APPROXIMATION-SOLVABILITY OF NONLINEAR FUNCTIONAL AND DIFFERENTIAL EQUATIONS

PURE AND APPLIED MATHEMATICS

A Program of Monographs, Textbooks, and Lecture Notes

MONOGRAPHS AND TEXTBOOKS IN
PURE AND APPLIED MATHEMATICS

Additional Volumes in Preparation

APPROXIMATION-SOLVABILITY OF NONLINEAR FUNCTIONAL AND DIFFERENTIAL EQUATIONS

Wolodymyr V. Petryshyn

Rutgers University
New Brunswick, New Jersey

CRC Press
Taylor & Francis Group
Boca Raton London New York

CRC Press is an imprint of the
Taylor & Francis Group, an **informa** business

CRC Press
Taylor & Francis Group
6000 Broken Sound Parkway NW, Suite 300
Boca Raton, FL 33487-2742

First issued in paperback 2019

ISBN-13: 978-0-8247-8793-6 (hbk)
ISBN-13: 978-0-367-40257-0 (hbk)

Library of Congress Cataloging-in-Publication Data

Petryshyn, Wolodymyr V.
    Approximation–solvability of nonlinear functional and differential equations / Wolodymyr V. Petryshyn.
       p. cm.—(Monographs and textbooks in pure and applied mathematics ; 171)
    Includes bibliographical references and index.
    ISBN 0-8247-8793-5
    1. Nonlinear functional analysis. 2. Mappings (Mathematics)
3. Topological degree. I. Title. II. Series.
    QA321.5.P48   1993
    515′.7—dc20
                                         92-36577
                                         CIP

Visit the Taylor & Francis Web site at
http://www.taylorandfrancis.com

and the CRC Press Web site at
http://www.crcpress.com

In memory of my parents
Vasyl and Maria

Preface

During the last 20 or 25 years the main thrust of the development of nonlinear functional analysis and its applications has been in the direction of breaking out of the classical framework into a much wider field of noncompact operators, such as operators of monotone and accretive type, operators of set and ball-condensing type, and operators of approximation-proper (A-proper) type. There exist extensive treatments of the theories and applications of the first two classes of operators.

The purpose of this book is to present a comprehensive theory of mappings of A-proper type and its application to differential equations in the context of constructive functional analysis, which came about as the answer to the following problem posed by the author some 25 years ago: For what type of linear and nonlinear mapping T is it possible to construct a solution of a given equation (*) $Tx = f$ as a strong limit of solutions x_n of finite dimensional approximate equations (**) $T_n(x_n) = f_n$? In a series of papers in the 1960s the author studied this problem, and the notion that evolved from these investigations in 1967 is that of an A-proper mapping. It turned out that the A-properness (and the pseudo-A-properness) of T not only is intimately connected with the constructive solvability (solvability) of equation (*), but in view of the fact that these classes of mappings are very extensive, their theory generalizes and unifies earlier results concerning Galerkin-type methods (**) for linear and nonlinear equations (*), with the more recent results in the theory of operators of monotone and accretive type, operators of type (S) and (S$_+$) [or type (α)], P$_\gamma$-compact and ball-contractive vector fields, and their combinations.

In view of this, the mappings of A-proper type became a subject of extensive study by pure and applied mathematicians in the United States, various countries in Western and Eastern Europe, the former Soviet Union (and in particular, Russia and Ukraine), Israel, China, and other countries, where I visited and gave many lectures during the last dozen years. Many of my listeners urged me to write a book on maps of A-proper type.

Keeping this in mind, in 1987 I gave two courses at Rutgers University, Topics in Analysis and Numerical Functional Analysis, for advanced graduate students and postdoctoral fellows, which were almost entirely devoted to the presentation of the theory of A-proper and pseudo-A-proper mappings and its application to the solvability of ordinary and partial differential equations, including those to which the classical results based on the Leray–Schauder degree theory are not applicable. This book reproduces, with extensive additions, the material presented in these courses.

Chapter I offers a comprehensive introduction, at a fairly elementary level, to the general theory of A-proper and pseudo-A-proper maps presented in Chapters II to V, including examples and applications to the solvability of differential equations of order 2.

Chapter II is devoted to the development of the linear theory of A-proper mappings and its application to the variational solvability of linear elliptic PDEs of order $2m$. As is shown, the abstract results in this chapter are the best possible.

The first part of Chapter III establishes the existence of fixed points and eigenvalues for P$_\gamma$-compact maps, including the classical results, while the second part provides a number of surjectivity theorems for pseudo-A-proper and weakly A-proper mappings that unify and extend the earlier results on monotone type and accretive mappings. This chapter also shows how the Friedrichs linear extension theory can be generalized to the extensions of densely defined nonlinear operators in a Hilbert space. Applications of both theories to ODEs and PDEs are given.

Chapter IV presents the generalized topological degree theory for A-proper mappings developed by Browder and Petryshyn. The degree theory is applied to local bifurcation and asymptotic bifurcation problems, among others.

In Chapter V the abstract results of Chapters II to IV are applied to the solvability of boundary value problems of ODEs and PDEs and to bifurcation and asymptotic bifurcation problems. For the convenience of the reader we include an Appendix which contains the statements of some basic theorems from real function theory and measure/integration theory. As to references, because of the great number of mathematicians working in this broad area, we have explicitly mentioned some papers only of those authors to whom direct reference is made in the book.

The book should be of interest to graduate students in pure and applied mathematics, mathematical physicists, and abstract numerical analysts. It should also be of interest to researchers in nonlinear functional analysis and its applications.

Both the level of difficulty and the prerequisites increase slowly from the beginning to the end of the book. This makes it suitable for a variety of courses taught to different kinds of students:

1. A one-semester course on current nonlinear functional analysis (Chapters I and III).
2. A one-semester course on the constructive solvability of linear equations (Chapters I and II).
3. A one-semester course on the topological degree theory and its application.
4. A full-year course on nonlinear functional analysis and partial differential equations since the mid-1960s.

It is assumed that the reader is familiar with functional analysis and some fundamentals of classical nonlinear analysis. This book should be considered a necessary companion to the monograph written earlier by the author on the topological degree in the solvability of semilinear densely defined operator equations.

I am grateful to my wife, Arcadia, and my friends for their encouragement. I would like to thank Robert Wilson and the Graduate Committee at Rutgers for giving me the opportunity to complete this monograph. I am also indebted to my former doctoral student Beatriz Lafferriere for providing me with her lecture notes of my courses given in 1987. Finally, I wish to thank Mary Jablonski and Barbara Mastrian for providing timely and efficient typing of many parts of the manuscript.

Wolodymyr V. Petryshyn

Contents

I
Solvability of Equations Involving A-Proper and Pseudo-A-Proper Mappings

Let X, Y be Banach spaces with their respective duals denoted by X^* and Y^*, D a subset of X, $T: D \subseteq X \to Y$ a possibly nonlinear mapping, $\Gamma = \{X_n, V_n; E_n, W_n\}$ an approximation scheme for the equation

$$T(x) = f \qquad (x \in D, f \in Y) \tag{1.1}$$

and a sequence of finite-dimensional equations

$$T_n(x_n) = W_n f \qquad (x_n \in D_n, W_n f \in E_n) \tag{1.2}$$

approximating equation (1.1) by means of the scheme Γ, where $T_n: D_n \subseteq X_n \to E_n$ is a finite-dimensional approximation to T determined by Γ. The main purpose of Chapter I is to introduce the notion and give some examples of an *A-proper* mapping $T: D \subseteq X \to Y$ with respect to a suitable scheme Γ and to show that, on the one hand, this notion is well suited to provide a new approach to the study of the constructive solvability of equation (1.1) and, on the other hand, the class of A-proper mappings is large enough to allow us to unify and properly extend (in a constructive way) the recent existence results for mappings of strongly monotone and accretive type and for ball-condensing and P_1-compact vector fields. The second purpose is to introduce a related notion of a pseudo-A-proper map $T: D \subset X \to Y$ which is intimately connected with the solvability (i.e., existence of solutions) of equation (1.1). Thus, we shall see, under some mild conditions on X, X^*, and T, the monotone maps $T: X \to X^*$ and

1

accretive maps $T: X \to X$ are pseudo-A-proper, but they are certainly not A-proper. The two classes of maps are essentially in the same relation, as are the notions of approximation solvability and solvability of equation (1.1) as defined in the next section.

1. DEFINITIONS AND SOME FACTS ABOUT A-PROPER MAPPINGS AND THE APPROXIMATION-SOLVABILITY OF EQUATION (1.1)

In Section 1 we fix the terminology and notation, introduce the definitions of some basic concepts used in this book, give a historical background and some examples of A-proper mappings, and prove some of the properties that are relevant to this section.

1.1 Admissible Schemes and Approximation-Solvability of Operator Equations

Let $\{X_n\}$ and $\{E_n\}$ be sequences of oriented finite-dimensional spaces with $X_n \subset X$, V_n an injective map of X_n into X, and W_n a continuous linear map of Y onto E_n for each $n \in Z_+$. We use the same symbol, $\| \cdot \|$ to denote the norms $\| \cdot \|_X$, $\| \cdot \|_Y$, and $\| \cdot \|_{E_n}$ and "$\to$" and "$\rightharpoonup$" to denote strong and weak (or weak*) convergence, respectively. For the sake of simplicity of notation and exposition we define the concept of *approximation properness* (*A-properness*) of $T: D \subseteq X \to Y$ and of the *approximation solvability* of equation (1.1) in terms of a given *admissible* scheme for (X,Y) defined as follows:

Definition 1.1   The scheme $\Gamma = \{X_n, V_n; E_n, W_n\}$ is said to be *admissible* for (X,Y) provided that dim $X_n =$ dim E_n for each $n \in Z_+$, dist$(x,X_n) \to 0$ as $n \to \infty$ for each x in X, and $\{W_n\}$ is uniformly bounded.

Note that we do not require $\{E_n\}$ to be subspaces of Y, nor do we assume that $\{X_n\}$ is nested. Hence the finite element method can be used in the construction of admissible schemes (see [284,314,316]). It follows that the existence of such an admissible scheme for (X,Y) implies that X is separable, $\cup_n X_n$ is dense in X, but Y need not be separable for Γ to be admissible. The following examples of simple admissible schemes (others will be given later), which we letter for further reference, will be used.

(a)   *Injective scheme for* (X,X^*). If $Y = X^*$, $E_n = X_n^*$, and $W_n \equiv V_n^*: X^* \to X_n^*$, the injective scheme $\Gamma_I = \{X_n, V_n; X_n^* V_n^*\}$ is admissible for (X,X^*).

(b) *Projective scheme for* (X,X). If $E_n = X_n$ and $W_n \equiv P_n: X \to X_n$ is a linear projection with $\alpha = \sup_n\| P_n \| < \infty$, the projectional scheme $\{X_n, V_n; X_n, P_n\}$, which we denote by $\Gamma_\alpha = \{X_n, P_n\}$, is admissible for (X,X), and in fact, Γ_α is projectionally complete in the sense that $P_n(x) \to x$ for each x in X.

(c) *Projective scheme for* (X,Y). If $E_n = Y_n$ and $W_n \equiv Q_n: Y \to Y_n$ is a linear projection with $\| Q_n \| \leq \beta$ for some $\beta \geq 1$, the projectional scheme $\Gamma_P = \{X_n, V_n; Y_n, Q_n\}$ is admissible for (X,Y).

If additionally, we assume that $\text{dist}(y, Y_n) \to 0$ for each y in Y, then $Q_n(y) \to y$ for each y in Y, so in this case we say that Γ_P is *complete* for (X,Y). If we assume further that $P_n: X \to Y_n$ are uniformly bounded, the scheme $\Gamma_P = \{X_n, P_n; Y_n, Q_n\}$ is *projectionally complete* since in this case $P_n(x) \to x$ and $Q_n(y) \to y$ for each x in X and y in Y. In case X and Y are Hilbert spaces, the projections are always assumed to be orthogonal.

Remark 1.1 Example (a) shows that when X is separable, (X, X^*) always has an admissible scheme, while example (b) shows that there is an admissible scheme for (X,X) whenever X has a Schauder basis. Indeed, when X has a Schauder basis $\{\phi_i\} \subset X$, then setting $X_n = [\phi_1, \ldots, \phi_n]$, $P_n(x) = \sum_{i=1}^{n} (\Phi_i, x)\phi_i$ for $x \in X$, $Y_n = [\Phi_1, \ldots, \Phi_n]$, and $P_n^*(w) = \sum_{i=1}^{n} (w, \phi_i)\Phi_i$ for w in X^*, where $\{\Phi_i\} \subset X^*$ is such that $(\Phi_i, \phi_j) = \delta_i^j$ with $[Q]$ denoting the linear span of the set Q and (u, x) the value of u in X^* at x in X, it is known (see [160]) that $\Gamma_\alpha = \{X_n, P_n\}$ is projectionally complete for (X,X) with $\{X_n\}$ nested and $\Gamma_p = \{X_n, V_n; Y_n, P_n^*\}$ is admissible for (X, X^*) since $P_n(x) \to x$ for each x in X and $\sup_n\| P_n \| = \sup_n\| P_n^* \| = \alpha$ for some $\alpha \geq 1$ but Γ_p need not be projectionally complete for (X, X^*) since, in general, $P_n^*(w) \nrightarrow w$ in X^* for each w in X^*. However, when $\{\phi_i\} \subset X$ is a *shrinking* basis, that is, $\{\phi_i\}$ is such that

$$\alpha_n(u) = \sup\left\{\left|\left(u, x - \sum_{i=1}^{n} (\Phi_i, x)\phi_i\right)\right| : \| x \| \leq 1\right\} \to 0$$

$$\text{for all} \quad u \in X^*,$$

then $\Gamma_p = \{X_n, P_n; Y_n, P_n^*\}$ is projectionally complete for (X, Y^*). The same assertion is true if X is a reflexive space with a Schauder basis $\{\phi_i\} \subset X$ (see [160]).

To state precisely the results concerning the approximation solvability of equation (1.1) and the A-properness of $T_n = W_n T |_{D_n}: D_n \to Y_n$, where $D_n = D \cap X_n$ for each $n \in Z_+$, we need the following:

Definition 1.2 Equation (1.1) is said to be *strongly* (respectively, *feebly*) *approximation solvable* w.r.t. the admissible scheme $\Gamma = \{X_n, V_n; E_n, W_n\}$

if there exists $n_0 \in Z_+$ such that equation (1.2) has a solution $x_n \in D_n \equiv D \cap X_n$ for each $n \geq n_0$ such that $x_n \to x_0 \in D$ (respectively, $x_{n_j} \to x_0 \in D$ for some subsequence $\{x_{n_j}\}$ of $\{x_n\}$) and $T(x_0) = f$. We say that equation (1.1) is *uniquely approximation solvable* if the approximate solutions $x_n \in D_n$ and the limit solution $x_0 \in D$ are unique.

Note that the approximation solvability of equation (1.1) implies its solvability, but the converse need not be true.

1.2 A-Proper Mappings, Examples, and Some Properties

In a series of papers ([202–205]) the author investigated the type of mappings T for which equation (1.1) is strongly approximation solvable w.r.t. a suitable scheme Γ [i.e., for which it is possible to construct a solution $x \in D$ of equation (1.1) as a strong limit of solutions $x_n \in D_n$ of approximate equation (1.2)]. The result of this study was that in 1967 the author introduced in [206,207] a wide class of mappings, those mappings satisfying *condition (H)*, which proved to be very suitable for the study of the constructive solvability of equation (1.1). In the author's paper [208] and subsequently, mappings satisfying condition (H) have been referred to as *A-proper*. The theory of these mappings developed by the author and other writers and its applications was surveyed up to 1975 by the author in [225] under the assumption that T is A-proper w.r.t. a projectionally complete scheme $\Gamma_P = \{X_n, P_n; Y_n, Q_n\}$.

In this book the theory of A-proper mappings and of mappings of A-proper type, together with its applications, are presented in the framework of admissible schemes given by Definition 1.1, although many results are valid for more general schemes, as indicated below. Consequently, the results presented in this book are applicable to more general situations than those outlined in [225]. In addition, the book contains some further new results as well as some new and more general applications to linear and nonlinear differential and integral equations.

Definition 1.3 Let $\Gamma = \{X_n, V_n; E_n, W_n\}$ be an admissible scheme for (X,Y). A mapping $T: D \subseteq X \to Y$ is said to be *A-proper* w.r.t. Γ if and only if $T_n: D_n \subseteq X_n \to E_n$ is continuous for each $n \in Z_+$ and the following condition holds:

(H) If $\{x_{n_j} \mid x_{n_j} \in D_{n_j}\}$ is any bounded sequence such that $\| T_{n_j}(x_{n_j}) - W_{n_j}(g) \| \to 0$ as $j \to \infty$ for some g in Y, there exists a subsequence $\{x_{n_{j(k)}}\}$ of $\{x_{n_j}\}$, $x \in D$, such that $x_{n_{j(k)}} \to x$ as $k \to \infty$ and $T(x) = g$.

Let us add that the continuity assumption on T_n is a very weak requirement. It always holds when T is finitely continuous and, in particular, when T is continuous, demicontinuous, or weakly continuous. We recall that T is *demicontinuous* at $x \in D$ if $\{x_j\} \subset D$ and $x_j \to x$ in X imply $T(x_j) \rightharpoonup T(x)$ in Y; T is *weakly continuous* at $x \in D$ if $\{x_j\} \subset D$ and $x_j \rightharpoonup x$ in X imply that $T(x_j) \rightharpoonup T(x)$ in Y; T is *finitely continuous* if for any finite-dimensional subspace V of X and any sequence $\{x_j\} \subset D \cap V$ such that $x_j \to x \in D \cap V$ we have $T(x_j) \rightharpoonup T(x)$ in Y.

Although the fixed-point and eigenvalue theories for generalized projectionally compact (P_γ-*compact*) mappings are discussed in Chapter III, it should be mentioned here that the concept of an A-proper mapping evolved from the notion of a P_γ-compact mapping introduced by the author in [202] for $\gamma = 0$ and in [206] for $\gamma > 0$ for the constructive approach to fixed-point and eigenvalue problems for noncompact maps in Banach spaces and for surjectivity theorems for strongly monotone and strongly accretive maps acting in Hilbert and Banach spaces, respectively. For the present we note that the definition of P_γ-compactness used in [206] can be stated as follows.

Definition 1.4 A map $F: D \subseteq X \to X$ is called P_γ-*compact* w.r.t. $\Gamma_\alpha = \{X_n, P_n\}$ if there exists a constant $\gamma \geq 0$ such that for each λ dominating γ (i.e., $\lambda \geq \gamma$ if $\gamma > 0$ and $\lambda > \gamma$ if $\gamma = 0$) the map $T_\lambda \equiv \lambda I - F: D \to X$ is A-proper w.r.t. Γ_α.

Before giving some simple but important examples of A-proper mappings, we first prove some of their properties that will be useful in various applications. The following fact provides the motivation for the terminology "A-proper" and at the same time is useful in other respects.

Recall first that $T: D \subseteq X \to Y$ is called *proper* if $T^{-1}(G) \equiv \{x \in D: Tx \in G\}$ is compact in X whenever G is compact in Y.

Theorem 1.1 If $D \subset X$ is open and $T: \overline{D} \to Y$ is continuous and A-proper w.r.t. $\Gamma = \{X_n, V_n; E_n, W_n\}$, the restriction of T to every bounded closed subset of $\overline{D}$ is proper.

Proof. Let G be any compact set in Y and let M be any bounded closed subset of $\overline{D}$ such that $M \cap T^{-1}(G) \neq \emptyset$. Let $\{x_k\}$ be any sequence in $M \cap T^{-1}(G)$. Then $\{Tx_k\} \subset G$ and since G is compact, we may assume that $T(x_k) \to g$ for some g in Y. Let $\{\epsilon_k\}$ be a sequence of positive numbers such that $\epsilon_k \to 0$ as $k \to \infty$. Since $T: \overline{D} \to Y$ is continuous, to each k there exists $\delta_k > 0$ such that $\| Tx_k - Tx \| < \epsilon_k$ for all $x \in B(x_k, \delta_k) \cap \overline{D}$. Since D is open, there is $y_k \in D$ such that $\| x_k - y_k \| < \eta_k$ for each k, where

$\eta_k = \min\{\frac{1}{2}\epsilon_k, \frac{1}{2}\delta_k\}$. Since $y_k \in D$ and $\text{dist}(y_k, X_n) \to 0$ as $n \to \infty$ for each fixed k, there exists $n(k) \in Z_+$ and $v_{n(k)} \in X_{n(k)}$ such that $v_{n(k)} \in \overline{D}$ and $\| y_k - v_{n(k)} \| < \eta_k$. Thus $v_{n(k)} \in \overline{D}_{n(k)}$ and $\| x_n - v_{n(k)} \| \le \| x_k - y_k \| + \| y_k - v_{n(k)} \| < 2\eta_k = \min\{\epsilon_k, \eta_k\}$. Hence $v_{n(k)} \in B(x_k, \delta_k)$, so $\| Tx_k - Tv_{n(k)} \| < \epsilon_k$. Since M is bounded and $\| W_n \| \le \beta_0$ for some β_0 as a map from Y to E_n, it follows that as $k \to \infty$ we have

$$\| W_{n(k)}Tv_{n(k)} - W_{n(k)}g \| \le \| W_{n(k)}Tv_{n(k)} - W_{n(k)}Tx_k \|$$

$$+ \| W_{n(k)}Tx_k - W_{n(k)}g \| \le \beta_0\{\| Tv_{n(k)} - Tx_k \| + \| Tx_k - g \|\} \to 0.$$

Hence, by the A-properness of T, we may assume that $v_{n(k)} \to x$ with $x \in \overline{D}$ and $Tx = g$. But then $x_k \to x$ with $x \in M$ since M is closed and $\{x_k\} \subset M$. Q.E.D.

Corollary 1.1 Suppose that D, Γ, and T are as in Theorem 1.1. Then any bounded sequence $\{x_k\} \subset \overline{D}$ such that $Tx_k \to g$ for some g in Y has a convergent subsequence and, in particular, T maps closed bounded sets in $\overline{D}$ into closed sets in Y.

The following example from [225] shows that a continuous proper mapping need not be A-proper w.r.t. a given scheme. As we shall see in Chapter II, this example will also show that the adjoint T^* of an A-proper map $T \in L(X,Y)$ need not be A-proper w.r.t. to the dual scheme $\Gamma^*_\#$.

Example 1.1 Let H be a Hilbert space with the orthonormal basis $\{\phi_1, \phi_2, \ldots\}$, $H_n = [\phi_1, \ldots, \phi_n]$, $P_n: H \to H_n$ an orthogonal projection, and $T \in L(H,H)$ given by

$$T\phi_1 = \phi_2, \quad T\phi_2 = \phi_3, \ldots, \quad T\phi_i = \phi_{i+1}, \quad \ldots \tag{1.3}$$

It is easy to see that T defined by (1.3) is such that $N(T) = \{0\}$ and $R(T)$ is closed with $R(T) = [\phi_2, \phi_3, \ldots]$.

Proposition 1.1 Let $T \in L(H,H)$ be defined by (1.3). Then T restricted to closed bounded sets is proper, but T is not A-proper w.r.t. $\Gamma_H = \{H_n, P_n\}$.

Proof. It is easy to see that T defined by (1.3) is such that $N(T) = \{0\}$ and $R(T)$ is closed with $R(T) = [\phi_2, \phi_3, \ldots]$. Hence, by the result of Yood [310], T restricted to bounded closed sets is proper. To show that T is not A-proper w.r.t. Γ_H, it suffices to establish the existence of a bounded sequence $\{x_n \mid x_n \in X_n\}$ such that $P_nT(x_n) \to g$ for some g in H and $\{x_n\}$ has no convergent subsequence. Now, if $\{x_n\} \equiv \{\phi_n\}$, then $x_n \in X_n$ for each n, $\| x_n \| = 1$ and $P_nTx_n = 0$ since $Tx_n = T\phi_n = \phi_{n+1}$ and

$P_n\phi_{n+1} = 0$ for each $n \in Z_+$. Thus $P_n Tx_n \to 0$, but $\{x_n\} = \{\phi_n\}$ has no convergent subsequence.   Q.E.D.

Later we show that, in general, an A-proper map $T: D \to Y$ need not be proper if we drop the requirement that T be continuous. It is obvious that when $T: D \to Y$ is A-proper and $a \neq 0$, then aT is also A-proper, but the following example shows that the sum of two A-proper maps need not be A-proper.

Example 1.2   Let H and Γ_H be as in Example 1.1 and let $A \in L(H,H)$ be given by

$$A\phi_1 = 0, \quad A\phi_2 = \phi_1, \quad \ldots, \quad A\phi_i = \phi_{i-1}, \quad i \geq 2. \tag{1.4}$$

It is easy to see that A, given by (1.4), is such that $N(A) = [\phi_1]$, $R(A) = H$, and $H = N(T) \oplus R(T^*)$. For later use we recall (see [182,184]) that each $T \in L(H,H)$ with a closed range $R(T)$ has a unique *generalized inverse* $T^+: H \to H$ which (among others) is characterized by one of the following relations:

$$T^+T = P_{R(T^*)}, \qquad N(T^+) = N(T^*) \tag{1.5}$$

$$TT^+ = P_{R(T)}, \qquad N(T^{**}) = N(T), \tag{1.6}$$

where $P_{R(T^*)}$ and $P_{R(T)}$ are the orothgonal projections of H onto $R(T^*)$ and $R(T)$, respectively.

Proposition 1.2   The map $A \in L(H,H)$ given by (1.4) is A-proper w.r.t. Γ_H and so is the identity I on H, but $T = I + A$ is not A-proper w.r.t. Γ_H.

Proof.   Let $\{x_n \mid x_n \in H_n\}$ be any bounded sequence such that $P_n A(x_n) \to g$ for some g in H. Since $x_n \in H_n$ and $\{x_n\}$ is bounded, it follows that for all $n \in Z_+$ and some $M > 0$,

$$x_n = \sum_{i=1}^n \alpha_i^n \phi_i \quad \text{with} \quad \| x_k \| = \sum_{i=1}^n (\alpha_i^n)^2 \leq M.$$

Hence, by (1.4), we see that $Ax_n = \alpha_2^n \phi_1 + \cdots + \alpha_n^n \phi_{n-1}$ and $g_n \equiv P_n Ax_n = Ax_n = \sum_{i=2}^n \alpha_i^n \phi_{i-1}$ since $Ax_n \in H_n$. Thus $g_n \in R(A)$ for each n and $g_n = P_n Ax_n \to g$ with $g \in R(A) = H$. In view of this and (1.5), we have

$$A^+ g_n = A^+ Ax_n = P_{R(A^*)}(x_n) = \alpha_2^n \phi_2 + \cdots + \alpha_n^n \phi_n \to A^+ g.$$

Since $\{\alpha_1^n\} \subset [-M^{1/2}, M^{1/2}]$, there exists $\alpha_1 \in [-M^{1/2}, M^{1/2}]$ and a subsequence $\{n_j\}$ of $\{n\}$ such that $\alpha_1^{n_j} \to \alpha_1$ and

$$x_{n_j} = A^+(g_{n_j}) + \alpha_1^{n_j}\phi_1 \to A^+g + \alpha_1\phi_1 \equiv x.$$

This and (1.6) imply that $Ax = A(A^+g + \alpha_1\phi_1) = g$, because $\phi_1 \in N(A)$ (i.e., A is A-proper w.r.t. Γ_H).

It is obvious that I is A-proper w.r.t. Γ_H, but $T = I + A$ is not A-proper w.r.t. Γ_H. In fact, if $x_1 = \phi_1$ and $x_k = k^{-1/2}(\phi_1 - \phi_2 + \phi_3 - + - + \cdots - (-1)^k\phi_k)$ for $k \geq 2$, then $x_n \in H_n$ for each $n \in Z_+$, $\| x_n \| = 1$, and $x_n + P_nAx_n \to 0$ as $n \to \infty$. However, for $n > m$,

$$\| x_n - x_m \|^2 = 2\left[1 - \left(\frac{m}{m}\right)^{1/2}\right] > \left[1 - \left(\frac{m}{m}\right)^{1/2}\right].$$

Therefore, $\{x_n\}$ has no convergent subsequence (i.e., $I + T$ is not A-proper w.r.t. Γ_H).   Q.E.D.

The foregoing fact has an important bearing, for example, on the homotopy theorem in the generalized degree theory for A-proper mappings developed in Chapter IV as well as on the solvability of equation (1.1) when T is a uniform limit of A-proper mappings.

The following theorem shows that A-properness is invariant under compact perturbations. This fact will prove to be useful in various applications of the A-proper mapping theory to the (constructive) solvability of differential and integral equations.

Theorem 1.2 If $T: D \subseteq X \to Y$ is A-proper w.r.t. Γ and $C: D \to Y$ is compact, then $A \equiv T + C$ is A-proper w.r.t. Γ.

Proof. Now $T_n: D_n \subseteq X_n \to Y_n$ is clearly continuous for each $n \in Z_+$. So let $\{x_{n_j} \mid x_{n_j} \in D_{n_j}\}$ be any bounded sequence such that $\| W_{n_j}(T + C)(x_{n_j}) - W_{n_j}g \| \to 0$ for some g in Y. Since $\{x_{n_j}\}$ is bounded and C is compact, we may assume that $C(x_{n_j}) \to y$ for some y in Y. This and the uniform boundedness of $\{W_n\}$ imply that

$$T_{n_j}(x_{n_j}) - W_{n_j}(g - y) = A_n(x_{n_j}) - W_{n_j}(g) + W_{n_j}(y - C(x_{n_j})) \to 0$$

as $j \to \infty$. Hence, by the A-properness of T, there exist a subsequence $\{x_{n_{j(k)}}\}$ of $\{x_{n_j}\}$ and $x \in D$ such that $x_{n_{j(k)}} \to x$ and $Tx = g - y$. Since, by continuity of C, $C(x_{n_{j(k)}}) \to Cx = y$, we see that $Tx + Cx = g$.
                                                                                    Q.E.D.

Since the identity I is A-proper w.r.t. an admissible scheme $\Gamma_\alpha = \{X_n, P_n\}$ for (X,X) on any closed subset D of X, an immediate consequence of Theorem 1.1 is the following:

Corollary 1.2 Let Γ_α be admissible for (X,X), $D \subseteq X$ closed and $C: D \to X$ compact. Then $\lambda I - C: D \to X$ is A-proper w.r.t. Γ_α for each $\lambda \neq 0$ and, in particular, C is P-compact.

To deduce our next consequence of Theorem 1.1 and for subsequent applications we need the following classical result in nonlinear functional analysis, which is usually referred to as the Picard–Banach theorem or

Contraction mapping theorem Suppose that D is a closed subset of X and $T: D \to D$ is k-contractive with $k \in (0,1)$ (i.e., $\| Tx - Ty \| \le k \| x - y \|$ for $x,y \in D$). Then there exists a unique point z in D such that $Tz = z$. Moreover, if $x_0 \in D$ and $x_n = Tx_{n-1}$ for $n = 1,2, \ldots$, then $\lim_n x_n = z$ and we have the estimate

$$\| x_n - z \| \le k^n (1 - k)^{-1} \| x_1 - x_0 \| \qquad \text{for each} \quad n \in Z_+. \qquad (1.7)$$

In particular, if $D = X$, then $x - Tx = y$ has a unique solution for each $y \in X$.

Proof. For the sake of completeness we give a proof of this important result.

Now, if $n \ge 1$, then $\| x_{n+1} - x_n \| = \| Tx_n - Tx_{n-1} \| \le k \| x_n - x_{n-1} \|$ and it follows by induction that $\| x_{n+1} - x_n \| \le k^n \| x_1 - x_0 \|$. If $m,n \ge 1$, then

$$\| x_{n+m} - x_n \| \le \sum_{i=0}^{m-1} \| x_{n+i+1} - x_{n+i} \| \le \sum_{i=0}^{m-1} k^{n+1} \| x_1 - x_0 \|$$

$$\le k^n \| x_1 - x_0 \| \sum_{i=0}^{m-1} k^i \le k^n \| x_1 - x_0 \| (1 - k)^{-1}.$$

Thus $z = \lim_n x_n$ exists and $z \in D$ since D is closed. Letting $m \to \infty$ in the preceding estimate we obtain (1.7). Moreover, since T is continuous, $Tz = \lim_n Tx_n = \lim_n x_{n+1} = z$. If $w \in D$ and $Aw = w$, then $\| z - w \| = \| Az - Aw \| \le k \| z - w \|$, which implies that $z = w$ since $k < 1$.

The second part follows from the first if we let $D = X$ and observe that for each given y in X the map $T_y(x) = Tx + y$ is a k-contraction of X into X and z is a fixed point of T_y if and only if z is the solution of $x - Tx = y$. Q.E.D.

We can now deduce the second consequence of Theorem 1.1.

Corollary 1.3 Let D, C, and Γ_α be as in Corollary 1.2 with $\alpha = 1$. If $S: X \to X$ is k-contractive with $k \in (0,1)$, then $\lambda I - S - C: D \to X$ is A-proper w.r.t. Γ_1 for each $|\lambda| > k$ and, in particular, $S + C: D \to X$ is P_1-compact.

Proof. Although Corollary 1.3 can be deduced from more general results below, it is instructive to give it a simple and independent proof. In view of Theorem 1.1, it suffices to prove that $S_\lambda \equiv \lambda I - S: D \to X$ is A-proper w.r.t. Γ_1 for each fixed $|\lambda| > k$. Now $S_{\lambda n}: D_n \to X_n$ is obviously continuous. So let $\{x_{n_j} \mid x_{n_j} \in D_{n_j}\}$ be any bounded sequence such that $\lambda x_{n_j} - P_{n_j} S x_{n_j} - P_{n_j} g \to 0$ for some g in X. Since $P_{n_j}(g) \to g$ in X, it follows that $\lambda x_{n_j} - P_{n_j} S x_{n_j} \to g$. In virtue of the contractive mapping theorem and the fact that $|\lambda| > k$, it follows that $\lambda I - S$ is a homeomorphism of X onto X. Hence there exists a unique $x_0 \in X$ such that $\lambda x_0 - S x_0 = g$. This and the assumption that $\| P_n \| = 1$ imply that

$$\| S_{\lambda n_j}(x_{n_j}) - P_{n_j} S_{\lambda n_j} P_{n_j}(x_0) \|$$
$$\geq (|\lambda| - k) \| x_{n_j} - P_{n_j}(x_0) \| \qquad \text{for all } j \in Z_+.$$

Since the left-hand side in the inequality above converges to zero and $P_{n_j}(x_0) \to x_0$ as $j \to \infty$, it follows that $x_{n_j} \to x_0$ and, of course, $S_\lambda x_0 = g$ by the continuity of S_λ. Thus S_λ and $S_\lambda + C: D \to X$ are A-proper w.r.t. Γ_1 for each $|\lambda| > k$. $\hspace{2cm}$ Q.E.D.

Corollary 1.3 implies, in particular, that $T \equiv I - S - C: D \to X$ is A-proper w.r.t. Γ_1 whenever $S: X \to X$ is k-contractive with $k \in (0,1)$. However, it is unknown if $T: D \to X$ remains A-proper if it is assumed that S is defined and k-contractive only on D, so we pose the following question.

Question 1.1 Assume that S is defined and k-contractive with $k \in (0,1)$ on $\bar{B}(0,r)$ and T maps $\bar{B}$ into X or even into $\bar{B}$. Is T A-proper w.r.t. an admissible scheme $\Gamma_1 = \{X_n, P_n\}$?

As we shall see below, the answer is yes if we impose some additional condition on X or on T as we shall see from the results of Fitzpatrick [76] and Webb [297].

To state our next example we recall the following.

Definition 1.5 Let ϕ be a continuous function of R^+ to R^+ and $T: X \to X^*$ a mapping such that

$$(Tx - Ty, x - y) \geq \phi(\| x - y \|) \qquad \text{for all } x,y \in X. \qquad (1.8)$$

Then T is called *monotone* if $\phi(t) \equiv 0$, *strongly monotone* if $\phi(t) = mt^2$ for some $m > 0$, and *firmly monotone* if $\phi(0) = 0$ and $\phi(t) > 0$ for $t > 0$.

The importance of mappings of monotone type stems from the fact that various classes of concrete and abstract differential operators in divergence form give rise to equations involving operators of monotone type acting from suitable Sobolev spaces to their respective duals. In view of this, since 1960 the class of operators above has been studied extensively by many authors and an impressive theory with applications has been developed (see [16,26,27,44,75,113,154,178,211,281,252,311] and [36,119]). The useful part of this theory is a number of surjectivity theorems obtained under various growth conditions on $T: X \to X^*$. It will be shown in this book that most of these theorems will be deduced in case of separable spaces from the more general results for mappings $T: X \to Y$ of A-proper type (e.g., weakly A-proper, pseudo-A-proper, uniform limits of A-proper maps, etc.).

For the present we state the following basic result.

Proposition 1.3 Let X be a separable reflexive Banach space. If $T: X \to X^*$ is demicontinuous and firmly monotone and $C: X \to X^*$ is compact, then $A = T + C$ is A-proper w.r.t. the injective scheme $\Gamma_I = \{X_n, V_n; X_n^*, V_n^*\}$.

We omit the proof of Proposition 1.3 since in Section 2 we establish a very general theorem (Theorem 2.3) which also includes Proposition 1.4 below for a more general class of mappings of types (S) and (S$_+$) introduced by Browder [35]. See also Skrypnik [262] for an analogous notion of a map of type (α).

Definition 1.6 A mapping $T: X \to X^*$ is said to be of *type* (S) [respectively, of type (S$_+$)] if when $\{x_j\} \subset X$ is any sequence such that $x_j \to x$ in X and $\lim_j (Tx_j - Tx, x_j - x) = 0$ [respectively, $\lim \sup(Tx_j - Tx, x_j - x) \leq 0$], then $x_j \to x$ in X.

Before we state our next result we first recall that T is called *semibounded* if $\{Tx_j\}$ is bounded whenever the sequences $\{x_j\}$ and $\{(Tx_j, x_j)\}$ are bounded.

Proposition 1.4 Let X be a separable reflexive Banach space and $T: X \to X^*$ demicontinuous and semibounded. (i) If T is also of type (S), then T is A-proper w.r.t. Γ_I. (ii) Moreover, the maps T of type (S$_+$) form a convex subclass of the class of maps of type (S).

Let us add that each T of type (S) is clearly of type (S$_+$), but the converse is not true. We underline that one of the important features of the maps of type (S_+) is that if T_1 and T_2 are of type (S_+), so is the map $T_\lambda \equiv \lambda T_1 + (1 - \lambda)T_2$ for each $\lambda \in (0,1)$ and hence T_λ is A-proper. This need not be the case when T_1 and T_2 are only of type (S). As we shall see later, this fact is particularly important in the degree theory for A-proper maps. Assertion (i) was first proved in [210, Theorem 2.2], when T is of modified type (S) and continuous and, in particular, when $T: X \rightarrow X^*$ is bounded, continuous, and of type (S). Subsequently, the latter fact was proved in [36].

In 1973, Skrypnik devoted a major portion of his book [262] to a systematic study of maps $T: D \subseteq X \rightarrow Y$ of *type* (α) [i.e., T is such that if $\{x_j\} \subset D$ is such that $x_j \rightharpoonup x$ in X and $\lim_j (Tx_j - Tx, x_j - x) \le 0$, then $x_j \rightarrow x$ in X]. Clearly, if T is bounded, then T is type (α) if and only if T is of type (S_+). Thus a number of results in [262] will be deduced from the A-proper mapping theory developed in this book.

Remark 1.2   Before we continue with further examples, let us first show that Proposition 1.3 follows from Proposition 1.4. To that end we need the following result proved by Fitzpatrick and the author (see [231]).

Lemma 1.1   If $T: X \rightarrow X^*$ is monotone, then T is semibounded.

Proof.   Let $\{x_j\} \subset X$ be any bounded sequence such that $\{(Tx_j, x_j)\}$ is bounded. To show that $\{Tx_j\}$ is bounded, by the uniform boundedness principle, it suffices to show that $\{|(Tx_j, v)|\}$ is bounded for each v in X. Clearly, to prove the latter, it suffices to show that for each v in X there exists a constant C_v such that $(Tx_j, v) \le C_v$ for all $n \in Z_+$. Now, for any v in X, we have $(Tx_j - Tv, x_j - v) \ge 0$ for all $j \in Z_+$ and thus $(Tx_j, x_j) - (Tx_j, v) - (Tv, x_j) + (Tv, v) \ge 0$ for $j \in Z_+$. Letting

$$C_v = \sup_j \{(Tx_j, x_j) - (Tv, x_j) + (Tv, v)\},$$

we see that $C_v < \infty$ and $(Tx_j, v) \le C_v$ for all $j \in Z_+$.   Q.E.D.

Now, to deduce Proposition 1.3 from Proposition 1.4, it suffices to show that $A = T + C$ is semibounded and of type (S). The fact that A is semibounded follows from Lemma 1.1 and the boundedness of C. To show that A is of type (S), let $\{x_j\}$ be any bounded sequence in X such that $x_j \rightharpoonup x$ in X and $(Ax_j - Ax, x_j - x) \rightarrow 0$ as $j \rightarrow \infty$. To show that $x_j \rightarrow x$ in X, it suffices to show that from each subsequence of $\{x_j\}$ we can extract a further subsequence which converges to x. So let us take an arbitrary subsequence $\{x_k\}$ of $\{x_j\}$. Since C is compact, X is reflexive and $\{x_k\}$ is bounded, we may suppose that $C(x_k) \rightarrow y$ in X^* for some y in X^*.

Now

$$(Tx_k - Tx, x_k - x)$$

$$= (Ax_k - Ax, x_k - x) - (Cx_k - Cx, x_k - x) \to 0$$

since the two terms on the right converge to zero as $k \to \infty$. This and the firm monotonicity of T imply that $\phi(\| x_k - x \|) \to 0$, so $x_k \to x$ as $k \to \infty$ [i.e., $A = T + C$ is of type (S)]. Consequently, Proposition 1.3 follows from Proposition 1.4.

When X is a complex Banach space, then in analogy with Definition 1.5 we have the following:

Definition 1.7 If X is a complex space and $\phi \colon R^+ \to R^+$, as in Definition 1.5, $T \colon X \to X^*$ is called *firmly monotone* if

$$\text{Re}(Tx - Ty, x - y) \geq \phi(\| x - y \|) \qquad \text{for all} \quad x,y \in X \qquad (1.9)$$

and *firmly complex monotone* if

$$| (Tx - Ty, x - y) | \geq \phi(\| x - y \|) \qquad \text{for all} \quad x,y \in X. \qquad (1.10)$$

When $\phi(t) = mt^2$ for some $m > 0$, the mapping T satisfying (1.10) is usually called *complex monotone*. Since every firmly monotone map is firmly complex monotone, it suffices to prove our next assertion for the latter.

Proposition 1.5 Let X be a separable and complex reflexive Banach space. If $T \colon X \to X^*$ is firmly complex monotone and either continuous, weakly continuous, or bounded and demicontinuous, then T is A-proper w.r.t. Γ_I.

Proof. Although Proposition 1.5 can be deduced from a more general result in Chapter VI, it is instructive to give it a simple and independent proof, especially since that type of argument is used in other situations. Clearly, $T_n = V_n^* T |_{X_n} \colon X_n \to X_n^*$ is continuous for each $n \in Z_+$. So let $\{ x_{n_j} \mid x_{n_j} \in X_{n_j} \}$ be any bounded sequence such that $\| T_{n_j}(x_{n_j}) - V_{n_j}^*(g) \| \to 0$ for some g in X^*. Since X is reflexive and $\{x_{n_j}\}$ is bounded, by the Eberlein–Smulian theorem (see [108]) we may assume that $x_{n_j} \to x_0$ for some $x_0 \in X$.

Suppose first that T is continuous. Since $\text{dist}(x,X_n) \to 0$ for $x \in X$, there exists $y_n \in X_n$ such that $y_n \to x_0$ in X. By (1.10),

$$\phi(\| x_{n_j} - y_{n_j} \|) \leq |(Tx_{n_j} - Ty_{n_j}, x_{n_j} - y_{n_j})|$$

$$= |(T_{n_j}(x_{n_j}) - V_{n_j}^* g, x_{n_j} - y_{n_j})$$

$$+ (g - Ty_{n_j}, x_{n_j} - y_{n_j})|.$$

Since T is continuous and $x_{n_j} - y_{n_j} \rightharpoonup 0$, both terms on the right converge to zero. Hence $\phi(\| x_{n_j} - y_{n_j} \|) \to 0$ as $j \to \infty$. This implies that $\| x_{n_j} - y_{n_j} \| \to 0$. Indeed, if $\| x_{n_i} - y_{n_i} \| \to a > 0$ for some subsequence $\{i\}$ of $\{j\}$, then by the continuity of ϕ we would have $\phi(\| x_{n_i} - y_{n_i} \|) \to \phi(a) > 0$, a contradiction. Thus $x_{n_j} \to x_0$ in X.

Suppose that T is either weakly continuous or demicontinuous and bounded. Again, by (1.10), we have for each $j \in Z_+$,

$$\phi(\| x_{n_j} - x_0 \|) \le |(Tx_{n_j} - Tx_0, x_{n_j} - x_0)|$$

$$= |(V_{n_j}^* Tx_{n_j} - V_{n_j}^* g, x_{n_j} - y_{n_j}) + (g, x_{n_j} - y_{n_j})$$

$$+ (Tx_{n_j} - g, y_{n_j} - x_0) - (Tx_0, x_{n_j} - x_0)|.$$

Since $x_{n_j} - y_{n_j} \to 0$, $V_{n_j}^* Tx_{n_j} - V_{n_j}^* g \to 0$, $y_{n_j} - x_0 \rightharpoonup 0$, and $x_{n_j} - x_0 \rightharpoonup 0$, our conditions on T imply that each term on the right converges to zero. Hence $x_{n_j} \to x_0$ as $j \to \infty$, as above.

To complete the proof, we must show that $Tx_0 = g$. Let y be any element in X and let $y_n \in X_n$ be such that $y_n \to y$ in Y. Since $Tx_{n_j} \rightharpoonup Tx_0$ in either case, it follows that

$$(g - Tx_0, y) = \lim_j (g - Tx_{n_j}, y_{n_j}) = \lim_j (V_{n_j}^* g - T_{n_j}(x_{n_j}), y_{n_j}) = 0$$

[i.e., $(g - Tx_0, y) = 0$ for each $y \in X$]. Hence $Tx_0 = g$.   Q.E.D.

If T is demicontinuous but not bounded, the simple proof above fails. In Chapter VI we provide a direct (but more complicated proof) of Proposition 1.5 for the case when T is complex monotone and demicontinuous but not necessarily bounded. The indirect proof that uses the surjectivity theorem of Browder [26] is quite simple and is given in Chapter III.

The initial examples given above indicate that the class of A-proper mappings is quite large. Additional and more general classes of A-proper mappings $T: D \subseteq X \to Y$ are given in Section 2 and in Chapters III and IV. It will be shown in Section 2 that certain A-proper mappings remain A-proper even when perturbed by noncompact mappings.

1.3   Approximation-Solvability as Related to A-Stable and K-Coercive A-Proper Mappings

In this section we show that there exists an intimate relationship between the approximation solvability of the equation

$$T(x) = f \qquad (x \in X, f \in Y) \tag{1.11}$$

and the A-properness of $T: X \to Y$. In fact, our first result in this section characterizes A-proper maps on the whole of X.

Theorem 1.3 Suppose that $\Gamma = \{X_n, V_n; E_n, W_n\}$ is an admissible scheme for (X, Y), $T: X \to Y$ such that $T_n: X_n \to Y_n$ is continuous for each $n \in Z_+$ and there exist $n_0 \in Z_+$ and a continuous function $\alpha: R^+ \to R^+$ such that

$$\| T_n(x) - T_n(y) \| \geq \alpha(\| x - y \|) \tag{1.12}$$
$$\text{for all} \quad x, y \in X_n \quad \text{and} \quad n \geq n_0,$$

where $\alpha(0) = 0$, $\alpha(r) > 0$, and $r > 0$ and $\alpha(r) \to \infty$ when $r \to \infty$.

Then equation (1.11) is uniquely approximation solvable w.r.t. Γ for each $f \in Y$ if and only if T is A-proper w.r.t. Γ and injective.

Proof. Suppose first that T is A-proper w.r.t. Γ and injective. By (1.12), T_n is an injective continuous mapping of X_n into Y_n for each $n \geq n_0$, and hence, by the Brouwer theorem of invariance of domain, the range $R(T_n)$ is an open set in Y_n. Furthermore, it follows from (1.12) and the continuity of T_n that $R(T_n)$ is also closed in Y_n for each $n \geq n_0$. Indeed, if $\{y_m\} \subset R(T_n)$ and $y_m \to y$ in Y_n, there exists a sequence $\{x_m\} \subset X_n$ such that $y_m = T_n(x_m)$ and, by (1.12), we find that $\alpha(\| x_{m+p} - x_m \|) \leq \| y_{m+p} - y_m \| \to 0$ as $m \to \infty$ (uniformly w.r.t. $p \geq 1$). We claim that $\{x_m\}$ is a Cauchy sequence in X_n. Indeed, if not, there would exist $a > 0$ and $\{m_k\}$, $\{p_k\}$ in Z_+ with $m_k \to \infty$ as $k \to \infty$ such that $\| x_{m_k + p_k} - x_{m_k} \| \to a$. Hence, by the continuity of α, we have $\alpha(\| x_{m_k + p_k} - x_{m_k} \|) \to \alpha(a) > 0$, a contradiction. Hence $x_m \to x$ for some x in X_n and $T_n(x_m) \to T_n(x) = y$ [i.e., $R(T_n)$ is closed in Y_n for each $n \geq n_0$]. Since $R(T_n)$ is a nonempty set in Y_n, which is both open and closed in Y_n, it follows that $R(T_n) = Y_n$ for each $n \geq n_0$.

Thus, for each $n \geq n_0$ and for each given f in Y, there exists a unique $x_n \in X_n$ such that $T_n(x_n) = W_n(f)$. For the sequence $\{x_n\}$ thus determined, (1.12), and the fact that $\| W_n T(0) \| \leq K_0$ for some $K_0 > 0$ and all $n \geq n_0$ imply that

$$\| T_n(x_n) \| \geq \| T_n(x_n) - T_n(0) \| - \| T_n(0) \| \geq \alpha(\| x_n \|) - K_0.$$

Since $\| T_n(x_n) \| = \| W_n f \| \leq K_1$ for some $K_1 > 0$ and all $n \geq n_0$, it follows from this and the last inequality that $\alpha(\| x_n \|) \leq K_1 + K_0$. This and the condition that $\lim_{r \to \infty} \alpha(r) = \infty$ imply that $\{x_n\}$ is bounded with $x_n \in X_n$ for $n \geq n_0$. Since $T_n(x_n) - W_n f = 0 \to 0$ as $n \to \infty$ and T is A-proper w.r.t. Γ, there exists a subsequent $\{x_{n_j}\}$ and $x \in X$ such that $x_{n_j} \to x$ and $T(x) = f$. Now we claim that $x_n \to x$ in X. If not, there exist $\epsilon > 0$ and a subsequence $\{x_{n_k}\}$ such that $\| x_{n_k} - x \| \geq \epsilon$ for all $k \in Z_+$. But $T_{n_k}(x_{n_k}) - W_{n_k}(t) = 0 \to 0$ as $k \to \infty$, so again, by the A-properness of T, there exists a subsequence $\{x_{n_{k(j)}}\}$ of $\{x_{n_k}\}$, $\bar{x} \in X$, such that $x_{n_{k(j)}} \to \bar{x}$ as $j \to \infty$ and $T(\bar{x}) = f$. Since T is one-to-one, we see that $x = \bar{x}$, in contradiction to the assumption that $\| x_{n_k} - x \| \geq \epsilon$ for all $k \in Z_+$. Consequently, the entire sequence $\{x_n\}$ converges strongly to x; that is, equation (1.11) is uniquely approximation solvable w.r.t. Γ for each f in Y.

Converse. Since T_n is continuous, T will be A-proper if we show that condition (H) holds. So let $\{x_{n_j} \mid x_{n_j} \in X_{n_j}\}$ be any bounded sequence such that $T_{n_j}(x_{n_j}) - W_{n_j}(g) \to 0$ for some g in Y. Since, by hypothesis, equation (1.11) is uniquely approximation solvable w.r.t. Γ for each f in Y, it follows that T is injective and there exists $n_0 \geq 1$ and a unique $z_n \in X_n$ such that $T_n(z_n) = W_n(g)$ for each $n \geq n_0$, $z_n \to z$ in X, and $T(z) = g$. Hence $T_{n_j}(z_{n_j}) = W_{n_j}(g)$ for $n_j \geq n_0$ and $T_{n_j}(x_{n_j}) - T_{n_j}(z_{n_j}) = T_{n_j}(x_{n_j}) - W_{n_j}(g) \to 0$ as $j \to \infty$. If we apply the inequality (1.12) to the element $x = x_{n_j}$ and $y = z_{n_j}$, we find that

$$\alpha(\| x_{n_j} - z_{n_j} \|) \leq \| T_{n_j}(x_{n_j}) - T_{n_j}(z_{n_j})\| \to 0 \qquad \text{as} \quad j \to \infty,$$

Hence $\| x_{n_j} - z_{n_j} \| \to 0$ as $j \to \infty$, and consequently, $x_{n_j} \to z$ in X with $T(z) = g$.   Q.E.D.

Remark 1.3 Going over the proof of Theorem 1.3, we see that it suffices to assume that the function $\alpha(t)$ be defined and continuous for $t > 0$ and that (1.12) hold in the form

$$\| T_n(x) - T_n(y)\| \geq \alpha(\| x - y \|)$$
$$\text{for all} \quad x,y \in X_n \quad \text{with} \quad x \neq y \quad \text{and} \quad n \geq n_0. \tag{113}$$

Moreover, the assumption "$\alpha(t) \to \infty$ as $t \to \infty$" was not used in the proof of the A-properness of T. Since the inequality (1.13) plays an important role in the study of the constructive solvability of equation (1.11) and the stability of the approximate method

$$T_n(x) = W_n f \qquad (x_n \in X_n, \ T_n = W_n T \mid_{X_n}, \ W_n f \in E_n), \tag{1.14}$$

we shall refer to maps $T: X \to Y$ for which (1.13) holds as *approximation stable (a-stable)*.

Corollary 1.4 If $T: X \to Y$ is continuous and a-stable, then equation (1.11) is uniquely approximation solvable for each f in Y if and only if T is A-proper w.r.t. Γ.

Proof. Since $T_n = W_n T \mid_{X_n}: X_n \to Y_n$ is obviously continuous for each $n \in Z_+$, to deduce Corollary 1.4 from Theorem 1.3 it suffices to show that T is injective. Suppose that $u \neq v$ and $T(u) = T(v)$. Since $\text{dist}(x, X_n) \to 0$ for each $x \in X$, there exist $u_n, v_n \in X_n$ such that $u_n \to u$ and $v_n \to v$ in X. Thus (1.13), the continuity of T, and the uniform boundedness of $\{W_n\}$ imply that

$$\alpha(\| u_n - v_n \|) \le \| W_n T(u_n) - W_n T(v_n) \| \to 0 \qquad \text{as} \quad n \to \infty.$$

Thus $0 \ge \alpha(\| u - v \|) > 0$ since $u \ne v$. This contradiction shows that T is one-to-one.   Q.E.D.

When $T: X \to X$ is linear and Γ is a projective admissible (but not necessarily complete) scheme $\Gamma_P = \{X_n, V_n; Y_n, Q_n\}$, we deduce from Corollary 1.4 the following simple but important theorem, which characterizes the class of bounded linear A-proper mappings.

Theorem 1.4   If $T \in L(X,Y)$, then equation (1.11) is uniquely approximation solvable w.r.t. Γ_P for each f in Y if and only if T is A-proper w.r.t. Γ_P and injective.

In order to have a self-contained exposition in Chapter II of the results for equation (1.11) involving linear A-proper mappings $T \in L(X,Y)$, we omit the proof of Theorem 1.4, which is based on Corollary 1.4 (and thus ultimately on the Brouwer theorem on invariance of domain) since we deduce it in Chapter II from a more general theorem for linear A-proper maps whose proof is based on a simple finite-dimensional result from linear algebra rather than on the sophisticated Brouwer's theorem.

However, to get acquainted with the technique, the reader should do

Exercise 1.1   Prove Theorem 1.4 by using Corollary 1.4.

The importance and consequences of Theorem 1.4 are considered in Chapter II, which is devoted to the study of linear A-proper mappings. In Section 3 we use it to solve PDEs.

In view of Propositions 1.3 and 1.5, an immediate consequence of Theorem 1.3 is the following constructive surjectivity result.

Corollary 1.5   Let X be a separable reflexive Banach space, $T: X \to X^*$ a demicontinuous (continuous or weakly continuous) map and assume also that $\phi(t)$ in Definition 1.5 is such that $\alpha(t) \equiv \phi(t)/t \to \infty$ as $t \to \infty$.

(A1) If X is real and T is firmly monotone, then

$$T(x) = f \qquad (x \in X, f \in X^*) \tag{1.15}$$

is strongly approximation solvable w.r.t. Γ_I for each f in X^* and, in particular, T is bijective.

(A2) If X is complex and T is firmly complex monotone and bounded when T is demicontinuous, the conclusion of (A1) holds.

Proof. By Propositions 1.3 and 1.5, T is A-proper w.r.t. Γ_l. Furthermore, since $V_n x = x$ for all $x \in X_n$, it follows from (1.8) when X is real that

$$\phi(\| x - y \|) \leq (Tx - Ty, V_n(x - y)) = (T_n(x) - T_n(y), x - y)$$

$$\leq \| T_n(x) - T_n(y) \| \, \| x - y \|.$$

Hence setting $\alpha(t) = \phi(t)/t$ for $t > 0$, we see that $\alpha(t)$ is a continuous function from $(0,\infty)$ into $(0,\infty)$ such that $\alpha(t) > 0$ for $t > 0$, $\alpha(t) \to \infty$ if $t \to \infty$ and

$$\| T_n(x) - T_n(y) \| \geq \alpha(\| x - y \|)$$

$$\text{for all} \quad x,y \in X_n, x \neq y, n \in Z_+ .$$

The same facts are also true when X is complex and T is firmly complex monotone. Consequently, the assertions of Corollary 1.5 follow from Theorem 1.3.   Q.E.D.

Theorem 1.3 points to the importance of the problem of finding conditions which, assuming (1.13), would imply that $T: X \to Y$ is A-proper w.r.t. Γ. The partial answer to this problem is given by Theorem 1.5 below. We discuss this problem and a more general problem of finding sufficient conditions for a given map $T: D \subset X \to Y$ to be A-proper more comprehensively in subsequent sections.

Definition 1.8 A mapping $T: D \subseteq X \to Y$ is said to be *pseudo-A-proper* if $T_n: D_n \to Y_n$ is continuous and if T satisfies condition (h): For any bounded sequence $\{x_{n_j} \mid x_{n_j} \in D_{n_j}\}$ such that $T_{n_j}(x_{n_j}) - W_{n_j}g \to 0$ for some g in Y, there exists $x \in D$ such that $Tx = g$.

Chapter IV is devoted to the study of the solvability of equations involving pseudo-A-proper mappings, where it is shown that it is indeed a very large class of maps for which existence results can be established. For the present we have the following:

Theorem 1.5 Suppose that $T: X \to Y$ is continuous and a-stable w.r.t. $\Gamma = \{X_n, V_n; E_n, W_n\}$. Then the following assertions are equivalent:

(A1) T is A-proper w.r.t. Γ.
(A2) Equation (1.11) is uniquely approximation solvable for each $f \in Y$.
(A3) T is surjective.
(A4) T satisfies condition (h).

Proof. To prove Theorem 1.5 it suffices to show that, under our hypotheses, (A1) $\Rightarrow$ (A2) $\Rightarrow$ (A3) $\Rightarrow$ (A4) $\Rightarrow$ (A1).

First note that since $T: X \to Y$ is continuous and a-stable, the proof of Corollary 1.4 shows that T is one-to-one. Furthermore, by Corollary 1.4 and Definition 1.2, to prove Theorem 1.5 it suffices to show that (A4) $\Rightarrow$ (A1).

So let $\{x_{n_j} \mid x_{n_j} \in X_{n_j}\}$ be any bounded sequence such that $T_{n_j}(x_{n_j}) - W_{n_j}(g) \to 0$ for some g in Y. Hence, by condition (h), there exists $x \in X$ such that $Tx = g$. Since Γ is admissible, there exists $y_n \in X_n$ such that $y_n \to x$ in X. If we apply the inequality (1.13) to the elements $x = x_{n_j}$ and $y = y_{n_j}$, we find that

$$\alpha(\| x_{n_j} - y_{n_j} \|) \leq \| T_{n_j}(x_{n_j}) - T_{n_j}(y_{n_j}) \|$$

$$\leq -\| T_{n_j}(x_{n_j}) - W_{n_j}(g)\|$$

$$+ \| W_{n_j}(g) - W_n T(y_{n_j})\| \to 0$$

since the first term on the right approaches zero by assumption, while the second also converges to zero because $y_{n_j} \to x$, T is continuous, $\{W_n\}$ are uniformly bounded, and $Tx = g$.   Q.E.D.

Remark 1.4 We would like to emphasize the practical usefulness of Theorem 1.5. It allows us to construct a strongly convergent sequence of approximate solutions $x_n \in X_n$ of equation (1.11) if (1.13) holds and we know (no matter how) that T is either surjective or T satisfies condition (h). The various classes of mappings for which this is the case are discussed in Section 2 and Chapter III.

Our next result shows that equation (1.11) admits constructive solvability without the condition that T is a-stable. To that end we first recall:

Definition 1.9 If $K: X \to Y^*$ is some map such that $Kx \neq 0$ when $x \neq 0$, then $T: X \to Y$ is called *K-coercive* when

$$(Tx, Kx)/\| Kx \| \to \infty \quad \text{as} \quad \| x \| \to \infty. \tag{1.16}$$

It is obvious that $T: X \to Y$ is K-coercive if there exists a function $c(r)$ of R^+ into R^1 with $c(r) \to \infty$ as $r \to \infty$ such that

$$(Tx, Kx) \geq c(\| x \|) \| Kx \| \quad \text{for all} \quad x \in X. \tag{1.17}$$

Our next result establishes the approximation solvability of (1.11) involving K-coercive A-proper mappings which will prove to be useful in various applications to differential equations not necessarily in divergence form.

Theorem 1.6 Let K be a map of X into Y^* with $Kx \neq 0$ when $x \neq 0$, K_n a map of X_n into $D(W_n^*)$, and $\Gamma = \{X_n, V_n; E_n, W_n\}$ an admissible scheme for (X, Y) such that

$$(W_n g, K_n(x)) = (g, Kx) \qquad \text{for all} \quad x \in X_n, \quad g \in Y, \quad n \in Z_+.$$

$$(1.18)$$

Let $M_n: X_n \to E_n$ be a linear isomorphism such that

$$(M_n(x), K_n(x)) > 0 \qquad \text{for} \quad x \in X_n \quad \text{with} \quad x \neq 0. \qquad (1.19)$$

If $T: X \to Y$ is A-proper w.r.t. Γ and K-coercive, equation (1.11) is feebly approximation-solvable for each f in Y. Moreover, if for a given $f \in Y$, (1.11) has at most one solution, it is strongly approximation solvable.

We omit the proof of Theorem 1.6 since it will be deduced as a special case from a more general theorem in Chapter III.

Let us add that, by virtue of Theorem 1.2, Theorem 1.6 allows us to study the approximation solvability of the perturbed equations of the form

$$Tx = Ax + Cx = f \qquad (f \in Y), \qquad (1.20)$$

where $C: X \to Y$ is compact and $A: X \to Y$ is A-proper. In this case the conditions of Theorem 1.6 are implied, for example, by the assumption:

A is K-coercive and there exists a constant $c_0 > 0$ such that $(Cx, Kx) \geq -c_0 \| Kx \|$ for $x \in X$. $\qquad (1.21)$

2. FURTHER EXAMPLES OF A-PROPER MAPPINGS

In recent years the solvability of operator equations involving mappings of *strongly monotone* and *accretive types* and the mappings of *condensing type* have been studied by many authors. Various surjectivity and fixed-point theorems, mapping theorems, and topological degree theories have been obtained for these mappings and then applied successfully to existence problems for certain classes of nonlinear ordinary and partial differential equations and integral and integrodifferential equations.

The purpose of this section is to show that if these mappings act in spaces having admissible schemes, they form proper subclasses of A-proper mappings. Consequently, the surjectivity and fixed-point theorems obtained in a constructive way, together with mapping theorems and the generalized degree theory for A-proper mappings, will unify and extend some of the basic results for these special classes of mappings.

2.1 Ball-Contractive Perturbation of Accretive Mappings

Let us recall that the set $G \subset X$ is precompact if and only if to each $\epsilon > 0$ there are finitely many balls of radius ϵ such that their union covers G. If G is only bounded, there is a positive lower bound for such numbers ϵ. These facts suggest the following notions introduced in [103] and [144], respectively.

Definition 2.1 Let G be any bounded set in X. Then the *ball measure of noncompactness of G relative to X* is defined by $\beta_X(G) = \inf\{r > 0 \mid G$ admits a finite r-ball covering with centers in $X\}$. The *set measure of noncompactness* of G is defined by $\gamma(G) = \inf\{d > 0 \mid G$ admits a finite covering by sets of diameter $\leq d\}$.

The following lemma concerning the properties of β_X and γ will be useful in the sequel.

Lemma 2.1 Let Φ denote either β_X or γ, and let G and Q be any bounded set in X. Then:

(i) $G \subset Q$ implies that $\Phi(G) \leq \Phi(Q)$; $\Phi(G \cup Q) = \max\{\Phi(G), \Phi(Q)\}$.

(ii) $\Phi(\overline{G}) = \Phi(G)$ and $\Phi(G) = 0$ if and only if $\overline{G}$ is compact.

(iii) $\Phi(\lambda Q) = |\lambda| \phi(Q)$ and $\Phi(G + Q) \leq \Phi(G) + \Phi(Q)$ for each $\lambda \in R$, where $G + Q = \{x + y \mid x \in G, y \in Q\}$.

(iv) $\Phi(co(G)) = \Phi(G)$, where $co(G)$ denotes the convex hull of G.

(v) $\Phi(N_\epsilon(G)) \leq \Phi(G) + 2\epsilon$, where $N_\epsilon(G) = \{x \in X \mid \text{dist}(x, G) < \epsilon\}$.

For the proof of these properties, see [59,101,191,253].

Remark 2.1 It should be noted, however, that despite Lemma 2.1, the measures β_X and γ differ in one crucial way. If $G \subset Q \subset X$ and Q inherits its metric from X, then the set measure of G, as a subset of Q, is the same as the set measure of noncompactness of G as a subset of X; and this is reflected in the notation $\gamma(G)$, which does not explicitly mention the encompassing space. On the other hand, it can easily happen that $\beta_Q(G) \neq \beta_X(G)$. For example, if $\Gamma = \{x \in X : \|x\| = 1\} \subset H$, then $\beta_\Gamma(\Gamma) = \sqrt{2}$, while $\beta_B(\Gamma) = 1$, where $B = \overline{B}(0,1) \subset H$. If X is a Banach space, then $\beta_X(B) = 1$, while $\gamma(B) = 2$. In general, one can easily show that if X is a Banach space and $G \subset X$ bounded, then $\beta_X(G) \leq \gamma(G)$ and $\gamma(G) \leq 2\beta_X(G)$. Since, for any $G \subset X$, we shall only consider its ball measure of noncompactness relative to X, instead of $\beta_X(G)$ we will just write $\beta(G)$. See [192,291] for proof and other comments.

Closely associated with β and γ are the concepts of k-ball and k-set contractions defined as follows.

Definition 2.2 A continuous bounded map $T: D \subseteq X \to Y$ is called *k-ball contractive* (respectively, *k-set-contractive*) if and only if $(\beta(T(G)) \le k\beta(G)$ [respectively, $\gamma(T(G)) \le k\gamma(G)$] for each bounded $G \subset D$ and some $k \ge 0$.

Notice that in the above we used β (respectively, γ) to refer to the ball measure (respectively, set measure) of noncompactness in X and in Y even though they are different. We hope that no confusion will result from this usage.

It follows from the definition that $T: D \subset X \to Y$ is compact iff T is 0-ball contractive iff T is 0-set contractive. It is also easy to show that if $T_j: X \to Y$ is k_j-ball contractive ($j = 1,2$) and $T_3: Y \to Z$ is k_3-ball contractive, then $T_3 T_1$ is $k_3 k_1$-ball contractive and $T_1 + T_2$ is $(k_1 + k_2)$-ball contractive. The same is true for k-set contractions. An immediate consequence of the remarks above is the following simple result, which illustrates the difference between the two classes.

Lemma 2.2 If $S: D \subset X \to Y$ is k-contractive and $C: D \to Y$ compact, then $T = S + C$ is k-set contractive. T is k-ball contractive provided that S is defined and k-contractive on all of X.

Proof. Let $G \subset D$ by any bounded set and $\gamma(G) = d$. Then given $\epsilon > 0$, we can write $G = \cup_{i=1}^m A_i$, $\text{diam}(A_i) \le d + \epsilon$. Thus $S(G) = \cup_{i=1}^m S(A_i)$, and since S is k-contractive, $\text{diam}(S(A_i)) \le k(d + \epsilon)$. Hence $\gamma(S(G)) \le kd$ since $\epsilon > 0$ is arbitrary, so $S: D \to Y$ and $T = S + C: D \to Y$ is k-set contractive since C is compact.

Suppose now that $S: X \to Y$ is k-contractive and let $G \subset D$ be any bounded set with $\beta(G) = r$. Then given $\epsilon > 0$, we can choose $\{x_1, \ldots, x_k\} \subset X$ with $G \subset \cup_{i=1}^k B(x_i, r + \epsilon)$. This and our condition on S imply that $S(G) \subset \cup_{i=1}^k B(Sx_i, k(r + \epsilon))$. Thus $\beta(S(G)) \le k(r + \epsilon)$, so $\beta(S(G)) \le k(r + \epsilon)$ since $\epsilon > 0$ is arbitrary. Q.E.D.

It follows from the proof of the second part of Lemma 2.2 that if S is defined only on the proper subset D of X, we are unable to show that $S: D \to Y$ is k-ball contractive since there is no guarantee that for an arbitrary $G \subset D$, the centers of the balls defining $\beta(G)$ lie in D.

In general, one cannot say much about the precise relation between k-ball and k-set contractions except for a general fact that follows from Remark 2.1: If $T: D \subseteq X \to Y$ is k-set contractive, then T is $2k$-ball contractive, and conversely, if $T: D \subseteq X \to Y$ is k-ball contractive, then T is $2k$-set contractive. Lemma 2.2 shows that in practice to verify that a map is k-ball contractive requires that it be everywhere defined, or at least defined on some larger subset than D, which, as we shall see, is usually the case in various applications. This point is illustrated by Lemma 2.3 below concerning semicontractive maps introduced by Browder [31] and Kirk [132]. See [290] (and [224]) for the proof of Lemma 2.3.

Definition 2.3   A continuous map $T: D \subset X \to Y$ is k-semicontractive if there exists a constant $k \in (0,1]$ and a continuous map $V: X \times X \to Y$ such that $T(x) = V(x,x)$ for $x \in D$ and for each fixed $x \in X$, $V(\cdot,x): X \to Y$ is k-contractive and $V(x,\cdot): X \to Y$ is compact.

Lemma 2.3   If $T: D \subset X \to Y$ is a continuous k-semicontractive map defined above, then T is k-ball contractive.

Proof.   Let $G \subset D$ be any bounded set with $\beta(G) = r$. Then given $\epsilon > 0$, we can choose $\{x_1, \ldots, x_k\} \subset X$ with $G \subset \cup_{j=1}^k B(x_j, r + \epsilon)$. For each x in X, $V(x,G)$ is a precompact set in Y, so $\cup_{n=1}^k V(x_n,G)$ is also precompact. Therefore, for the given $\epsilon > 0$, there exist points $z_1, \ldots,$ z_p in Y such that $\cup_{n=1}^k V(x_n,G) \subset \cup_{j=1}^p B(z_j,\epsilon)$. Now, given any x in G we choose n such that $\| x - x_n \| \leq r + \epsilon$, and observe that $\| V(x,x) - V(x_n,x)\| \leq k \| x_n - x \| \leq k(r + \epsilon)$. Moreover, we may choose j such that $\| V(x_n,x) - z_j \| < \epsilon$. Thus

$$\| T(x) - z_j \| = \| V(x,x) - z_j \| \leq \| V(x,x) - V(x_n,x)\|$$
$$+ \| V(x_n,x) - z_j \|$$
$$\leq k(r + \epsilon) + \epsilon.$$

It follows from this that $T(G) \subset \cup_{j=1}^p B(z_j, kr + k\epsilon + \epsilon)$. Since $\epsilon > 0$ can be chosen arbitrarily small, we get $\beta(T(G)) \leq k\beta(G)$ (i.e., T is k-ball contractive).   Q.E.D.

In what follows we shall also need a slight generalization of k-ball contractions with $k < 1$ introduced in [94,291] and given by:

Definition 2.4   A continuous bounded map $T: D \subset X \to Y$ is called *ball condensing* iff $\beta(T(G)) < \beta(G)$ for each bounded set $G \subset D$ with $\beta(G) \neq 0$.

Similarly, one defines a set-condensing map. It follows that every k-ball-contractive map with $k < 1$ is ball condensing, but the converse is not true (see [191]).

The main reason for considering the ball measure of noncompactness is contained in the following result, which enables us to show that ball-condensing vector fields and the k-ball-contractive perturbations of strongly accretive maps form subclasses of A-proper mappings. We need [62]:

Proposition 2.1 Let $\Gamma_\alpha = \{X_n, P_n\}$ be an admissible scheme for the Banach space X and let $G \subset X$ be bounded. Then

$$\beta(G) \leq \beta(\bigcup_{n \geq 1} P_n G)$$

$$= \lim_m \beta(\bigcup_{n \geq m} P_n G) \leq \beta(G)\alpha, \qquad \text{where} \quad \alpha = \sup_n \| P_n \|.$$

Proof. The second equality is obvious since $\bigcup_{n=1}^m P_n(G)$ is precompact. Let $r > \beta(\bigcup_{n \geq m} P_n G)$ and choose $\{x_1, \ldots, x_k\} \subset X$ such that $\bigcup_{n \geq m} P_n G \subset \bigcup_{i=1}^k B(x_i, r)$. Then $G \subset \bigcup_{i=1}^k B(x_i, r)$ since $P_n x \to x$ for each x in X, and therefore $\beta(G) \leq r$. Hence

$$\beta(G) \leq \lim_m \beta(\bigcup_{n \geq m} P_n G).$$

Now, let $G \subset \bigcup_{i=1}^k B(x_i, r)$ with $r > \beta(G)$. Given $\epsilon > 0$, we can choose $m \in Z_0$ such that $\| P_n x_i - x_i \| \leq \epsilon$ for $n \geq m$ and $i = 1, 2, \ldots, k$ since $P_n x \to x$ for each x in X. It follows from this and the fact that $\sup_n \| P_n \| = \alpha$ that

$$\bigcup_{n \geq m} P_n(G) \subset \bigcup_{i=1}^k B(x_i, r\alpha + \epsilon).$$

This implies that $\lim_m \beta(\bigcup_{n \geq m} P_n G) \leq \beta(G)\alpha$. \hfill Q.E.D.

We will mostly be interested in the following corollary of Proposition 2.1, proved by Webb [291].

Corollary 2.1 If $\{y_n\}$ is any bounded sequence in X, then $\beta(\{P_n y_n\}) \leq \beta(\{y_n\}) \sup_n \| P_n \|$.

We are now in the position to prove the first basic result in this section (see [171,220,221,275,294]).

Theorem 2.1 Suppose that the projectional scheme $\Gamma_P = \{X_n, V_n; Y_n, Q_n\}$ is complete for (X, Y) with $\eta = \sup_n \| Q_n \|$, $D \subseteq X$ is closed and $T: D \to Y$ is a continuous map such that:

(2.1) There exist a constant $\mu_0 > 0$ and an $n_0 \in Z_+$ such that $\beta(\{Q_n T x_n\}) \geq \mu_0 \beta(\{x_n\})$ for each bounded sequence $\{x_n \mid x_n \in D_n\}$ with $n \geq n_0$.

If $F: D \to Y$ is a k-ball contraction such that $k < \mu_0\eta^{-1}$, then $T_t \equiv T + tF: D \to Y$ is A-proper w.r.t. Γ_P for each $t \in [-1,1]$.

If $\mu_0 = \eta = 1$, then the same conclusion holds when F is ball condensing.

Proof. Note first that $T_{tn}: D_n \subset X_n \to Y_n$ is obviously continuous for each $n \in Z_+$ and $t \in [-1,1]$. Now, let $t \in [-1,1]$ be fixed and let $\{x_{n_j} \mid x_{n_j} \in D_{n_j}\}$ be any bounded sequence such that $Q_{n_j}T_t(x_{n_j}) - Q_{n_j}(g_t) \to 0$ as $j \to \infty$ for some $g_t \in Y$, where without loss of generality we assume that $n_j \geq n_0$ for all $j \in Z_+$. Since $Q_{n_j}(g_t) \to g_t$ in Y, we see that

$$g'_{n_j} \equiv Q_{n_j}T(x_{n_j}) + tQ_{n_j}F(x_{n_j}) \to g_t \text{ in } Y.$$

Suppose first that $t = 0$. Then $Q_{n_j}T(x_{n_j}) \to g_0$ in Y, and therefore, in view of (2.1), $\beta(\{x_{n_j}\}) = 0$ since $\beta(\{Q_{n_j}Tx_{n_j}\}) = 0$. Hence there exist a subsequence $\{x_{n_{j(k)}}\}$ and $x \in X$ such that $x_{n_{j(k)}} \to x$ in X with $x \in D$ since D is closed. Thus the continuity of T and the completeness of Γ_P imply that $Tx = g_0$ (i.e., T is A-proper).

Suppose now that $t \neq 0$. Since $\{g'_{n_j}\}$ is precompact and

$$Q_{n_j}T(x_{n_j}) = g'_{n_j} - tQ_{n_j}F(x_{n_j}) \qquad \text{for each } j \in Z_+, \tag{2.2}$$

it follows from (2.2), (2.1), Corollary 2.1, and the condition on F that $\mu_0\beta(\{x_{n_j}\}) \leq \beta(\{Q_{n_j}T(x_{n_j})\}) \leq |t| \; \beta(\{Q_{n_j}F(x_{n_j})\}) \leq k\eta\beta(\{x_{n_j}\})$. Thus $\beta(\{x_{n_j}\}) = 0$, so $\{x_{n_j}\}$ has a convergent subsequence $\{x_{n_{j(k)}}\}$ with $x_{n_{j(k)}} \to x$ for some $x \in D$. Hence, by the continuity of T and F and the completeness of Γ_P, we see that $Tx + tFx = g_t$ (i.e., T_t is A-proper w.r.t. Γ_P for each $t \in [-1,1]$).

To prove the second part, note that if $\mu_0 = \eta = 1$, it follows from (2.2), (2.1), and Corollary 2.1 that

$$\beta(\{x_{n_j}\}) \leq \beta(\{Q_{n_j}Tx_{n_j}\}) \leq |t| \; \beta(\{Q_{n_j}Fx_{n_j}\}) < \beta(\{x_{n_j}\}).$$

This again implies that $\{x_{n_j}\}$ has a convergent subsequence $\{x_{n_{j(k)}}\}$ such that $x_{n_{j(k)}} \to x \in D$ and $T_t(x) = g$. \hfill Q.E.D.

Corollary 2.2 When $Y = X$, $Y_n = X_n$, and $Q_n = P_n$ for each $n \in Z_+$ and $T = I$, then $T_t = I + tF: D \to X$ is A-proper w.r.t. $\Gamma_\alpha = \{X_n, P_n\}$ for each $t \in [-1,1]$ if F is either k-ball contractive with $k < \alpha^{-1}$ or $\alpha = 1$ and F is ball-condensing.

Proof. If $Y = X$, $\Gamma_P = \Gamma_\alpha = \{X_n, P_n\}$ and $T = I$, the hypothesis (2.1) is trivially satisfied with $\mu_0 = 1$ and $n_0 = 1$, so Corollary 2.2 follows from Theorem 2.1. \hfill Q.E.D.

Our next result shows that certain A-proper mappings remain A-proper even when perturbed by noncompact mappings.

Theorem 2.2 Suppose that the scheme Γ_P in Theorem 2.1 is also nested, $T: X \to Y$ is continuous and A-proper w.r.t. Γ_P, and there exist $n_0 \in Z_+$ and $\mu_0 > 0$ such that

$$\| Q_n T(x) - Q_n T(y) \| \geq \mu_0 \| x - y \| \tag{2.3}$$
$$\text{for all} \quad x, y \in X_n \quad \text{and} \quad n \geq n_0.$$

If $F: D \subseteq X \to Y$ is k-ball contraction with $k < \mu_0 \eta^{-1}$ and with D some closed set, then $T_t = T + tF: D \to Y$ is A-proper w.r.t. Γ_P for each $t \in [-1,1]$. If $\mu_0 = \eta = 1$, the same conclusion holds when F is ball-condensing.

Proof. Theorem 2.2 follows from Theorem 2.1, Corollary 1.4, and the following:

Lemma 2.4 If T satisfies the conditions of Theorem 2.2, the hypothesis (2.1) of Theorem 2.1 holds: that is, $\beta(\{Q_n T x_n\}) \geq \mu_0 \beta(\{x_n\})$ for each bounded sequence $\{x_n \mid x_n \in D_n\}$ with $n \geq n_0$.

Proof. For $n \geq n_0$ let $\{x_n \mid x_n \in D_n\}$ be any bounded sequence and suppose that $\{Q_n T x_n\}$ is covered by finitely many balls $B(y_j, r)$ with $\{y_1, \ldots, y_k\} \subset Y$. Since $T: X \to Y$ is continuous and A-proper w.r.t. Γ_P for which (2.3) holds, Corollary 1.4 implies that T is bijective, and consequently, there exist $\{u_1, \ldots, u_k\} \subset X$ such that $T(u_j) = y_j$ for $1 \leq j \leq k$. By continuity of T and the denseness of $\cup_n X_n$ in X, there exist $z_j \in X_{n_j}$ such that $\| T z_j - T u_j \| < \epsilon$ for $1 \leq j \leq k$ and any given $\epsilon > 0$. Let $n_1 \geq n_0$ be such that $\| Q_n T z_j - T z_j \| < \epsilon$ for $n \geq n_1$ and $1 \leq j \leq k$. Then, for those n such that $\{Q_n T x_n\}$ belongs to $B(y_j, r)$ and exceeds n_j (with j fixed) we have that $z_j \in X_n$ since $\{X_n\}$ is nested and

$$\mu_0 \| x_n - z_j \| \leq \| Q_n T x_n - Q_n T z_j \| \leq \| Q_n T x_n - y_j \|$$
$$+ \| T u_j - T z_j \| + \| T z_j - Q_n T z_j \|$$
$$\leq r + 2\epsilon.$$

It follows that the set $\{x_n \mid n \geq \max(n_1, n_2)\}$ is covered by the balls $B(z_j, (r + 2\epsilon)/\mu_0)$, $1 \leq j \leq k$. As ϵ is arbitrary, we have shown that $\beta(\{x_n\}) \leq \mu_0^{-1} \beta(\{Q_n T x_n\})$ for $n \geq n_0$, as required. Q.E.D.

Corollary 2.3 Suppose that all the conditions of Theorem 2.2 hold except for the requirement that T is A-proper, which is replaced by the hypothesis that T is either surjective or satisfies condition (h). Then the conclusions of Theorem 2.2 concerning the A-properness of $T + tF: D \to Y$ hold.

Proof. Since $T: X \to Y$ is continuous and a-stable with $\alpha(t) = \mu_0 t$, in virtue of Theorem 1.5, the hypothesis that T is either surjective or T satisfies condition (h) implies that T is A-proper w.r.t. Γ_P. Hence Corollary 2.3 follows from Theorem 2.2.   Q.E.D.

To illustrate the type of operators for which Theorem 2.2 or Corollary 2.3 applies, we need the following

Definition 2.5   Let $K: X \to Y^*$ be a (possibly nonlinear) mapping. An operator $T: X \to Y$ is said to be *strongly K-monotone* or *strongly accretive* if there exists a constant $m > 0$ such that

$$(Tx - Ty, K(x - y)) \geq m \| x - y \| \| K(x - y)\| \qquad (2.4)$$
$$\text{for all} \quad x, y \in X.$$

It will be shown later that depending on some additional conditions on K and/or X and Y, T satisfying (2.4) is bijective or satisfies condition (h). Hence, for each such mapping T, Corollary 2.3 implies:

Corollary 2.4   Suppose that $T: X \to Y$ is continuous and strongly K-monotone and the scheme Γ_P is such that

$$Q_n^* K(x) = K(x) \qquad \text{for all} \quad x \in X_n \quad \text{and} \quad n \in Z_+. \qquad (2.5)$$

If $F: D \subseteq X \to Y$ satisfies the conditions of Theorem 2.2 with $\mu_0 = m$, then $T + tF: D \to Y$ is A-proper w.r.t. Γ_P for each $t \in [-1,1]$.

Proof.   First note that by virtue of (2.5), it follows from (2.4) that $\| Q_n Tx - Q_n Ty \| \geq m \| x - y \|$ for all $x,y \in X_n$ [i.e., T satisfies (2.3) of Theorem 2.2 with $\mu_0 = m$ and $n_0 = 1$]. Since we may assume that T is either surjective or satisfies condition (h), Corollary 2.4 follows from Corollary 2.3.   Q.E.D.

Thus, for example, if X is a separable reflexive Banach space, $Y = X^*$ and $\Gamma_P = \{X_n, V_n; R(P_n^*), P_n^*\}$ with $P_n^*(w) \to w$ in X^* for each w in X^*, then Corollary 2.4 applies if we set $K = I$ and note that in this case $T: X \to X^*$ is strongly monotone and thus A-proper and surjective. When $Y = X$ and $K = J$, following the arguments of Webb [295], it will be shown in Chapter III that even a demicontinuous strongly accretive map is A-proper without recourse to differential equations.

2.2   Mappings of Modified Type (KS) and (KS$_+$)

It was noted in Section 1.2 that, among the operators $T: X \to X^*$ of monotone type, bounded and continuous mappings of type (S) [and of type (S$_+$)] introduced by Browder [35] play a very important role in the solvability of PDEs that are in divergence form. A somewhat more restrictive class of maps $T: X \to X^*$ was studied earlier by Pohodjayev (see *Condition 1* in [239]). However, when T is not bounded and, more important, when T maps X into Y and $Y \neq X^*$, condition (S) cannot be used directly to establish the A-properness of T. In [210] the author extended the notions of [35] by modifying the condition (S) of Browder in such a way as to be applicable to continuous mappings $T: X \to Y$ and to be more in consonance with the notion of the A-properness w.r.t. a given scheme Γ. This extension, which is carried out in this section, will also allow us to apply the abstract approximation-solvability and existence results to differential equations, which need not be in divergence form. The basic result in this section is Theorem 2.3, which establishes the A-properness of a mapping $T: X \to Y$ of modified types (KS) and (KS$_+$) under suitable conditions on the mapping $K: X \to Y^*$ and the admissible scheme $\Gamma = \{X_n, V_n; E_n, W_n\}$. When T is continuous, Propositions 1.3 and 1.4, as well as other results, will be deduced from Theorem 2.3.

Definition 2.6   Let $K: X \to Y^*$ be a suitable (in general nonlinear) operator. A mapping $T: X \to Y$ is said to be of the *modified type* (KS) [respectively, *modified type* (KS$_+$)] if for any sequence $\{x_{n_j} \mid x_{n_j} \in X_{n_j}\}$ for which $x_{n_j} \rightharpoonup x$ in X and $\lim_j (Tx_{n_j} - Ty_{n_j}, K(x_{n_j} - y_{n_j})) = 0$ [respectively, $\overline{\lim}_j (Tx_{n_j} - Ty_{n_j}, K(x_{n_j} - y_{n_j})) \leq 0$] we have $x_{n_j} \to x$ in X, where $\{y_{n_j} \in X_{n_j}\}$ is such that $y_{n_j} \to x$ in X as $j \to \infty$.

Theorem 2.3   Let Γ be an admissible scheme for (X,Y) with X reflexive, $K: X \to Y^*$ continuous and $K_n: X_n \to D(W_n^*)$ such that

$$\{K_n(z_n)\} \text{ is bounded whenever } \{z_n \mid z_n \in X_n\} \text{ is bounded,} \qquad (2.6)$$

$$Kz_n \rightharpoonup K(0) = 0 \text{ in } Y^* \text{ whenever } z_n \rightharpoonup 0 \text{ in } X \text{ and } \overline{R(K)} = Y^*, \qquad (2.7)$$

$$(W_n g, K_n x) = (g, Kx) \text{ for all } x \in X_n, g \in Y, n \in Z_+. \qquad (2.8)$$

Then, if $T: X \to Y$ is continuous and of modified type (KS), T is A-proper w.r.t. Γ. Moreover, the continuous maps $T: X \to Y$ of modified type (KS$_+$) form a convex subclass of the class of modified type (KS).

Proof.   Since $T_n: X_n \to Y_n$ is obviously continuous for each $n \in Z_+$, to prove the A-properness of $T: X \to Y$ it suffices to show that T satisfies condition (H). So let $\{x_{n_j} \mid x_{n_j} \in X_{n_j}\}$ be any bounded sequence such that $T_{n_j}(x_{n_j}) - W_{n_j}g \to 0$ for some g in Y. Now, since $\{x_{n_j}\}$ is bounded and X is reflexive, we may assume that $x_{n_j} \rightharpoonup x_0$ in X for some $x_0 \in X$. Let

$\{y_n \mid y_n \in X_n\}$ be a sequence such that $y_n \to x_0$ in X. Then $x_{n_j} - y_{n_j} \in X_{n_j}$, $x_{n_j} - y_{n_j} \to 0$ in X since $y_{n_j} \to x_0$ and by virtue of (2.8) and a simple calculation, one gets the equality

$$(Tx_{n_j} - Ty_{n_j}, K(x_{n_j} - y_{n_j}))$$

$$= (T_{n_j}(x_{n_j}) - W_{n_j}g, K_{n_j}(x_{n_j} - y_{n_j})) + (g - Ty_{n_j}, K(x_{n_j} - y_{n_j})).$$

Since $Ty_{n_j} \to Tx_0$ in X, $K(x_{n_j} - y_{n_j}) \to 0$ in Y^* by (2.7), it follows from this and (2.6) that both terms on the right-hand side of the equality above converge to zero. Hence $(Tx_{n_j} - Ty_{n_j}, K(x_{n_j} - y_{n_j})) \to 0$ with $x_{n_j} \rightharpoonup x_0$ and $y_{n_j} \to x_0$ as $j \to \infty$. Since T is of modified type (KS), it follows that $x_{n_j} \to x_0$ in X and, consequently, $Tx_{n_j} \to Tx_0$ in Y by the continuity of T. To show that $Tx_0 = g$, note that for any fixed $w \in R(K)$ there exists $x \in X$ and $z_n \in X_n$ such that $w = Kx$, $z_n \to x$ in X, and $Kz_n \to Kx$ in Y^*. Thus, using (2.6)–(2.8), we see that for each $w \in R(K)$,

$$(g - Tx_0, w) = (g - Tx_0, Kx) = \lim_j (g - Tx_{n_j}, Kz_{n_j})$$

$$= \lim_j (W_{n_j}(g) - Tn_j(x_{n_j}), K_{n_j}(z_{n_j})) = 0.$$

Since $R(K)$ is dense in Y^* and $(g - Tx_0, w) = 0$ for each $w \in R(K)$, it follows that $Tx_0 = g$ (i.e., T is A-proper).

To prove the second part of our assertion, note first that, by its definition, every map of modified type (KS$_+$) is necessarily of modified type (KS). Thus we need only to prove the convexity of this class. Let $T_1, T_2 : X \to Y$ be of modified type (KS$_+$), and for each fixed $\lambda \in (0,1)$ let $T_\lambda \equiv \lambda T_1 + (1 - \lambda)T_2$. Let $\{x_{n_j} \mid x_{n_j} \in X_{n_j}\}$ and $\{y_{n_j} \mid y_{n_j} \in X_{n_j}\}$ be such that $x_{n_j} \rightharpoonup x_0$ in X, $y_{n_j} \to x_0$ in X, and $\lim \sup(T_\lambda(x_{n_j}) - T_\lambda(y_{n_j}), K(x_{n_j} - y_{n_j})) \le 0$. Since λ and $(1 - \lambda)$ are positive, it follows from the equality

$$(T_\lambda(x_{n_j}) - T_\lambda(y_{n_j}), K(x_{n_j} - y_{n_j})) = \lambda(T_1(x_{n_j}) - T_1(y_{n_j}), K(x_{n_j} - y_{n_j}))$$

$$+ (1 - \lambda)(T_2(x_{n_j}) - T_2(x_{n_j}), K(x_{n_j} - y_{n_j}))$$

that for an infinite subsequence $\{x_{n_{j(k)}}\}$ we have either

$$\lim \sup(T_1(x_{n_j}) - T_1(y_{n_j}), K(x_{n_j} - y_{n_j})) \le 0$$

or

$$\lim \sup(T_2(x_{n_j}) - T_2(y_{n_j}), K(x_{n_j} - y_{n_j})) \le 0.$$

In either case we see that $x_{n_{j(k)}} \to x$ as $k \to \infty$. This allows that T_λ is of the modified type (KS$_+$).   Q.E.D.

An immediate corollary of Theorem 2.3 is the following practically useful result.

Corollary 2.5 Suppose that (X, Y), Γ, K, and K_n satisfy the conditions of Theorem 2.3 and that $T: X \to Y$ is a continuous map such that:

There exists a compact map $C: X \to Y$ and a functional $\phi: X \to R^1$ with $\phi(0) = 0$ which is weakly upper semicontinuous at $x = 0$ [i.e., if $z_j \to 0$ in X, then $\lim_j \sup \phi(z_j) \leq \phi(0)$] and such that for each sufficiently large $n \in Z_+$,

$$(Tx - Ty, K(x - y)) + (Cx - Cy, K(x - y)) \qquad (2.9)$$
$$+ \phi(x - y) \geq c(\| x - y \|) \qquad \forall \ x, y \in X_n,$$

where $c: R^+ \to R^+$ is continuous and $t \to 0$ whenever $c(t) \to 0$.

Then T is of modified type (KS_+) and, in particular, T is A-proper.

Proof. Let $\{x_{n_j} \mid x_{n_j} \in X_{n_j}\}$ be any sequence such that $x_{n_j} \rightharpoonup x_0$ in X, $\{y_{n_j} \mid y_{n_j} \in X_{n_j}\}$ such that $y_{n_j} \to x_0$ in X and

$$\lim_j \sup a_{n_j} \equiv \lim_j \sup (Tx_{n_j} - Ty_{n_j}, K(x_{n_j} - y_{n_j})) \leq 0.$$

To show that $x_{n_j} \to x_0$ in X, note first that since C is compact, we may assume that $Cx_{n_j} \to g$ for some g in Y. Since $Cy_{n_j} \to Cx_0$ in X, $x_{n_j} - y_{n_j} \to 0$ in X and, by (2.9),

$$c(\| x_{n_j} - y_{n_j} \|) \leq (Tx_{n_j} - Ty_{n_j}, K(x_{n_j} - y_{n_j}))$$
$$+ (Cx_{n_j} - Cy_{n_j}, K(x_{n_j} - y_{n_j})) + \phi(x_{n_j} - y_{n_j}),$$

the properties of C, ϕ, and $\{a_{n_j}\}$ imply that by taking the limit superior in the last inequality, we get

$$\lim_j \sup c(\| x_{n_j} - y_{n_j} \|) \leq 0 + 0 + \lim_j \sup \phi(x_{n_j} - y_{n_j}) \leq \phi(0) = 0.$$

Since $c(\| x_{n_j} - y_{n_j} \|) \geq 0$ for each j, it follows that $\lim_j c(\| x_{n_j} - y_{n_j} \|) = 0$, from which we obtain the convergence $x_{n_j} \to x$ in X; that is, T is of modified type (KS_+) and in particular, T is A-proper w.r.t. Γ by Theorem 2.3. Q.E.D.

Remark 2.2 To apply Theorems 1.6 and 2.3 and Corollary 2.5 and subsequent more general results of nature similar to equation (1.11) involving maps T from X to either X^*, X, or Y, we have to choose K, K_n, M_n, and Γ in some specified way so that conditions (1.18) [i.e., (2.8)] and possibly (2.6) and (2.7) are satisfied. The possible choices are illustrated by the following discussion.

Case 1. Suppose that $Y = X^*$, so that $T: X \to X^*$. Then $Y^* = X^{**}$, so $X \subset Y^*$ with $X = Y^*$ when X is reflexive. If for Γ we choose $\Gamma_1 =$

$\{X_n, V_n; X_n^*, V_n^*\}$ (i.e., $E_n = X_n^*$ and $W_n = V_n^*$), then $D(W_n^*) = X_n$, so in this case the simplest choice for K and K_n that satisfies condition (1.18) or (2.8) is $K = I$ and $K_n = I_n$, the identities on X and X_n, respectively. If X is also reflexive, the choice of K and K_n above also satisfies (2.6) and (2.7) of Theorem 2.3. If $\{\phi_1, \ldots, \phi_n\}$ is a basis in X_n and $\{\Phi_1, \ldots, \Phi_n\}$ is a basis in X_n^* such that $(\Phi_i, \Phi_j) = \delta_i^j$, then the linear isomorphism $M_n: X_n \to X_n^*$ defined by $M_n(\phi_j) = \Phi_j$ for $1 \le j \le n$ satisfies (1.19) of Theorem 1.6 with $K_n = I_n$.

It is not hard to show (see Chapter III for more general results) that a bounded continuous map $T: X \to X^*$ which is of type (S) is also of modified type (S). Hence Proposition 1.4 follows from Theorem 2.3 when T is continuous and bounded.

Case 2. Suppose that $Y = X$, so that $T: X \to X$. Then $Y^* = X^*$ and if for Γ we choose the scheme $\Gamma_1 = \{X_n, P_n\}$ (i.e., $E_n = X_n$ and $W_n = P_n$), then $D(W_n^*) = X^*$, so in this case [assuming, additionally, that X^* is of type (H)] one possible choice for K and K_n that satisfies (1.18) or (2.8) is $K = J$ and $K_n = P_n^* J|_{X_n}$, where J is some single-valued duality map of X to X^* corresponding to a given gauge function $\mu: R^+ \to R^+$ (see Section 2.3). Unfortunately, as we shall see later, except for some special spaces (e.g., Hilbert and l^p spaces) J is not weakly continuous at $x = 0$, so that in general the first part of (2.7) is not satisfied.

Of course, as will be seen in Section 3, it is sometimes possible to choose suitable linear maps K and K_n for which (2.6)–(2.8) hold. In many applications to differential equations, such a choice is often possible (e.g., see Section 3).

Case 3. If $T: X \to Y$ with Y being neither X nor X^*, and if for Γ we choose the projective scheme $\Gamma_P = \{X_n, V_n; Y_n, Q_n\}$, then from the point of view of applications, the most useful choice for K and K_n, which satisfies (1.18) or (2.8) and (2.6)–(2.7), is a bounded, linear, injective map $K: X \to Y^*$ with $R(K)$ dense in Y^* such that $Q_n^* K x = K x$ for all $x \in X_n$ and $n \in Z_+$. We shall see how this choice is made when the abstract results are applied to differential boundary value problems in Section 3 and other chapters.

2.3 Duality Mappings

An important example of an A-proper mapping from a reflexive space X into X^* is given by so-called duality mapping. This concept was introduced in [14] and later generalized and studied extensively in [36,222] and by other authors (see [128,244] for further references). In this section we define a duality mapping and prove those of its properties that we shall use. Further properties will be discussed in subsequent chapters as they are needed.

We recall that a continuous function $\phi: R^+ \to R^+$ is called a *gauge function* if ϕ is strictly increasing, $\phi(0) = 0$ and $\phi(t) \to \infty$ as $t \to \infty$.

Definition 2.7   A map $J: X \to X^*$ is called a *duality mapping* with gauge function ϕ if to each x in X:

$$(Jx,x) = \| x \| \, \phi(\| x \|) \quad \text{and} \quad \| Jx \| = \phi(\| x \|). \qquad (2.10)$$

If $\phi(t) \equiv t$, then J is called a *normalized duality* map.

It follows from the Hahn–Banach theorem that $J(x) \neq \emptyset$ for each $x \in X$. Further, it is not hard to show that for each fixed $x \in X$, $J(x)$ is a closed convex set that lies on $\partial B(0,r)$ with $r = \phi(\| x \|)$. Thus, in general, J is a multivalued map of X into X^*.

Lemma 2.5   Let $\Gamma_1 = \{X_n, P_n\}$ be a projectionally complete scheme for (X,X) and $J: X \to X^*$ a duality map with gauge function ϕ. Then $P_n^* Jx \subset Jx$ for each $x \in X_n$ and $n \in Z_+$. Moreover, if X^* is strictly convex, then J is single-valued with $P_n^* Jx = Jx$ for all $x \in X_n$ and J is continuous from the strong topology of X to the weak* topology of X^*.

Proof.   For any given $n \in Z_+$, let $x \in X_n$ be any fixed element and let $w \in J(x)$. We claim that $P_n^* w \in Jx$. To see this, note first that

$$(P_n^* w, x) = (w, P_n x) = (w, x) = \| w \| \cdot \| x \|. \qquad (2.11)$$

It follows from (2.11) that $\| w \| \leq \| P_n^* w \|$. Hence $\| w \| = \| P_n^* w \|$ since $\| P_n \| = 1$ for each $n \in Z_+$, so $\| P_n^* w \| = \phi(\| x \|)$. This together with (2.11) shows that $P_n^* w \in J(x)$ (i.e., $P_n^* Jx \in Jx$).
    Now, to prove the second part, observe that since $J(x) \subset \partial B(0,r) \subset X^*$ with $r = \phi(\| x \|)$ and X^* is strictly convex, it follows that $J(x)$ consists of a single point and thus J is single valued and $P_n^*(w) = Jx$ (i.e., $P_n^* Jx = Jx$ for $x \in X_n$). Finally, we claim that $Jx_j \xrightarrow{*} Jx_0$ [i.e., $(Jx_j, x) \to (Jx_0, x)$ for each $x \in X$, whenever $x_j \to x_0$ in X]. To prove this it suffices to show that every subsequence of $\{Jx_j\}$ contains a further subsequence that converges weak* to Jx_0. Since X is separable and $\{Jx_j\}$ is bounded in X^*, it follows that each subsequence of $\{Jx_j\}$ contains a weak* convergent subsequence. Let us denote this last subsequence by $\{Jx_i\}$ and its weak* limit by w_0. Since

$$\| x_0 \| \, \phi(\| x_0 \|) = \lim_i \| x_i \| \, \phi(\| x_i \|) = \lim_j (Jx_i, x_i) = (w_0, x_0), \qquad (2.12)$$

we see that $\phi(\| x_0 \|) \leq \| w_0 \|$. On the other hand, since $Jx_i \xrightarrow{*} w_0$, it follows that $\| w_0 \| \leq \lim \inf \| Jx_i \| = \lim \phi(\| x_i \|) = \phi(\| x_0 \|)$ and therefore $\| w_0 \| = \phi(\| x_0 \|)$. This and (2.12) imply that $w_0 = Jx_0$.   Q.E.D.

To state our first result we recall that X is said to have property (H) iff X is strictly convex and such that if $x_j \rightharpoonup x$ and $\| x_j \| \to \| x \|$ then $x_j \to x$ in X. It is known (see [5]) that Hilbert spaces, uniformly convex Banach spaces, and locally uniformly convex Banach spaces are spaces with property (H).

The following proposition for duality mappings J, which will prove to be important in a number of situations, will be particularly useful in the generalized degree theory for A-proper maps from subsets of X to X^* in a similar way as the identity I is useful in the Leray–Schauder degree theory for compact vector fields from subsets of X into X (see Chapter IV and [222]).

Proposition 2.2 If X is a separable reflexive space with X and X^* having property (H), then the duality map $J: X \to X^*$ with gauge function ϕ is continuous and A-proper w.r.t. $\Gamma_I = \{X_n, V_n; X_n^*, V_n^*\}$. If $\Gamma_\alpha = \{X_n, P_n\}$ is projectionally complete, J is also A-proper w.r.t. $\Gamma_P = \{X_n, P_n; R(P_n^*), P_n^*\}$.

Proof. Since X is reflexive and X^* is strictly convex, it follows from Lemma 2.5 that J is single-valued and $Jx_j \rightharpoonup Jx$ in X^* if $x_j \to x$ in X. Moreover, $\| Jx_j \| = \phi(\| x_j \|) \to \phi(\| x \|)$ because $x_j \to x$ in X. Thus $Jx_j \to Jx$ in X^* since X^* has property (H) (i.e., J is continuous). To show that J is A-proper w.r.t. Γ_I and Γ_P, by Proposition 1.4, it suffices to prove that J is of type (S). We will prove more, namely that J is of type (S$_+$), since this fact will prove to be useful later.

So let $\{x_j\}$ be any sequence in X such that $x_j \rightharpoonup x$ in X and $\lim \sup(Jx_j - Jx, x_j - x) \le 0$. Since, by (2.10),

$$(Jx_j - Jx, x_j - x) \ge (\phi(\| x_j \|) - \phi(\| x \|))(\| x_j \| - \| x \|)$$

and ϕ is strictly increasing, it follows that J is monotone and $\lim \inf(Jx_j - Jx, x_j - x) \ge 0$. Consequently, $(Jx_j - Jx, x_j - x) \to 0$ and therefore $(\phi(\| x_j \|) - \phi(\| x \|))(\| x_j \| - \| x \|) \to 0$ as $j \to \infty$. This implies that $\| x_j \| \to \| x \|$. Since we also have that $x_j \rightharpoonup x$ in X and X has property (H), it follows that $x_j \to x$ in X. Q.E.D.

Since $(Jx, x)/\| x \| = \phi(\| x \|) \to \infty$ as $\| x \| \to \infty$, as immediate consequence of Proposition 2.2, Theorem 1.6, and Case 1 of Remark 2.2 is the following result, which will prove useful in our study of accretive maps.

Proposition 2.3 If X is as in Proposition 2.2, then the duality map J with gauge function ϕ is a bijection of X onto X^*.

Remark 2.3 Let us add that the additional assumption in Propositions 2.2 and 2.3 that X is such that X and X^* have property (H) is not really restrictive since, by the results of Asplund [5] and Kadec [124], every separable reflexive space has an equivalent norm such that in the new norm both X and X^* are locally uniformly convex and, in particular, they have property (H). Consequently, we shall always assume this to be the case when we deal with duality mappings defined on separable reflexives spaces.

3. A-PROPERNESS AND THE CONSTRUCTIVE SOLVABILITY OF SOME ORDINARY AND PARTIAL DIFFERENTIAL EQUATIONS

In this section we show how the notion of an A-proper mapping enters naturally into the problem of constructive solvability of boundary value problems for ordinary and partial differential equations. Section 3.1 is devoted to the introduction of the notions of some functional spaces and to the statements of some basic facts that are useful in the study of differential and integral equations. In Section 3.2 the results for A-proper mappings obtained in the preceding sections are used to construct strong solutions for ordinary and partial differential equations of the second order. Section 3.3 is devoted to the construction of a weak solution of PDE of order $2m$, which is given in the divergence form. An interesting feature of our discussion in this section is that the existence of a weak solution in a given subspace $V \subset W_2^m$ ($V \supseteq \mathring{W}_2^m$) for a linear boundary value problem is deduced via A-proper mapping theory without the customary usage of the Gårding inequality and Fredholm alternative.

3.1 Basic Functional Spaces and Some of Their Properties

In this section we introduce the notation and state those facts from the theory of Sobolev spaces which are needed in the formulation of boundary value problems for differential equations. Discussions of the results stated in this section appear in [1,73,87,277,314].

Let Q be a domain in R^n and let the points of Q be denoted by $x = (x_1, \ldots, x_n)$. For any $m \in Z_+$, let $C^m(Q)$ be the vector space of all m-times continuously differentiable functions $u: Q \to R^1$. We set $C^0(Q) = C(Q)$, and let $C^\infty(Q) \equiv \cap_{m=0}^\infty C^m(Q)$. The subspace $C_0^\infty(Q)$ consists of those functions in $C^\infty(Q)$ that have compact support in Q. If $\alpha = (\alpha_1, \ldots, \alpha_n)$ is a multi-index with $\alpha_j \in Z_+$, we denote by $D^\alpha = \partial^{\alpha_1}/\partial x_1^{\alpha_1} \cdots \partial^{\alpha_n}/\partial x_n^{\alpha_n}$ the differential operator of order $|\alpha| = \alpha_1 + \cdots + \alpha_n$. Since Q is open, $u \in C^m(Q)$ need not be bounded on Q. If $u \in C(Q)$ is bounded and uniformly continuous on Q, it possesses a unique bounded continuous extension to $\overline{Q}$. Accordingly, $C^m(\overline{Q}) = \{u \in C^m(Q) : D^\alpha$ is bounded and uniformly continuous on Q for $|\alpha| \leq m\}$. With the usual linear structure, $C^m(\overline{Q})$ is a Banach space with the norm given by $\| u \|_{m,Q} = \sum_{|\alpha| \leq m} \sup_{x \in Q} | D^\alpha u(x) |$.

We say that X is *continuously embedded* into Y, and write $X \subset Y$, if X is a vector subspace of Y and $\| u \|_Y \leq K_0 \| u \|_X$ for all $u \in X$ and some constant $K_0 > 0$. The embedding theorem for the spaces $C^m(\overline{Q})$ asserts that if Q is bounded, then the embedding $C^{m+1}(\overline{Q}) \subset C^m(Q)$ is compact for $m = 0, 1, \ldots$ [i.e., every bounded set in $C^{m+1}(\overline{Q})$ is precompact in $C^m(\overline{Q})$]. According to the *Ascoli–Arzela theorem*, a set $G \subset C(\overline{Q})$ is precompact iff G is bounded and equicontinuous.

Suppose now that $Q \subset R^n$ is a measurable set with respect to the Lebesgue measure in R^n. For example, a closed or open set in R^n is measurable. Following the usual convention, we say that an assertion P is true for almost every x in Q [i.e., for $x \in Q$ (a.e.)] if the set $Q_0 \subset Q$ of points x at which P is not true is a set of meas$(Q_0) = 0$. Two measurable functions u_1 and u_2 defined on Q are said to be *equivalent* iff $u_1(x) = u_2(x)$ for $x \in Q$ (a.e.). If u_1 is integrable (in the sense of Lebesgue) on Q, every function u_2 on Q that is equivalent to u_1 is also integrable and $\int_Q u_1(x)\, dx = \int_Q u_2(x)\, dx$. Thus we shall make no distinction between equivalent functions. To formulate integral equations and differential boundary value problems as operator equations in suitable functional (e.g., Sobolev) spaces, the following known facts for integrable functions will prove to be useful.

Suppose that $u_j : Q \subset R^n \to R^1$ is a sequence of integrable functions such that $\lim_j u_j(x) \neq \pm\infty$ for $x \in Q$ (a.e.). Then:

> If to each $\epsilon > 0$ there exists $\delta(\epsilon) > 0$ independent of j such   (3.1a)
> that $\int_G | u_j(x)|\, dx < \epsilon \; \forall j \in Z_+$ and $\forall G \subset Q$ with meas(G)
> $< \delta(\epsilon)$ (i.e., the integrals are absolutely continuous in G,
> uniformly in j), *Vitali's theorem* asserts that $\lim_j \int_Q u_j(x)\, dx$
> $= \int_Q u(x)\, dx$.

> If $\lim_j \int_G u_j(x)\, dx \neq \pm\infty$ for all measurable sets $G \subset Q$, the   (3.1b)
> *Vitali–Hahn–Saks theorem* asserts that $\lim_j \int_Q u_j(x)\, dx = \int_Q u(x)\, dx$ and the integrals are absolutely continuous in G,
> uniformly in j.

If $p \in [1,\infty)$ is any integer, then $L_p \equiv L_p(Q)$ denotes the Banach space of measurable functions $u : Q \to R^1$ such that $| u |^p$ are integrable and the norms of u are given by $\| u \|_p = \{\int_Q | u(x)|^p\, dx\}^{1/p}$. L_p is separable for $p \in [1,\infty)$ and uniformly convex (and thus reflexive) for $1 < p < \infty$. But L_1 and L_∞ are not reflexive with L_∞ also not separable, where L_∞ is the

Banach space of all essentially bounded functions $u: Q \to R^1$ with the norm $\| u \|_\infty = \text{ess sup}_{x \in Q} | u(x) |$. One of the most important inequalities in the L_p-spaces is

Generalized Hölder inequality   If $u_i \in L_{p_i}$, $1 \leq i \leq k$, $1 < p_1, \ldots, p_k < \infty$ and $\sum_i p_i^{-1} = 1$, then $| \int_Q \prod_i u_i \, dx | \leq \prod_i \| u_i \|_{p_i}$.

We will also use the following inequalities for reals $\lambda_1, \ldots, \lambda_k$:

$$a(\sum_i | \lambda_i |^s)^{1/s} \leq \sum_i | \lambda_i | \leq b(\sum_i | \lambda_i |^s)^{1/s}, \qquad 1 \leq s < \infty \qquad (3.2a)$$

$$(\sum_i | \lambda_i |)^r \leq c(\sum_i | \lambda_i |^r), \qquad\qquad 0 < r < \infty, \qquad (3.2b)$$

with positive constants a, b, c depending only on k, s, r. It is known that if $p \in [1,\infty)$ and $q > 1$ is such that $1/p + 1/q = 1$, then the dual L_p^* of the space L_p can be identified with $L_q(Q)$ and thus $u_j \rightharpoonup u$ in L_p provided that $\lim_j \int_Q u_j v \, dx = \int_Q uv \, dx$ for each v in L_q. Further, if $u_j \rightharpoonup u$ in L_p and $u_j(x) \to v(x)$ for $x \in Q$ (a.e.), then $u = v$. Also, if $u_j \rightharpoonup u$ in L_p, there exists a subsequence $\{u_{j_i}\}$ such that $u_{j_i}(x) \to u(x)$ for $x \in Q$ (a.e.). Finally, let us add that if $\text{vol}(Q) = \int_Q 1 \, dx < \infty$ and $1 \leq p \leq q$, then $L_q \subset L_p$ and $\| u \|_p \leq \text{vol}(Q)^{1/p - 1/q} \| u \|_q$ for $u \in L_q$.

To study the solvability of nonlinear differential and integral equations, we first recall some relevant abstract results.

Definition 3.1 $f: Q \times R^k \to R^1$ satisfies the *Carathéodory condition* if for $x \in Q$ (a.e.) the function $u \equiv (u_1, \ldots, u_k) \to f(x, u_1, \ldots, u_k)$ is continuous for each $u \in R^k$ and for each $u \in R^k$ the function $x \to f(x, u_1(x), \ldots, u_k(x))$ is measurable on Q (as a function of x).

The following lemma plays a key role in the theory of nonlinear differential and integral equations.

Lemma 3.1   Suppose that $f: Q \times R^k \to R^1$ satisfies the Carathéodory condition and for q, $p_i \geq 1$, some $h \in L_q$ and $K_0 > 0$,

$$| f(x, u_1, \ldots, u_k)| \leq K_0 \left(\sum_{i=1}^k | u_i |^{p_i/q} \right) + h(x))$$

$$\forall \, x, u \in G \times R^k. \qquad (3.3)$$

Then the Nemytskii operator F defined by

$$F(u)(x) = f(x, u_1(x), \ldots, u_k(x)) \qquad [u = (u_1, \ldots, u_k)] \qquad (3.4)$$

is continuous and bounded as a map from $L_{p_1} \times \cdots \times L_{p_k}$ into L_q.

The proof of Lemma 3.1 depends on the Hölder inequality, (3.2), and the theorems of Vitali and Vitali–Hahn–Saks. Another result that will prove to be useful is the following:

Lemma 3.2 Let Q be bounded and $f_n: Q \to R^1$ a sequence of measurable functions such that $f_n(x) \to f(x)$ for $x \in Q$ (a.e.). Then $f_n \to f$ in L_p ($p \geq 1$) iff, given any subsequence $\{f_{n'}\}$ of $\{f_n\}$, there exists a further subsequence $\{f_{n''}\}$ of $\{f_{n'}\}$ and an L_p-function g (depending on $\{f_{n''}\}$) such that $|f_{n''}(x)| \leq g(x)$ for all n'' and $x \in Q$ (a.e.).

If $m \in Z_+$ and $p \in [1,\infty)$, we let $W_p^m(Q) = \{u \in L_p: D^\alpha u \in L_p$ for $|\alpha| \leq m\}$, where $D^\alpha u$ denote the generalized derivatives of u. This "Sobolev space" is a Banach space under the norm

$$\| u \|_{m,p} = (\sum_{|\alpha| \leq m} \| D^\alpha u \|_p^p)^{1/p} \qquad (u \in W_p^m). \tag{3.5}$$

The space W_p^m is separable if $1 \leq p < \infty$ and uniformly convex (and thus reflexive) if $1 < p < \infty$. When $m = 0$ in (3.5), $\| \cdot \|_{0,p}$ is the ordinary L_p-norm, so in this case we set $\| \cdot \|_{0,p} = \| \cdot \|_p$. We denote by $\mathring{W}_p^m$ the completion of $C_0^\infty(Q)$ in W_p^m and say that $u \in \mathring{W}_p^m$ satisfies the homogeneous Dirichlet boundary conditions in generalized sense. This is justified by the fact that if the boundary ∂Q is smooth and $u \in W_p^m(Q) \cap C^m(\overline{Q})$, then $u \in \mathring{W}_p^m$ iff $D^\alpha u = 0$ on ∂Q for $|\alpha| < m$. On the other hand, the condition $u \in \mathring{W}_p^m$ applies to a wider class of not so regular functions u. It is useful to note that if Q is bounded, then

$$\| D^m u \|_p \equiv (\sum_{|\alpha| = m} \| D^\alpha u \|_p^p)^{1/p},$$
$$\text{(where} \quad D^k u = \{D^\alpha u: |\alpha| = k\}), \tag{3.5a}$$

is an equivalent norm on $\mathring{W}_p^m$. This follows from

Friedrichs' inequality If $Q \subset R^n$ is bounded, then for $|\beta| \leq m$,

$$\int_Q | D^\beta u |^p \, dx \leq K_0(\sum_{|\alpha| = m} \int_Q | D^\alpha u |^p \, dx) \quad \forall u \in \mathring{W}_p^m.$$

When $p = 2$, the Sobolev space W_2^m, which is sometimes denoted by $H^m(Q)$, becomes a Hilbert space with the inner product

$$(u,v)_{m,2} = \sum_{|\alpha| \leq m} (D^\alpha u, D^\alpha v) = \sum_{|\alpha| \leq m} \int_Q D^\alpha u D^\alpha v \, dx.$$

The space $\mathring{W}_2^m = \mathring{H}^m$ has the inner product $(u,v)_{2,m} = \sum_{|\alpha| = m} (D^\alpha u, D^\alpha v)$.

38 **Chapter I**

One of the most important results in the theory of Sobolev spaces is the following embedding theorem, which plays a key role in the solvability of differential boundary value problems.

Sobolev embedding theorem Let $Q \subset R^n$ be a bounded domain with a sufficiently smooth boundary ∂Q (see [1,87]) and let $1 \le p < \infty$. Then:

(a) If $0 \le j < k$, then $W_p^k(Q) \subset W_r^j(Q)$ for $r \le (n - p)/(n - p(k - j))$ with a continuous embedding mapping. The embedding is compact if $r < np/(n - p(k - j))$ and $r < \infty$.

(b) If $0 \le j < k$ and $j < k - n/p$, then $W_p^k(Q) \subset C^j(\overline{Q})$ and the embedding map is always compact.

Remark 3.0 The conclusions in (a) and (b) are valid for $\mathring{W}_p^k$ and $\mathring{W}_r^j$ without the regularity assumption on ∂Q.

When Q is bounded and ∂Q is regular in the sense of Calderon [43] (see also [304]), one also has the following inequality of the Sobolev type, which will prove to be useful in what follows.

$$\int_Q |u|^p \, dx \le K_1 \left\{ \left[\sum_{i=1}^n \left(\frac{\partial u}{\partial x_i} \right)^2 \right]^{p/2} + \left| \int_Q u \, dx \right|^p \right\}$$
$$\text{for} \quad u \in W_p^1(Q) \tag{3.6}$$

with $K_1 = K_1(n,p,Q)$. When $p = 2$, (3.6) is called the inequality of Poincaré.

Let us add that when $n = 1$ and $\overline{Q} = [a,b]$, then $W_p^m([a,b])$ is the space of all absolutely continuous (a.c.) functions $u(x)$ on $[a,b]$ whose derivatives $D^j u$, $1 \le j \le m - 1$ are a.c. and $D^m u \in L_p([a,b])$, $\mathring{W}_p^m = \{u \in W_p^m : D^j(a) = D^j(b) = 0, 0 \le j < m - 1\}$, and $W_p^m([a,b]) \subset C^{m-1}([a,b])$ is compact. Moreover, the Poincaré inequality becomes

$$\int_a^b |u|^2 \, dx \le \frac{b - a}{2} \left(\int_a^b |u'|^2 \, dx + \left| \int_a^b u \, dx \right| \right)$$
$$\text{for} \quad u \in W_2^1. \tag{3.6a}$$

It follows from the *bounded convergence theorem* that if Q is bounded and $\{f_j(x)\}$ is a sequence of measurable functions such that $\| f_j(x)\|_\infty \le K_0$ for all $j \in Z_+$ and $f_j(x) \to f(x)$ for $x \in Q$ (a.e.), then $f_j \to f$ in L_p for any $p \in (0,\infty)$. As was shown in [304], this theorem admits a useful extension.

Lemma 3.3 If $r \in (0, \infty]$, $\{f_j(x)\}$ is a sequence of measurable functions and $K_0 > 0$ such that $\| f_j \|_\infty \leq K_0$ for all $j \in Z_+$ and $f_j(x) \to f(x)$ for $x \in Q$ (a.e.), then $f_n \to f$ in L_p for any $p \in (0, r)$.

We complete this section with the following result, which will prove to be essential in our study of differential boundary value problems.

Let $Q \subset R^n$ be a bounded domain with ∂Q so that the Sobolev embedding theorem holds on Q. For $m \in Z_+$, let s_m be the cardinal number of the set $S_m = \{\alpha \mid \alpha \in Z_+^n : |\alpha| \leq m\}$, $\xi(u) = \{D^\alpha u : |\alpha| \leq m\}$, $\eta(u) = \{D^\alpha u : |\alpha| \leq m - 1\} \in R^{s_{m}-1}$, $\{(u) \equiv D^m u = \{D^\alpha u : |\alpha| = m\} \in R^{s'_m}$ with $s'_m = s_m - s_{m-1}$ and for each $\xi \in R^{s_m}$ we write $\xi = (\eta, \zeta)$.

Proposition 3.1 Suppose that $b: Q \times R^{s_{m}-1} \times R^{s'_m} \to R^1$ satisfies the Carathéodory condition and for a given $1 < p < \infty$ there exist $K_0 > 0$ and $h \in L_q(Q)$ with $q = p(p - 1)^{-1}$ such that

$$(A) \mid b(x, \eta, \xi) \mid \leq K_0\{h(x) + |\eta|^{p-1} + |\xi|^{p-1}\} \quad \text{for} \quad x \in (Q) \text{ (a.e.)},$$

and $(\eta, \xi) \in R^{s_{m}-1} \times R^{s'_m}$.

Let V be any closed subspace of $W_p^m(Q)$ with $V \supseteq \mathring{W}_p^m$. Then:

(a) The operator $B: V \to L_q$ defined by $B(u) = b(\cdot, \xi(u)(\cdot))$ is bounded and continuous.

(b) For each fixed v in V, the map $B_v: V \to L_q$ defined by $B_v(u) = B(\cdot, \eta(u)(\cdot), D^m v)$ is completely continuous; that is, if $\{u_j\} \subset V$ is such that $u_j \rightharpoonup u_0$ in V, then $B_v(u_j) \to B_v(u_0)$ in L_q.

Proof. Proposition 3.1 can be deduced from Lemma 3.1, but for the sake of completeness, we shall prove it by using essentially the same type of arguments as one does in proving Lemma 3.1.

(a) In view of (A), the inequality (3.2b) with $r = q$ implies that $\mid b(x, \xi(u)(x))\mid^q \leq \text{const}\{ \mid h \mid^q + \mid \eta(u)\mid^p + \mid \xi(u)\mid^p\}$. Hence, integrating the last inequality over Q and again using (3.2b) with $r = 1/q$, we get that $\| Bu \|_q \leq K_1 \phi(\| u \|_{m,p})$ for some $K_1 > 0$ and $\phi(t) = \| h \|_q + t^{p/q}$. Thus B maps bounded sets in V into bounded sets in L_q. To show that $B: V \to L_q$ is continuous, let $\{u_j\} \subset V$ be such that $u_j \to u_0$ in V. Then $\xi(u_j) \to \xi(u_0)$ in L_p spaces. Passing an infinite subsequence, we may assume without loss of generality that $\xi(u_j)(x) \to \xi(u_0)(x)$ for $x \in Q$ (a.e.) and thus, by the Carethéodory condition on b, $\bar{b}_j(x) \equiv b(x, \xi(u_j)(x)) - b(x, \xi(u_0)(x)) \to 0$ for $x \in Q$ (a.e.).

In view of (A) and (3.2a), we see that for any measurable subset G of Q,

$$I_j(G) \equiv \int_G |\bar{b}_j(x)|^q \, dx$$

$$\leq K_1 \int_G \{| b(x,\xi(u_j)(x))|^q + | b(x,\xi(u_0)(x)|^q\} \, dx$$

$$\leq \text{const} \left\{ \int_G [2 | h |^q + \sum_{|\alpha|\leq m} | D^\alpha u_0 |^p] \, dx \right.$$

$$\left. + \int_G \sum_{|\alpha|\leq m} | D^\alpha u_j |^p \, dx \right\}$$

The last integrals in the inequality above converge as $j \to \infty$ since $\xi(u_j) \to \xi(u_0)$ in $L_p(Q)$ and, in particular, in $L_p(G)$. Hence, by the Vitali–Hahn–Saks theorem, these integrals are absolutely continuous in G, uniformly in j. The other integrals are also absolutely continuous in G. Hence, to any given $\epsilon > 0$, there exists $\delta(\epsilon) > 0$ such that $I_j(G) < \epsilon$ for all $G \subset Q$ with $\text{meas}(G) < \delta(\epsilon)$ and all j. In view of this, Vitali's theorem implies that $\bar{b}_j = B(u_j) - B(u_0) \to 0$ in L_q. Actually, we have shown that every subsequence of $\{B(u_j)\}$ itself has a subsequence converging strongly in L_q to $B(u_0)$. Hence $\{B(u_j)\}$ itself so converges.

(b) It follows from (a) that $B_v: V \to L_q$ is obviously bounded and continuous. Since $\eta(u_j) \to \eta(u_0)$ in L_p whenever $u_j \to u_0$ in V, the same arguments show that B_v is completely continuous. Alternatively, B_v is completely continuous since the Sobolev embedding theorem ensures the complete continuity of $u \to (\eta(u),\zeta(v))$ of V into $(L^p(Q))^{sm}$, and because the Nemytskii operator, induced by b, is continuous from L_p^{sm} into L_q.

Q.E.D.

3.2 Construction of Strong Solutions of Linear and Semilinear Ordinary and Partial Differential Equations of Order 2

The main purpose of this section is to show how the theory of A-proper maps is used in the construction of strong solutions for Dirichlet boundary value problems for linear and semilinear equations of the second order. The restriction to partial differential equations (PDEs) of the second order is motivated by the use of Sobolevskii's inequality [271] for PDEs. Non-linear equations of higher order and with different boundary conditions will be treated in subsequent sections and chapters after the theory of mappings of A-proper type is more fully developed. Nevertheless, the examples in this section are so chosen that other known abstract results need not apply to them. In dealing with semilinear equations, we allow the nonlinearities to depend on the second-order derivatives such that the degree theories for compact or condensing vector fields will not be available.

Problem (A)

For the sake of simplicity we first consider a simple boundary value problem for an ordinary differential equation (ODE) of the form

$$-D^2u - g(x,u,Du,D^2u) + h(x,u,Du) = f(x), \qquad (3.7)$$
$$x \in (0,1), \ u(0) = u(1) = 0.$$

We let $X = \{u: u \in W_2^2([0,1]), u(0) = u(1) = 0\}$, $Y = L^2([0,1])$, and for $u \in X$, let $\| u \|_X = \| D^2u \|_2$. Then X is a Banach space whose norm $\| \cdot \|_X$ is equivalent to the W_2^2-norm. Indeed, since the smallest eigenvalue of $Ku = -D^2u$ equals π^2 and $u(0) = u(1) = 0$, we have

$$\| u \|_2^2 \le \pi^{-2} \| D^2u \|_2^2 \qquad \text{for} \quad u \in X. \qquad (3.8)$$

On the other hand, using Poincaré's inequality (3.6a) for the function $Du \in X$ with $a = 0$ and $b = 1$, we get

$$\| Du \|_2^2 \le 2^{-1} \| D^2u \|_2^2 \qquad \text{for} \quad u \in X. \qquad (3.9)$$

Hence $\| \cdot \|_X$ and $\| \cdot \|_{2,2}$ are equivalent and X is a Hilbert space with respect to the inner product

$$[u,v] = (D^2u, D^2v) \qquad (u,v \in X), \qquad (3.10)$$

where $(\cdot,\cdot)$ denotes the inner product in $Y = L^2$. Furthermore, K is an isometric isomorphism of X onto Y. In what follows we refer to any function u in X that satisfies (3.7) almost everywhere as a *strong solution* of (3.7). Since X is a separable Hilbert space, there exists a sequence $\{X_n\} \subset X$ of finite-dimensional subspaces such that $\text{dist}(u,X_n) = \inf_{x \in X_n} \| u - x \|_X \to 0$ as $n \to \infty$ for each u in X. If we set $Y_n = K(X_n) \subset Y$ and let $P_n: X \to X_n$ and $Q_n: Y \to Y_n$ denote the orthogonal projections, then $\Gamma_K = \{X_n, P_n; Y_n, Q_n\}$ is projectionally complete for (X,Y) and

$$Q_n Ku = Ku \qquad \text{for all} \quad u \in X_n \quad \text{and} \quad n \in Z_+. \qquad (3.11)$$

To apply Theorem 1.6 to the boundary value problem (3.7), we impose the following conditions on the functions g and h.

(a1) $g: [0,1] \times R^1 \to R^1$ satisfies the Carathéodory conditions and there exist $a_1, a_2, a_3, a_4 \in R^+ \backslash \{0\}$ such that

$$| g(x,q,s,r)| \leq a_1 + a_2 | q | + a_3 | s | + a_4 | r | \qquad (3.12)$$
$$\text{for} \quad x \in [0,1] \quad \text{and} \quad q,s,r \in R,$$

$$[g(x,q,s,r_1) - g(x,q,s,r_2)][r_1 - r_2] \geq -c_0 | r_1 - r_2 |^2 \qquad (3.13)$$
$$\text{for} \quad x \in [0,1] \quad \text{and} \quad q,s,r_1,r_2 \in R.$$

(a2) $h: [0,1] \times R^2 \to R^1$ satisfies the Carathéodory conditions and there exist $c_1, c_2, c_3 \in R^+ \backslash \{0\}$ with $1 - c_0 - c_2/\pi^2(c_3/2) > a_2/\pi^2 + a_3/2$ and

$$| h(x,q,s)| \leq c_1 + c_2 | q | + c_3 | s | \qquad (3.14)$$
$$\text{for} \quad x \in [0,1] \quad \text{and} \quad q,s \in R.$$

We define the operators $F,C: X \to Y$ by $F(u)(x) = g(x,u,Du,D^2u)$ and $C(u)(x) = h(x,u,Du)$ for $x \in [0,1]$ and $u \in X$. In view of Lemma 3.1, the conditions (a1) and (a2) imply that F and C are well defined, bounded, and continuous as mappings from X to Y. Furthermore, since by the Sobolev embedding theorem, X is compactly embedded into $C^1([0,1])$, it follows that C is compact and $g(x,u_j,Du_j,D^2u_0) \to g(x,u_0,Du_0,D^2u_0)$ in Y whenever $\{u_j\} \subset X$ is such that $u_j \rightharpoonup u_0$ in X by Proposition 3.1.

Thus, to show that $A \equiv K - F + C: X \to Y$ is A-proper w.r.t. Γ_P, it suffices to prove that $T \equiv K - F$ is A-proper w.r.t. Γ_K. Since, by virtue of (3.11), Γ_P, K, and $K_n \equiv K |_{X_n}: X_n \to Y_n$ satisfy conditions (2.6)–(2.8) of Theorem 2.3, by Corollary 2.5, it suffices to show that (2.9) holds for $T = K - F$. Now, it follows from (3.13) and a simple calculation that for all u and v in X, we have

$$(Tu - Tv, Ku - Kv) \geq (1 - c_0)\| Ku - Kv \|^2 + ([g(x,u,Du,D^2v)$$
$$- g(x,v,Dv,D^2v)], Ku - Kv).$$

Thus (2.9) of Corollary 2.5 holds on all of X with $C \equiv 0$, $c(t) \equiv (1 - c_0)t^2$, and

$$\phi(u - v) \equiv -\int_0^1 [g(x,u,Du,D^2v) - g(x,v,Dv,D^2v)][D^2u - D^2v] \, dx.$$

The definition of ϕ above implies that $\phi(0) = 0$ and that if $\{u_j\} \subset X$ is such that $u_j \rightharpoonup u_0$ in X, then $\phi(u_j - u_0) \to 0$ as $j \to \infty$ because $g(x,u_j,Du_j,D^2u_0) \to g(x,u_0,Du_0,D^2u_0)$ in $Y = L_2$ and $D^2u_j \rightharpoonup D^2u$ in Y, as was shown above.

We are now in the position to prove the following:

Theorem 3.1 Suppose that the functions g and h satisfy conditions (a1) and (a2), respectively. Then the boundary value problem (3.7) is feebly approximation solvable w.r.t. Γ_P for each $f \in L_2$ and, in particular, (3.7) has a strong solution in $X \subset W_2^2$ for each f in L_2. It is strongly approximation solvable for a given $f \in L_2$ if it has at most one solution in X.

Proof. The preceding discussion implies that (3.7) is equivalent to the solvability of the operator equation

$$Ku - Fu + Cu = f \quad (u \in X, f \in Y). \tag{3.15}$$

Since, as was shown above, the mapping $A \equiv K - F + C : X \to Y$ is A-proper w.r.t. $\Gamma_K = \{X_n, P_n; K(X_n), Q_n\}$, continuous, and bounded, to apply Theorem 1.6 to (3.15) it suffices to show that A is K-coercive. Now, the last inequality also yields the following estimates:

$$(Au, Ku) \geq (1 - c_0)\| u \|_X^2$$
$$- (g(x, u, Du, 0), Ku) | - \| F(0) \| \|u\|_X - |(Cu, Ku)|.$$

But the hypotheses (a1) and (a2), together with Hölder's inequality and the fact that

$$\| u \|_2 \leq \frac{1}{\pi^2} \| D^2 u \|_2 \quad \text{and} \quad \left| \int_0^1 Du D^2 u \, dx \right| \leq \frac{1}{2} \| D^2 u \|_2$$

for each $u \in X$, imply that

$$(Au, Ku) \geq \left(1 - c_0 - \frac{a_2 + c_2}{\pi^2} - \frac{a_3 + c_3}{2} \right) \| Ku \|_2^2$$
$$- (a_1 + c_1 + \| F(0) \|) \| Ku \|_2,$$

so we see that $(Au, Ku)/\| Ku \|_2 \to \infty$ as $\| u \|_X \to \infty$.   Q.E.D.

    Let us add that the same method is applicable to boundary value problems for semilinear ODEs of higher order and with different boundary conditions provided that the linear part is a homeomorphism between two suitable Sobolev spaces.

Remark 3.1 The conclusions of Theorem 3.1 certainly hold if instead of (3.13) in (a1) we assume that the function g is Lipschitzian in the highest-order derivative, that is, if

$$| g(x, q, s, r_1) - g(x, g, s, r_2) | \leq c_0 | r_1 - r_2 | \tag{3.16}$$
$$\text{for} \quad x \in [0, 1] \quad \text{and} \quad q, s, r_1, r_2 \in R.$$

Remark 3.2 If $g(x,q,s,r) \equiv 0$ for $(x,q,s,r) \in [0,1] \times R^3$, the mapping $K^{-1}C: X \to X$ is compact, so in this case one can use the classical results to obtain the surjectivity or even constructive surjectivity theorem for the boundary value problem (3.7).

Problem (B)

Let $Q \subset R^n$ be a bounded domain so that the Sobolev embedding theorem holds on Q. To study the solvability of the Dirichlet problem for second-order semilinear equations of the form

$$Lu - \mathcal{N}(x,\delta u,D^2 u) = f(x) \qquad [u \in D(L) \equiv W_2^2 \cap \mathring{W}_2^1], \qquad (3.17)$$

where $D^2 u = \{D^\alpha u: |\alpha| = 2\}$, $\delta u = \{D^\alpha u: |\alpha| \le 1\}$, and $f \in L^2$, we will employ the inequality of Sobolevskii [271] (see also [145]), which states that if

$$Lu = -\sum_{i,j=1}^n a_{ij}(x) \frac{\partial^2 u}{\partial x_i \, \partial x_j} + \sum_{i=1}^n a_i(x)u + a_0(x)u \qquad (3.18)$$

$$[x \in D(L)]$$

is a strongly elliptic operator with $a_{ij}(x) \in C^1(Q)$ and $a_i(x) \in C(Q)$, and if K is any other second-order operator defined on $D(L)$ with the same properties, there exist constants $\tau_1 > 0$ and τ_2 (depending on L, K, and ∂Q but not on u) such that

$$(Lu,Ku) \ge \tau_1 \| u \|_{2,2}^2 - \tau_2 \| u \|_2^2 \qquad \text{for all} \quad u \in W_2^2 \cap \mathring{W}_2^1. \qquad (3.19)$$

We shall use (3.19) when $K = -\Delta$, with Δ being the Laplacian on $D(L)$. We recall that L is said to be *strongly elliptic* provided that there exists a constant $\mu > 0$ such that

$$\sum_{i,j=1}^n a_{ij}(x)s_i s_j \ge \mu \left(\sum_{i=1}^n s_i^2 \right) \qquad (3.20)$$

$$\text{for} \quad x \in Q \quad \text{and} \quad s = (s_1, \ldots, s_n) \in R^n.$$

It is well known that K is a self-adjoint positive definite map of $D(L) \subset L^2$ into L^2 and that K is a homeomorphism as a mapping $W_2^2 \cap \mathring{W}_2^1$ onto L^2. Hence there exist constants $m_1 > 0$ and $m_2 > 0$ such that

$$m_1 \| Ku \|^2 \le \| u \|_{2,2}^2 \le m_2 \| Ku \|^2 \qquad \text{for} \quad u \in W_2^2 \cap \mathring{W}_2^1. \qquad (3.21)$$

Let c be the smallest positive number such that

$$\left(\sum_{|\alpha|=2} \| D^\alpha u \|_2^2 \right)^{1/2} \le c \| Ku \| \qquad \text{for} \quad u \in W_2^2 \cap \mathring{W}_2^1. \qquad (3.22)$$

Now assume that $\mathcal{N}$ satisfies the following conditions:

(b1) $\mathcal{N}: Q \times R^{S_1} \times R^{S_2} \to R^1$ satisfies the Carathéodory condition and there exist $K_0 > 0$ and $h \in L^2$ such that

$$| \mathcal{N}(x,\eta,\zeta) | \le K_0\{h(x) + | \eta | + | \zeta | \}$$

$$\text{for} \quad (x,\eta,\zeta) \in Q \times R^{s_1} \times R^{s_2}.$$

(b2) There exists $\gamma \in (0,\tau_1 m_1/c)$ such that for $(x,\eta) \in Q \times R^{s_1}$,

$$| \mathcal{N}(x,\eta,\zeta) - \mathcal{N}(x,\eta,\zeta')| \le \gamma | \zeta - \zeta' |_t \quad \text{for} \quad \zeta,\zeta' \in R^{s_2}.$$

If we set $X = W_2^2 \cap \mathring{W}_2^1$, $Y = L^2(Q)$, and define $N: X \to Y$, $N(u)(x) = \mathcal{N}(x,\delta u(x),D^2u(x))$ for $x \in Q$ and $u \in X$, then by Proposition 3.1, condition (b1) implies that N is a bounded continuous mapping of X into Y. It is easy to show that the map $L: X \to Y$ defined by the right-hand side of (3.18) lies in $L(X,Y)$. Hence the solvability of (3.17) is equivalent to the solvability of the operator equation

$$Lu + Nu = f \quad (u \in X, f \in Y). \tag{3.23}$$

In what follows we refer to the solution of (3.23) as a *strong solution* of (3.17); that is, u is a function in $W_2^2 \cap \mathring{W}_2^1$ that satisfies (3.17) for $x \in Q$ (a.e.).

In order to apply Theorem 1.6 to (3.23), we have to show that $T \equiv L + N: X \to Y$ is A-proper w.r.t. to a suitable scheme Γ for (X,Y).

Let $\{X_n\}$ be a sequence of finite-dimensional subspaces of X such that dist$(u,X_n) \to 0$ for each u in X, $Y_n = K(X_n)$, and $Q_n: Y \to Y_n$ an orthogonal projection for each $n \in Z_+$. Then $\Gamma_K = \{X_n,V_n;Y_n,Q_n\}$ is a complete projection scheme for (X,Y) with the property that

$$Q_nKu = Ku \quad \text{for all} \quad u \in X_n \quad \text{and each} \quad n \in Z_+. \tag{3.24}$$

The projection method for (3.23) consists of finding an element $u_n \in X_n$ for each sufficiently large $n \in Z_+$ from the finite-dimensional quasilinear equation

$$Q_nL(u_n) + Q_nN(u_n) = Q_nf \quad (u_n \in X_n, Q_nf \in Y_n) \tag{3.25}$$

or, equivalently, from the nonlinear algebraic system

$$(Lu_n,Ku) + (N(u_n),Ku) = (f,Ku) \quad \text{for} \quad u \in X_n \tag{3.26}$$

such that $\{u_n\}$ (or at least a subsequence of $\{u_n\}$) converges to some u_0 in X and u_0 is a solution of (3.23).

Before we continue with our discussion of the solvability of (3.23), the following remark seems to be in order.

Remark 3.3 Suppose that $\mathcal{N}(x,\eta,\zeta) \equiv 0$, so that (3.17) reduces to the linear equation

$$Lu = f \qquad [u \in D(L) = X, f \in L^2], \tag{3.27}$$

while the approximate equation (3.25) reduces to

$$Q_n L(u_n) = Q_n f \qquad (u_n \in X_n, Q_n f \in Y_n). \tag{3.28}$$

Since the injection of X into $Y = L^2$ is compact, it follows that the functional $\phi: X \to R$ defined by $\phi(u) = \| u \|_2^2$ is such that if $u_j \rightharpoonup u_0$ in X, then $\phi(u_n) \to \phi(0) = 0$. In view of this, property (3.24), satisfied by Γ_K, and inequality (3.19), Corollary 2.5 in Section 2.2 (with $C \equiv 0$) implies that $L: X \to Y$ is A-proper w.r.t. Γ_K. Hence Theorem 1.4 in Section 1 implies the validity of the following constructive theorem.

Theorem 3.2 Let L be a strongly elliptic operator defined on $W_2^2 \cap \mathring{W}_2^1 \equiv X$ and let Γ_K be the complete projection scheme for (X,L^2) constructed above.

If L is one-to-one, equation (3.23) is uniquely approximation solvable w.r.t. Γ_K for each f in L^2.

Note (i) Theorem 3.2 strengthens the corresponding result of Ladyženskaya [145], who established the approximation solvability of (3.27) under the assumption that (3.27) is uniquely solvable for each f in L^2 and that L satisfies the inequality

$$(Lu,u) \geq c \| u \|_{2,1}^2 \qquad \text{for all} \quad u \in X \quad \text{and some} \quad c > 0. \tag{3.29}$$

(ii) Theorem 3.2 provides a new proof for the existence of solutions in $W_2^2 \cap \mathring{W}_2^1$ of equation (3.27). It does not use the classical Fredholm alternative, which was used in all studies known to the author concerning the solvability of (3.27) in $W_2^2 \cap \mathring{W}_2^1$.

To continue our discussion of the solvability of the semilinear equation (3.23), we first prove the following:

Lemma 3.4 If L is strongly elliptic and the function $\mathcal{N}$ satisfies conditions (b1) and (b2), then $T = L + N: X \to Y$ is A-proper w.r.t. Γ_K.

Proof. Since $K: X \to Y$ is a linear homeomorphism and (3.24) holds, in view of Corollary 2.5, to prove Lemma 3.4 it suffices to show the existence of a functional $\phi: X \to R$ with $\phi(0) = 0$, which is weakly upper semicontinuous at 0 and such that

$$(Tu - Tv, Ku - Kv) \tag{3.30}$$
$$\geq c(\| K(u - v) \|) - \phi(u - v) \qquad (u,v \in X),$$

where $c: R^+ \rightarrow R^+$ is continuous and $t \rightarrow 0$ whenever $c(t) \rightarrow 0$.

Now, if we define $B: X \times X \rightarrow Y$ by $B(u,v) = -\mathcal{N}(x,\delta u, D^2 v)$, then our growth condition (b1) on $\mathcal{N}$ implies that B is bounded and continuous, and (b2), together with the Hölder's inequality, imply that

$$| (B(u,u) - B(u,v), Ku - Kv)| \leq \gamma c \| Ku - Kv \|^2 \qquad \text{for} \quad u,v \in X.$$

Thus (3.19) and (3.21) show that (3.30) holds with $c(t) = (\tau_1 m_1 - \gamma c) t^2$ and

$$\phi(u - v) = \tau_2 \| u - v \|_2^2 + (B(u,v) - B(v,v), Ku - Kv)$$
$$\text{for} \quad u,v \in X.$$

The definition above of ϕ implies that $\phi(0) = 0$ and that if $\{u_j\} \subset X$ is such that $u_j \rightharpoonup u_0$ in X, then $\phi(u_j - u_0) \rightarrow 0$ as $j \rightarrow \infty$. Indeed, to prove the last assertion, suppose that $\{u_j\} \subset X$ is such that $u_j \rightharpoonup u_0$ in X. Then $Ku_j \rightarrow Ku_0$ in Y and $u_j \rightarrow u_0$ in W_2^1 by the Sobolev embedding theorem. Hence $B(u_j,u_0) \rightarrow B(u_0,u_0)$ in Y since $\mathcal{N}(x,\delta u_j, D^2 v) \rightarrow \mathcal{N}(x,\delta u_0, D^2 v)$ in Y for any fixed v in X by Proposition 3.1, and therefore, $\phi(u_j - u_0) = (B(u_j,u_0) - B(u_0,u_0), Ku_j - Ku_0) \rightarrow 0$. This proves Lemma 3.4.

Q.E.D.

We are now in the position to prove our approximation solvability and/ or existence result for (3.17).

Theorem 3.3 Suppose that L is strongly ellipitic and the function $\mathcal{N}$ satisfies conditions (b1) and (b2). Suppose further that $T = L + N: X \rightarrow Y$ is K-coercive, that is,

$$\frac{(Tu, Ku)}{\| Ku \|} \rightarrow \infty \qquad \text{as} \quad \| u \|_{2,2} \rightarrow \infty.$$

Then equation (3.23) or (3.17) is feebly approximation solvable in X w.r.t. Γ_K for each f in L^2 [and, in particular, (3.17) has a strong solution for each f in L^2]. Moreover, if for a given f in L^2, (3.17) has at most one strong solution, then (3.17) is strongly approximation solvable [i.e., the solutions u_n of (3.25) exist for all $n \geq n_0$, $u_n \rightarrow u$ in the W_2^2-norm and we have $Tu = f$].

Proof. Theorem 3.3 follows from Theorem 1.6 in Section 1 if in the latter we take for Γ the scheme $\Gamma_K = \{X_n, V_n; Y_n, Q_n\}$. Because of (3.24), we see that if we set $K_n = K|_{X_n}: X_n \rightarrow Y_n$ for each $n \in Z_+$, conditions (1.18) and (1.19) of Theorem 1.6 hold with $E_n = Y_n$ and $W_n = Q_n$ since Y is a

Hilbert space and Q_n is an orthogonal projection of Y onto $Y_n = K(X_n)$.
                                                                                    Q.E.D.

Note that if $\tau_2 = 0$, then since $c(t) = (\tau_1 m_1 - c)t^2$ in (3.30) it is easy to show that T is K-coercive if

$$\left(\int_Q | \mathcal{N}(x,u,u,0)|^2 \right)^{1/2} \leq a_1 \| u \|_{2,2} + a_2 \qquad \text{for} \quad u \in X$$

and a_1 is sufficiently small.

Problem (C)

As our next example we consider

$$Lu - \mathcal{N}_1(x,\delta u,\Delta u) = f(x) \qquad (u \in W_2^2 \cap \mathring{W}_2^1), \tag{3.31}$$

where L is the same as in problem (B) and the function $\mathcal{N}_1(x,\eta,r)$ satisfies the following conditions:

(c1) $\mathcal{N}_1 : Q \times R^{s_1} \times R^1 \to R^1$ satisfies the Carathéodory condition and there exists $K_0 > 0$ and $h \in L^2$ such that

$$| \mathcal{N}_1(x,\eta,r)| \leq K_0 \{ h(x) + | \eta |$$
$$+ | r |\} \qquad \text{for} \quad (x,\eta,r) \in Q \times R^{s_1} \times R^1.$$

(c2) There exists $\gamma \in (0,\tau_1 m_1)$ such that for $(x,\eta) \in Q \times R^{s_1}$

$$(\mathcal{N}_1(x,\eta,r_1) - \mathcal{N}_1(x,\eta,r_2))(r_1 - r_2)$$
$$\geq -\gamma | r_1 - r_2 |^2 \qquad \text{for} \quad r_1,r_2 \in R^1.$$

The same arguments as those in the discussion of problem (B) show that $\mathcal{N}_1 : X \equiv W_2^2 \cap \mathring{W}_2^1 \to Y \equiv L^2$, defined by $N_1 u = -\mathcal{N}_1(x,\delta u(x),\Delta u(x))$ for $x \in Q$ and $u \in X$, is continuous, bounded, and $T = L - N_1 : X \to Y$ is A-proper w.r.t. Γ_K since

$$(Tu - Tv, Ku - Kv) \geq (\tau_1 m_1 - \gamma) \| Ku - Kv \|^2 - (V_1(u,v)$$
$$- V_1(v,v), Ku - Kv) \qquad \text{for} \quad u,v \in X,$$

where $V_1 : X \times X \to Y$ is given by $V_1(u,v) = N(x,\delta u,\Delta v)$.

Hence, in view of the above, Theorem 1.6 implies:

Theorem 3.4   Suppose that L is strongly elliptic, N_1 satisfies (c1) and (c2), and $T = L + N_1$ is K-coercive. Then the conclusions of Theorem 3.3 hold for equation (3.31).

4. EXISTENCE THEOREMS AND PSEUDO-A-PROPER MAPPINGS

In the preceding section we studied the approximation solvability of equation (1.1) involving continuous or at least bounded and finitely continuous mappings $T: X \rightarrow Y$ which are A-proper with respect to the scheme $\Gamma = \{X_n, V_n; E_n, W_n\}$. Among the results we have shown is that if, in addition to the A-properness of T, we know that equation (1.1) has at most one solution for a given vector f in Y, then this equation is strongly approximation solvable with respect to Γ [i.e., we obtained the constructive solvability of equation (1.1)].

Although the class of A-proper maps is quite large, there are existence theorems of operator equations (e.g., with weakly closed or with monotone-type operators) for which the A-proper mappings theory is not directly applicable. This is not surprising since A-properness is connected with constructive solvability and one cannot expect this property of all equations for which existence of solutions can somehow be established.

However, it was noted in [213] that a closer look at the A-proper mapping theory suggests that if one is interested primarily in its existence rather than its constructive aspect, the same approach can still be used to obtain existence theorems for a much wider class of *pseudo-A-proper* maps and for *uniform limits* of such maps. In this section we present basic existence theorems for equations involving the foregoing classes of mappings and then indicate some special cases obtained earlier by other authors. For the sake of simplicity we define this class of maps in terms of projectionally complete scheme $\Gamma_P = \{X_n; P_n; Y_n; Q_n\}$ for (X, Y), although the same goes through for more general schemes, including those used in [210].

4.1 Surjectivity Theorems

Let $D \subset X$ be a subset and let $T: D \subset X \rightarrow Y$ be a given map. As before, we associate with the equation

$$Tx = f (x \in D, f \in Y) \tag{4.1}$$

a sequence of approximate equations

$$T_n(x) = Q_n f \quad (x \in D_n, Q_n f \in Y_n). \tag{4.2}$$

Definition 4.1 A mapping $T: D \subset X \rightarrow Y$ is said to be *pseudo-A proper* with respect to Γ_p iff $T_n: D_n \subset X_n \rightarrow Y_n$ is continuous and satisfies condition (h): If $\{x_{n_i} \mid X_{n_i} \in D_{n_i}\}$ is any bounded sequence such that

$T_{n_j}(x_{n_j}) \to g$ for some g in Y as $j \to \infty$, then there exists an $x \in D$ such that $Tx = g$.

Mappings satisfying condition (h) have been introduced in [203] and studied further in [213,220,225,233] and [307]. As was indicated in [233], where $Y = X$, a pseudo-A-proper mapping is related to the notion of a G-operator introduced by De Figueiredo [72]. For further studies, see [171–175,296].

In what follows K will be a mapping (in general nonlinear) from X to Y^* such that $K(0) = 0$, $Kx \neq 0$, if $x \neq 0$ and

$$Q_n^* Kx = Kx \qquad \text{for all} \quad x \in X_n \quad \text{and all} \quad n \in Z^+. \qquad (4.3)$$

The mapping M_n will always be a linear isomorphism of X_n onto Y_n such that

$$(M_n(x), Q^* Kx) > 0 \qquad \text{for} \quad x \in X_n \quad \text{with} \quad x \neq 0. \qquad (4.4)$$

Our first basic existence result in this section is the following:

Theorem 4.1 Let K and M_n satisfy (4.3) and (4.4), respectively.

(B1) If D is a bounded open subset of X with $0 \in D$, $T: \overline{D} \to Y$ pseudo-A proper, and f an element in Y such that

$$(Tx, Kx) \geq (f, Kx) \qquad \text{for} \quad x \in \partial D, \qquad (4.5)$$

then equation (4.1) has a solution in $\overline{D}$ for each $f \in Y$ satisfying (4.5).

(B2) If $T: X \to Y$ is pseudo-A proper and K-coercive, then equation (4.1) is solvable for each f in Y (i.e., T is surjective).

Proof. (B1) For each fixed n and every $x \in \overline{D}_n$, consider the mapping $A_n(x) \equiv T_n(x) - Q_n f$, where $T_n \equiv Q_n T|_{\overline{D}_n}: \partial \overline{D}_n \subset X_n \to Y_n$. Our conditions imply that for all $x \in \partial D_n \subset X_n \cap \partial D$ and each n, we have

$$\begin{aligned}(A_n x, K_n x) &= (Q_n Tx, K_n x) - (Q_n f, K_n x) \\ &= (Tx, Kx) - (f, Kx) \geq 0.\end{aligned} \qquad (4.6)$$

Now since M_n is a linear isomorphism of X_n onto Y_n, $V_n \equiv M_n(D_n)$ is a bounded open set in Y_n with $0 \in V_n$, $V_n \cap \partial V_n = \emptyset$, and M_n maps ∂D_n homomorphically onto ∂V_n. Let G_n be the mapping from $\overline{V}_n$ into Y_n defined by $G_n = I_n - A_n L_n$ with $L_n = M_n^{-1}$ and I_n the identity in Y_n. Since for each fixed n, $y_n \in \overline{V}_n$ is a fixed point of G_n if and only if $x_n = L_n(y_n)$ is a solution of equation (4.2) [i.e., of $T_n(x_n) = Q_n f$], to establish the solvability of the latter equation for each n, in view of the finite-dimensional Leray–Schauder theorem, it suffices to show that G_n satisfies the condition

(LS): $G_n(y) \neq \lambda y$ for all $\lambda > 1$ and $y \in \partial V_n$. If this is not true, $G_n(y_0) = \lambda_0 y_0$ for some $\lambda_0 > 1$ and $y_0 \in \partial V_n$. Let $x_0 \in \partial D_n$ be such that $y_0 = M_n x_0$. This and (4.6) imply that

$$\lambda_0(M_n x_0, K_n x_0) = (y_0, K_n x_0) = (G_n y_0, K_n x_0) - (A_n L_n y_0, K_n x_0)$$
$$= (M_n x_0, K_n x_0) - (A_n x_0, K_n x_0) \leq (M_n x_0, K_n x_0).$$

Since $(M_n x_0, K_n x_0) > 0$ by (4.4), our assumption that $\lambda_0 > 1$ is false. Hence for each n there exists $x_n \in D_n$ such that $T_n(x_n) = Q_n f$. Since $\{x_n \mid x_n \in \overline{D}_n\}$ is bounded, T is pseudo-A proper on $\overline{D}$, and $T_n(x_n) = Q_n f \rightarrow f$ in Y as $n \rightarrow \infty$, there exists an element x^* in $\overline{D}$ such that $T(x^*) = f$.

(B2) To prove (B2), we first show that our K-coerciveness condition satisfied by T implies condition (4.5) for each f and suitable $D = B(0,r_f)$. Indeed, if f is any vector in Y, then since $c(r) \rightarrow \infty$ as $r \rightarrow \infty$, there exists a number $r_f > 0$ such that $\| f \| < c(r_f)$. Hence for all $x \in \partial B(0,r_f)$, condition (4.5) holds since

$$(Tx, Kx) - (f, Kx) \geq (c(r_f) - \| f \|) \| Kx \| > 0 \qquad \text{for} \quad x \in \partial B(0,r_f).$$

Now, since for each $f \in Y$ there exists $r_f > 0$ such that (4.5) holds on $\partial B(0,r_f)$, the assertion (B2) follows from (B1) for $D = B(0,r_f)$.   Q.E.D.

Remark 4.1 We remark in passing that Theorem 4.1 represents essentially a global existence result in the sense that if T is pseudo-A proper on $\overline{D}$ and if $T_{n_j}(x_{n_j}) \rightarrow g$ in Y for some bounded sequence $\{x_{n_j} \mid x_{n_j} \in \overline{D}_{n_j}\}$, the equation $Tx = g$ is necessarily solvable in $\overline{D}$. Clearly, if for a given f in Y, the equations $Tx = f$ is not solvable in $\overline{D}$ and if T is pseudo-A proper on $\overline{D}$, we cannot find a bounded sequence $\{x_{n_j} \mid x_{n_j} \in \overline{D}_{n_j}\}$ such that $T_{n_j}(x_{n_j}) \rightarrow f$ in Y as $j \rightarrow \infty$. On the other hand, the proof of Theorem 4.1 suggests the possibility of obtaining local existence results for T, which is only pointwise pseudo-A proper in the following sense:

Definition 4.1' $T: D \subset X \rightarrow Y$ is said to be *pseudo-A proper at f in Y* if for any bounded sequence $\{x_{n_j} \mid x_{n_j} \in D_{n_j}\}$ such that $T_{n_j}(x_{n_j}) \rightarrow f$ in Y, there exists an $x_0 \in \overline{D}$ such that $Tx_0 = f$.

Looking over the proof of Theorem 4.1, we see that it is also implies the validity of the following local result.

Theorem 4.1' Let D, Γ_p, K, K_n, and M_n satisfy the conditions of Theorem 4.1, and let $T: \overline{D} \rightarrow Y$ be pseudo-A proper at f in Y and $(Tx, Kx) \geq (f, Kx)$ for $x \in \partial D$; then the equation $Tx = f$ is solvable in $\overline{D}$.

For various applications of Theorems 4.1 and 4.1', see [225] and [233].

Now we apply Theorem 4.1 to obtain the basic surjectivity theorem for $T: X \to Y$, which is a uniform limit on bounded subsets of X of a special sequence of pseudo-A-proper mappings and which is not coercive.

Theorem 4.2 Let $K: X \to Y^*$, $K_n: X_n \to Y'_n$ ($\equiv Q_n^* Y$), and let $M_n: X_n \to Y_n$ be a linear isomorphism such that for all $n \in Z^+$ and all $x \neq 0$ in X_n and g in Y,

$$(Q_n g, K_n x) = (g, Kx) \quad \text{and} \quad (M_n(x), K_n(x) > 0. \tag{4.7}$$

Let $T: X \to Y$ be a finitely continuous mapping and suppose also that there exist a bounded finitely continuous mapping $F: X \to Y$ such that

(C1) If $\{x_k\} \subset X$ is bounded and $Tx_k \to g$, there exists $x \in X$ such that $Tx = g$.

(C2) $T_\mu \equiv T + \mu F$ is pseudo-A proper for each $\mu > 0$.

(C3) F is positively homogeneous of order $\gamma \geq 1$ [i.e., $F(tx) = t^\gamma Fx$ for $x \in X$, $t \geq 0$, and some $\gamma \geq 1$].

(C4) $(T_\mu(x) - T_\mu(0), Kx) \geq (\mu b \parallel x \parallel^\gamma - c) \parallel Kx \parallel$ for each $\mu > 0$ and all $x \in X$ with $b > 0$ and $c \geq 0$ some constants independent of x and μ.

(C5) $\parallel Tx \parallel \to \infty$ as $\parallel x \parallel \to \infty$.

Then under the conditions noted above, T is surjective.

Proof. To apply Theorem 4.1 in the proof of Theorem 4.2, we first note that for each given $f \in Y$ and $\mu > 0$ and all $x \in X$, we have, in view of (C4) and the equality $F(0) = 0$, the relation

$$
\begin{aligned}
(T_\mu x - f, Kx) &= (T_\mu x - T_\mu(0), Kx) + (T_\mu(0) - f, Kx) \\
&\geq (\mu b \parallel x \parallel^\gamma - c - \parallel T(0) - f \parallel) \parallel Kx \parallel.
\end{aligned}
\tag{4.8}
$$

Hence, since $\parallel x \parallel^\gamma \to \infty$ as $\parallel x \parallel \to \infty$, to each given f in Y and $\mu > 0$ there corresponds a positive number $r_{\mu f} > 0$ such that the right-hand side of the inequality (4.8) is positive for all $x \in \partial B(0, r_{\mu f})$ and therefore

$$(T_\mu(x), K(x)) \geq (f, K(x)) \quad \text{for all} \quad x \quad \text{in} \quad \partial B(0, r_{\mu f}).$$

Since by our assumption, T_μ is a finitely continuous and pseudo-A-proper mapping of X into Y, Theorem 4.1 implies that in view of the last inequality, to each fixed $\mu_k > 0$ (with $\mu_k \to 0$ as $k \to \infty$) and f in Y. There exists a vector $x_k \in \overline{B}(0, r_{\mu_k f})$ such that

$$T\mu_k(x_k) = f \quad \text{or} \quad T(x_k) = f - \mu_k F(x_k). \tag{4.9}$$

The equation (4.9) and (C4) imply that for each k we have

$$(T_{\mu_k}(x_k) - T_{\mu_k}(0), K(x_k))$$
$$= (f - T(0), K(x_k) \geq (\mu_k b \| x_k \|^\gamma - c) \| K(x_k) \|.$$

Hence, using the Schwartz–Buniakovskii inequality, we get the relation $c + \| f - T(0) \| \geq b(\| y_k \|^\gamma)$ with $y_k = \mu_k^{1/\gamma} x_k$ for each k, from which it follows that $\{y_k\}$ is bounded. Since F is bounded and $\mu_k F(x_k) = F(y_k)$, it follows that $\{F(y_k)\}$ is also bounded, and so is the sequence $\{T(x_k)\} = \{f - F(y_k)\}$. Hence (C5) implies that $\{x_k\}$ is bounded. This and the fact that F is bounded and $\mu_k \to 0$ as $k \to \iota$ imply that

$$T(x_k) = f - \mu_k F(x_k) \to f \quad \text{in} \quad y \quad \text{as} \quad k \to \infty,$$

from which, on account of (C1), we see that there exists an element x_0 in X such that $T(x_0) = f$. Since f was arbitrary, it follows that $T(X) = Y$.   Q.E.D.

For further study of general pseudo-A-proper mappings, see the author's paper [233] as well as some work of Milojević [171–173] and Webb [296].

4.2 Special Cases

Let X and Y be real reflexive Banach spaces with a projectionally complete scheme $\Gamma_P = \{X_n, P_n; Y_n, Q_n\}$, and let $D \subset X$ be an open subset. A mapping $T: \overline{D} \to Y$ is said to be *weakly closed* if $\{x_n\} \subset \overline{D}$ is a sequence such that $x_n \rightharpoonup x_0$ weakly in X and $Tx_n \rightharpoonup h$ weakly in Y, then $x_0 \in \overline{D}$ and $Tx_0 = h$. If D is also assumed to be convex, every bounded weakly closed maping $T: \overline{D} \subset X \to Y$ is pseudo-A proper w.r.t. Γ_P. This and the pseudo-A-properness of a weakly continuous mapping will be given by:

Proposition 4.1 Let (X,Y) and Γ_p be as above, K a demicontinuous mapping of X into Y^*, and K_n a mapping of X_n into $Y'_n (\equiv Q^* Y)$ such that

$$(Q_n g, K_n x) = (g, Kx) \qquad \forall\, x \in X_n, \quad q \in Y. \tag{4.10}$$

(a) If D is an open subset of X and T a bounded weakly closed mapping of $\overline{D}$ into Y, then T is pseudo-A proper w.r.t. Γ_P.

(b) If D is an open convex subset of X and T a weakly continuous mapping of $\overline{D}$ into Y, then T is pseudo-A proper w.r.t. Γ_P.

For the simple proof of Proposition 4.1 the reader is referred to the article [233].

In view of Proposition 4.1, Theorem 4.3 implies the validity of the following corollary, which we state here as an illustration of the generality of Theorem 4.1, which is discussed in [233].

Corollary 4.1 Suppose that in addition to (4.10) we also assume that there exists a linear isomorphism M_n of X_n onto Y_n such that

$$(M_n x, K_n x) > 0 \qquad \text{for} \quad x \in X_n, \quad x \neq 0, \quad g \in Y. \tag{4.11}$$

(a) If D is an open bounded subset of X with $0 \in D$ and T a finitely continuous, bounded, and weakly closed map of $\overline{D}$ into Y such that $(Tx, Kx) \geq (f, Kx)$ for all $x \in \partial D$ and some f in Y, then there exists $x_0 \in \overline{D}$ such that $Tx_0 = f$.

(b) If $T: X \to Y$ is weakly continuous and K-coercive, then T maps X onto Y.

An easy consequence of Theorem 4.2 is the following result for weakly closed mappings, which will prove to be useful for various applications.

Corollary 4.2 Suppose that conditions (4.10) and (4.11) of Corollary 4.1 hold. Let F be a bounded finitely continuous mapping of X into Y such that F is positively homogeneous of order $\gamma \geq 1$ and for some $b > 0$:

$$(Fx, Kx) \geq b \| x \|^\gamma \| Kx \| \qquad \text{for all} \quad x \text{ in } X. \tag{4.12}$$

If T is a finitely continuous weakly closed mapping of X into Y such that $T + \mu F$ is psuedo-A proper for each $\mu > 0$,

$$(Tx, Kx) \geq (T(0), Kx) - |(T(0), Kx)| \qquad \text{for all} \quad x \in X, \tag{4.13}$$

and $\| Tx \| \to \infty$ as $\| x \| \to \infty$, then T maps X onto Y.

Let us also add that, as is easy to prove, if $Y = X^*$ and $T: X \to X^*$ is demicontinuous and monotone, then T is pseudo-A proper. Using some results of Reich [248], an important example of a pseudo-A-proper mapping was provided by Webb [296], who gave a direct proof of the following:

Proposition 4.2 Suppose that X and X^* are uniformly convex and that $T: X \to X$ is a demicontinuous mapping which is accretive [i.e., $(Tx - Ty, J(x - y)) \geq 0$ for all $x, y \in X$]. Then T is pseudo-A proper w.r.t. $\Gamma_1 = \{X_n, P_n\}$. Moreover, if $\{x_n\}$ is any bounded sequence in X and $Tx_n \to g$ for some g in X, there exists an element x in X such that $Tx = g$.

This proposition will be used in subsequent discussion.

II
Equations Involving Linear A-Proper Mappings

Let $T \in L(X,Y)$, $\Gamma = \{X_n, V_n; E_n, W_n\}$ be an admissible scheme for the equation

$$Tx = f \quad (x \in X, f \in Y) \tag{1.1}$$

and consider a sequence of finite-dimensional linear equations

$$T_n(x) = W_n f \quad (x \in X_n, W_n f \in Y_n, T_n = W_n T \,|_{X_n}) \tag{1.2}$$

approximating (1.1). Section 1.1 of 1 contains some basic theorems, which provide the necessary and sufficient conditions for (1.1) to be uniquely approximation solvable w.r.t. a given scheme [i.e., for the solution x_0 of (1.1) to be constructed as a strong limit of solutions $x_n \in X_n$ of (1.2)]. These theorems are then used in Sections 1.2 and 2 to deduce, as special cases, both some earlier results and some new ones concerning the convergence of various methods of *Galerkin* and *moments* type used in the constructive solvability of (1.1) involving bounded and unbounded operators. Use of the generalized Friedrichs extension of densely defined nonsymmetric operators plays an essential role in the solvability of (1.1), involving unbounded abstract and differential operators. The existence of such extensions is proved in Section 1.2.

Fredholm alternatives for (1.1) involving A-proper maps and their special cases are treated in Section 3. The results obtained in Sections 1 to 3 are then applied to the constructive solvability of linear boundary value

55

problems for ordinary and partial differential equations, which are more general than those treated in Section I.3. Section 5 contains results on the relation of the A-properness of T to the stability of the method (1.2) in the sense of Mikhlin [166], with application to the Galerkin and Petrov–Galerkin methods. The rate of convergence of the errors and the residuals is studied in Section 6.

Since our primary goal in this chapter is to study linear equations, we will give detailed proofs of the results presented in [204,208,215,216,223, 225]. The relation of our results to those of other authors is treated briefly. For further references, see [68,73,117,123,139,141,145,162,164,165,241, 242,271,283,315].

1. APPROXIMATION-SOLVABILITY AND BOUNDED A-PROPER MAPPINGS

This section contains some basic theorems concerning the relationship between the A-properness of $T \in L(X,Y)$ and the unique approximation solvability of equation (1.1).

1.1. Basic Constructive Theorems

Our first result is the following theorem, which includes the characterization Theorem I.1.4 for $T \in L(X,Y)$.

Theorem 1.1 Let $T \in L(X,Y)$, and let $\Gamma = \{X_n, V_n; E_n, W_n\}$ be a given admissible scheme for (X,Y). If T is injective and A-proper w.r.t. Γ, equation (1.1) is uniquely approximation solvable w.r.t. Γ for each f in Y. The converse holds if Γ is an admissible projection scheme $\Gamma_p = \{X_n, V_n, Y_n, Q_n\}$.

Moreover, with $\| W_n \| \leq \beta$, the following error estimate holds:

$$\| x_n - x_0 \| \leq (1 + \beta/\gamma \| T \|) \inf\{\| x_n - x \| : x \in X_n\}, \qquad n \geq n_0.$$

$$(E1)$$

Proof. Suppose that T is injective and A-proper w.r.t. Γ. We first show that T is a-stable in the sense that there exist a constant $\gamma > 0$ and $n_0 \in Z_+$ such that

$$\| T_n(x) \| \geq \gamma \| x \| \qquad \text{for all} \quad x \in X_n \quad \text{and} \quad n \geq n_0. \qquad (1.3)$$

Indeed, if (1.3) were not true, there would exist a sequence $\{x_n \mid x_n \in X_n\}$ which, by linearity of T, we may assume that $\{x_n\} \subset \partial B(0,1)$ such that $T_n(x_n) \to 0$ as $n \to \infty$. Since T is A-proper and $\{x_n \mid x_n \in X_n\}$ is bounded, there exist a subsequence $\{x_{n_j}\}$ and $x \in X$ such that $x_{n_j} \to x$ and

$Tx = 0$ with $\| x \| = 1$, in contradiction to the injective property of T. Hence (1.3) is true. Thus since $T_n: X_n \to E_n$ is linear and dim $X_n = $ dim E_n, it follows from (1.3) that T_n is a one-to-one map of X_n onto Y_n and so, for each $n \geq n_0$, equation (1.2) has a unique solution $x_n \in X_n$ for each $f \in Y$ since $W_n(Y) = E_n$.

Now since $\{W_n\}$ is uniformly bounded and $\{x_n \mid x_n \in X_n\}$ satisfies (1.2), it follows from this and (1.3) that $\{x_n\}$ is bounded and $T_n(x_n) - W_nf = 0 \to 0$. Hence the A-properness of T w.r.t. Γ implies the existence of a subsequence $\{x_{n_j}\}$ and $x_0 \in X$ such that $x_{n_j} \to x_0$ in X as $j \to \infty$ and $Tx_0 = f$ [i.e., x_0 is a solution of (1.1) that is unique since T is one-to-one]. This and the A-properness of T imply that $x_n \to x_0$, since otherwise there would exist a subsequence $\{x_{n_k}\}$ of $\{x_n\}$ and some $\epsilon > 0$ such that $\| x_{n_k} - x_0 \| \geq \epsilon$ for all $k \in Z_+$. But $T_{n_k}(x_{n_k}) - W_{n_k}(f) = 0 \to 0$, so, again, by the A-properness of T, there exist a subsequence $\{x_{n_{k(i)}}\}$ and $x' \in X$ such that $x_{n_{k(i)}} \to x'$ as $i \to \infty$ and $T(x') = f$ with $x' = x_0$ since x_0 is the unique solution of (1.1), contradicting the assumption that $\| x_{n_k} - x_0 \| \geq \epsilon$ for all $k \in Z_+$. Thus equation (1.1) is uniquely approximation solvable w.r.t. Γ for each $f \in Y$.

Converse. Suppose that (1.1) is uniquely approximation solvable w.r.t. $\Gamma_P = \{X_n, V_n, Y_n, Q_n\}$ for each f in Y. Hence $T^{-1} \in L(Y, X)$, $T_n^{-1} \equiv (Q_nT|_{X_n})^{-1}: Y_n \to X_n$ exists for each $n \geq n_0$ and $x_n = T_n^{-1}Q_n(f) \to x \equiv T^{-1}f$ in X for each $f \in Y$. Hence, by the principle of uniform boundedness (see [66]), the norms of $T_n^{-1}Q_n: Y \to X$ are uniformly bounded by some $K_0 > 0$ for each $n \geq n_0$. Since $Q_nf = f$ for $f \in Y_n$, it follows that the norms $T_n^{-1}: Y_n \to X_n$ are also uniformly bounded by K_0 for $n \geq n_0$. Thus for each $n \geq n_0$ and all $x \in X_n$, $\| x \| = \| T_n^{-1}T_n(x) \| \leq K_0 \| T_n x \|$ [i.e., (1.3) holds with $\gamma = K_0^{-1}$]. Now, to show that T is A-proper w.r.t. Γ_P, let $\{x_{n_j} \mid x_{n_j} \in X_n\}$ be any bounded sequence such that $g_{n_j} \equiv Q_{n_j}T(x_{n_j}) - Q_{n_j}g \to 0$ in Y for some g in Y. Since (1.1) is uniquely approximation solvable for each $f \in Y$, there exists a unique $u_n \in X_n$ such that $T_n(u_n) = Q_ng$ and $u_n \to u = T^{-1}g$ in X. Hence, since (1.3) holds, for all sufficiently large j we have

$$\| x_{n_j} - u_{n_j} \| = \| T_{n_j}^{-1}(T_{n_j}(x_{n_j})) \| < K_0 \| T_{n_j}(x_{n_j}) - Q_{n_j}g \|.$$

But $g_{n_j} \to 0$ and therefore $x_{n_j} \to u$ since $u_{n_j} \to u$ and $\| x_{n_j} - u_{n_j} \| \to 0$ as $j \to \infty$. Thus T is injective and A-proper w.r.t. Γ_P.

To get (E1) note that there is $\tilde{x}_n \in X_n$ for each $n \in Z_+$ such that $\| x_0 - \tilde{x}_n \| = \text{dist}(x_0, X_n) \to 0$. But then $Tx_n \to Tx_0 = f$, $x_n - \tilde{x}_n \in X_n$, and hence by (1.3) and (1.2) and the fact that $\| W_n \| \leq \beta$, we have

$$\gamma \| x_n - \tilde{x}_n \| \leq \| T_n(x_n) - W_nT\tilde{x}_n \|$$

$$= \| W_nf - W_nT\tilde{x}_n \| \leq \beta \| T \| \| x_0 - \tilde{x}_n \|.$$

This and $\| x_n - x_0 \| \leq \| x_n - x_n \| + \| x_n - x_0 \|$ yield (E1).   Q.E.D.

Since (1.3) plays a key role in the study of (1.1), we refer to $T \in L(X,Y)$ satisfying (1.3) as *a-stable* (see Theorem 1.3).

Theorem 1.1 highlights the importance of finding conditions that would imply the A-properness of T. We do this in Theorems 1.2, 1.3, and 1.4.

Theorem 1.2 T is A-proper w.r.t. Γ and injective iff T is a-stable and pseudo-A-proper.

Proof.   It follows from the proof of the sufficiency part of Theorem 1.1 that if T is injective and A-proper w.r.t. Γ, then T is obviously pseudo-A-proper and a-stable. Suppose, conversely, that T is pseudo-A-proper and a-stable. First, to show that T is injective, let $Tu = Tv$ for $u \neq v$. Since dist$(x, X_n) \to 0$ for each $x \in X$, there exist u_n and v_n in X_n such that $u_n \to u$ and $v_n \to v$ in X. It follows from the a-stability of T [i.e., from (1.3)] that

$$c \parallel u_n - v_n \parallel \leq \parallel T_n(u_n - v_n) \parallel \leq \beta \parallel Tu_n - Tv_n \parallel \to 0$$

since $\parallel W_n \parallel \leq \beta$, $Tu_n \to Tu$, and $Tv_n \to Tv$ in Y. Thus $\parallel u_n - v_n \parallel \to 0$, and therefore $u = v$. This contradiction shows that T is one-to-one. To prove that T is A-proper, let $\{x_{n_j} \mid x_{n_j} \in X_{n_j}\}$ be any bounded sequence such that $T_{n_j}(x_{n_j}) - W_{n_j}g \to 0$ for some g in Y. It follows from this and the pseudo-A-properness of T w.r.t. Γ that there exists $x \in X$ such that $Tx = g$. Let $u_n \in X_n$ be such that $u_n \to x$ in X. Then, by virtue of the a-stability of T, for each $j \in Z_+$ sufficiently large, we have

$$\gamma \parallel x_{n_j} - u_{n_j} \parallel \leq \parallel T_{n_j}(x_{n_j}) - T_n(u_{n_j}) \parallel \leq \parallel T_{n_j}(x_{n_j}) - W_{n_j}g \parallel$$

$$+ \parallel W_{n_j}(g - Tu_{n_j}) \parallel \to 0 \qquad \text{as} \quad j \to \infty.$$

Hence $x_{n_j} \to x$ and $Tx = g$; that is, T is A-proper and injective.

$$\text{Q.E.D.}$$

A consequence of Theorems 1.1 and 1.2 is the following useful result, which is an improvement of Theorem I.1.5, where T was nonlinear.

Theorem 1.3   The following assertions are equivalent:

(A1) T is injective and A-proper w.r.t. Γ.
(A2) $R(T) = Y$ and T is a-stable w.r.t. Γ.
(A3) Equation (1.1) is uniquely approximation solvable w.r.t. Γ.

The key observation that should be made about the proof of Theorems 1.1 and 1.2 is that it does not use the notion of the adjoint T^* of T, which is a highly linear concept. Consequently, the same type of argument applies to equations involving linear and nonlinear A-proper mappings. The practical usefulness of Theorem 1.3 stems from the fact that if we know (no matter how) that T is surjective, then under condition (1.3) the map T is injective and A-proper and thus equation (1.1) is uniquely approximation solvable w.r.t. Γ for each f in Y.

If we strengthen somewhat the conditions on (X,Y) and on Γ, the surjectivity hypothesis in (A2) can be eliminated.

Theorem 1.4 Let $\Gamma_P = \{X_n, V_n, Y_n, Q_n\}$ be a projectional scheme that is complete for (X,Y) and suppose that $\{X_n\}$ and $\{Y_n\}$ are nested. Suppose further that one of the following holds:

  (i) X is reflective and $R(Q_n^*) \subset R(Q_m^*)$ for $n \leq m$.
  (ii) $Q_n^* w \to w$ in Y^* for each w in Y^*.

Then T is A-proper w.r.t. Γ_P iff condition (1.3) holds. In particular, equation (1.1) is uniquely approximation solvable w.r.t. Γ_P for each $f \in Y$ iff (1.3) holds.

Proof. By Theorem 1.3 it suffices to show that T is pseudo-A-proper w.r.t. Γ_P if either (i) or (ii) holds.

  (i) Let $\{x_{n_j} \mid x_{n_j} \in X_{n_j}\}$ be any bounded sequence such that $T_{n_j}(x_{n_j}) - Q_{n_j}g \to 0$ for some g in Y. Since X is reflexive and $\{x_{n_j}\}$ is bounded, we may assume that $x_{n_j} \rightharpoonup x$ in X, so $T x_{n_j} \rightharpoonup Tx$ in Y. Since Γ_P is complete and $Q_n y \to y$ for each y in Y and $W \equiv \cup_n R(Q_n^*)$ is total. To see the latter, let $y \in Y$ be such that $(w,y) = 0$ for all $w \in W$. This implies that $(Q_m^* u, y) = 0$ for each u in Y^*. Hence $0 = (Q_m^* u, y) = (u, Q_m y) \to (u,y) = 0$ for each u in Y^*, from which it follows that $y = 0$ (i.e., W is total). Let w be any element in W. Then w lies in some $R(Q_m^*)$ and hence, by the second assumption in (i), for all $n_j \geq m$ we have

$$0 = \lim_j (T_{n_j}(x_{n_j}) - Q_{n_j}g, w) = \lim_j (T x_{n_j} - g, w) = (Tx - g, w).$$

Thus $Tx = g$, so T is pseudo-A-proper.

  (ii) First note that since $Q_n y \to y$ for each y in Y and $\{Q_n\}$ is uniformly bounded, it follows from this and the continuity of T that if $u_j \to u$ in X, then $Q_j T u_j - Tu = Q_j T u_j - Q_j Tu + Q_j Tu - Tu \to 0$ in Y. Now let x be any element in X. Since $\operatorname{dist}(x, X_n) \to 0$, there exists $x_n \in X_n$ such that $x_n \to x$ and so, by (1.3), $\| Q_n T x_n \| \geq \gamma \| x_n \|$ for $n \geq n_0$. Thus, by the preceding discussion, taking the limit in the last inequality as $n \to \infty$, we get $\| Tx \| \geq \gamma \| x \|$ for each $x \in X$. This implies that T has a continuous inverse T^{-1} defined on $R(T) \subset Y$ with $R(T)$ closed.

Suppose now that $\{x_{n_j} \mid x_{n_j} \in X_{n_j}\}$ is any bounded sequence such that

$T_{n_j}(x_{n_j}) - Q_{n_j}g \to 0$ for some g in Y. Let $w \in N(T^*)$. Then since $\{x_{n_j}\}$ is bounded, $Q^*_{n_j}(w) \to w$ in Y^* and $T^*Q^*_{n_j}(w) \to T^*(w) = 0$ in X^*, it follows that $(T^*Q^*_{n_j}w, x_{n_j}) \to 0$ as $j \to \infty$. In view of this and the equality $(T^*Q^*_{n_j}w, x_{n_j}) = (w, Q_{n_j}Tx_{n_j} - Q_{n_j}g) + (w, Q_{n_j}g)$, the passage to the limit in the latter as $n_j \to \infty$ yields the relation $(w,g) = 0$ for each $w \in N(T^*)$. Thus $g \in R(T)$ since $R(T)$ is closed, so there exists $x \in X$ such that $Tx = g$ [i.e., T is pseudo-A-proper w.r.t. Γ_P if (ii) holds].

The last assertion follows from the facts proved above and Theorem 1.2.   Q.E.D.

If X is reflexive, the additional condition in (i) imposed on Γ_P is certainly satisfied when Y has a Schauder basis. If X is not necessarily reflexive, the additional condition in (ii) imposed on Γ_P holds if Y has a shrinking basis.

We conclude this section by singling out the following special case of Corollary I.2.5, which will prove to be useful in Section 2.

Proposition 1.1   Let X be reflexive, $K \in L(X,Y^*)$ injective with $\overline{R(K)} = Y^*$ and $\Gamma_P = \{X_n, V_n; Y_n, Q_n\}$ an admissible projective scheme for (X,Y) such that $Q^*_n Kx = Kx$ for $x \in X_n$ and $n \in Z_+$. If there exist constants $a_1 > 0$ and a_2 and $n_0 \in Z_+$ such that

$$(Tx, Kx) \geq a_1 \| x \|^2 - a_2(Cx, Kx) \quad \text{for} \quad x \in X_n \quad \text{and} \quad n \geq n_0, \quad (1.4)$$

where $C \in L(X,Y)$ is compact, then T is A-proper w.r.t. Γ_P.

1.2   Special Cases: Galerkin and Petrov–Galerkin Methods

In this section and in Section 2 we indicate how Theorems 1.1 to 1.4 of Section 1.1 can be used to deduce as special cases some earlier results, as well as some new ones, concerning the convergence of various methods of Galerkin type used in the solvability of equation (1.1) involving special classes of bounded and unbounded linear operators. The study of these methods was initiated in [123,165,199,241] and pursued further in [139,157,215,225,283,284] and others. We survey these results as they appeared in their chronological order. It is worthwhile noting that practically all of these authors assumed the existence of solutions and were concerned only with the convergence problems. In contradistinction, our results in Section 1.1 in most cases provide constructive existence theorems. We mention explicitly only those authors to whom our results are directly related. For references, see [139,157].

Galerkin Method

Let $M \in L(X,X)$ and let $\Gamma_\alpha = \{X_n, P_n\}$ be projectionally complete for (X,X) with $\alpha = \sup_n \| P_n \|$. The Galerkin method for the equation

$$x - Mx = f \qquad (x \in X, f \in X) \qquad (1.5)$$

consists of solving the finite-dimensional equations

$$x_n - P_n M x_n = P_n f \qquad (1.6)$$

for the approximate solutions $x_n \in X_n$ and then showing that $\{x_n\}$ converges strongly to a solution of (1.5). Depending on the operator M and the scheme Γ_α, Theorems 1.1 to 1.4 imply the following constructive results for equation (1.5).

Corollary 1.1 If $T \equiv I - M$ is injective, then equation (1.5) is uniquely approximation solvable w.r.t. Γ_α for each $f \in Y$ provided that one of the following two conditions holds:

(a) M is k-ball contractive with $k < \alpha^{-1}$ and, in particular, $M = S + C$ with C compact and $\| S \| < \alpha^{-1}$.

(b) $\{X_n\}$ is nested, $\alpha = 1$, $P_n^* w \to w$ in X^* for each w in X^* and $M = F + C - A$, where F is k-ball contractive with $k < 1$, C compact and A accretive [i.e., $(Ax,w) \geq 0$ for each $x \in X$ and $w \in Jx$ with J being a normalized duality map of X into X^*].

Proof. (a) Since, by Corollary I.2.2, $T = I - M$ is A-proper w.r.t. Γ_α when M is k-ball contractive with $k < \alpha^{-1}$, the assertion of Corollary 1.1 in this case follows from Theorem I.1.4 (or Theorem 1.1) when $Y = X$, $Y_n = X_n$, and $Q_n = P_n$ for each $n \in Z_+$.

(b) Since $\| P_n \| = 1$, it follows from Lemma I.2.5 that $P_n^* Jx \subset Jx$ for each $x \in X_n$ and $\| P_n^* w \| = \| w \|$ for $w \in Jx$. Hence $\| P_n (I + A)(x) \| \geq \| x \|$ for all $x \in X_n$. In view of this and the assumption that $P_n^*(u) \to u$ in X^* for each u in X^*, it follows from Theorem 1.4 that $I + A: X \to X$ is A-proper w.r.t. Γ_α. Hence, by Theorem I.2.2, the map $I + A - F$ is A-proper w.r.t. Γ_α since F is k-ball contractive with $k < 1$. This and the compactness of C imply that $T = I - (F + C - A) = I + A - F - C$ is A-proper w.r.t. Γ_α, so the assertion of Corollary 1.1, in case (b), also follows from Theorem 1.4. Q.E.D.

Another corollary of Theorem 1.4, which includes the results proved in [162,164,204,242] for the case when X is a Hilbert space, is the following result for equation (1.1).

Corollary 1.2   Suppose that $\Gamma_\alpha = \{X_n, P_n\}$ and J are the same as in Corollary 1.1(b) and $T \in L(X,X)$ is such that any one of the following conditions holds:

$$(Tx, w) \geq c \| x \|^2 \qquad \text{for all} \quad x \in X, \; w \in J(x) \quad \text{and some} \quad c > 0.$$

$$(1.7)$$

X is reflexive with X^* strictly convex and T is injective with   (1.8)
$\lim \inf(Tx_j J x_j) \geq \delta > 0$ whenever $\{x_j\} \subset \partial B(0,1)$ and $x_j \rightharpoonup 0$.

$T = A + C$, T injective, C compact, and either (1.7) or (1.8)   (1.9)
holds for A.

Then equation (1.1) is uniquely approximation solvable w.r.t. Γ_α for each f in Y.

     The proof is similar to that of Corollary 1.1(b) and therefore is left as an exercise for the interested reader.

Petrov–Galerkin Method

     Let $\{X_n\}$ and $\{Y_n\}$ be two distinct sequences in X so that $\Gamma_P = \{X_n, P_n; Y_n, Q_n\}$ is projectionally complete for (X,X). The generalization of the Galerkin method suggested in [199], when M is compact, consists in constructing a solution of (1.5) as a limit of solutions $x_n \in X_n$ of the equations

$$Q_n x_n - Q_n M x_n = Q_n f \qquad (x_n \in X_n, \; Q_n f \in Y_n), \qquad (1.10)$$

which are different from (1.6) since $Q_n x_n \neq x_n$. In this abstract setting and under the assumption that (1.5) is uniquely solvable for each f in X, the unique approximation solvability of (1.5) w.r.t. Γ_P for each f in X was proved by Polskii [242] under the following:

Condition (A)   There exists $\Theta > 0$ and $n_0 \in Z_+$ such that $\| x \| \leq \Theta \| Q_n x \|$ for all $x \in X_n$ and $n \geq n_0$.

     It was noted in [241] that condition (A) is also a necessary condition for (1.5) to be uniquely approximation solvable w.r.t. Γ_P for each f in X in the sense that if condition (A) does not hold, one can give an example of M and f for which (1.10) is either not solvable or has infinitely many solutions or $\{x_n\}$ does not converge. It was shown by Vainikko [283] (even in a more general setting) that condition (A) is both necessary and sufficient for the convergence of the Petrov–Galerkin method.

     It turns out that the basic result in [242] can easily be embedded into the general theory of A-proper mappings. It follows from Theorem I.1.4 (or Theorem 1.1) and the following lemma deduced from Theorem 1.3.

Lemma 1.1 Suppose that $M \in L(X,X)$ is compact. Then $T = I - M$ is A-proper w.r.t. Γ_P if and only if condition (A) holds.

Proof. Suppose that $T = I - M$ is A-proper w.r.t. $\Gamma_P = \{X_n, P_n; Y_n, Q_n\}$. Then, by Theorem I.1.2, the identity $I = T + M$ is A-proper w.r.t. Γ_P. This and the injective property of I imply the existence of $\gamma > 0$ and $n_0 \in Z_+$ such that $\| Q_n x \| \geq \gamma \| x \|$ for all $x \in X_n$ and $n \geq n_0$ [i.e., condition (A) holds with $\Theta = \gamma^{-1}$].

Conversely, suppose that condition (A) holds. Since I is also surjective, Theorem 1.3 shows that I is A-proper w.r.t. Γ_P. Hence, again by Theorem I.1.2, $T = I - M$ is A-proper w.r.t. Γ_P.   Q.E.D.

By virtue of Lemma 1.1, Theorem 1.1. implies the following result for the Petrov–Galerkin method.

Corollary 1.3 Let $\Gamma_P = \{X_n, P_n; Y_n, Q_n\}$ be projectionally complete for (X,X) and $M \in L(X,X)$ compact. Then the following assertions are equivalent:

(A1) Equation (1.5) is uniquely approximation solvable w.r.t. Γ_P.
(A2) $T = I - M$ is injective and A-proper w.r.t. Γ_P.
(A3) T is injective and condition (A) holds.

An analogous characterization of the unique approximation solvability of Polskii [242] obtained for (1.1) with $T \in L(H,H)$ bijective is deduced even for $T \in L(X,X)$ from Theorem 1.1 and the following consequence of Theorem 1.3.

Lemma 1.2 Let $\Gamma_\alpha = \{X_n, P_n\}$ be projectionally complete for (X,X) and $T \in L(X,X)$. Then T is injective and A-proper w.r.t. Γ_α if and only if T is bijective and satisfies condition (A), $\tau = \lim \inf \tau_n > 0$, where $\tau_n = \inf\{\| P_n z \| : z \in T(X_n), \| z \| = 1\} > 0$ for $n \geq n_0$.

Proof. Let T be injective and A-proper. Then, by Theorem 1.1, T is bijective and there is $n_0 \in Z_+$ and $\gamma > 0$ such that $\| P_n T x \| \geq \gamma \| x \|$ for $x \in X_n$ and $n \geq n_0$. This and $\| Tx \| \leq \| T \| \| x \|$ for $0 \neq x \in X_n$ imply that $\| P_n T x \| / \| Tx \| \geq \gamma / \| T \|$ for $n \geq n_0$ and $0 \neq x \in X_n$. This shows that T is bijective and satisfies condition (A).

Converse. Let T be bijective and satisfy condition (A). Then $\| P_n T x \| \geq \tau_n \| Tx \|$ for all $x \in X_n$ and $n \geq n_0$. Since $\| Tx \| \geq m \| x \|$ for $x \in X$

and some $m > 0$, we see that $\| P_n Tx \| \geq \tau_n m \| x \|$ for $x \in X_n$ and $n \geq n_0$. Since $\tau > 0$, the definition of τ implies that for any given $\epsilon > 0$ with $\tau - \epsilon > 0$, there exists $n(\epsilon) \in Z_+$ such that $m\tau_n \geq m(\tau - \epsilon)$ for $n \geq n(\epsilon)$. Thus $\| P_n Tx \| \geq m(\tau - \epsilon)\| x \|$ for $x \in X_n$ and $n \geq \max\{n(\epsilon),n_0\}$ (i.e., T is a-stable and bijective). Hence by Theorem 1.3, T is injective and A-proper w.r.t. Γ_α.   Q.E.D.

2. EQUATIONS INVOLVING UNBOUNDED LINEAR OPERATORS

In this section we show how the theory of A-proper maps developed in Section 1 can be used to study the approximation solvability of equations involving various classes of unbounded densely defined linear operators. The relation of results presented here to those obtained earlier by the writer and other authors concerning the methods of Ritz, Galerkin, moments, and least squares will be indicated. The basic virtue of the theory of A-proper mappings, as used here, lies not only in the fact that it presents a general and unified approach to problems treated differently by a number of authors, but as will be seen later, it represents a new approach to the approximation solvability of equations involving abstract and differential operators.

2.1 Solvable Extensions of Densely Defined K-p.d. Operators

Let H be a real Hilbert space. It was shown by Friedrichs [88] (see also [165]) that if T is a densely defined linear operator that is symmetric and positive definite, T has a unique self-adjoint extension which is positive definite and bijective. However, if $L_2([0,1])$ is a real space and we consider the simple boundary value problem

$$-D((p(x)D^2u) + g(x)Du = f(x), \qquad (2.1)$$
$$u(0) = Du(0) = D^2u(1) = 0 \qquad (f \in L_2),$$

where $p \in C^1[0,1]$ with $p(x) \geq p_0 > 0$ for $x \in [0,1]$ and $g \in C[0,1]$ with $g(x) \geq 0$ for $x \in [0,1]$, then T defined on

$$D(T) = C_0^3 = \{u \in C^3[0,1] \mid u(0) = Du(0) = D^2u(1) = 0\}$$

$$\text{by} \quad Tu = -(D(pD^2u)) + gDu$$

is not symmetric or positive definite, and hence Friedrichs' theory is not applicable to (2.1). To treat the boundary value problem (2.1) (and more general ones) in a way analogous to [88,165], it was shown by the author in [215,216] that Friedrichs' procedure admits an extension to *K-positive definite* (*K*-p.d.) and *K-symmetric* maps given by

Definition 2.1   A densely defined map $T: D(T) \subset H \to H$ is called *K*-p.d. if there exist a closable operator K with $D(K) \supseteq D(T)$ and $KD(T)$ dense in H and constants $\alpha_1 > 0$, $\alpha_2 > 0$ such that

$$(Tx,Kx) \geq \alpha_1 \| x \|^2 \quad \text{and} \quad \| Kx \|^2 \leq \alpha_2(Tx,Kx) \qquad \text{for} \quad x \in D(T);$$

$$(2.2)$$

T is called *K-symmetric* iff $(Tx,Ky) = (Kx,Ty)$ for $x,y \in D(T)$.

Note that if we set $Ku = Du$ for $u \in D(K) = \{u \in C^1[0,1] \mid u(0) = 0\}$, it is not hard to show that T, defined by the left-hand side of (2.1), is *K*-p.d. and *K*-symmetric with $Ku = Du$ on $D(K)$ in the sense of Definition 2.1.

Since the Friedrichs-type extensions of *K*-p.d. and *K*-symmetric operators, as well as of *K*-p.d. but not necessarily *K*-symmetric operators, play an essential role in the approximation solvability of equations involving such operators, we use this section to outline the basic results from [215,216].

Lemma 2.1   If *T* is *K*-p.d. and *K*-symmetric, then:

(a) *T* has a bounded inverse defined on $R(T)$ and *T* is closable.
(b) $|(Tx,Ky)|^2 \leq (Tx,Kx)(Ty,Ky)$ for $x,y \in D(T)$.

Exercise 2.1   Prove Lemma 2.1.

Let f be any element in H and let $F(x)$ be the functional

$$F(x) = (Tx,Kx) - 2(f,Kx) \qquad [x \in D(T)]. \tag{2.3}$$

It is not hard to see that the problem of solving the equation

$$Tx = f \qquad [x \in D(T), f \in H] \tag{2.4}$$

is equivalent to the problem of minimizing the functional $F(x)$ and to solve the latter it is, in general, necessary to extend somewhat the set $D(T)$ on which F is defined and with it the operator T. It was shown by the author in [216] that, using essentially the same arguments as in [165,215], the variational problem is solvable and that T possesses a unique closed and continuously invertible K_0-p.d. and K_0-symmetric extension, which can be described briefly as follows.

Let H_0 be the completion of $D(T)$ with respect to

$$[x,y] = (Tx,Ky), \quad |x| = [x,x]^{1/2} \quad [x,y \in D(T)]. \tag{2.5}$$

Lemma 2.2   (a) H_0 can be regarded as a subset of H.
  (b) K has a bounded extension K_0: $H_0 \to H$ with $K \subset K_0 \subset \overline{K}$.
  (c) $|x| \geq \alpha_1^{1/2} \|x\|$ and $\|K_0 x\| \leq \alpha_2^{1/2} |x|$ for $x \in H_0$.

Exercise 2.2   Prove Lemma 2.2.

Having constructed H_0, it is now easy to solve the variational problem. In fact, by Lemma 2.2, $(f, K_0 x)$ is a bounded linear functional of $x \in H_0$ and hence, by the Riesz theorem, to each f in H there exists a unique $w \in H_0$ such that

$$(f, K_0 x) = [w,x] \quad \text{for all} \quad x \in H_0. \tag{2.6}$$

Consequently, the functional

$$F(x) = [x,x] - 2[w,x]^2 = |x - w|^2 - |w|^2 \tag{2.7}$$

which, by (2.3), is valid for all x in $D(T)$, can be extended to all of H_0. Considered on H_0, F attains its infimum d at $x = w$ with

$$d = \inf\{F(x) \mid x \in H_0\} = -|w|^2. \tag{2.8}$$

Note that, in general, w, which minimizes F on H_0, does not necessarily lie in $D(T)$, so that equation (2.4) may not have a solution unless T could be somewhat extended. Theorem 2.1 shows that for a K-p.d. and K-symmetric T, such an extension is always possible.

Theorem 2.1   If T is K-p.d. and K-symmetric, then T has a unique K_0-p.d. and K_0-symmetric extension $T_0 \supseteq T$ such that T_0 is continuously invertible and $D(T_0)$ consists of all elements in H_0 realizing the infimum of $F(x)$ on H_0 as f ranges through all of H.

Proof.   We outline the proof of Theorem 2.1. As before, for each f in H there exists a unique $w = Gf \in H_0$ such that

$$(f, K_0 x) = [Gf, x] \quad \text{for all} \quad x \in H_0. \tag{2.9}$$

By Lemma 2.2, G defined by (2.9) is a bounded linear map from H to H_0 and from H to H with $|G| \leq \alpha_2^{1/2}$ and $\|G\| \leq \alpha_2^{1/2}/\alpha_1^{1/2}$. It follows from (2.9) and the denseness of $KD(T)$ that G is injective. Thus $T_0 \equiv G^{-1}$ exists and is a closed operator such that $T_0 \supseteq T$ and T_0 is K_0-p.d. and K_0-symmetric. Furthermore, T_0 is the only extension with $D(T_0) \equiv R(G^{-1})$

$\subset H_0$ such that if T^1 is any other K_0-p.d. and K_0-symmetric extension of T with $D(T^1) \subset H_0$, then $T_0 \supseteq T^1$ (see [215,216] for details).   Q.E.D.

The operator T_0 will be called the *generalized Friedrichs extension* of T. For $K = I$, Theorem 2.1 furnishes the self-adjoint extension of Friedrichs. An element $u_0 \in D(T_0)$ such that $T_0(u_0) = f$ will be called a *strong solution* of equation (2.4).

An immediate consequence of Lemma 2.2 and Theorem 2.1 is the following result, which is sometimes useful in applications.

Corollary 2.1 If T is K-p.d. and K-symmetric and K is closed with $D(K) = D(T)$, then $K = K_0$, $H_0 = D(T)$, $T = T_0$, $R(T) = R(K) = H$ and $|x|$, $\| Tx \|$, $\| Kx \|$ define equivalent norms on $D(T)$.

Corollary 2.1 implies that T and K form an acute angle [271]; that is,

$$(Tx, Kx) \geq \delta \| Tx \| \| Kx \| \qquad \text{for all} \quad x \in D(T) \quad \text{and some} \quad \delta \in (0,1].$$
$$(2.10)$$

The approximation solvability of (2.4) with T having an acute angle will be studied in Section 2.2.

If we apply the discussion above to (2.1) with $Ku = Du$, then H_0 in this case is the completion of C_0^3 with respect to

$$[u,u] = \int_0^1 \{pu''v'' + gu'v'\} \, dx, \quad |u|_0 = [u,u]^{1/2}$$
$$(u' \equiv Du, \ u'' \equiv D^2 u)$$
$$(2.11)$$

Because $|u|_0^2 \geq p_0\|(u'')^2\|_2^2$ and $u(0) = u'(0) = 0$, it is easy to show that the H_0-norm is equivalent to $W_2^2(0,1)$-norm and that $H_0 = \{u \in W_2^2 \mid u(0) = u'(0) = 0\}$. Since, as was noted above, T defined on C_0^3 by the left side of (2.1) is K-p.d. and K-symmetric, we have the following result (see [51]):

Proposition 2.1 T defined on C_0^3 by $Tu = -D(pD^2u) + gDu$ has a unique generalized Friedrichs extension T_0 with $D(T_0) \subset H_0$. Thus (2.1) has a unique strong solution $u \in D(T_0)$ for each $f \in L_2$. Moreover, $D(T_0) = \{u \in H_0 \mid u'' \text{ is a.c. on } [0,1], u''' \in L_2(0,1), \text{ and } u''(1) = 0\}$.

Note that the natural boundary condition $u''(1) = 0$ is lost in the construction of H_0 but is regained in $D(T_0)$.

Proof. The first two assertions follow from Theorem 2.1. To prove the last assertion, note that for each $f \in L_2$ there exists a unique $u = T_0^{-1}f \in D(T_0)$ such that $[u,v] = (f,Kv)$ for each $v \in H_0$, that is,

$$\int_0^1 \{p(x)u''(x)v''(x) - [g(x)u'(x) - f(x)]v'(x)\}\, dx = 0.$$

The second term can be transformed in the following way:

$$\int_0^1 [g(x)u'(x) - f(x)]v'(x)\, dx$$

$$= \int_0^1 \left[\frac{d}{dx} \left(\int_0^x \{g(s)u'(s) - f(s)\}\, ds \right) \right] v'(x)\, dx$$

$$= \int_0^x \{g(s)u'(s) - f(s)\}\, ds\, v'(x)\Big|_0^1$$

$$- \int_0^1 \left(\int_0^x \{g(s)u'(s) - f(s)\}\, ds \right) v''(x)\, dx.$$

Since $v'(0) = 0$ and $v'(1) = \int_0^1 v''(x)\, dx$, combining the above, we get

$$\int_0^1 \{p(x)u'' + \int_0^1 [g(s)u'(s) - f(s)]\, ds$$

$$- \int_0^x [g(s)u'(s) - f(s)]\, ds\}v''(x)\, dx = 0.$$

But for $z \in L_2(0,1)$, if we set $w(t) = \int_0^t z(s)\, ds$ and $v(x) = \int_0^x w(t)\, dt$, then $v \in H_0$ and $v''(x) = z$, so that the map $B: H_0 \to L_2$, given by $Bv = v''$, is bijective. Therefore, the expression in the braces of the last equality equals 0 for $x \in [0,1]$ (a.e.), that is,

$$p(x)u''(x) + \int_0^1 [g(s)u'(s) - f(s)]\, ds - \int_0^x [g(s)u'(x) - f(s)]\, ds = 0.$$

Since $p(1)u''(1) = 0$ and $p(x) \geq p_0$ for all $x \in [0,1]$, it follows from the last equality that u'' is a.c. on $[0,1]$, $u''(1) = 0$, $u''' \in L^2(0,1)$, and $-(p(x)u'') + g(x)u'(x) = f(x)$ for $x \in [0,1]$ (a.e.). We have so far shown that $D(T_0) \subseteq M \equiv \{u \in H_0 \,|\, u''$ is a.c. on $[0,1]$, $u''' \in L^2(0,1)$ and $u''(1) = 0\}$. Conversely, let $v \in M$ be such that $f = -(pv'')' + gv' \in L_2$, and let $u \in D(T_0)$ be such that $T_0 u = f$. Then for each z in H_0, $[v,z] = \int_0^1 (pv''z'' + gv'z')\, dx = \int_0^1 fz'\, dx = [u,z]$. Thus $v = u \in D(T_0)$ and hence $D(T_0) = M$.   Q.E.D.

Remark 2.1 The last assertion of Proposition 2.1 shows that for some differential K-p.d. and K-symmetric operators one can determine the nature of the domain $D(T_0)$, which, as we shall see later, is important in proving an estimate for the residual $Tu_n - f$, where u_n is a Galerkin approximate to u_0 and obtain the convergence of u_n to u and of all of the derivatives of u_n up to the order of T.

Using T_0 we can now construct a *solvable extension* L_0 for a K-p.d. but not necessarily K-symmetric operator L and use this fact in the approximation solvability of the equation

$$Lx = f \qquad [x \in D(L), f \in H]. \tag{2.12}$$

Thus the following extension obtained in [216] of the results of Friedrichs [88], Lax and Milgram [150], and the writer [215] will prove to be useful.

Theorem 2.2 Let T be K-p.d. and K-symmetric and let L be a linear map with $D(L) = D(T)$ such that for some $\eta_1 > 0$ and $\eta_2 > 0$,

$$(Lx, Kx) \geq \eta_1 |x|^2 \qquad \text{for} \quad x \in D(L) \tag{2.13}$$

$$|(Lx, Ky)| \leq \eta_2 |x||y| \qquad \text{for } x,y \in D(L). \tag{2.14}$$

Then L has a unique closed extension $L_0 \supseteq L$ of the form $L_0 = T_0 W_0$ with $D(L_0) \subset H_0$ such that L_0 is continously invertible and L_0 satisfies (2.13) and (2.14) with K_0 replacing K, where W_0 is an extension of $W = T_0^{-1}L$ such that $W \subset W_0 \subset \overline{W}$ and $R(W_0) = D(T_0)$.

Proof. Let T_0 be the generalized Friedrichs extension of T. Then $W = T_0^{-1}L: D(L) \subset H_0 \to H_0$ is bounded since if $v = Wx$ with $x \in D(L)$, (2.14) and the properties of T_0 imply that

$$|v|^2 = |Wx|^2 = |T_0^{-1}Lx|^2 = (Lx, K_0v) \leq \eta_2 |x||v|$$

[i.e., $|Wx| \leq \eta_2 |x|$ for $x \in D(L)$]. It follows from (2.13) that

$$(Lx, Kx) = [T_0^{-1}Lx, v] = [Wx, x] \geq \eta_1 |x|^2 \qquad \text{for all} \quad x \in D(L).$$

Since $D(L)$ is dense in H_0 and W is continuous, it follows that W has a unique bounded closure $\overline{W}: H_0 \to H_0$ such that $[\overline{W}x, x] \geq \eta_1 |x|^2$ for $x \in H_0$. Hence, by the Lax–Milgram lemma [150] (or by Corollary 1.2 if H is separable, $X = H_0$ and $J = I$), $\overline{W}$ is a homeomorphism.

Let W_0 be such that $W \subset W_0 \subset \overline{W}$ and $R(W_0) = D(T_0)$ and set $L_0 = T_0 W_0$, where $D(L_0) = D(W_0)$. We see that $L_0 \supseteq L$ since, for $x \in D(L)$, $W_0 x = Wx = T_0^{-1}Lx$ and $L_0 x = T_0 W_0 x = Lx$. Further, L_0 maps $D(L_0)$ onto H and is injective. Indeed, $R(W_0) = D(T_0)$ and T_0 maps $D(T_0)$ onto H and $L_0 x = 0$ implies that $W_0 x = 0$ and, consequently, $x = 0$. Next, $L_0: D(L_0) \subset H_0 \to H$ is closed. Indeed, let $x_n \to x$ in H_0 and $L_0 x_n \to f$ in H. Then $W_0 x_n \to \overline{W}x$ in H_0 and $L_0 x_n = T_0 W_0 x_n \to f$ in H with $W_0 x_n \to \overline{W}x$ in H by Lemma 2.2(c). This and the closedness of T_0 imply that

$\overline{W}x \in D(T_0) = R(W_0)$ [i.e., $x \in D(W_0) = D(L_0)$ and $L_0x = T_0W_0x = f$]. Since L_0 is onto H, injective, and closed, as a map of $D(L_0) \subset H_0 \to H$, there is a constant $c > 0$ such that $\| L_0x \| \geq c \, | \, x \, | \geq c\alpha_1^{1/2}\| \, x \, \|$ for $x \in D(L_0)$. Hence L_0^{-1} is continuous and therefore closed as a map from H to H. Hence $L_0 = (L_0^{-1})^{-1}$ is closed as a map from $D(L_0) \subset H$ onto H. Now it is obvious that L_0 satisfies (2.13) and (2.14) for $x,y \in D(L_0)$ with K_0 replacing K. Since the equation $(L_0x,K_0y) = [W_0x,y]$ would be true for any L_0 satisfying the conditions of the theorem, L_0 is uniquely determined by these conditions.   Q.E.D.

We add that the interesting feature of the proof of Theorem 2.2 is that it depends only on the linear operator T_0 and the bijectivity of $\overline{W}$ and not on the linearity of L. Hence, as we will see later, Theorem 2.2 remains valid for densely defined nonlinear operators satisfying generalized monotonicity conditions, so that a Friedrichs-type extension admits generalization to nonlinear operators.

The reader is invited to show that the following useful corollary follows from Theorem 2.2, which was first proved in [216] when $K = I$.

Corollary 2.2 Suppose that L in Theorem 2.2 is given by $L = T + S$ with $D(S) \supset D(T)$ and L satisfies condition (2.13).

(a) If $| \, (Sx,Ky) \, | \leq \eta_3 \, | \, x \, | \, | \, y \, |$ for $x,y \in D(T)$ and some $\eta_3 > 0$, then $L_0 = T_0(I + N_0)$, where N_0 is a certain extension of $N = T_0^{-1}S$ in H_0.

(b) If $\| Sx \| \leq \eta_4 \, | \, x \, |$ for $x \in D(T)$ and some $\eta_4 > 0$, then $D(L_0) = D(T_0)$ and $L_0 = T_0 + S_0$, where S_0 is an extension of S to H_0.

Finally, Theorem 2.2 and Corollary 2.1 imply:

Corollary 2.3 If K is Theorem 2.2 is closed with $D(K) = D(T)$ and L: $D(L) = D(T) \subset H \to H$ is closable and satisfies (2.13), then $L = L_0$ and $H_0 = D(L)$.

Remark 2.2 The extension results presented above are also valid for complex Hilbert spaces. If H is complex, (2.13) is replaced by $|(Lx,Kx)| \geq \eta_1 \, | \, x \, |^2$ for $x \in D(L)$ and the K-symmetry of T follows, by polarization, from the fact that T is K-p.d. (see [215]).

We complete this section with the following lemma, which will prove to be useful in our study of the generalized moments method.

Lemma 2.3 Suppose that L is as in Theorem 2.2. If $\{\phi_i\} \subset D(L)$ is a sequence of linearly independent elements such that $\{L\phi_i\}$ is complete in H, then $\{\phi_i\}$ is complete in H_0.

Proof. Let v be any element in $D(L) = D(T)$. Since $\{L\phi_i\}$ is complete in H, there exists $n \in Z_+$ and numbers $\{\beta_1, \ldots, \beta_n\}$ such that $\| Lv - \sum_{j=1}^{n} \beta_j L\phi_j \| < (2\eta_1)^{-1}\alpha_2^{-1/2}\epsilon$, where $\epsilon > 0$ is an arbitrary small given number. Set $v_n = \sum_{j=1}^{n} \beta_j \phi_j$ and note that since, by Lemma 2.2, $\| Kx \| \leq \alpha_2^{1/2} | x |$ for $x \in D(L)$, it follows from (2.13) and the inequality above that

$$| v - v_n |^2 \leq (1/\eta_1) \| L(v - v_n) \| \, \| K(v - v_n) \|$$

$$< \eta_1^{-1} \cdot \tfrac{1}{2}\eta_1 \alpha_2^{-1/2} \epsilon \cdot \alpha_2^{1/2} | v - v_n |$$

(i.e., $| v - v_n | < \tfrac{1}{2}\epsilon$). Suppose now that u is any element in H_0. By definition of H_0, there exists $v \in D(T)$ such that $| u - v | < \tfrac{1}{2}\epsilon$. Now for the element v we choose the numbers n, $\{\beta_1, \ldots, \beta_n\}$, such that $| v - \sum_{j=1}^{n} \beta_j \phi_j | < \tfrac{1}{2}\epsilon$. Hence, by the triangle inequality,

$$\left| u - \sum_{j=1}^{n} \beta_j \phi_j \right| \leq | u - v | + \left| v - \sum_{j=1}^{n} \beta_j \phi_j \right| < \epsilon.$$

Q.E.D.

2.2 Generalized Method of Moments for the Approximation-Solvability of Equations Involving Unbounded K-p.d. Maps

In this section we apply the extension results of Section 2.1 and of Section 1 to the constructive solvability of equations involving perturbations of unbounded K-p.d. operators acting in separable Hilbert spaces.

Let $T: D(T) \subset H \to H$ be K-p.d. and K-symmetric and H_0 the auxiliary space constructed above. Our problem is to study the approximation solvability of the equation of the form

$$Lx + Mx = f \quad (f \in H), \tag{2.15}$$

where L satisfies the conditions of Theorem 2.2 (or weaker) and M is such that $D(M) \supseteq D(L) = D(T)$ and $T_0^{-1}M: D(T) \subset H_0 \to H_0$ is bounded. Now, if $\{X_n\} \subset H_0$ is a sequence of finite-dimensional spaces such that $\mathrm{dist}(h, X_n) \to 0$ for each h in H_0, $Y_n = K(X_n) \subset H$ for $n \in Z_+$, and P_n: $H_0 \to X_n$ and $Q_n: H \to Y_n$ are orthogonal projections in H_0 and H, respectively, then the projectional scheme $\Gamma_K = \{X_n; P_n; Y_n, Q_n\}$ is admissible for (H_0, H).

As in [141], assuming that $\{X_n\} \subset D(L)$, the generalized moments method (GM-method) for the approximation solvability of (2.15) consists of finding $x_n \in X_n$ for $n \geq n_0$ from the condition that

$$(Lx_n + Mx_n - f, Kx) = 0 \qquad \text{for each} \quad x \in X_n \qquad (n \geq n_0 \geq 1).$$
$$(2.16)$$

Of course, if one chooses a basis $\{\phi_i\} \subset X_n$ and sets $x_n = \sum_{i=1}^{n} a_i^n \phi_i$, one determines $\{a_i^n\}$ from the equivalent algebraic system

$$\sum_{i=1}^{n} \{(L\phi_i, K\phi_j) + (M\phi_i, K\phi_j)\} a_i^n = (f, K\phi_j) \qquad (1 \leq j \leq n). \qquad (2.17)$$

Since $Y_n = K(X_n)$, (2.16) or (2.17) can be put in the form

$$Q_n L x_n + Q_n M x_n = Q_n f. \qquad (2.18)$$

It follows easily from Lemma 2.2 and Theorem 2.2 that (2.16) can be written in the form

$$P_n W x_n + P_n T_0^{-1} M x_n = P_n T_0^{-1} f \qquad (2.19)$$

or in the form

$$P_n \overline{W} x_n + P_n F x_n = P_n T_0^{-1} f, \qquad (2.20)$$

where $\overline{W}$ and F are the closures in H_0 of W and $T_0^{-1}M$, respectively. Thus the GM-method for equation (2.15) reduces to the projection method (2.20) for the equation

$$\overline{W}x + Fx = T_0^{-1} f \qquad (x \in H_0, T^{-1}f \in H_0) \qquad (2.21)$$

involving bounded maps to which the results of Section 1 apply.

It is clear that every solution of (2.15) is a solution of (2.21), but the converse is not true in general. However, if $D(M) \supseteq H_0$, then $F = T_0^{-1}M$, and since $L_0 = T_0 W_0$ with $R(W_0) = D(T_0)$, it follows that in this case a solution $x_0 \in H_0$ of (2.21) lies in $D(L_0)$ and is thus a *strong solution* of (2.15).

In what follows we regard a solution $x_0 \in H_0$ of (2.21) as a *weak solution* of (2.15). This is equivalent to the requirement that x_0 in H_0 satisfies

$$l(x_0, x) + m(x_0, x) = (f, K_0 x) \qquad \text{for all} \quad x \in H_0, \qquad (2.22)$$

where $l(y, x)$ and $m(y, x)$ are bilinear forms on H_0 such that $l(y, x) = (Lx, Ky)$ and $m(y, x) = (My, Kx)$ for $y, x \in D(L)$.

Remark 2.3 If we do not assume that $\{X_n\}$ lies in $D(L)$, the approximate equations (2.20), or rather their equivalent and practically more useful versions of the form

$$l(x_n, x) + m(x_n x) = (f, K_0 x) \qquad (x \in X_n), \qquad (2.23)$$

must be used for the construction of the approximates $x_n \in X_n$. This is usually done when one is solving differential boundary value problems and the coordinate functions $\{\phi_i\}$ are assumed to satisfy the boundary conditions only in a generalized sense. In this case one usually constructs a weak solution of (2.15).

If we assume that F is k-ball contractive with $k < \eta_1$, the inequality $[\overline{W}x,x] \geq \eta_1 |x|^2$ for $x \in H_0$ and Theorem I.2.2 imply that $\overline{W} + F: H_0 \to H_0$ is A-proper w.r.t. $\Gamma_1 = \{X_n, P_n\}$. Hence Theorem 1.1 implies the following new basic result for the GM-method, which as will be seen, properly contains most of the earlier results for projective-type methods obtained for the case when F is 0-ball contractive (i.e., F is compact).

Theorem 2.3 If F is k-ball contractive with $k < \eta_1$ and $\overline{W} + F$ is injective, there exists $n_0 \in Z_+$ such that (2.16) or (2.23) has a unique solution $x_n \in X_n$ for each $n \geq n_0$, $x_n \to x_0$ in H_0 and x_0 is the unique (possibly weak) solution of (2.15).

Moreover, the following estimate holds:

$$|x_n - x_0| \leq (1 + |A|/\gamma|) P_n x_0 - x_0|, \qquad n \geq n_0, \qquad \text{(E2)}$$

where $A = \overline{W} + F$ and $\gamma > 0$ is such that $\|P_n A x\| \geq \gamma |x|$ for $x \in X_n$, $n \geq n_0$.

Remark 2.4 The map $F: H_0 \to H_0$ is obviously k-ball contractive with $k < \eta_1$ if its norm $|F| < \eta_1$. The last condition follows easily from the assumption that $\|Mx\| \leq b |x|$ for all $x \in D(T)$ and some $b > 0$ such that $b(\|KT_0^{-1}\|)^{1/2} \leq b\alpha_2^{1/2} < \eta_1$ since $|T_0^{-1}Mx|^2 = (Mx, K(T_0^{-1}Mx)) \leq \|Mx\| \|K(T_0^{-1}Mx)\| \leq b\alpha_2^{1/2} |T_0^{-1}Mx| |x|$ for $x \in D(T)$ and $\|KT_0^{-1}y\| \leq \alpha_2^{1/2} \|y\|$ for all y in H.

If $M \equiv 0$ or $D(M) \supseteq H_0$, Theorem 2.2 implies that x_0 constructed by Theorem 2.3 does lie in $D(L_0)$. Since $[Wx,y] = (Lx, Ky)$ for $x,y \in D(L)$ and $W \equiv T_0^{-1}L: D(L) \subset H_0 \to H_0$ is bounded by (2.14), we may weaken condition (2.13) if we strengthen the requirement on F. Indeed, in virtue of Proposition 1.1, the following variant of Theorem 2.3 is useful in applications at least in those cases when Sobolev-type embedding theorems apply.

Theorem 2.4 If F in Theorem 2.3 is compact, its conclusion still holds even when the inequality (2.13) is replaced by the weaker assumption:

(c1) There is a compact map $C: H_0 \to H_0$ and constants $\eta_5 > 0$ and η_6 such that $(Lx, Kx) \geq \eta_5 |x|^2 - \eta_6[Cx,x]$ for $x \in D(L)$.

It was noted above that if $D(M) \supseteq D(T)$, then in general the weak solution $x_0 \in H_0$ of (2.15) constructed by Theorem 2.3 need not lie in $D(L_0)$. Moreover, it was shown by Mikhlin [165] that unless L is of a very special type and $\{\phi_i\}$ are chosen in a special way, the residuals $\| Lx_n + Mx_n - f \|$ will not converge to 0. The next theorem shows that both of these deficiencies can be eliminated if we strengthen the conditions on L and K. It also has the interesting feature that the theory of A-proper mappings is applicable directly to equation (2.15) without using the auxiliary equation (2.21).

Theorem 2.5 Let T be K-p.d. and K-symmetric with K closed and $D(K)$ $= D(T)$ and K closable with $D(L) = D(T)$. Suppose that any one of the following two conditions holds:

(i) L satisfies (2.13), $M: H_0 \to H$ is k-ball-contractive with $k < \eta_1 \alpha_2^{-1/2}$ and $L + M$ is injective.

(ii) L satisfies (c1), $M: H_0 \to H$ is compact and $L + M$ is injective.

Then equation (2.16) or (2.18) has a unique solution $x_n \in X_n$ for each large $n \in Z_+$, $x_n \to x_0$ in H_0, $x_0 \in D(L)$, $\| Lx_n + Mx_n - f \| \to 0$, and

$$| x_0 - x_n | \le (1 + | A |_{H_0 H}/\gamma)| P_n x_0 - x_0 |, \qquad n \ge n_0, \tag{E3}$$

where $A = L + M$, $| A |_{H_0 H}$ is the norm of $A: H_0 \to H$, and $\| Q_n A x \| \ge \gamma | x |$ for $x \in X_n$.

Proof. Note first that in view of our conditions on K, Corollary 2.1 implies that $K_0 = K$, $H_0 = D(T)$, with $| x |$, $\| Tx \|$, and $\| Kx \|$ providing equivalent norms on $D(T)$. Further, since $L: D(L) \subset H \to H$ is closable, $L: H_0 \to H$ is bounded. To prove the latter, by the closed graph theorem, it suffices to show that L is closed as a map from H_0 to H. But since L is defined on all of H_0, it suffices to show that $L: H_0 \to H$ is closable; that is, if $\{x_j\} \subset H_0$ is such that $| x_j - 0 | \to 0$ and $\| Lx_j - g \| \to 0$, then $g = 0$. Since $| x | \ge \alpha_1^{1/2} \| x \|$ for $x \in D(L)$, it follows that $\| x_j - 0 \| \to 0$ and $\| Lx_j - g \| \to 0$. Hence $g = 0$, because L is closable in H.

Suppose now that (i) holds. It follows from (2.13) and (2.2) that $\| Q_n Lx \| \ge \eta_1 \alpha_2^{-1/2} | x |$ for all $x \in X_n$ and all $n \in Z_+$. Hence our condition on M in (i) and Theorem I.2.2 imply that $L + M: H_0 \to H$ is A-proper w.r.t. Γ_K. On the other hand, if (ii) holds, the compactness of $C: H_0 \to H_0$ and the continuity of $T: H_0 \to H$ imply that $\hat{C} \equiv TC: H_0 \to H$ is compact. Thus (c1) can be put in the form

$$(Lx, Kx) \ge \eta_5 | x |^2 - \eta_6(\hat{C}x, Kx) \qquad \text{for} \quad x \in H_0.$$

Consequently, by Proposition 1.1, $L: H_0 \to H$ is also A-proper w.r.t. Γ_K, and so is $L + M$, since $M: H_0 \to H$ is compact. Now, part one of Theorem 2.5 follows from Theorem 1.1, while the last assertion follows from the fact that $H_0 = D(L)$ and $L,M: H_0 \to H$ are continuous. The estimate (E3) follows from (E1) of Theorem 1.1.   Q.E.D.

A consquence of Theorem 2.5(i) is the following convergence result for equation (2.15), with L forming an acute angle. Following Sobolevskii [271], we say that two densely defined operators L and T form an *acute angle* iff $D(L) = D(T)$, L and T vanish only at the zero element and

$$(Lx,Tx) \geq \delta \parallel Lx \parallel \parallel Tx \parallel \qquad \text{for all} \quad x \in D(L) \quad \text{and some} \quad \delta \in (0,1).$$
$$(2.24)$$

Lemma 2.4 If L and T form an acute angle and are closed with $R(L) = R(T) = H$, there exist constants $\gamma_1 > 0$ and $\gamma_2 > 0$ such that

$$\parallel Lx \parallel \leq \gamma_1 \parallel Tx \parallel, \qquad \parallel Tx \parallel \leq \gamma_2 \parallel Lx \parallel \qquad \text{for} \quad x \in D(T). \qquad (2.25)$$

Proof. Introduce in $D(T)$ a new inner product and norm by

$$[x,y]_D = (Tx,Ty), \qquad |x|_1^2 = [x,x]_D, \qquad x,y \in D(T). \qquad (2.26)$$

Since T^{-1} is closed and defined on all of H, it is bounded. Hence $D(T)$ is a Hilbert space with respect to the metric (2.26), which we denote by H_1. Now, as in the proof of Theorem 2.5, one shows that $L: H_1 \to H$ is bounded; that is, $\parallel Lx \parallel_- < \gamma_1 |x|_1 = \gamma_1 \parallel Tx \parallel$ for all x in H_1 and some $\gamma_1 > 0$. Interchanging the role of T and L, we obtain the second inequality in (2.25).

Now, if L and T satisfy conditions of Lemma 2.4, then (2.24) implies that $(Lx,Tx) \geq \delta\gamma_2^{-1} |x|_1^2$ for $x \in D(L)$ (i.e., L is K-p.d. with $K = T$ and $\eta_1 = \delta\gamma_2^{-1}$).   Q.E.D.

In view of Lemma 2.4, Theorem 2.5(i) implies the validity of the following new result for equation (2.15).

Corollary 2.4 Suppose that $L: D(L) \subset H \to H$ forms an acute angle with T and that T and L are closed and surjective. If $M: H_1 \to H$ is k-ball contractive with $k < \eta_1\gamma_2^{-1}$ and $L + M$ is injective, equation (2.15) is uniquely approximation solvable w.r.t. Γ_T, $\parallel Lx_n + Mx_n - f \parallel \to 0$ and the estimate (E3) holds.

Special Cases

The following are the most important special cases of the GM-method formulated and studied earlier by a variety of principles and methods.

(i) *Galerkin–Bubnov method*. When $K = I$ and $L = T$, then T_0 is self-adjoint and positive definite, H_0 is the energy space H_T, $\overline{W} = I$, and the GM-method reduces to the well-known Galerkin–Bubnow or Galerkin–Krylov method. In this abstract form, Theorem 2.3 was proved by Mikhlin [165] under the assumption that $T_0^{-1}M$ is compact in H_T, although beginning with Galerkin the method has been used extensively earlier in the study of the solvability of differential equations (see [143,156,162] for numerous references). For more recent study of the method, see [157,225].

(ii) *Method of moments*. When $L = T$ and $K = T$, then H_0 is the space with the scalar product $[u,v] = (Tu,Tv)$ and the GM-method reduces to the method of moments whose study was initiated by Krylov [143], Kravchuk (see [141]), and studied further by a number of authors (e.g., [88,165, 215,315]). When M is compact, Theorem 2.3 was proved by the author [225]. If $L = T$ and $K \neq T$, then L is K-p.d. and K-symmetric and the method has been studied by Martyniuk [162] (see also [68,146]) under the assumption that $R(T) = H$ and by the author [215] in its present setting. When $L \neq T$ and $K \neq T$, the GM-method was studied in [145,215,266, 271,315]. The GM-method is related (see Remark 2.5 below) to the Galerkin–Petrov method initiated by Petrov [199] and studied extensively by Polskii [241,242], Vainikko [282,283], Luchka [156], and others (see [68,157,225] for further references).

(iii) *Generalized Ritz method*. When $M = 0$, $K = I$, and $L = T$, then T_0 is self-adjoint and positive definite, $H_0 = H_T$, and the GM-method reduces to the method of Ritz studied by a number of authors—in particular, by [109,123,165,167,215] and others. For the extensive literature on this method, see [165]. If $K \neq I$, the generalized Ritz method has been studied in [162,215] and others. When $K \neq I$ and $L \neq T$ the method has been studied in [73,215]. When $D(L) = D(T)$, L and T form an acute angle with $R(T) = R(L) = H$ and $M = 0$, Corollary 2.4 was stated without proof by Sobolevskii [271] under the additional condition that T is self-adjoint and Q_n commutes with T. In their present forms, Theorms 2.3, 2.4, and 2.5 and Corollary 2.4 are new.

(iv) *Method of least squares*. When $M = 0$ and $L = T = K$, the GM-method reduces to the method of least squares, which has been studied by a number of authors (see [139,157,165,225] for references).

Advantages of the GM-Method

It is obvious that the first positive feature of the GM-method is that it presents a unified approach to a number of special methods and that it is applicable to a more general class of equations than any of its special cases. Another of its advantages is that the wide freedom in the choice of T and K makes the GM-method computationally more flexible and useful. When applied to differential equations it gives a better characterization of convergence than the method of Galerkin.

2.3 Projection Methods in the Approximation-Solvability of Equations Involving Unbounded Operators in Banach Spaces

In this section we indicate briefly how the results of Section 1.1 apply to the approximation solvability of

$$Lx = f \qquad [x \in D(L), f \in Y], \tag{2.27}$$

where L is an unbounded densely defined linear operator in a Banach space X such that $\{Z_n\} = \{LX_n\}$ is complete for some sequence $\{X_n\} \subset D(L)$; that is, dist$(g,Z_n) = \inf_z \{\| g - z \| : z \in Z_n\} \to 0$ for each g in Y. Let $\{Y_n\} \subset Y$ be finite-dimensional spaces, dim Y_n = dim Z_n, and Q_n: $Y \to Y_n$ projections such that $Q_n g \to g$ in Y for each g in Y. If U_n: $Z_n \to Y$ denotes an inclusion map for each $n \in Z_+$, then $\Gamma_L = \{Z_n, U_n; Y_n, Q_n\}$ is a complete projectional scheme for $\{Y,Y\}$. The general projection method considered in [241] (see also [204]) consists in finding an approximate solution $x_n \in X_n$ from the equation

$$Q_n Lx_n = Q_n f \qquad (x_n \in X_n, Q_n f \in Y_n). \tag{2.28}$$

If we set $Lx = y \in Y$ and $Lx_n = y_n \in Z_n$, equations (2.27) and (2.28) take the respective forms

$$Ty = f \tag{2.29}$$

$$Q_n Ty_n = Q_n f \qquad (y_n \in Z_n, Q_n f \in Y_n), \tag{2.30}$$

where $T = I$, the identity map on Y. Now, by Theorem 1.1, there exists $n_0 \in Z_+$ such that (2.30) has a unique solution for $n \geq n_0$ such that $y_n = \bar{Q}_n^{-1} Q_n f$ converges to y in Y for each f in Y if and only if the map $T = I$ is A-proper w.r.t. Γ_L, where $\bar{Q}_n$ is the restriction of Q_n to Z_n. Since, obviously, I is bijective, Theorem 1.3 shows that I is A-proper w.r.t. Γ_L if and only if there exist $n_0 \in Z_+$ and $\gamma > 0$ such that

$$\| Q_n(w) \| \geq \gamma \| w \| \qquad \text{for all} \quad w \in Z_n \quad \text{and} \quad n \geq n_0. \tag{2.31}$$

It follows from these remarks that if we set $\tau_n = \inf \{\| Q_n(w) \| \mid w \in Z_n, \| w \| = 1\}$, then Polskii's *condition* (A): lin inf $\tau_n = \tau > 0$ is both necessary and sufficient for (2.30) to have a unique solution $y_n \in Z_n$ for each large n such that $y_n = \bar{Q}_n^{-1} Q_n f \to y \ (= Lx)$ in Y for each f in Y.

The discussion above and the argument used in proving (E1) in Theorem 1.1 imply the validity of the following.

Theorem 2.6 Suppose that the scheme $\Gamma_L = \{Z_n, U_n; Y_n, Q_n\}$ is complete for (Y, Y). Then the following assertions are equivalent:

(A1) There exists an $n_0 \in Z_+$ such that (2.30) is uniquely solvable for each $n \geq n_0$ and $y_n = Lx_n \to y = Lx$ for every f in Y.

(A2) The identity map I on Y is A-proper w.r.t. Γ_L.

(A3) There exist $n_0 \in Z_+$ and $\gamma > 0$ such that (2.31) holds.

(A4) Condition (A) holds.

Moreover, if any one of the four conditions holds, then

$$\| Lx_n - Lx \| \leq (1 + \beta/\eta) \inf_{u \in Z_n} \| f - u \|, \qquad n \geq n_0, \tag{E4}$$

where $\| Q_n \| \leq \beta$ and $\| Q_n u \| \geq \eta \| u \|$ for all $u \in Z_n$ and $n \geq n_0$.

Remark 2.5 If there exists a bounded operator L^{-1}, then $\{x_n\}$ converges to x and we have the error estimate

$$\| x_n - x \| \leq \| L^{-1} \| (1 + \beta/\eta) \inf_{u \in Z_n} \| f - u \|, \qquad n \geq n_0. \tag{E4$'$}$$

Remark 2.6 We complete this section with the following general observation. The methods of Sections 2.2 and 2.3 have been studied (and applied) so extensively under various names that it is almost impossible not to miss some of the contributors. Since Section 1.2 and Section 2 of this chapter are used primarily to show how the abstract results connected with these methods can be deduced from the theory of A-proper mappings developed in Section 1.1, we make no claim on the completeness of the special results and/or the literature surveyed in these sections.

3. GENERALIZED FREDHOLM ALTERNATIVE

To begin, we first recall (see [66,310]) that $T \in L(X, Y)$ is said to be *Fredholm* iff $d(T) \equiv \dim N(T) < \infty$, $R(T)$ is closed, and $\dim(Y/R(T)) \equiv d(T^*) = \dim N(T^*) < \infty$. If T is Fredholm, its *index* is defined by $i(T) = d(T) - d(T^*)$. In what follows we let

$$N(T)^\perp = \{u \in X^* : (u, x) = 0 \ \forall \ x \in N(T)\},$$

$$N(T^*)^\perp = \{y \in Y : (w, y) = 0 \ \forall \ w \in N(T)\}.$$

If $C \in L(X, X)$ is compact, $T = I - C$ and $T^* = I^* - C^*$, then the classical Fredholm alternative asserts:

Proposition 3.1   Either the equations $Tx = f$ and $T^*u = w$ are uniquely solvable for each f in X and w in X^*, or $N(T)$ and $N(T^*)$ have the same finite dimension and $Tx = f$ is solvable iff $f \in N(T^*)^\perp$, while $T^*u = w$ is solvable iff $w \in N(T)^\perp$.

It is well known that Proposition 3.1 plays an essential role in the theory of the solvability of Fredholm integral equations of the second kind, and the latter theory is again important in the solvability of certain boundary value problems for ordinary and partial differential equations.

3.1   Generalized Alternative for A-Proper Mappings

Since, as we have seen in Chapter I, not every linear differential equation can be reduced to the form to which Proposition 3.1 is applicable, it is interesting to know if and in what way Proposition 3.1 can be extended to A-proper mappings $T \in L(X,Y)$. The results presented here were obtained by the author in [208,223] (see also [78]) in less general form than those given in this section. Our first result in this direction is the following:

Lemma 3.1   Suppose that $T \in L(X,Y)$ is A-proper w.r.t. the admissible scheme $\Gamma = \{X_n, V_n; E_n, W_n\}$. Then $d(T) < \infty$ and $R(T)$ is closed in Y.

Proof.   Suppose that $N(T) \neq \{0\}$. If $N(T)$ were not finite-dimensional, there would exist a sequence $\{x_j\} \subset S_1 \equiv \{x \in N(T) : \| x \| = 1\}$ such that $\| x_i - x_j \| \geq \frac{1}{2}$ for $i \neq j$ and $Tx_j = 0$ for each $j \in Z_+$. Hence, by Corollary I.1.1, $\{x_j\}$ has a convergent subsequence. This contradiction shows that $d(T) < \infty$.

To prove that $R(T)$ is closed, let $\{x_n\} \subset X$ be such that $Tx_n \to y$ for some y in Y. We claim that $y \in R(T)$. Since $d(T) < \infty$, for each x_n there exists $z_n \in N(T)$ such that $d_n = \text{dist}(x_n, N(T)) = \| x_n - z_n \|$ for each $n \in Z_+$. If the sequence $\{x_n - z_n\}$ contains a bounded subsequence $\{x_{n_j} - z_{n_j}\} \equiv \{u_j\}$, then $Tu_j = Tx_{n_j} \to y$ as $j \to \infty$. Hence again, by Corollary I.1.1, there is a subsequence $\{u_{j_k}\}$ such that $u_{j_k} \to u$ for some u in X and $T(u_{j_k}) \to Tu = y$ [i.e., in this case $y \in R(T)$]. Thus, to complete the proof, it will be sufficient to show that the assumption that $\| x_n - z_n \| \to \infty$ leads to a contradiction. Now, assuming that $\| x_n - z_n \| \to \infty$ and noting that $T(x_n - z_n) = Tx_n \to y$, we see that $Tw_n \to 0$, where $w_n = (x_n - z_n)/\| x_n - z_n \|$ for each $n \in Z_+$. Hence, as before, Corollary I.1.1 implies the existence of a subsequence $\{w_{n_j}\}$ such that $w_{n_j} \to w$ in X and $Tw_{n_j} \to Tw = 0$ [i.e., $\| w \| = 1$ and $w \in N(T)$]. Now, if we fix j so that

$$\| w_{n_j} - w \| = \| (x_{n_j} - z_{n_j})/\| x_{n_j} - z_{n_j} \| - w \| < \tfrac{1}{2},$$

then

$$\| x_{n_j} - \tilde{w}_{n_j} \| < \tfrac{1}{2} \| x_{n_j} - z_{n_j} \| = \tfrac{1}{2} d_{n_j} \quad \text{with} \quad \tilde{w}_{n_j}$$

$$= z_{n_j} + \| x_{n_j} - z_{n_j} \| w \in N(T),$$

which contradicts the definition of d_{n_j}. Q.E.D.

A consequence of Theorem I.1.1 and Lemma 3.1 is the following:

Theorem 3.1 If $T \in L(X,Y)$ is A-proper w.r.t. Γ, then T is Fredholm with $i(T) \geq 0$.

Proof. It follows from Lemma 3.1 that $d(T) < \infty$ and $R(T)$ is closed. Hence, by the well-known results on topological complements (see [48,66]), there exists a closed subspace X_1 of X such that $X = N(T) \oplus X_1$, $T(X_1) = R(T)$, and $T_1 = T|_{X_1}: X_1 \to Y$ is injective. Suppose, contrary to our assertion, that $d(T^*) > d(T)$. Then there exists a subspace N' of Y with dim $N' = d(T)$, an isomorphism Π of $N(T)$ onto N', and a linear map U of X into Y given by $U = T + \Pi P$, where P is the projection of X onto $N(T)$, such that $N' \cap R(T) = \{0\}$ and $U(X) = R(T) \oplus N'$. Moreover, U is injective. Indeed, if $Ux = 0$, then since $x = x_0 + x_1$ with $x_0 \in N(T)$ and $x_1 \in X_1$, we see that $Ux = Tx_1 + \Pi x_0 = 0$ implies that $Tx_1 = 0$ and $\Pi x_0 = 0$ since $N' \cap R(T) = \{0\}$. Hence $x_0 = 0$ and $x_1 = 0$ (i.e., $x = 0$). Thus $U = T + C$ is A-proper w.r.t. Γ since T is A-proper and $C = \Pi P: X \to N'$ is compact. Since U is also injective, it follows from Theorem I.1.1 that $U(X) = Y$. Hence $Y = R(T) \oplus N'$ and therefore $d(T^*) = $ dim $N' = d(T)$ [i.e., one always has the relation $d(T^*) \leq d(T)$ or $i(T) \geq 0$]. Q.E.D.

It should be noted that the proof of Theorem 3.1, which is based on Theorem I.1.1, does not utilize the adjoint T^* of T. If one assumes that Γ is a projectionally complete scheme $\Gamma_P = \{X_n,P_n;Y_n,Q_n\}$ for (X,Y) with the additional property that its "adjoint" scheme $\Pi_P^* = \{R(Q_n^*),Q_n^*;R(P_n^*),P_n^*\}$ is also projectionally complete for (Y^*,X^*) (which would be the case if X and Y had shrinking Schauder bases), then it makes sense to talk about the A-properness of $T^*: Y^* \to X^*$ w.r.t. Γ_P^*. However, the following example shows that the adjoint T^* of an A-proper map T need not be A-proper w.r.t. Γ_P^*.

Proposition 3.2 Let $T \in L(H,H)$ be the map of Example I.1.2 given by $T\phi_1 = 0$, $T\phi_2 = \phi_1, \ldots, T\phi_i = \phi_{i-1}$ for $i \geq 2$. Then T is A-proper w.r.t. $\Gamma_H = \{X_n,P_n\}$ but its adjoint $T^*: H \to H$ is not A-proper w.r.t. $\Gamma_H^* = \Gamma_H$. Note that $N(T) = [\phi_1], N(T^*) = \{0\}$, so $i(T) = d(T) - d(T^*) = 1$.

Proof. Proposition I.1.2 shows that T is A-proper w.r.t. Γ_H. To show that T^* is not A-proper w.r.t. $\Gamma_H^* = \Gamma_H$, observe first that T^* is given by (1.3) in Example I.1.1 (i.e., by $T\phi_i = \phi_{i+1}$ for $i \geq 1$). Hence, by Proposition I.1.1, T^* is not A-proper w.r.t. $\Gamma_H^* = \Gamma_H$. Q.E.D.

To state conditions on the A-proper map $T \in L(X,Y)$ under which T^* would also be A-proper, we first need the following:

Lemma 3.2 Let $\Gamma_P = \{X_n, P_n; Y_n, Q_n\}$ and Γ_P^* be projectionally complete for (X,Y) and (Y^*, X^*), respectively. Then there exists $a_0 > 0$ and $n_0 \in Z_+$ such that

(a) $\| Q_n TP_n(x) \| \geq a_0 \| x \|$ for all $x \in X_n$ and each $n \geq n_0$ if and only if there exist $a_1 > 0$ and $n_1 \in Z_+$ such that

(b) $\| P_n^* T^* Q_n^*(g) \| \geq a_1 \| g \|$ for all $g \in R(Q_n^*) \equiv Y_n'$ and $n \geq n_1$.

Proof. (a) $\Rightarrow$ (b). It follows from the definition of the norm in Y^* that if $g \in Y^*$, then for any $k > 1$ there exists $y \in Y$ with $\| y \| = 1$ such that $|(g,y)| \geq (1/k)\| g \|$. Hence if $g_n \in Y_n'(\subset Y^*)$ with $\| g_n \| = 1$ for each n, there exists $y_n \in Y_n$ with $\| y_n \| \leq \beta$ for all $n \in Z_+$ such that $|(g_n, y_n)| \geq \frac{1}{2}$ for $n \in Z_+$. Indeed, by the preceding observation, letting $k = 2$, we see that for each $g_n \in Y_n'$ there exists $z_n \in Y$ with $\| z_n \| = 1$ such that

$$\tfrac{1}{2} \leq |(g_n, z_n)| = |(Q_n^* g_n, z_n)| = |(g_n, y_n)| \qquad \text{for all} \quad n,$$

where $\| y_n \| = \| Q_n z_n \| \leq \| Q_n \| \| z_n \| \leq \beta$ for all $n \in Z_+$. Now, if (b) does not hold, there exists a sequence $\{g_n \in Y_n' : \| g_n \| = 1\}$ such that $P_n^* T^* Q_n^*(g_n) \to 0$ as $n \to \infty$. But for each g_n there is $y_n \in Y_n$ with $\| y_n \| \leq \beta$ for all n such that $|(g_n, y_n)| \geq \frac{1}{2}$ for $n \in Z_+$. Since, by (a), $\| Q_n TP_n(x) \| \geq a_0 \| x \|$ for $x \in X_n$ and $n \geq n_0$, the map $T_n = Q_n TP_n|_{X_n} : X_n \to Y_n$ is an isomorphism. Hence for each $n \geq n_0$ there exists $x_n \in X_n$ such that $T_n(x_n) = y_n$ with $\| x_n \| \leq \beta a_0^{-1}$ for $n \in Z_+$. Thus

$$\tfrac{1}{2} \leq |(g_n, y_n)| = |(g_n, T_n(x_n))| = |(P_n^* T^* Q_n^*(g_n), x_n)| \to 0$$

since $P_n^* T^* Q_n^*(g_n) \to 0$ as $n \to \infty$, a contradiction.

(b) $\Rightarrow$ (a). By the Hahn–Banach theorem, for $x \in X$ with $x \neq 0$, there exists $g \in X^*$ with $\| g \| = 1$ such that $(g,x) = \| x \|$. This implies that for each $x_n \in X_n (\subset X)$ with $\| x_n \| = 1$ there exists $g_n \in R(P_n^*) \equiv X_n' (\subset X^*)$ with $\| g_n \| \leq \alpha$ such that $(g_n, x_n) = 1$ for each $n \in Z_+$. Now, if (a) were not true, there would exist $\{x_n \in X_n : \| x_n \| = 1\}$ such that $Q_n TP_n(x_n) \to 0$ as $n \to \infty$. In view of (b), to each $f_n \in X_n'$ with $n \geq n_1$ there exists $g_n \in Y_n'$ such that $f_n = T_n'(g_n)$, where $T_n' \equiv P_n^* T^* Q_n^*|_{Y_n'}$. Hence $1 = (f_n, x_n) = (T_n' g_n, x_n) = (g_n, Q_n TP_n x_n) \to 0$ since $Q_n TP_n x_n \to 0$ as $n \to \infty$, a contradiction. Q.E.D.

Theorem 3.2   Suppose that Γ_P and $\Gamma_{\tilde{P}}^*$ are as in Lemma 3.2. Then $T \in L(X,Y)$ is injective and A-proper w.r.t. Γ_P iff T^* is injective and A-proper w.r.t. $\Gamma_{\tilde{P}}^*$.

Proof.   Suppose that T is injective and A-proper; then, by Theorem 1.3, T is continuously invertible and Lemma 3.2(a) holds. Hence T^* is also continuously invertible and, by Lemma 3.2, T^* is a-stable w.r.t. $\Gamma_{\tilde{P}}^*$ [i.e., Lemma 3.2(b) holds]. Thus, by Theorem 1.3, T^* is a A-proper w.r.t. $\Gamma_{\tilde{P}}^*$ and injective. The converse is proved similarly.   Q.E.D.

The following result provides a complete generalization (partially in constructive form) of the classical Fredholm alternative.

Theorem 3.3   (generalized Fredholm alternative)   Let Γ_P and $\Gamma_{\tilde{P}}^*$ be as in Lemma 3.2 and let $T \in L(X,Y)$ be A-proper w.r.t. Γ_P.

(A1) Either the equation

$$Tx = f \qquad (x \in X, f \in Y) \tag{3.1}$$

is uniquely approximation solvable w.r.t. Γ_P for each $f \in Y$, in which case the adjoint equation

$$T^*u = w \qquad (u \in Y^*, w \in X^*) \tag{3.2}$$

is also uniquely approximation solvable w.r.t. $\Gamma_{\tilde{P}}^*$ for each w in X^*, or $N(T) \neq \{0\}$ and $d(T) < \infty$; in the latter case $d(T) \geq d(T^*)$ and (3.1) is solvable iff $f \in N(T^*)$, while (3.2) is solvable iff $w \in M(T)$.

(A2) dim $N(T) = $ dim $N(T^*)$ [i.e., $i(T) = 0$] iff T^* is A-proper w.r.t. $\Gamma_{\tilde{P}}^*$.

Proof.   (A1) The first part of Theorem 3.3 (A1) follows from Theorems 3.2 and 1.3 and Lemma 3.1, while the second part of (A1) follows from Theorem 3.1 and the known facts (see [66]) that $R(T)$ is closed in Y iff $R(T^*)$ is closed in X^* and that in this case (3.1) [respectively, (3.2)] is solvable iff $f \in N(T^*)$ [respectively, $w \in N(T)$].

(A2) To prove Theorem 3.3 (A2), suppose first that $d(T) = d(T^*) = m$. It follows from Theorem 3.1 that m is finite and from Theorem 3.2 that it is sufficient to consider the case when $m \geq 1$. Let $x_1, \ldots, x_m$ and $y_1^*, \ldots, y_m^*$ be the bases for $N(T)$ and $N(T^*)$, respectively. Then there exist elements $x_1^*, \ldots, x_m^*$ in X^* and $y_1, \ldots, y_m$ in Y such that $(x_i^*, x_j) = (y_i^*, y_j) = \delta_{ij}$ for $1 \leq i, j \leq m$. Now consider the map $A = T + C$ of X into Y, where

$$Cx = \sum_{i=1}^{m} (x_i^*, x)y_i \qquad (x \in X).$$

Then $C: X \to Y$ is clearly compact, so by Theorem I.1.2, A is A-proper w.r.t. Γ_P. Moreover, A is injective. Indeed, if

$$Ax = Tx + Cx = Tx + \sum_{i=1}^{m} (x_i^*,x)y_i = 0, \qquad (3.3)$$

then since $(y_i^*,Tx) = (T^*y_i^*,x) = 0$ and $(y_i^*,y_j) = \delta_{ij}$, it follows from (3.3) that $(x_i^*,x) = 0$ for $1 \le i \le m$. Hence, again by (3.3), $Tx = 0$ [i.e., $x \in N(T)$], so $x = \sum_{i=1}^{m} d_i x_i$. This implies that $d_i = (x_i^*,x) = 0$ for $1 \le i \le m$. Thus A is injective. Hence, by Theorem 3.2, $A^* = T^* + C^*: Y^* \to X^*$ is injective and A-proper w.r.t. Γ_P^*. Since C^* is compact, $T^* = A^* - C^*$ is A-proper w.r.t. Γ_P^*.

Suppose, conversely, that $T^*: Y^* \to X^*$ is A-proper w.r.t. Γ_P^*. Then, by Theorem 3.1, T^* is Fredholm with $i(T^*) \ge 0$. But as is well known, $i(T^*) = -i(T)$ and $i(T) \ge 0$ since T is A-proper. Consequently, $i(T) = 0$.   Q.E.D.

Exercise 3.1   Prove directly that dim $N(T) =$ dim $N(T^*)$ if T^* is A-proper without using the fact that $i(T^*) = -i(T)$.

3.2 Special Cases of Generalized Alternative

In this section we discuss some special cases of Theorem 3.3. In particular, in addition to others, it will be shown that the classical Fredholm alternative follows from Theorem 3.3 for Banach spaces with projectionally complete schemes.

Suppose first that $Y = X$, $\Gamma_P = \Gamma_\alpha = \{X_n,P_n\}$ with $\alpha \equiv \sup_n \| P_n \|$, and $\Gamma_P^* = \Gamma_\alpha^* = \{X_n',P_n^*\}$. We assume that Γ_α and Γ_α^* are projectionally complete for (X,X) and (X^*,X^*) respectively.

Case 1.   If $T = I - C$ with $C \in L(X,X)$ compact, then T is A-proper w.r.t. Γ_α and so is $T^* = I^* - C^*: X^* \to X^*$ w.r.t. Γ_α^* since C^* is compact.

Case 2.   If $T = I - S - C$ with $\| S \| < 1$, then T is A-proper w.r.t. Γ_1 by Corollary I.1.3 and so is $T^* = I^* - S^* - C^*$ w.r.t. Γ_1^* since $\| S^* \| = \| S \| < 1$ and C^* is compact.

Case 3.   If $T = L + C$ with C compact and $(Lx,Jx) \ge c \| x \|^2$ for all $x \in X$ and some $c > 0$ (i.e., L is c-accretive), where J is the normalized duality mapping from X to X^*, and $\{X_n\}$ is nested, then Theorem 1.4 implies that $L: X \to X$ is A-proper w.r.t. Γ_1 since $\| P_n L P_n x \| \ge c \| x \|$ for all x in X_n and $n \in Z_+$ and so is the map $L + C$. Since L is also injective, Theorem 3.2 implies that $L^*: X^* \to X^*$ is injective and A-proper w.r.t. Γ_1^*, so the map $T^* = L^* + C^*$ is also A-proper w.r.t. Γ_1^* since C^* is compact.

The discussion above implies the following consequence of Theorem 3.3.

Corollary 3.1 If T, Γ, and Γ^* are as in Case 1, 2, or 3, the following alternative holds. Either the equations

$$Tx = f \quad (f \in X), \qquad T^*u = w \quad (w \in X^*) \tag{3.4}$$

are uniquely approximation solvable for each f in X and w in X^*, or $N(T)$ and $N(T^*)$ have the same finite dimension and $Tx = f$ is solvable iff $f \in N(T^*)^\perp$, while $T^*u = w$ is solvable iff $w \in N(T)^\perp$.

To obtain other special cases we need the following result obtained by Nussbaum [194].

Proposition 3.3 If $F \in L(X,Y)$ is k-ball contractive, then $F^* \in L(Y^*,X^*)$ is k-set contractive for the same k. F is k-set contractive if T^* is k-ball contractive for the same k.

Proof. Suppose that T is k-ball contractive. To show that T^* is k-set contractive, it suffices to show that if $S \subset Y^*$ is a set of diameter $\leq d$, then $T^*(S)$ can be covered by a finite number of sets of diameter $\leq kd + \epsilon$ for any $\epsilon > 0$.

Let $B = \{x \in X : \|x\| \leq 1\}$ and consider $T(B)$. Since $\chi(B) \leq 1$ and T is k-ball contractive, $T(B) \subset \cup_{i=1}^m B(y_i, k + \epsilon/2d)$ for any $\epsilon > 0$. Let $M > 0$ be such that $\|y_i\| \leq M$ for $1 \leq i \leq m$ and $\|y^*\| \leq M$ for $y^* \in S$, so that $|(y^*(y_i)| \leq M^2$ for $y^* \in S$. Decompose the closed interval $[-M^2, M^2]$ into the union of disjoint intervals Δ_i, $1 \leq i \leq p$, of length $< \epsilon/2$. Corresponding to this decomposition, we divide S into equivalence classes S_j, $1 \leq j \leq q$, as follows: Given y_1^* and $y_2^* \in S$, write $y_1^* \sim y_2^*$ iff for each i, $1 \leq i \leq m$, $y_1^*(y_i)$ and $y_2^*(y_i)$ lie in the same interval $\Delta_{j(i)}$, $1 \leq j(i) \leq p$. It is clear that each y^* in S lies in some S_j. We claim that $\text{diam}(T^*(S_j)) \leq kd + \epsilon$. For take y_1^* and y_2^* in S_j. We have

$$\|T^*(y_1^*) - T^*(y_2^*)\| = \|y_1^*T - y_2^*T\| = \sup_{x \in B}|y_1^*(Tx) - y_2^*(Tx)|$$

$$= \sup_{y \in T(B)}|y_1^*(y) - y_2^*(y)|.$$

If $y \in T(B)$, then y lies in $B(y_i, k + \epsilon/2d)$ for some i, $1 \leq i \leq n$. It follows that

$$|y_1^*(y) - y_2^*(y)| \leq |y_1^*(y - y_i) - y_2^*(y - y_i)|$$
$$+ |y_1^*(y_i) - y_2^*(y_i)| = |(y_1^* - y_2^*)(y - y_i)| + |y_1^*(y_i) - y_2^*(y_i)|.$$

Since y_1^* and y_2^* lie in the same equivalence class, $|y_1^*(y_i) - y_2^*(y_i)| < \epsilon/2$. Since $\|y_1^* - y_2^*\| < d$ and $\|y - y_i\| \le k + \epsilon/2d$, $|(y_1^* - y_2^*)(y) - y_i)| \le d(k + \epsilon/2d) = kd + \epsilon/2$. Adding, we see that $|y_1^*(y) - y_2^*(y)| \le kd + \epsilon$, and because y was an arbitrary point in $T(B)$, $\|T^*(y_1^*) - T^*(y_2^*)\| \le kd + \epsilon$. This shows that diam$(T^*(S_j)) \le kd + \epsilon$, and since $T^*(S) \subset \cup_{j=1}^q T(S_j)$, we have covered $T^*(S)$ by a finite number of sets of diameter $\le kd + \epsilon$ and proved the first part of Proposition 3.3.

The second part follows from the first. Suppose that $T^* \in L(Y^*,X^*)$ is k-ball contractive. By the first part, $B^{**} \in L(X^{**},Y^{**})$ is k-set contractive. Let J_X be the isometric embedding of X into X^{**} and J_Y of Y into Y^{**}. Take a bounded set A in X. Since J_X is an isometry, $\gamma(J_X(A)) = \gamma(A)$, and since T^{**} is k-set contractive, $\gamma(T^{**}(J_X(A))) \le k\gamma(J_X(A)) = \gamma(A)$. But $T^{**}J_X = J_Y T$, so $\gamma(J_Y T(A)) \le k\gamma(A)$, and since J_Y is an isometry, $\gamma(T(A)) \le k\gamma(A)$. $\hspace{2cm}$ Q.E.D.

In view of Proposition 3.3, Theorem 3.3 implies the following:

Proposition 3.4 Let $\{X_n\}$ be nested, L c-accretive and, F k-ball contractive with $k < c$. Then:

(a) Either the equation

$$Lx + Fx = f \qquad (f \in X) \tag{3.5}$$

is uniquely approximation solvable w.r.t. Γ_1 for each f in X, in which case the adjoint equation

$$(L^* + F^*)(u) = w \qquad (w \in X^*) \tag{3.6}$$

is also uniquely approximation solvable w.r.t. Γ_1^* for each $w \in X^*$, or $N(L + F)$ and $N(L^* + F^*)$ have the same finite dimension and (3.5) is solvable iff $f \in N(L + F)^*)^\perp$, while (3.6) is solvable iff $w \in N(L + F)^\perp$.

Proof. We first show that $T = L + F$ is A-proper w.r.t. Γ_1. This follows from Corollary I.3.2 because L is bijective, $\|P_n LPx\| \ge c\|x\|$ for all $x \in X_n$ and $n \in Z_+$, and F is k-ball-contractive with $k < c$. In view of this and Theorem 3.3, to prove Proposition 3.4, it suffices to exclude the strict inequality $d(T^*) < d(T)$ since we know that $d(T^*) \le d(T)$. Let m and n be the respective dimensions of $N(T^*)$ and $N(T)$, let $y_1^*, \ldots, y_m^*$ and $x_1, \ldots, x_n$ be the bases for $N(T^*)$ and $N(T)$, respectively, and assume that $m < n$. As before, there exists a set of elements $x_1^*, \ldots, x_n^*$ in X^* and $y_1, \ldots, y_m$ in X such that $(x_i^*, x_j) = \delta_{ij}$ for $1 \le i \le n$ and $(y_i^*, y_j) = \delta_{ij}$ for $1 \le i, j \le m$. Consider the mapping $A = T^* + V$ of X^* into X^*, where

$$V(u) = \sum_{i=1}^m (u, y_i) x_i^* \qquad (u \in X^*).$$

First note that A is injective. Indeed, if $Au = 0$, that is,

$$T^*u + Vu = T^*u + \sum_{i=1}^{m} (u,y_i)x_i^* = 0, \tag{3.7}$$

then since $(T^*u,x_i) = (u,Tx_i) = 0$ and $(x_i^*,x_j) = \delta_{ij}$, it follows from (3.7) that $(u,y_i) = 0$ for $1 \le i \le m$. Hence, again by (3.7), $T^*u = 0$ [i.e., $u \in N(T^*)$] and therefore $u = \sum_{i=1}^{m} d_i x_i^*$. Since $x_1^*, \ldots, x_m^*$ are linearly independent, $d_i = (u,y_i) = 0$ for $1 \le i \le m$ (i.e., $u = 0$). Thus $T^* + V$ is an injective map of X^* into X^*. Since L is continuously invertible, the map $L^*: X^* \to X^*$ is also continuously invertible and $\| L^{*-1} \| = \|(L^{-1})^*\| = \| L^{-1}\| \le 1/c$. Now u in X^* is a solution of $Au = g$ for any g in X^* iff $u + L^{*-1}(F^* + V)u = L^{*-1}g$. But by Proposition 3.3, F^* is k-set contractive, so $L^{*-1}F^* + L^{*-1}V$ is also k/c-set contractive with $k/c < 1$. Since $I + L^{*-1}(F^* + V)$ is obviously injective, it follows (see [194,219,291]) that $I + L^{*-1}(F^* + V)$ is bijective [i.e., $(T^* + V)u = g$ is uniquely solvable for each g in X^*]. In particular, there exists a unique $u^* \in X^*$ such that $T^*u^* + vu^* = x_{m+1}^*$. The last equation leads to the impossible equality

$$1 = (x_{m+1}^*,x_{n+1}) = (T^*u^*,x_{n+1}) + \sum_{i=1}^{m} (u,y_i)(x_i^*,x_{m+1}) = 0 + 0 = 0.$$

Thus we must conclude that $m = n$.   Q.E.D.

Exercise 3.2   Show that if $C: X \to X$ is k-set contractive with $k < 1$ and $I - C$ is injective, then $I - C$ is bijective.

Proposition 3.4 implies the validity of the following:

Corollary 3.2   If L is c-accretive and F is k-ball-contractive with $k < c$, then $(L + F)^*$ is A-proper w.r.t. $\Gamma_1^* = \{X_n',P_n^*\}$.

It should be added that the basic Proposition 3.3 was improved by the author in [219], where it was shown that if F is ball condensing, then F^* is k-set contractive with $k = x(T(B))$, and if F^* is ball condensing, then F is k-set contractive with $k = F^*(B^*)$. For the case when $Y = X$ and $L = I$, linear results obtained in [291] are analogous to those in [219].

4. CONSTRUCTIVE SOLVABILITY OF LINEAR ELLIPTIC BOUNDARY VALUE PROBLEMS

To apply the A-proper mapping theory to the solvability of the general variational boundary value problems for nonlinear elliptic partial differential operators, we first show how this theory can be used to obtain in a simple way the constructive existence results for linear elliptic equations under a more restrictive condition on strong ellipticity (see [211]).

Let Q be a bounded domain in R^n with boundary ∂Q so smooth that the Sobolev embeddings theorem holds on Q. Let $C_0^\infty(Q)$ be the family of real- or complex-valued infinitely differentiable functions with compact support in Q. For a multi-index $\alpha = (\alpha_1, \ldots, \alpha_n)$ we denote by D^α the derivative $D^\alpha = D_1^{\alpha_1} \cdots D_n^{\alpha_n}$ of order $|\alpha| = \alpha_1 + \cdots + \alpha_n$. For any real number p with $1 < p < \infty$, let $L_p(Q)$ be the Banach space with the norm $\| \cdot \|_p$ and for any integer $m \geq 0$ let

$$W_p^m \equiv W_p^m(Q) = \{u \mid u \in L_p(Q), D^\alpha u \in L_p(Q) \quad \text{for} \quad |\alpha| \leq m\}$$

be the Sobolev space with the norm $\| u \|_{m,p} = \{\sum_{|\alpha| \leq m} \| D^\alpha u \|_p^p\}^{1/p}$, where $D^\alpha u$ denotes the distribution derivatives of u. W_p^m is a uniformly convex and separable Banach space which for every p includes $C_0^\infty(Q)$. We let $\mathring{W}_p^m$ denote the closure of $C_0^\infty(Q)$ in W_p^m. Let $\langle u,v \rangle = \int_Q uv \, dx$ [and $\langle u,v \rangle = \int_Q u\bar{v} \, dx$ if $L_p(Q)$ is complex] denote the natural pairing between $u \in L_p$ and $v \in L_q$ with $q = p(p-1)^{-1}$. When $p = 2$, $\langle \cdot,\cdot \rangle$ denotes the inner product in $L_2(Q)$. In case $p = 2$, W_2^m is a Hilbert space with the inner product $(u,v)_m = \sum_{|\alpha| \leq m} \langle D^\alpha u, D^\alpha v \rangle$ and norm $\| u \|_{m,2} = (u,u)_m^{1/2}$.

Generalized abstract boundary value problem   Let V be any closed subspace of W_2^m such that $\mathring{W}_2^m \subseteq V$. To study the variational solvability of the generalized abstract boundary value problem for linear elliptic equations of order $2m$ ($m \geq 1$), we consider a bilinear form $B[v,u]$ (i.e., $B[v,u]$ that is linear in v and conjugate linear in u when W_2^m is complex) defined on $V \times V$ by

$$B[v,u] = \sum_{|\alpha|,|\beta| \leq m} \langle D^\alpha v, A_{\alpha\beta}(x)D^\beta u \rangle \qquad (u,v \in V), \tag{4.1}$$

where the complex-valued coefficients $A_{\alpha\beta}$ of $B[u,v]$ satisfy the conditions:

(c1) $A_{\alpha\beta}(x) \in L_\infty(Q)$ for all α and β with $|\alpha| \leq m$ and $|\beta| \leq m$.
(c2) There exists a constant $\mu_0 > 0$ such that for $x \in Q$ (a.e.)

$$\text{Re}(\sum_{|\alpha| = |\beta| = m} A_{\alpha\beta}(x)\zeta_\alpha\zeta_\beta) \geq \mu_0 \sum_{|\alpha| = m} |\zeta_\alpha|^2$$

$$\text{for all} \quad \zeta = \{\zeta_\alpha : |\alpha| = m\} \in R^{s_m'},$$

where s_m' is the number of derivatives of order m. The form $B[v,u]$ is uniformly strongly elliptic in the foregoing sense.

It follows from (c1), (4.1), and the Buniakovskii–Schwarz inequality that $B[v,u]$ is a continuous bilinear form on $V \times V$.

Definition 4.1   Let F be a given bounded linear functional on V (i.e., $F \in V^*$). By the *generalized abstract problem* (GA problem) we mean a problem of finding an element $u \in V$ such that

$$B[\psi, u] = F(\psi) \qquad \text{for all} \quad \psi \in V. \tag{4.2}$$

To apply the results of Section 3 to the solvability of (4.2), we first have to select an admissible scheme Γ_V for maps from V to V. Since V is a separable Hilbert space we may choose a sequence $\{X_n\}$ of finite-dimensional subspaces of V such that $\text{dist}(u, X_n) = \inf\{\| u - \phi \|_{m,2} \mid \phi \in X_n\} \to 0$ as $n \to \infty$ for each $u \in V$. If P_n denotes the orthogonal projection of V onto X_n, then $\Gamma_V = \{X_n, P_n\}$ is an admissible scheme since Γ_V is projectionally complete.

For a given $F \in V^*$ we associate with (4.2) a sequence of finite-dimensional linear algebraic equations

$$B[\phi_j, u_n] \equiv \sum_{|\alpha|,|\beta| \leq m} \langle D^\alpha \phi_j, A_{\alpha\beta} D^\beta u_n \rangle = F(\phi_j) \qquad (1 \leq j \leq n) \tag{4.3}$$

for determination of the nth-order Galerkin approximate solution $u_n \in X_n$, where $\{\phi_j \mid 1 \leq j \leq n\}$ is a basis in X_n for each $n \in Z_+$. For completeness we recall:

Definition 4.2   The GA problem (4.2) is said to be *uniquely approximation solvable* w.r.t. Γ_V iff there exists an integer $n_0 \in Z_+$ such that (4.3) has a unique solution $u_n \in X_n$ for each $n \geq n_0$ such that $u_n \to u$ in V and u is the unique solution of (4.2).

We call the reader's attention to the difference between the solvability (i.e., the existence of solutions) of (4.2) and the approximation solvability (i.e., the constructive solvability) of (4.2).

It will be seen below that a simple consequence of the approximation-solvability theory and of the generalized Fredholm alternative for equations involving A-proper mappings developed by the author in [208] and in Section 3 is the following partially constructive basic theorem for solvability of the GA problem (4.2).

Theorem 4.1   Let V be any closed subspace of W_2^m with $\mathring{W}_2^m \subset V$ and suppose that the coefficients $A_{\alpha\beta}(x)$ of the bilinear form $B[v, u]$ defined on $V \times V$ by (4.1) satisfy conditions (c1) and (c2).

(A1) If the nullspace $N(B) = \{u \in V \mid B[\phi, u] = 0 \ \forall \ \phi \in V] = \{0\}$, then the GA problem (4.2) is uniquely approximation solvable w.r.t. Γ_V

for each $F \in V^*$ and, in particular, for each $F \in V^*$ there exists a unique $u \in V$ such that (4.2) holds.

(A2) In general, the following alternative holds: Either the equations

$$B[\phi,u] = F(\phi) \quad \forall \, \phi \in V, \qquad \overline{B[v,\psi]} = F'(\psi) \quad \forall \, \psi \in V, \qquad (4.4a)$$

are uniquely approximation solvable w.r.t. Γ_V for each pair $F, F' \in V^*$, or the homogeneous equations

$$B[\phi,u] = 0 \quad \forall \, \phi \in V, \qquad \overline{B[v,\psi]} = 0 \quad \forall \, \psi \in V, \qquad (4.4b)$$

have the same finite number, say $d \geq 1$, of linearly independent solutions $u_1, \ldots, u_d \in N(B)$ and $v_1, \ldots, v_d \in N(B^*)$ and then equations (4.4a) are solvable iff $F(v_j) = 0$ and $F'(u_j) = 0$ for $1 \leq j \leq d$.

Proof. (A1) for each fixed $u \in V$, $B([v,u]$ is a bounded linear functional of v in V. Hence, by the well-known theorem of Riesz, there exists a unique element $L(u) \in V$ such that

$$B[\phi,u] = (\phi,Lu)_m \quad \forall \, \phi \in V. \qquad (4.5)$$

It is obvious that L is a bounded linear mapping of V into V. Moreover, L is one-to-one since $N(L) = N(B) = \{0\}$. Now, by the same Riesz theorem, to each $F \in V^*$ there corresponds a unique element $w_F \in V$ such that $F(\phi) = (\phi,w_F)_m$ for all $\phi \in V$. Thus the solvability of (4.2) is equivalent to that of

$$Lu = w_F \quad (u \in V, w_F \in V), \qquad (4.6)$$

while the solvability of (4.3) is equivalent to that of

$$P_n L(u_n) = P_n w_F \quad (u_n \in X_n, P_n w_F \in X_n). \qquad (4.7)$$

Hence the approximation solvability of the GA problem (4.2) is equivalent to that of Eq. (4.6).

It follows from Theorem 1.4 that Eq. (4.6) is uniquely approximation solvable w.r.t. Γ_V for each $w_F \in V$ if and only if L is one-to-one and A-proper w.r.t. Γ_V. Thus, to complete the proof of (A1) it suffices to show that L is A-proper.

To that end we first recall the definition of an approximation-proper (A-proper) mapping, which we state here in terms of an admissible scheme $\Gamma = \{X_n,V_n;E_n,W_n\}$ for Banach spaces (X,Y) since this more general situation will be encountered in Section 2.

Definition 4.3 A not necessarily linear map $T: X \to Y$ is said to be A-*proper* w.r.t. Γ iff $T_n = W_n T|_{X_n}: X_n \to E_n$ is continuous and T satisfies condition (H): If $\{x_{n_j} | x_{n_j} \in X_{n_j}\}$ is any bounded sequence such that

$T_{n_j}(x_{n_j}) - W_{n_j}(g) \to 0$ for some g in Y, there exists a subsequence $\{x_{n_{j(k)}}\}$, $x \in X$, such that $x_{n_{j(k)}} \to x$ and $Tx = g$.

Now, as noted above, to complete the proof of (A1) of Theorem 4.1, we need the following:

Lemma 4.1 If the coefficients $A_{\alpha\beta}$ of $B[v,u]$ satisfy (c1) and (c2), then $L: V \to V$ given by (4.5) is A-proper w.r.t. Γ_V.

Proof. Since $L_n = P_n L |_{X_n}: X_n \to X_n$ is continuous for each $n \in Z_+$, to show that L is A-proper w.r.t. Γ_V, it suffices to prove that L satisfies condition (H). Now L can be written in the form $L = L_1 + L_2$, where

$$(v, L_1 u)_m = B_1[v,u] \equiv \sum_{|\alpha| = |\beta| = m} \langle D^\alpha v, A_{\alpha\beta}(x) D^\beta u \rangle \qquad (u,v \in V)$$

(4.8)

$$(v, L_2 u)_m = B_2[v,u] \equiv \sum_{\substack{|\alpha| \le m, |\beta| \le m \\ |\alpha + \beta| < 2m}} \langle D^\alpha v, A_{\alpha\beta}(x) D^\beta u \rangle \qquad (u,v \in V)$$

(4.9)

with $L_2 \in L(V,V)$ completely continuous by virtue of the complete continuity of the embedding of W_2^m into W_2^j for $0 \le j \le m - 1$ and an easily established inequality $|B_2(u,u)| \le m_0 \|u\|_{m,2} \|u\|_{m-1,2}$ for all $u \in V$ and some $m_0 > 0$. Now, Theorem I.1.2 shows that A-properness is invariant under compact perturbations. Hence it suffices to show that L_1 is A-proper. So let $\{u_{n_j} | u_{n_j} \in X_{n_j}\}$ be any bounded sequence such that $P_{n_j} L_1(u_{n_j}) - P_{n_j} g \to 0$ [i.e., $P_{n_j} L_1(u_{n_j}) \to g$ for some g in V]. For simplicity of notation set $u_{n_j} \equiv u_j$ for each $j \in Z_+$ and note that because $\{u_j\}$ is bounded we may assume that $u_j \to u_0$ for some $u_0 \in V$, where "$\to$" denotes the weak convergence in V. Since $D^\alpha u_j \to D^\alpha u$ in L_2 for each α with $|\alpha| < m$ by the Sobolev embedding theorem, to prove that $u_j \to u_0$ in V, it suffices to show that $D^\alpha u_j \to D^\alpha u_0$ in L_2 for each α with $|\alpha| = m$. Now, since dist$(u,X_n) \to 0$ for each $u \in V$, there exists $w_j \in X_j$ such that $w_j \to u_0$ in V as $j \to \infty$. This and the continuity of L_1 imply that

$$\lim_j (u_j - w_j, L_1 u_j - L_1 w_j)_m = \lim_j (u_j - w_j, P_j L_1 u_j - P_j L_1 w_j)_m = 0$$

since $P_j L_1 u_j - P_j L_1 w_j \to g - L_1 u_0$ and $u_j - w_j \to 0$ as $j \to \infty$ and hence $\text{Re}(u_j - w_j, L_1(u_j - w_j))_m \to 0$. In view of this and the inequality

$$\text{Re}(L_1 u, u)_m \ge \mu_0 \left(\sum_{|\alpha| = m} \|D^\alpha u\|_2^2 \right) \qquad \text{for all} \quad u \in V, \qquad (4.10)$$

which is implied by (c2), it follows that $D^\alpha(u_j - w_j) \to 0$ as $j \to \infty$ for

$|\alpha| = m$ and, consequently, $D^\alpha u_j \to D^\alpha u_0$ in L_2 for each α with $|\alpha| = m$. Thus L_1 is A-proper w.r.t. Γ_V and so is $L = L_1 + L_2$ because L_2 is completely continuous. This proves (A1).

(A2) To prove (A2), let us first note that since $B[v,u]$ is a bounded bilinear functional on $V \times V$, its adjoint B^* is also a bounded bilinear functional on $V \times V$ which is defined by $B^*[v,u] = \overline{B[u,v]}$ for $u,v \in V$ and the linear operator associated uniquely with B^* is the adjoint L^* of L with $N(L^*) = N(B^*)$.

To continue with the proof, note that since L is A-proper w.r.t. Γ_V, a simple consequence of the author's generalized alternative ([208]) implies that either (4.9) is uniquely approximation solvable w.r.t. Γ_V for each $F \in V^*$, in which case the adjoint problem

$$B^*[v,\psi] = F'(\psi) \qquad \forall\, \psi \in V \tag{4.2*}$$

is also uniquely approximation solvable w.r.t. Γ_V for each $F' \in V^*$, or $N(B) = N(L) \neq \{0\}$ and L (or equivalently, B) is Fredholm with $\mathrm{ind}(L) = \dim N(L) - \dim N(L^*) \geq 0$; moreover, $\mathrm{ind}(L) = 0$ iff L^* is also A-proper w.r.t. Γ_V.

Now since $L^* = L_1^{**} + L_2^*$, where L_2^* is completely continuous and L_1^* satisfies the inequality (4.10), it follows that L^* is also A-proper w.r.t. Γ_V. Thus L (or $B[v,u]$) is Fredholm with $\mathrm{ind}(L) = 0$ and the alternative stated in (A2) follows from this and the preceding remarks.   Q.E.D.

Remark 4.1 Going over the proof of Theorem 4.1, we see that its assertions (A1) and (A2) remain valid if instead of condition (c2) we assume the weaker condition

(c2′) There exists a constant $\mu' > 0$ such that

$$\Big| \sum_{|\alpha|=|\beta|=m} \langle D^\alpha u, A_{\alpha\beta} D^\beta u \rangle \Big| \geq \mu' \Big(\sum_{|\alpha|=m} \| D^\alpha u \|_2^2 \Big) \qquad \forall\, u \in V.$$

Remark 4.2 It should be added that the existence part of Theorem 4.1 (A1) and (A2) can be deduced from the classical results on elliptic equations as presented, for example, in [2,277]. However, its constructive aspect appears to be new. Moreover, our proof of Theorem 4.1 is much simpler than that of any proof known to the author. It should be pointed out, however, that the strong ellipticity assumption (c2) used in this paper is somewhat different from the classical condition, which requires

(c2$_c$) $\mathrm{Re}\Big(\sum_{|\alpha|=|\beta|=m} A_{\alpha\beta}(x)\xi^\alpha \xi^\beta \Big) \geq \mu_0 \Big(\sum_{|\alpha|=m} |\xi^\alpha|^2 \Big)$

for $x \in Q$ (a.e.) and for all $\xi = (\xi_1, \ldots, \xi_n) \in R^n$.

Special Cases of the GA Problem

It was shown by Agmon [2] (see also Treves [277]) that the GA problem (4.2) actually solves some boundary value problem for an elliptic operator of order $2m$ given in the variational or divergence form:

$$Au(x) = \sum_{|\alpha|,|\beta| \le m} (-1)^{|\alpha|} D^\alpha (A_{\alpha\beta}(x) D^\beta u) \tag{4.11}$$

with boundary conditions determined by the form $B[v,u]$ and the choice of the subspace V. We illustrate this by the following special cases of the GA problem.

Generalized Dirichlet problem   The simplest special case of the GA problem is the generalized Dirichlet problem (GD problem), which for each bounded linear functional $F \in V^* = (W_2^m)^*$ calls for determination of a function $u \in W_2^m$ such that

$$B[\phi,u] = F(\phi) \qquad \text{for all} \quad \phi \in \mathring{W}_2^m. \tag{4.12}$$

When $\{f,g\}$ is a given pair with $f \in L_2(Q)$ and $g \in W_2^m(Q)$, a special case of the GD problem (4.12) is the problem of finding a solution $u \in W_2^m(Q)$ of

$$\begin{cases} B[\phi,u\} = \langle\phi,f\rangle & \forall \phi \in \mathring{W}_2^m(Q) \\ u - g \in \mathring{W}_2^m(Q). \end{cases} \tag{4.13}$$

Indeed, if we set $F(\phi) = \langle\phi,f\rangle - B[\phi,g]$ for $\phi \in \mathring{W}_2^m$, then clearly, F is a bounded linear functional on $\mathring{W}_2^m$ and $u \in W_2^m(Q)$ is a solution of (4.13) if and only if $u = v + g$ and $v \in \mathring{W}_2^m$ is a solution of

$$B[\phi,v] = \langle\phi,f\rangle - B[\phi,g], \qquad \forall \phi \in \mathring{W}_2^m. \tag{4.14}$$

Now it follows from (4.1) and (c1) that each $u \in W_2^m(Q)$ defines a bounded linear functional $F_u \in (\mathring{W}_2^m(Q))^*$ if we set

$$F_u(\phi) = B[\phi,u], \qquad \phi \in \mathring{W}_2^m, \tag{4.15}$$

and this functional is usually denoted by

$$F_u = B[\cdot,u] = \sum_{|\alpha|,|\beta| \le m} (-1)^{|\alpha|} D^\alpha (A_{\alpha\beta} D^\beta u). \tag{4.16}$$

Thus a solution $u \in W_2^m$ of the Dirichlet boundary value problem (4.13) can be regarded as a generalized solution of

$$\sum_{|\alpha|,|\beta| \le m} (-1)^{|\alpha|} D^\alpha (A_{\alpha\beta} D^\beta u) = f \qquad (x \in Q),$$

$$u - g \in \mathring{W}_2^m(Q). \tag{4.17}$$

In case $g = 0$ we get a solution that satisfies the homogeneous boundary condition $u \in \mathring{W}_2^m$.

We summarize the discussion above into

Corollary 4.1 Let $B[v,u]$ be the Dirichlet bilinear form on $\mathring{W}_2^m \times \mathring{W}_2^m$ given by (4.1) with $A_{\alpha\beta}(x)$ satisfying (c1) and (c2) or (c2'). If $N(B) = \{0\}$, the GD problem (4.12) is uniquely approximation solvable for each $F \in (\mathring{W}_2^m)^*$. In general, $B[v,u]$ is Fredholm of index zero; more precisely, the alternative (A2) holds for (4.2) and its adjoint (4.2*) with $V = \mathring{W}_2^m(Q)$.

When Corollary 4.1 is applied to the Dirichlet boundary value problem (4.14) with $F(\phi) = \langle \phi, f \rangle - B[\phi, g]$ for all $\phi \in \mathring{W}_2^m$ with $f \in L_2$ and $g \in W_2^m$, its assertions are valid for the special Dirichlet problem (4.13). In particular, (4.13) has a unique solution $u \in W_2^m$ for each pair $(f, g) \in L_2 \times W_2^m$ if $N(B) = \{0\}$, which is also a generalized solution of (1.17); if $N(B) \neq \{0\}$, then (4.13) has a solution $u \in W_2^m$ if and only if $\langle f, v_j \rangle - B[v_j, g] = 0$, where $v_1, \ldots, v_d \in \mathring{W}_2^m$ are such that $B[v_j, \psi] = 0$ for all $\psi \in \mathring{W}_2^m$.

Remark 4.3 Note that the interesting feature of our approach is the (constructive) solvability of the generalized Dirichlet boundary value problem and its special cases is that we do not use the Gårding inequality, and thus the leading coefficients $A_{\alpha\beta}(x)$, with $|\alpha| = |\beta| = m$, need not be continuous on $\overline{Q}$. However, as pointed out in Remark 1.2, the strong ellipticity assumption (c2) used in this paper is somewhat different from the classical condition (c2$_c$).

It is known (see [104]) that if (c2$_c$) holds and $A_{\alpha\beta}(x)$ are continuous on $\overline{Q}$ for $|\alpha| = |\beta| = m$, we can show (the proof is not easy) that $B[u,u]$ satisfies the Gårding inequality on $\mathring{W}_2^m(Q)$; that is, there exist constants $c_0 > 0$ and c_1 such that

(G) $\mathrm{Re}(Lu, n)_m \geq c_0 \|u\|_{m,2}^2 - c_1 \|u\|_2^2$ for all $u \in \mathring{W}_2^m(Q)$.

It follows from (G) that L is also A-proper w.r.t. Γ_V and hence Corollary 4.1 also holds in this case.

The main point is that when (c2) holds, we obtain the solvability of the GA problem (4.2) in a rather simple way and for V not necessarily equal to $\mathring{W}_2^m(Q)$. The more important reason for our discussion of the solvability of (4.2) under condition (c2) is that it leads naturally (by way of arguments and conditions) to the (constructive) variational solvability of the nonlinear problem (1) if its leading part satisfies the nonlinear version of (c2).

The Neumann problem   Another special case of the GA problem is the Neumann problem, which is associated with the choice $V = W_2^m(Q)$. We consider, then, the following continuous linear functional on $W_2^m(Q)$:

$$F(\phi) = \int_Q \phi \bar{f} \, dx + \sum_{j=1}^{m-1} \int_{\partial Q} \left(\frac{\partial}{\partial n}\right)^j \phi \bar{g}_j \, d\sigma, \tag{4.18}$$

where, say, $f \in L_2(Q)$, $g_j \in L_2(\partial Q)$ $(0 \le j \le m - 1)$, and $\partial/\partial n$ denotes the normal derivative. When $N(B) = \{0\}$, the equation

$$B[\phi,u] = F(\phi) \qquad \forall \; \phi \in W_2^m(Q) \tag{4.19}$$

is uniquely approximation solvable and, in particular, there exists a unique $u \in W_2^m(Q)$ satisfying (4.19). In general, we have the alternative (A2) for (4.19).

To write down explicitly the boundary conditions implied by (4.19) one assumes that u, ∂Q, and $A_{\alpha\beta}(x)$ are sufficiently smooth so that integration by parts and the use of Green's formula is possible. It was shown by Agmon [2] (see also Treves [277]) that there are linear differential operators N_j, defined in the neighborhood of ∂Q, of order $\le 2m - 1 - j$, respectively, such that for $u,v \in C^{2m}(Q)$,

$$B[v,u] = \langle v, Au \rangle + \sum_{j=0}^{m-1} \int_{\partial Q} \frac{\partial^j v}{\partial n^j} \overline{N_{2m-1-j}(x,D)u} \, d\sigma, \tag{4.20}$$

where

$$A(x,D)u = \sum_{|\alpha|,|\beta| \le m} (-1)^{|\alpha|} D^\alpha (A_{\alpha\beta} D^\beta u)$$

and the operators N_{2m-1-j} can be explicitly calculated for given $B[v,u]$ and Q (see [2,277]). Thus, if $F(\phi)$ is given by (4.18), then for u to be a solution of $Au = f$, we see that the boundary conditions implied by the equation (4.19) are

$$N_{2m-1-j}(x,D)u = g_j \quad \text{on} \quad \partial Q, \qquad j = 0,1, \dots, m - 1. \tag{4.21}$$

Note that N_{2m-1-j} contains the term $\partial^{2m-1-j} u/\partial n^{2m-1-j}$ with a nonvanishing coefficient. Thus the boundary conditions for u are specified by m differential operators on ∂Q, having orders $m, m + 1, \dots, 2m - 1$. Note that in this case the boundary conditions actually depend on the choice of the form B. These are the so-called *natural* boundary conditions. For further discussion, see [2] and [277].

5. STABILITY OF PROJECTIVE METHODS AND A-PROPERNESS

The study of the stability of projective methods was initiated by Mikhlin [166], who concentrated his attention on the Ritz method (see [168]); the stability of the Galerkin and Galerkin–Petrov methods was considered in [166] and later more completely in [282] and by others (see [225] and [279]). All results for A-proper maps presented here are new.

Consider the approximation solvability of the equation

$$Tx = f \qquad (x \in X, f \in Y). \tag{5.1}$$

In this section we indicate the relation of the notion of the A-properness of T with respect to $\Gamma = \{X_n, P_n; Y_n, Q_n\}$ to the notion of the stability of the method

$$T_n(x_n) = Q_n f \qquad (x_n \in X_n, T_n = Q_n T \mid X_n, n \geq n_0), \tag{5.2}$$

as defined by Mikhlin [166] with a slight variant due to [282] and [225]. For each fixed n ($\geq n_0$), along with equation (5.1) we consider the perturbed equation

$$(T_n + F_n)(y_n) = Q_n f + h_n \qquad (y_n \in X_n, h_n \in Y_n), \tag{5.3}$$

where $F_n: X_n \to Y_n$ is a linear operator perturbation of T_n and h_n is a perturbation of $Q_n f$.

Definition 5.1 Suppose that equation (5.2) is uniquely solvable for each $n \geq n_0$ and each $f \in Y$. Then the projection method (5.2) is said to be *stable from space X to Y* if there exist nonnegative constants p, q, r independent of n and f such that for $\| F_n \| \leq r$ and arbitrary h_n in Y_n the perturbed equation (5.3) has a unique solution $y_n \in X_n$ for $n \geq n_0$ and the following inequality holds:

$$\| x_n - y_n \| \leq p \| x_n \| \| F_n \| + q \| h_n \| \qquad (n \geq n_0). \tag{5.4}$$

Theorem 5.1 Suppose that $T \in L(X, Y)$ is surjective. Then the projective method (5.2) is stable in the sense of Definition (5.1) if and only if T is one-to-one and A-proper with respect to Γ.

Proof. Suppose that T is one-to-one and A-proper. Then, by Theorem 1.1, equation (5.2) is uniquely solvable for each $n \geq n_0$ and each $f \in Y$, so it makes sense to talk about the stability of (5.2) as given by Definition 5.1. Moreover, there exists $c > 0$ such that $\| T_n(x) \| \geq c \| x \|$ for $x \in X_n$ and $n \geq n_0$. Let r be any fixed number in $[0, c)$ and let $\| F_n \| \leq r$. Then for all x in X_n and $n \geq n_0$ we have $\|(T_n + F_n)x\| \geq (c - r)\|x\|$. Hence for each fixed $n \geq n_0$ and any $h_n \in Y_n$, equation (5.3) has a unique solution $y_n \in X_n$. The discussion above implies the equality

$$x_n - y_n = (T_n + F_n)^{-1}(F_n(x_n) - h_n),$$

from which (5.4) follows with $p = q = (c - r)^{-1}$.

To prove the converse, suppose that (5.2) is stable in the sense of Definition 5.1. Take $F_n = 0$ and for arbitrary $f \in Y$ and $h_n \in Y_n$ let x_n and y_n be the solutions for $n \geq n_0$ of the corresponding equations (5.2) and (5.3). Hence it follows from (5.4) that $\|x_n - y_n\| \leq q\|h_n\| = \|T_n(x_n - y_n)\|$ and consequently, $\|T_n(x)\| \geq q^{-1}\|x\|$ for all $x \in X_n$ and each $n \geq n_0$. In view of this and the surjectivity of T, it follows from Theorem 1.3 that T is one-to-one and A-proper. Q.E.D.

Note that if X is reflexive and $Q_n Q_m = Q_n$ for $m \geq n$ or $Q_n^* u \to u$ in Y^* for each u in Y, the hypotheses in Theorem 5.1 that T is surjective can be omitted.

Theorem 5.1 forms a basis for the study of stability of the numerical realization of the approxmation methods of the Galerkin type when the latter are applied to the solvability of equation (5.1) involving A-proper mappings.

Galerkin method   We illustrate the assertion above by discussing the stability of the Galerkin method when the latter is applied to the equation

$$Tx = f \quad [f \in H, T \in L(H,H)], \tag{5.5}$$

where H is a separable real Hilbert space with a linearly independent complete system $\{\phi_i\} \subset H$ and where T is one-to-one A-proper with respect to $\Gamma_0 = \{X_n, P_n; X_n, P_n\}$ with P_n the orthogonal projection of H onto $X_n = [\phi_1, \ldots, \phi_n]$ for each $n \geq 1$.

If the approximate solution $x_n \in X_n$ is taken in the form

$$x_n = \sum_{i=1}^{n} a_i^n \phi_i, \tag{5.6}$$

then in the Galerkin method the unknowns a_i^n are defined by the algebraic system

$$\sum_{i=1}^{n} (T\phi_i, \phi_j)a_i^n = (f, \phi_j) \quad (1 \leq j \leq n) \tag{5.7}$$

which can be written in the form

$$G_n a^{(n)} = f^{(n)}, \tag{5.8}$$

where $a^{(n)} = (a_1^n, \ldots, a_n^n)^T$, $f^{(n)} = ((f,\phi_1), \ldots, (f,\phi_n))^T$, $G_n = ((T\phi_i,\phi_j))$ for $1 \le i, j \le n$. Let γ_{ij} denote the errors arising in the computation of $(T\phi_i,\phi_j)$, let $\Gamma_n = (\gamma_{ij})$ be the error matrix, and let $\delta^{(n)}$ be the corresponding error in $f^{(n)}$. Then instead of the exact Galerkin process (5.7) we solve the "nonexact" process

$$(G_n + \Gamma_n)b^{(n)} = f^{(n)} + \delta^{(n)}, \tag{5.9}$$

and obtain the "nonexact" solution

$$y_n = \sum_{i=1}^{n} b_i^n \phi_i. \tag{5.10}$$

Following Mikhlin [166] (with variant due to Vainikko [282]) we say that the Galerkin method is stable if there exist nonnegative constants r, p, q independent of n and of f such that $\| \Gamma_n \| \le r$ and arbitrary $\delta^{(n)} \in R^n$ the perturbed equation (5.9) has a unique solution $b^{(n)} \in R^n$ for $n \ge n_0$ and

$$\| a^{(n)} - b^{(n)} \| \le p \| x_0 \| \| \Gamma_n \| + q \| \delta^{(n)} \|, \tag{5.11}$$

where x_0 is the unique solution of equation (5.5) that exists by Theorem 1.1. By the norms of the vectors and matrices we mean their norms are elements and operators in the space R^n.

Now, as was mentioned before, equation (5.7) may be written in X_n as

$$P_n T x_n = P_n f \quad (x_n \in X_n, P_n f \in X_n). \tag{5.12}$$

To use Theorem 5.1 we have to write equation (5.8) as an operator equation in X_n. For this purpose we use the approach of [139]. Let $\bar{S}_n: H \to R^n$ be the linear map defined by $\bar{S}_n(x) = ((x,\phi_1), \ldots, (x,\phi_n))^T$ for each x in H, and let S_n be the restriction of $\bar{S}_n$ to X_n. It follows that $\| S_n \| = \| \bar{S}_n \|$, S_n^{-1} exists and $S_n^{-1}\bar{S}_n = P_n$. Moreover, if $u_n = \Sigma_{i=1}^n c_i^n \phi_i$ is any element in X_n, then $S_n(u_n) = R_n c^{(n)}$, where R_n is the Gramm matrix given by $R_n = ((\phi_i,\phi_j))$ for $1 \le i, j \le n$ and each $n \ge 1$. Since $c^{(n)} = R_n^{-1} S_n(u_n)$, the discussion above implies that equation (5.9) can be written in X_n as

$$P_n T y_n + F_n y_n = P_n f + h_n, \tag{5.13}$$

where $h_n = S_n^{-1}\delta^{(n)} \in X_n$ and $F_n = S_n^{-1}\Gamma_n R_n^{-1} S_n: X_n \to X_n$.

Now, since R_n is a positive symmetric matrix, its eigenvalues λ_i^n are positive and we order them by $0 < \lambda_n^n \le \cdots \le \lambda_n^n$. Following accepted terminology we say that the system $\{\phi_i\}$ is *strongly minimal* if there exists a constant $\lambda_0 > 0$ such that $\inf_n \lambda_1^n = \lim_n \lambda_1^n \ge \lambda_0$. Of course, an orthogonal system in H or even "almost orthogonal system $\{\phi_i\}$ in H" is strongly minimal.

Theorem 5.2 If $T \in L(H,H)$ is one-to-one and A-proper with respect to Γ_0 and if $\{\phi_i\}$ is strongly minimal in H, equation (5.5) is uniquely approximation solvable and the Galerkin method (5.6)–(5.7) is stable.

Proof. It follows from the results in [139] and [166] and the strong minimality of $\{\phi_i\}$ in H that $\| S_n^{-1} \| \leq \lambda_0^{-1/2}$ and $\| R_n^{-1} S_n \| \leq \lambda_0^{-1/2}$ for each $n \geq 1$. Hence

$$\| F_n \| \leq \| S_n^{-1} \| \| \Gamma_n \| \| R_n^{-1} S_n \| \leq \lambda_0^{-1} \| \Gamma_n \|$$

and $\quad \| h_n \| \leq \lambda_0^{-1/2} \| \delta^{(n)} \|$

for each n. Now, since T is one-to-one and A-proper with respect to Γ_0, there exists a constant $c > 0$ and an integer $n_0 \geq 1$ such that $\| T_n(x) \| \geq c \| x \|$ for all $x \in X_n$ and $n \geq n_0$. Consequently, if we assume that $\| \Gamma_n \| < r = \frac{1}{2}(c/\lambda_0)$, then $\| F_n \| < \frac{1}{2}c$, so by Theorem 5.1, equation (5.13), or equivalently equation (5.9), has a unique solution $b^{(n)}$ for $n \geq n_0$. For x_n and y_n given by (5.6) and (5.10), respectively, we have the inequality (5.4), which in our case reduces to

$$\| x_n - y_n \| \leq \frac{2}{c} \| x_n \| \| F_n \| + \frac{2}{c} \| h_n \|. \tag{5.14}$$

Since x_n is the solution of equation (5.2) and $x_n \to x_0$ in H, there exists a constant $\sigma > 0$ indendent of n or f such that $\| x_n \| \leq \sigma \| x_0 \|$. Now, since $(R_n e^{(n)}, e(n)) \geq \lambda_0 \| e^{(n)} \|^2$ for all $e^{(n)}$ in R_n and for each n and

$$\| x_n - y_n \|^2 = (R_n(a^{(n)} - b^{(n)}), a^{(n)} - b^{(n)}),$$

the inequality (5.11) follows from the discussion above and the inequality (5.14) with $p = \sigma c^{-1} \lambda_0^{-1/2}$ and $q = 2c^{-1}$. Hence the Galerkin method (5.6)–(5.7) is stable.   Q.E.D.

Remark 5.1 Since $T = I - C$ is A-proper when C is compact, Theorem 5.2 implies the stability of the Galerkin method

$$x_n - P_n C x_n = P_n f \qquad (x_n \in X_n, P_n f \in X_n), \tag{5.15}$$

where $Y_n = X_n$ and $Q_n = P_n$ for each $n \in Z^+$, obtained in [117].

Petrov–Galerkin method   We complete this section with the following problem concerning the stability of the Petrov–Galerkin method when applied to the solvability of the equation

$$x - C(x) = f \qquad (x \in H, f \in H), \tag{5.16}$$

where C is a compact linear mapping from a separable Hilbert space H into itself.

Let $\{\varphi_i\}$ and $\{\psi_i\}$ be two distinct bases in H, let $X_n = \{\varphi_1, \varphi_2, \ldots, \varphi_n\}$, $Y_n = \{\psi_1, \psi_2, \ldots, \psi_n\}$ for each $n \in Z^+$, and let $\Gamma_1 = \{X_n, P_n; Y_n, Q_n\}$ be a projectionally complete scheme for the pair (H, H). The Petrov–Galerkin method suggested by Petrov in [199] consists in constructing a solution $x \in H$ of equation (5.15) as a limit of solutions $x_n \in X_n$ of the equations

$$Q_n x_n - Q_n C x_n = Q_n f \qquad (x_n \in X_n, \, Q_n f \in Y_n), \tag{5.17}$$

which are different from equations (5.15) since $Q_n x_n \neq x_n$. In this abstract setting and under the assumption that Eq. (5.16) is uniquely solvable for each f in H, Polskii [241] and the author [215] established the unique approximation solvability of (5.16) with eespect to the scheme

$$\Gamma_1 = \{X_n, P_n; Y_n, Q_n\}.$$

Problem 5.1 Using Theorem 5.1 when $T = I - C$, show that the Petrov–Galerkin method (5.17) is stable in the sense of Definition 5.1.

III

Fixed-Point and Surjectivity Theorems for P_γ-Compact and A-Proper-Type Maps

In Chapter III we first present constructive fixed-point and surjectivity theorems for *generalized projectionally compact* (P_γ-*compact*) mappings whose study was initiated by the author in [202,206] and studied further by the writer (see [203,213,221]) and other authors (see [46,51,52,60,61, 64,70,125,201,236,307]). For historical reasons let me add that the first such theorems for P_0-compact mappings were obtained by the author in [202], where it was shown that in addition to classical and other fixed-point theorems, the theorem included the surjectivity results of Minty, Browder, and Shinbrot for maps $\lambda I - F$ with F monotone decreasing and either continuous, demicontinuous, or weakly continuous. This fact provided the main initial motivation for the study of P_γ-compact and later of A-proper mappings.

Since the results presented in this chapter, as well as the generalized degree theory for A-proper mappings that will be developed in Chapter IV, depend essentially on the Brouwer finite-dimensional topological degree theory, in Section 1 we state the basic properties of this degree theory and indicate those of its consequences which will be used in our study. The interested reader may find the justification of the definition and the proofs of the results stated in Section 1.1 in the earlier studies [19,112, 250] or in the more recent [62,155,314].

In Section 2 we first use the Galerkin method to establish a number of fixed-point theorems for mappings that are P_1-compact at 0. These theorems are then used to obtain, mostly in a constructive way, various fixed-point theorems for compact perturbation of ball-condensing and contractive mappings, for generalized contractions, and for strictly semicontractive mappings. Some constructive surjectivity theorems for P$_\gamma$-compact vector fields are treated in Section 3.

1. OUTLINE OF THE BROUWER DEGREE THEORY AND SOME OF ITS CONSEQUENCES

In Section 1.1 we define the Brouwer degree for a continuous mapping and state those of its properties that we will need, and in Section 1.2 we deduce some basic finite-dimensional fixed-point theorems.

1.1 Brouwer Degree and Some of Its Properties

Let D be a bounded open subset of the Euclidean space R^n. Then for each continuous mapping $T: \overline{D} \to R^n$ and each point $y \notin T(\partial D)$ there is defined an integer $\deg(T,D,y)$, called the *Brouwer degree of T on D over y*, which in principle is the algebraic count of the number of solutions $x \in D$ of the equation $Tx = y$ and which, roughly speaking, may be defined as follows. If ψ is any function in $C^1(\overline{D})$ such that $\| T(x) - \psi(x) \| < \text{dist}(y,T(\partial D))$, we define the degree of T by the equality

$$\deg(T,D,y) = \deg(\psi,D,y). \tag{1.1}$$

It can be shown that this definition does not depend on the smooth function ψ, which approximates T. Moreover, if y is not a critical value of ψ, then $\deg(\psi,D,y)$ is given by the simple formula

$$\deg(\psi,D,y) = \sum_{x \in \psi^{-1}(y)} J_\psi(x), \tag{1.2}$$

where $J_\psi(x)$ is the Jacobian determinant of ψ at x [i.e., $J_\psi(x) = \det \psi'(x)$]. However, if y is a critical point of ψ, one can define $\deg(T,D,y)$ as an integral involving an averaging kernel in the sense of Heinz [112] (see also [155] for details). The Brouwer degree thus defined has the following useful properties.

Theorem 1.1 (i) If $\deg(T,D,y) \neq 0$, then $T^{-1}(y) \neq \varnothing$.
　　(ii) If $T = I$, then $\deg(I,D,y) = 1$ if $y \in D$ and $\deg(I,D,y) = 0$ if $y \notin \overline{D}$.
　　(iii) If $H: [0,1] \times \overline{D} \to R^n$ is continuous and $y \notin H(t,\partial D)$ for $t \in [0,1]$, then $H(t,\cdot)$ is constant in $t \in [0,1]$.

(iv) If $D \supset \cup_{j=1}^{m} D_j$, $\overline{D} = \cup_{j=1}^{m} \overline{D}_j$, D_j open, $D_i \cap D_j = \varnothing$ for $j \neq i$ and $y \notin T(\partial D)$, then $\deg(T,D,y) = \sum_{j=1}^{m} \deg(T,D,y)$.

(v) If $T_1|_{\partial D} = T|_{\partial D}$ and $y \notin T(\partial D)$, then $\deg(T_1,D,y) = \deg(T,D,y)$.

(vi) $\deg(T,D,y) = \deg(T - y,D,0)$ and $\deg(T,D,y) = 0$ if $y \notin T(\overline{D})$.

(vii) If $D_1 \subset D$ is open and $y \notin T(D \backslash D_1)$, then $\deg(T,D,y) = \deg(T,D_1,y)$.

(viii) There exists a uniform neighborhood U of T such that if $T_1 \in U$, then $y \notin T_1(\partial D)$ and $\deg(T,D,y) = \deg(T_1,D,y)$.

(ix) $\deg(T,D,y)$ is constant on components of $R^n \backslash T(\partial D)$ and hence we may write $\deg(T,D,A)$ for $\deg(T,D,y)$ with $y \in A$ if A is a connected subset of $R^n \backslash T(\partial D)$.

In view of the definition of $\deg(T,D,y)$, the proofs of the foregoing properties follow immediately from (1.1) and the similar properties possessed by $\deg(\psi,D,y)$ when $\psi \in C^1(\overline{D})$. For a detailed proof of Theorem 1.1, see [112,155].

In what follows we shall need the following additional properties of $\deg(T,D,y)$, whose proofs, which can also be found in [112,155], are much more complicated than those listed in Theorem 1.1.

Theorem 1.2 (x) Let D_1 be an open bounded set containing $T(\overline{D})$, $\Delta = D_1 \backslash T(\partial D)$ and suppose that the components of Δ are Δ_j ($j = 1, 2, \ldots$). If $G: D_1 \to R^n$ is continuous and $y \notin G(T(\partial D)) \cup G(\partial D_1)$, then

$$\deg(G \cdot T,D,y) = \sum_j \deg(G,\Delta_j,y) \, \deg(T,D,\Delta_j). \qquad (1.3)$$

(xi) Suppose that D is also symmetric w.r.t. $0 \in D$ (i.e., $x \in D \Rightarrow -x \in D$), $T: D \to R^n$ is odd [i.e., $T(-x) = -T(x)$ for $x \in D$] and $0 \notin T(\partial D)$; then $\deg(T,D,0)$ is an odd integer.

(xii) If D is an open subset of R^n (not necessarily bounded) and $T: D \to R^n$ is continuous and injective, then $T(D)$ is open.

First note that since Δ can have at most countably many components Δ_j and since the set $T^{-1}(y)$ is compact and is covered by disjoint open sets Δ_j (and hence it meets only a finite number of the Δ_j's), it follows that the summation in (1.3) is finite. Property (x) is usually called the "multiplication" theorem, (xi) is referred to as an "odd mapping" or "Borsuk" theorem, and (xii) is usually called an "invariance of domain" theorem.

Theorem 1.3 If $D \subset R^n$ is open and bounded, $T: \overline{D} \to R^n$ is continuous and injective, and $y \in T(D)$, then $\deg(T,D,y) = \pm 1$.

Remark 1.1 When we come to define a topological degree for certain maps acting in infinite-dimensional spaces, we shall need to apply the theory outlined in this section in finite-dimensional spaces other than R^n.

Suppose that V is a real normed space of dimension n. Then V can be identified with R^n once a basis has been chosen. Thus the degree for mappings $T: \overline{D} \subset V \to V$ can be defined provided that we can show that the degree we have defined in R^n is independent of the basis chosen, which, as is well known, is the case. Thus if $(v_1, \ldots, v_n)$ is a basis in V, $(e_1, \ldots, e_n)$ a basis in R^n, and $h: V \to R^n$ a linear homeomorphism given by $h(v_j) = e_j$ for $j = 1, \ldots, n$, then for $y \notin T(\partial D)$ we define the degree of T on $D \subset V$ over $y \in V$ by the equality

$$\deg(T,D,y) = \deg(G,h(D),h(y)) \qquad (G = hTh^{-1}). \qquad (1.4)$$

It is easy to show that the definition above is independent of h (i.e., of the chosen basis). Hence we see that the degree theory developed above is valid when R^n is replaced by another normed space of dimension n.

If V_1 and V_2 are two normed spaces of the same finite dimension, we can reasonably expect to be able to apply the theory to continuous mappings $T: \overline{D} \subset V_1 \to V_2$. We obtain a uniquely defined degree by following the preceding definitions provided that V_1 and V_2 have specified orientations (i.e., are "oriented"). With this additional assumption the Brouwer degree is applicable to mappings between two oriented normed spaces of the same finite dimension. For further comments, see [155].

1.2 Brouwer and Other Finite-Dimensional Fixed-Point Theorems

To illustrate the power of the degree theory, we now deduce from Theorem 1.1 the well-known Brouwer and other fixed-point theorems that we shall need later.

Theorem 1.4 (Brouwer) If $K \subset V$ is compact and convex and $T: K \to K$ is continuous, then T has at least one fixed point in K.

Proof. Suppose first that $K = \overline{B}(0,r)$. Without loss of generality we may assume that $x \neq T(x)$ for $x \in \partial K$. Hence the homotopy $H: [0,1] \times K \to V$ defined by $H(t,x) = x - tTx$ is such that $0 \notin H(1,\partial K)$ and $\| H(t,x) \| \geq (1 - t)r > 0$ for $t \in [0,1)$ and $x \in \partial K$. Thus, by (iii) of Theorem 1.1, $\deg(1 - T,K,0) = \deg(I,K,0) = 1$ by (ii) and hence, by (i), there exists $x \in B(0,r)$ such that $Tx - x = 0$.

Now, if K is compact and convex, then we can choose $B(0,r)$ such that $B(0,r) \supset K$ and use Tietze's theorem (see [65,261]) to extend T to $\tilde{T}$ on

$\overline{B}(0,r)$ so that $\tilde{T}(\overline{B}(0,r)) \subset \text{co}(T(K)) \subset K \subset B(0,r)$. By the preceding result, there exists an $x \in \overline{B}(0,r)$ such that $\tilde{T}(x) = x$. Since $\tilde{T}(x) \in K$ we see that $x \in K$ and $Tx = x$.   Q.E.D.

An analytic proof of Theorem 1.4, which does not depend on the degree theory, can be found in [129]. A consequence of Theorem 1.4 is the following result, which we will find to be useful (see [202]).

Theorem 1.5 Let $T: \overline{B}(0,r) \to V$ be continuous and $\mu > 0$ any constant. Then there exists at least element x in $\overline{B}(0,r)$ such that

$$Tx - \mu x = 0 \tag{1.5}$$

provided that T satisfies either the condition $(\pi_\mu^\leq)$ or $(\pi_\mu^\geq)$, where:

$(\pi_\mu^\leq)$ If for some $x \in \partial B(0,r)$ the equation $Tx = \alpha x$ holds, then $\alpha \leq \mu$.
$(\pi_\mu^\geq)$ If for some $x \in \partial B(0,r)$ the equation $Tx = \alpha x$ holds, then $\alpha \geq \mu$.

Proof. Assume first that T satisfies $(\pi_\mu^\leq)$ and consider the map $Ax = Tx - \mu x + x$ and the retraction R of X onto $\overline{B}(0,r)$ defined by $Rx = x$ if $\| x \| \leq r$ and $Rx = rx/\| x \|$ if $\| x \| \geq r$. If for x in $\overline{B}(0,r)$ we define the map $A_1 x = RA(x)$, it follows that A_1 is a continuous map of $\overline{B}(0,r)$ into $\overline{B}(0,r)$. Hence, by Theorem 1.4, A_1 has a fixed point u in $\overline{B}(0,r)$. But then u is also a fixed point of A. Indeed, if $\| u \| < r$, it follows that $Au = u$, since in this case the assumption of the equality $Au = (\| Au \|/r)u$ would yield the contradiction of $\| u \| < r$. If $\| u \| = r$ and u is not a fixed point of A, then $\lambda \equiv \| Au \|/r > 1$ and, in view of our definition of A, we would get $Tu = (\lambda - 1 + \mu)u$ for $\| u \| = r$ and $\alpha \equiv \lambda - 1 + \mu > \mu$, which is excluded by condition $(\pi_\mu^\leq)$. Thus $Au = u$ and therefore u satisfies equation (1.5).

To prove Theorem 1.5 under condition $(\pi_\mu^\geq)$, define the operator B by $Bx = 2\mu x - Tx$ and note that B satisfies $(\pi_\mu^\leq)$ iff T satisfies $(\pi_\mu^\geq)$. Hence, by the first part of Theorem 1.5, there exists $u \in \overline{B}(0,r)$ such that $Bu = u$ [i.e., u satisfies (1.5)].   Q.E.D.

Remark 1.2 If T satisfies the conditions of Theorem 1.5 with $\mu = 1$ (or $\mu = 0$), then there exists $u \in \overline{B}(0,r)$ such that $Tu = u$ (or $Tu = 0$). In particular, condition $(\pi_\mu^\leq)$ is implied by the assumption that $T(\partial B(0,r)) \subset \overline{B}(0,r)$.

Remark 1.3 Note that $(\pi_\mu^\leq)$ holds iff (I): "$Tx \neq \eta x$ for all $x \in \partial B(0,r)$ and any $\eta > \mu$," while $(\pi_\mu^\geq)$ holds iff (II): "$Tx \neq \eta x$ for all $x \in \partial B(0,r)$ and any $\eta < \mu$." When $\mu = 1$, (I) is usually referred to as the *Leray–Schauder condition*.

Let us note that many other less general conditions satisfied by T on $\partial B(0,r)$, which were used to prove the existence of zeros or of fixed points of T, imply either $(\pi_\mu^\leq)$ or $(\pi_\mu^\geq)$, as the following shows.

Corollary 1.1 If $T: \overline{B}(0,r) \to V$ is such that either

$$\| x - Tx \|^2 \geq \| Tx \|^2 - \| x \|^2, \qquad x \in \partial B(0,r) \tag{1.6}$$

or

$$\| x - Tx \|^2 \leq \| Tx \|^2 - \| x \|^2, \qquad x \in \partial B(0,r), \tag{1.7}$$

then there exists a point $u \in \overline{B}(0,r)$ such that $Tu = u$.

Proof. Suppose that $Tx = \alpha x$ for some $x \in \partial B(0,r)$. Then (1.6) implies that $(1 - \alpha)^2 \geq \alpha^2 - 1$. Hence $\alpha \leq 1$ [i.e., T satisfies $(\pi_1^\leq)$]. On the other hand, (1.7) implies that $(1 - \alpha)^2 \leq \alpha^2 - 1$. Hence $\alpha \geq 1$ and T satisfies the condition $(\pi_1^\geq)$. Consequently, in both cases, the assertion of Corollary 1.1 follows from Theorem 1.5.

Remark 1.4 If T is of the form $Tx = Sx + x$, then (1.6) reduces to

$$\| Sx + x \|^2 \leq \| Sx \|^2 + \| x \|^2, \qquad x \in \partial B(0,r), \tag{1.8}$$

while (1.7) reduces to the condition

$$\| Sx + x \|^2 \geq \| Sx \|^2 + \| x \|^2, \qquad x \in \partial B(0,r). \tag{1.9}$$

In both cases Theorem 1.5 implies the existence of zeros of S.

We next state a fixed-point theorem for a general bounded open set $D \subset V$ with T satisfying the Leray–Schauder boundary condition, whose proof requires use of the degree theory (see [155,202]).

Theorem 1.6 Let D be a bounded open subset of V and $T: \overline{D} \to V$ continuous. If there exists $w \in D$ such that

(LS) $\quad Tx - w \neq \eta(x - w) \qquad$ for all $\quad x \in \partial D$ and all $\quad \eta > 1$

then T has at least one fixed point in $\overline{D}$.

Proof. Without loss of generality we may suppose that T has no fixed points on ∂D. Let $H(t,x) = x - tTx - (1 - t)w$ for $x \in \overline{D}$ and $t \in [0,1]$. Clearly, $H(1,x) \neq 0$ and $H(0,x) \neq 0$ for $x \in \partial D$. Now, if $H(t,x) = 0$ for some $t \in (0,1)$ and $x \in \partial D$, then $Tx - w = \eta(x - w)$ for $\eta = t^{-1} > 1$, in contradiction to (LS). Thus, by Theorem 1.1, $\deg(I - T,D,0) = \deg(I - w,D,0)$. Since $w \in D$, $\deg(I - w,D,0) = 1$ and so, by Theorem 1.1, there must be an $x \in D$ such that $x - Tx = 0$ (i.e., x is a fixed point of T). Q.E.D.

Remark 1.5 Note that when D is also convex, (LS) holds if $T(\partial D) \subset \overline{D}$.

2. CONSTRUCTIVE FIXED-POINT THEOREMS FOR P_γ-COMPACT MAPS

Let X be a real Banach space with a projectionally complete scheme $\Gamma_\alpha = \{X_n, P_n\}$, where $\alpha = \sup_n \| P_n \|$. In Section 2.1 we study the constructive existence of solutions (including fixed points) of mappings that are A-proper or P_1-compact at 0, while in Section 2.2 we establish a number of essentially constructive surjectivity theorems involving P_γ-compact mappings.

2.1 Zeros and Fixed Points of Pointwise P_γ-Compact Mappings

To state the first result in this section we recall (see Definition I.1.4) that $F: D \subset X \to X$ is *generalized projectionally compact* (P_γ-*compact*) if there exists a constant $\gamma \geq 0$ such that $pI - F$ is A-proper w.r.t. Γ_α for each p dominating γ (i.e., $p \geq \gamma$ if $\gamma > 0$ and $p > \gamma$ if $\gamma = 0$). As usual, F is P-compact iff F is P_0-compact. In what follows we write $\mu > \gamma$ whenever μ dominates γ. The following results were proved in [202,203,221].

Theorem 2.1 Suppose that $D = B(0,r)$, $F: \overline{D} \to X$ is P-compact and bounded and for some $\mu > \gamma$ the following is true:

$(\pi_\mu^\leq)$ If $Fx = \alpha x$ and $x \in \partial D$, then $\alpha \leq \mu$.

Then there exists $x_0 \in \overline{D}$ which satisfies the equation

$$Fx - \mu x = 0; \tag{2.1}$$

if $(\pi_\mu^<)$ holds on ∂D (i.e., if $Fx = \alpha x$ and $x \in \partial D \Rightarrow \alpha < \mu$), then (1.2) is feebly approximation solvable w.r.t. Γ_α, and strongly approximation solvable if x_0 is unique.

Proof. We may assume without loss of generality that (2.1) has no solutions on ∂D, since otherwise we are finished [i.e., we may assume that $(\pi_\mu^<)$ holds on ∂D]. Now the proof of Theorem 2.1 is based on Theorem 1.5 and the following:

Lemma 2.1 If F satisfies the conditions of Theorem 2.1 and $(\pi_\mu^<)$ holds on ∂D, then there is $n_0 \in Z_+$ such that if $n \geq n_0$ and $P_n F x = \beta x$ with $x \in \partial D_n$, then $\beta < \mu$ [i.e., $(\pi_\mu^<)$ holds on ∂D for $n \geq n_0$].

Proof. Now, if the assertion of Lemma 2.1 were not true for any $n_0 \in Z_+$, we could find sequences $\{n_j\} \subset Z_+$, $\{x_{n_j} \mid x_{n_j} \in D_{n_j}\}$, and $\{\beta_{n_j} \mid \beta_{n_j} \geq \mu\}$ with $n_j \to \infty$ as $j \to \infty$ such that

$$F_{n_j}(x_{n_j}) = \beta_{n_j} x_{j_j}. \tag{2.2}$$

To show that this assumption leads to a contradiction, observe first that $\{\beta_{n_j}\} \subset [\mu, a/r]$ since the boundedness of F implies that $\| F_n(x_n) \| = \| P_n F x_n \| \leq \alpha \| Fx \| \leq a$ and thus without loss of generality we may assume that $\beta_{n_j} \to \beta \in [\mu, a/r]$. Then (2.2) implies that $F_{n_j}(x_{n_j}) - \beta x_{n_j} = (\beta_{n_j} - \beta)(x_{n_j}) \to 0$ because $\{x_{n_j}\}$ is bounded. Since $\beta > \gamma$ and F is P_γ-compact, there exist a subsequence $\{x_{n_{j(k)}}\}$ and $x \in \bar{D}$ such that $x_{n_{j(k)}} \to x$ as $k \to \infty$ and $Fx - \beta x = 0$ with $x \in \partial D$, in contradiction to the assumption that $(\pi_\mu^<)$ holds on ∂D.

Proof of Theorem 2.1 continued. Since $F_n : \bar{D}_n \to X_n$ satisfies condition $(\pi_\mu^<)$ on ∂D_n for $n \geq n_0$, it follows from Theorem 1.5 that there exists $x_n \in \bar{D}_n$ for each $n \geq n_0$ such that $F_n(x_n) - \mu x_n = 0$. Again since $\mu > \gamma$, F is P_γ-compact and $F_n(x_n) - \mu x_n = 0 \to 0$ as $n \to \infty$, there exists a subsequence $\{x_{n_j}\}$ and $x_0 \in \bar{D}$ such that $x_{n_j} \to x_0$ as $j \to \infty$ and $F(x_0) - \mu x_0 = 0$ with $\| x_0 \| < r$ because of our additional assumption that $(\pi_\mu^<)$ holds on ∂D [i.e., (2.1) is feebly approximation solvable]. Clearly, as in Chapter I, one shows that if $x_0 \in D$ is the only solution of (2.1), then $x_n \to x_0$ in X [i.e., (2.1) is strongly approximation solvable w.r.t. Γ_α]. This means that in this case one can construct the solution of (2.1) by the Galerkin method.   Q.E.D.

Remark 2.1   A careful examination of the proof of Theorem 2.1 reveals two important facts:

(i) One may weaken the hypothesis that F is P_γ-compact to the requirement that F is P_γ-compact at 0 (i.e., $\lambda I - F : \bar{D} \to X$ is A-proper at 0 for each $\lambda > \gamma$).

To clarify this remark, let us introduce (see ([76,221])

Definition 2.1   Let $T : D \subset X \to X$ and $g \in Y$. Then T is said to be *A-proper at* g w.r.t. Γ_α if $T_n : D_n \to X_n$ is continuous and if $\{x_{n_j} \mid x_{n_j} \in D_{n_j}\}$ is any bounded sequence such that $P_{n_j} T(x_{n_j}) \to g$, then there exist a subsequence $\{x_{n_{j(k)}}\}$ and $x \in D$ such that $x_{n_{j(k)}} \to x$ and $Tx = g$.

It is easy to show that when $S: \overline{D} \to \overline{D}$ is contractive (i.e., k-contractive with $k < 1$), then $T = I - S$ is A-proper at 0 w.r.t. Γ_1, but to show that $T: \overline{D} \to X$ is A-proper some further conditions will be necessary.

(ii) The theorem remains valid if the boundedness assumption on F is replaced by the weaker condition:

(Δ) There exists $c > 0$ such that if $F_n(x) = \lambda x$ holds for $x \in \partial D_n$ and any $n \in Z_+$ with $\lambda > 0$, then $\lambda \leq c$.

Note that condition (Δ) is in no way a condition on the size of Fx or even on the size of $P_n Fx$. All it says is that when for some $x \in \partial D$ and any n the vector $P_n Fx$ is in the same direction as x, then $P_n Fx$ are uniformly bounded. Condition (Δ) certainly holds if $F(\partial D)$ is bounded or if $(Fx, Jx) \leq c \| x \|^2$ for all $x \in \partial D$ and some $c > 0$, where J is the normalized duality map.

In view of Remark 2.1, Theorem 1.2 yields the following practically useful and general fixed-point theorem, which was essentially proved by the author (see [203,221]).

Theorem 2.2   Suppose that $D = B(0,r)$ and $F: \overline{D} \to X$ is P_1-compact at 0 and satisfies conditions (Δ) and ($\pi_1^{\leq}$). Then F has a fixed point $x_0 \in \overline{D}$; if ($\pi_1^{<}$) holds on ∂D, then the equation

$$Fx - x = 0 \tag{2.3}$$

is feebly approximation solvable w.r.t. Γ_α and strongly approximation solvable if $x_0 \in D$ is unique fixed point of F.

Remark 2.2   The interesting feature of Theorem 2.2 is that its proof, which ultimately depends on the Brouwer fixed-point theorem, can be given without using any finite- or infinite-dimensional degree theory.

Remark 2.3   It is easy to show that the boundary condition ($\pi_1^{\leq}$) is implied by any one of the following:

(A1) $\| Fx - x \|^2 \geq \| Fx \|^2 - \| x \|^2$ for $x \in \partial B(0,r)$ (used in [6]).
(A2) $(Fx, Jx) \leq (x, Jx)$ for $x \in \partial B(0,r)$ (used in [136] when X is a Hilbert space and the normalized duality map is just $J = I$).
(A3) $Fx \in \overline{B}(0,r)$ for $x \in \partial B(0,r)$ (used in [250]).
(A4) X is a Hilbert space and there is a map $A: \overline{B}(0,r) \to X$ such that $(Ax,x) \leq \| x \|^2$ and $\| Fx - Ax \| \leq \| x - Ax \|$ for $x \in \partial B(0,r)$ (used in [225]).
   Consequently, since $F = S + C: \overline{D} \to X$ is P_1-compact and bounded

when S is ball-condensing and C compact (see Corollary I.2.1), the fixed-point theorems of Schauder [256], Rothe [250], Leray–Schauder [153], Altman [6], and Krasnoselskii [136] follow from Theorem 2.2 when $S \equiv 0$ and that of Sadovskii [253] when $C \equiv 0$. Moreover, the theorem of Kaniel [121] for quasicompact maps and of the author [202,203] for P-compact maps also follow from Theorem 2.2

Theorem 2.2 includes also the result of Krasnoselskii [136] concerning the convergence of the Galerkin method (see [139] for earlier contributions) when $S = 0$ and C is compact.

It is interesting to observe that if (A3) holds, then as the following shows, one can still weaken further the conditions on F imposed in Theorem 2.2.

Theorem 2.3 Suppose $F: D = \overline{B}(0,r) \subset X$ is such that $I - F$ is A-proper at 0 w.r.t. $\Gamma_1 = \{X_n, P_n\}$ and $F(\partial D) \subset \overline{D}$. Then (2.3) is feebly approximation solvable w.r.t. Γ_1. It is strongly approximation solvable if the fixed point is unique.

Proof. Since $\| P_n \| = 1$ for all $n \in Z_+$ and $Fx \in \overline{D}$ for $x \in \partial D$ (i.e., $\| Fx \| \le r$ if $\| x \| = r$), it follows that $F_n(x) \in \overline{D}_n$ for $x \in \partial D_n$ for each $n \in Z_+$. Hence, by Theorem 1.5 and Remark 1.2, there exists $x_n \in \overline{D}_n$ such that $x_n - F_n(x_n) = 0$. Since $I - F$ is A-proper at 0 and $x_n - F_n(x_n) = 0 \to 0$ as $n \to \infty$, there exist a subsequence $\{x_{n_j}\}$ and $x \in \overline{D}$ such that $x_{n_j} \to x \in \overline{D}$ and $x - Fx = 0$. If x is the only fixed point of F in $\overline{D}$, the entire sequence $\{x_n\}$ converges to x in X. Q.E.D.

Before we state the next theorem, we need the following known results, which we prove for the sake of completeness.

Lemma 2.2 Suppose that $\Gamma_\alpha = \{X_n, P_n\}$ is projectionally complete for (X,X). Then $P_n \to I$ uniformly on K iff K is precompact. Hence for $C: K \to X$, $P_n C$ converges on K uniformly to C iff $C(K)$ is precompact.

Proof. Suppose that K is precompact and assume that $b_n \equiv \sup_K \| P_n x - x \|$. Then there exist $\epsilon > 0$, a subsequence $\{P_j\}$ of $\{P_n\}$, and $\{x_j\} \subset K$ such that $\| P_j x_j - x_j \| \ge \epsilon$ for all $j \in Z_+$. We may assume without loss of generality that $x_j \to x_0$ for some $x_0 \in X$. This yields the contraction, since $\epsilon \le \| P_j x_j - x_j \| \le \alpha \| x_j - x_0 \| + \| P_j x_0 \| + \| x_0 - x_j \| \to 0$ as $j \to \infty$.

Suppose, conversely, that $b_n \to 0$ and the bounded set K is not precompact. Then there exist $\epsilon > 0$ and $\{x_n\} \subset K$ with $\| x_i - x_j \| \ge \epsilon$ for $i \ne j$. Choose n so that $b_n < \epsilon/4$. Then $\| P_n x_i - P_n x_j \| \ge \| x_i - x_j \| -$

$\epsilon/2 \geq \epsilon/2$ for $i \neq j$, in contradiction to the fact that $\{P_n x_j\}$, as a bounded sequence in X_n, has a convergent subsequence.   Q.E.D.

In view of Lemma 2.2, as an immediate corollary of Theorem 2.3, we get the following fixed-point theorem of Frum-Ketkov [89] with a correct proof for spaces with projectionally complete schemes $\Gamma_1 = \{X_n, P_n\}$ given by Nussbaum [191] (see also [221] for an account of papers devoted to Frum-Ketkov maps).

Theorem 2.4   Let $\Gamma_1 = \{X_n, P_n\}$ be projectionally complete for (X, X), $D = B(0, r)$, and let $F: \overline{D} \to X$ be a continuous mapping such that $F(\partial D) \subset \overline{D}$. Assume that there exists a compact set $K \subset X$ and a constant $k < 1$ such that $\mathrm{dist}(F(x), K) \leq k \, \mathrm{dist}(x, K)$ for $x \in \overline{D}$. Then the conclusion of Theorem 2.3 holds. Moreover, if Γ_1 is determined by a Schauder monotone basis, then the conclusion above holds if F is assumed to be only demicontinuous.

Proof.   In view of Theorem 2.3, to prove both assertions of Theorem 2.4, it suffices to show that $T = I - F: \overline{D} \to X$ is A-proper at 0 w.r.t. Γ_1. So let $\{x_{n_j} \mid x_{n_j} \in D_{n_j}\}$ be any sequence such that $g_{n_j} \equiv x_{n_j} - P_{n_j} F(x_{n_j}) \to 0$ as $j \to \infty$. Since K is compact, our condition on F implies the existence of $y_{n_j} \in K$ for each $j \in Z_+$ such that $\| F(x_{n_j}) - y_{n_j} \| < k\| x_{n_j} - y_{n_j} \|$. Hence the following inequalities hold for each $j \in Z_+$:

$$\| x_{n_j} - y_{n_j} \| \leq \| x_{n_j} - P_{n_j} F x_{n_j} \| + \| P_{n_j} F x_{n_j} - P_{n_j} y_{n_j} \|$$
$$+ \| P_{n_j} y_{n_j} - y_{n_j} \|$$
$$\leq \| g_{n_j} \| + \| F x_{n_j} - y_{n_j} \| + a_{n_j} \leq \| g_{n_j} \|$$
$$+ k\| x_{n_j} - y_{n_j} \| + a_{n_j}.$$

This implies that

$$(1 - k)\| x_{n_j} - y_{n_j} \| \leq \| g_{n_j} \| + a_{n_j} \to 0$$

since $\| g_{n_j} \| \to 0$ by hypothesis and $a_{n_j} = \| P_{n_j} y_{n_j} - y_{n_j} \| \to 0$ by Lemma 2.2. Now, since K is compact and $\{y_{n_j}\} \subset K$, there exist a subsequence, which we again denote by $\{y_{n_j}\}$ and $x \in K$ such that $y_{n_j} \to x$ and therefore it follows from the last inequality that $x_{n_j} \to x$ in X.

Suppose first that F is continuous. Then $F x_{n_j} \to Fx$ and $P_{n_j} F x_{n_j} \to Fx$ and, therefore, $0 = \lim_j g_{n_j} = \lim_j (x_{n_j} - P_{n_j} F x_{n_j}) = x - Fx$, so in this case, Theorem 2.4 is proved.

Suppose now that F is only demicontinuous and Γ_1 is determined by a monotone Schauder basis, say, $\{\phi_j\} \subset X$; that is, $X_n = \mathrm{span}\{\phi_1, \dots, \phi_n\}$ for each $n \in Z_+$, and for each $x \in X$, $P_n x = \sum_{j=1}^{r} (\Phi_j, x)\phi_j$, where $\{\Phi_j\} \subset X^*$ is such that $(\Phi_j, \phi_i) = \delta_{ij}$, $1 \le i, j \le n$, for each $n \in Z_+$. Then $X'_n = \text{span}\{\Phi_1, \ldots, \Phi_n\}$ for each $n \in Z_+$ is such that $P^*_n: X^* \to X'_n$, $X'_n \subset X'_{n+1}$ for each n, and $\cup_n X'_n$ is weak* dense in X^*. Since F is demicontinuous and $x_{n_j} \to x$, it follows that $Fx_{n_j} \rightharpoonup Fx$ in X and $P_{n_j}Fx_{n_j} = x_{n_j} - g_{n_j} \to x$ in X. Fix $k \in Z_+$, let $z \in X'_k$, and note that $(z, P_{n_j}Fx_{n_j}) = (z, Fx_{n_j})$ for $(z, P_{n_j}Fx_{n_j}) = (z, Fx_{n_j})$ for $n_j > k$. Hence $\lim_j(z, P_{n_j}Fx_{n_j}) = \lim_j(z, Fx_{n_j})$; that is, $(z, x) = (z, Fx)$ or $(z, Fx - x) = 0$ for any $z \in X'_k$ and any $k \in Z_+$. This implies that $x - Fx = 0$ (i.e., $I - F$ is A-proper at 0 w.r.t. Γ_1). Q.E.D.

To the best of my knowledge the second assertion of Theorem 2.4 appears to be new.

Using the Brouwer degree theory, Petryshyn and Tucker [236] obtained the following generalization of Theorem 2.2 (see also [221]).

Theorem 2.5 Let D be a bounded open subset of X with $0 \in D$. Suppose that $F: \overline{D} \to X$ is P_1-compact at 0 and satisfies conditions (Δ) and ($\pi_1^\le$) on ∂D. Then the conclusions of Theorem 2.2 hold for equation (2.3).

Proof. First we claim that there is $n_0 \in Z_+$ such $x - tP_nFx \ne 0$ for $t \in [0,1]$ and $x \in \partial D_n$ for each $n \ge n_0$. If this were not the case, there would exist sequences $\{n_j\} \subset Z_+$, $\{x_{n_j} \mid x_{n_j} \in \partial D_{n_j}\}$ and $\{t_j\} \subset [0,1]$ such that $x_{n_j} - t_jP_{n_j}Fx_{n_j} = 0$ for each $j \in Z_+$. Since $0 \in D$ and $\{x_{n_j}\} \subset \partial D$, there exist $\delta_1 > 0$ and $\delta_2 > \delta_1$ such that $\delta_1 \le \|x_{n_j}\| \le \delta_2$. Since without loss of generality we may assume that $t_j > 0$, and $t_j \to t_0 \in [0,1]$, we see that $P_{n_j}Fx_{n_j} = t_j^{-1}x_{n_j}$. Hence, by condition ($\Delta$), $t_j^{-1} \le c$ and $\|P_{n_j}Fx_{n_j}\| \le c\delta_2$. Hence

$$x_{n_j} - t_0P_{n_j}Fx_{n_j} = x_{n_j} - t_jP_{n_j}Fx_{n_j} + (t_j - t_0)P_{n_j}Fx_{n_j} \to 0$$

as $j \to \infty$. Since $c \ge t_0^{-1} > 1$ and F is P_1-compact at 0, there exists a subsequence $\{x_{n_{j(k)}}\}$ and $x_0 \in \partial D$ such that $x_{n_{j(k)}} \to x$ and $x_0 - t_0Fx_0 = 0$ or $Fx_0 - t_0^{-1}x_0 = 0$ with $t_0^{-1} > 1$, in contradiction to condition ($\pi_1^\le$). Hence $F_n(t,x) \equiv x - tP_nFx \ne 0$ for all $x \in \partial D_n$, $t \in [0,1]$ and each $n \ge n_0$. Therefore, by the homotopy part (iii) of Theorem 1.1, $\deg(I - F_n, D_n, 0) = \deg(I, D_n, 0) = 1$ for each $n \ge n_0$. Hence for each such n there exists $x_n \in \overline{D}_n$ such that $x_n - F_nx_n = 0$ and therefore since F is P_1-compact at 0, there exist a subsequence $\{x_{n_j}\}$ and $x \in \overline{D}$ such that $x_{n_j} \to x_0$ in X and $x_0 - Fx_0 = 0$. If $x_0 \in D$ is the only fixed point, then, as before, one shows, again using the P_1-compactness of F at 0, that $x_n \to x_0$ in X. Q.E.D.

Remark 2.4 The condition $(\pi_1^{\leq})$ holds if any one of the four conditions listed in Remark 2.3 holds on ∂D provided that in (A3) we assume additionally that D is convex. Since $F = S + C : \overline{D} \to X$ is P_1-compact and bounded when C is compact and S is ball condensing or k-ball contractive with $k < 1$, Theorem 2.5 is valid for these classes of mappings provided that $(\pi_1^{\leq})$ holds on ∂D and thus Theorem 2.5 extends the existence results of [185,249]. Moreover, we get the convergence of the Galerkin method, when applied to (2.3), provided that $(\pi_1^{<})$ holds on ∂D and F has at most one fixed point in D. Let us add that the remarks above also apply to the case when $F : \overline{D} \to X$ is k-set contractive with $k < \frac{1}{2}$. However, as we shall see in Chapter IV, using the Sadovskii–Nussbaum degree theory, one can get the existence of fixed points of k-set contractions F when $k < 1$, but it is unknown if one can use the Galerkin method to construct these fixed points.

Using the Schauder fixed-point theorem, it was shown by Schaefer [255] that if $C : X \to X$ is compact, then either C has a fixed point or the set $\{x \mid x - \lambda Cx = 0, 0 < \lambda < 1\}$ is unbounded. Martelli and Vignoli [159] extended this theorem to condensing mappings, while the author (see [221]) extended it further to 1-set-contractive mappings satisfying condition (C) (see Theorem 1.6 in Chapter IV).

Using Theorem 2.2, we now show that Schaefer's theorem admits extension to P_1-compact mappings.

Theorem 2.6 Let $F : X \to X$ be bounded, continuous, and P_1-compact at 0. Then either F has a fixed point or the set $\{x \mid x - \lambda Fx = 0, 0 < \lambda < 1\}$ is unbounded.

Proof. To prove Theorem 2.6, we need the following.

Lemma 2.3 Let $F : X \to X$ be bounded, continuous, and P_1-compact. If R is a continuous radial retraction of X onto $\overline{B}(0,r)$ given by $Rx = x$ if $\| x \| \leq r$ and $Rx = rx/\| x \|$ if $\| x \| \geq r$, then $RF : X \to X$ is also bounded, continuous, and P_1-compact. Further, if F is P_1-compact only at 0, then RF is also P_1-compact at 0.

Proof. Since RF is obviously bounded and continuous, it suffices to show that RF is P_1-compact. So let $\{x_{n_j} \mid x_{n_j} \in X_{n_j}\}$ be any bounded sequence such that $\lambda x_{n_j} - P_{n_j} RF(x_{n_j}) \to g$ for each $\lambda \geq 1$ and some g in X. Define the function $\phi(t)$ by $\phi(t) = 1$ if $0 \leq t \leq r$ and $\phi(t) = r/t$ if $t \geq r$. Since $\{Fx_{n_j}\}$ is bounded, there exists a subsequence which we again denote by $\{Fx_{n_j}\}$ such that $\| Fx_{n_j} \| \to \mu \geq 0$ as $j \to \infty$, where the number μ satisfies one of the following three cases:

Case 1. If $\mu < r$, then for $\epsilon = (r - \mu)/2$ there exists $n_1 \in Z_+$ such that $\mu - \epsilon < \| Fx_{n_j} \| \leq \epsilon + \mu < r$ for $n_j \geq n_1$ for all $j \in Z_+$. Hence $\lambda x_{n_j} - P_{n_j}RF(x_{n_j}) = \lambda x_{n_j} - P_nFx_{n_j} \to g$, and since F is P_1-compact, there exist a subsequence $\{x_{n_{j(k)}}\}$ and $x \in X$ such that $x_{n_{j(k)}} \to x$ as $k \to \infty$ and $\lambda x - Fx = g$. Since $\| Fx \| \leq r$, it follows that $\lambda x - RFx = g$.

Case 2. If $\mu = r$, then $\phi(\| Fx_{n_j} \|) \to 1$ as $j \to \infty$, and hence there exists a sequence $\{\epsilon_j\}$ with $\epsilon_j \to 0$ such that $\phi(\| Fx_{n_j} \|) = 1 + \epsilon_j$ for each $j \in Z_+$. It follows from this that $\lambda x_{n_j} - P_{n_j}Fx_{n_j} = \lambda x_{n_j} - P_{n_j}RF(x_{n_j}) + \epsilon_j P_{n_j}F(x_{n_j}) \to g$ as $j \to \infty$. Hence again there exists a subsequence $\{x_{n_{j(k)}}\}$ and $x \in X$ such that $x_{n_{j(k)}} \to x$ and $\lambda x - Fx = g = \lambda x - RFx$ since $\| Fx_{n_{j(k)}} \| \to \| Fx \| = \mu = r$ and $\phi(\| Fx_{n_{j(k)}} \|) \to 1$.

Case 3. If $\mu > r$, then $\phi(\| Fx_{n_j} \|) \to r/\mu = \beta < 1$ and therefore there exists a sequence $\{\epsilon_j\}$ with $\epsilon_j \to 0$ such that $\phi(\| Fx_{n_j} \|) = \beta + \epsilon_j$ for each $j \in Z_+$. This implies that $\lambda x_{n_j} - \beta P_{n_j}F(x_{n_j}) = \lambda x_{n_j} - P_{n_j}RF(x_{n_j}) + \epsilon_j P_{n_j}F(x_{n_j}) \to g$ as $j \to \infty$ [i.e., $\lambda \beta^{-1}x_{n_j} - P_{n_j}F(x_{n_j}) \to \beta^{-1}g$]. Hence since $\lambda \beta^{-1} > \lambda \geq 1$ and F is P_1-compact, there exist $\{x_{n_{j(k)}}\}$ and x in X such that $x_{n_{j(k)}} \to x$ and $\lambda \beta^{-1}x - Fx = \beta^{-1}g$. But then

$$\| Fx_{n_{j(k)}} \| \to \| Fx \| = \mu \quad \text{and} \quad \phi(\| Fx_{n_{j(k)}} \|) \to \phi(\| Fx \| = \frac{r}{\| Fx \|} = \beta.$$

This implies that $g = \beta(\lambda \beta^{-1}x - Fx) = \lambda x - \beta Fx = \lambda x - RFx$.

Going over the proof, we see that when F is P_1-compact at 0, then RT is also P_1-compact at 0, so that the second assertion is also true.

Q.E.D.

Proof of Theorem 2.6 continued. Suppose that the equation $x - Fx = 0$ has no solution in X, and for each fixed $n \in Z_+$, let R_n be the radial retraction of X onto $B_n = \overline{B}(0,n)$. Then $F_n = R_nF$ maps B_n into B_n and, by Lemma 2.3, F_n is P_1-compact at 0. Hence, by Theorem 2.2, there exists $x_n \in B_n$ such that $F_n(x_n) = x_n$ for each $n \in Z_+$, whence $Fx_n \in X \backslash B_n$ for all $n \in Z_+$ because if $Fx_n \in B_n$ for any $n = n_0$, then x_{n_0} would satisfy the equation $x_{n_0} = Fx_{n_0}$, which contradicts our assumption. Hence $F_nx_n = (n/\| Fx_n \|) Fx_n = x_n$ and $\| x_n \| = n$. Also, $\| Fx_n \| > n$ and $x_n - \lambda_n Fx_n = 0$ with $\lambda_n = n/\| Fx_n \| < 1$ for each $n \geq 1$. This proves Theorem 2.6.

Q.E.D.

Remark 2.5 Since the proof of Theorem 2.6 depends on Theorem 2.2, we see that Remark 2.2 applies (i.e., no finite- or infinite-degree theory is necessary). We shall see in Chapter IV that the proof of Theorem 2.6 is very simple and short when the generalized degree theory is used. The practical usefulness of Theorem 2.6 lies in the following. If we want to solve some differential or integral equation $x - Fx = y$ for any given y, we first rewrite this equation as an equivalent operator equation in a suitable Banach space X so that the map $x \to F_y(x) \equiv Fx - y$ is P_1-compact at 0 (or, equivalently, F is P_1-compact) and then show that if $x - \lambda F_y(x) = 0$ holds for $x \in X$ and some $\lambda \in (0,1)$ then $\| x \| \leq K_y$ for some constant $K_y > 0$ independent of λ. In this case, Theorem 2.6 guarantees the existence of a fixed point of F_y. The difficult part in this approach is to establish the a priori bound K_y on the solutions x of $x - \lambda F_y(x) = 0$, $0 < \lambda < 1$.

To obtain an important special case of Theorems 2.5 and 2.6, we first recall that $A: X \to X$ is called *accretive* if $(Ax - Ay, w) \geq 0$ for all $x, y \in X$ and some $w \in J(x - y)$, where J is the normalized duality mapping. The study of such mappings was initiated by [38,128] and was further continued by many authors. For relevant references, see Section 1 of Chapter IV, where we obtain some basic results for these mappings which are needed in our study. Among other results, it is shown that if A is a continuous accretive map, then $A + \lambda I$ is A-proper w.r.t. Γ_1 and bijective for each $\lambda > 0$. Moreover, $\| A_n x - A_n y \| \geq \lambda \| x - y \|$ for all $x, y \in X_n$ and $n \in Z_+$. Combining this with Theorem I.2.2, we see that if Γ_1 is nested, $S: X \to X$ is ball condensing and $C: X \to X$ compact, then $\lambda I + A - S - C$ is A-proper w.r.t. Γ_1 for each $\lambda \geq 1$ (i.e., $F \equiv S + C - A$ is P_1-compact).

In view of the discussion above, Theorems 2.5 and 2.6 apply to these mappings. We state explicitly only the following consequence of Theorem 2.6, which contains a number of special cases (see [159,221]).

Proposition 2.1 Let $\Gamma_1 = \{X_n, P_n\}$ be a nested scheme for (X, X), $y \in X$, $A: X \to X$ bounded, continuous, and accretive, $S: X \to X$ ball condensing, and $C: X \to X$ compact. Then $F \equiv S + C - A$ is P_1-compact and thus either F_y has a fixed point or the set $\{x \mid x - \lambda F(x) = 0, 0 < \lambda < 1\}$ is unbounded.

2.2 Fixed Points of Compact Perturbations of Maps of Lipschitz Type

In [137] it was shown by Krasnoselskii that if D is a closed, bounded convex subset of X, S contractive (i.e., k-contractive with $k < 1$) and C compact on D, then $T = S + C$ has a fixed point in D provided that the following rather restrictive condition holds:

$$Sx + Cy \in D \quad \text{for all} \quad x \text{ and } y \text{ in } D. \tag{2.4}$$

The recent contributions by many authors (see [297] for latest references) have extended the initial existence result of [137] in various directions (see [185,230,249] and to more general mappings, including those of semicontractive type introduced by Browder [31] (and Kirk [132] in a somewhat different manner). This aspect of the fixed-point theory will be studied in Chapter IV.

In this section we examine some of these classes from the viewpoint of A-properness and P$_\gamma$-compactness and in that context outline some of the results obtained in [203,213,221]; others are discussed in Chapter IV.

Let us first note that in view of Proposition I.1.2, Theorem 2.5 implies the following essentially constructive fixed-point theorem obtained in [236].

Corollary 2.1 Let $D \subset X$ be bounded and open with $0 \in D$, $C: \overline{D} \to X$ compact and $S: X \to X$ contractive. If $F = S + C$ satisfies condition $(\pi_1^\leqq)$, F has a fixed point $x_0 \in \overline{D}$; if $(\pi_1^\lessgtr)$ holds on ∂D, the equation

$$Sx + Cx - x = 0 \qquad (x \in D) \tag{2.5}$$

is feebly approximation solvable w.r.t. Γ_1 and strongly approximation solvable if x_0 is unique.

For maps F of the form $F = S + C$, with S contractive on all of X, Corollary 2.1 appears to be the most general result from which one can obtain the constructive existence of fixed points if one knows that F has at most one fixed point in D. However, if S is not defined on all of X, then even when $D = B(0,1)$ it has been known since the 1970s that $I - S: \overline{D} \to X$ is A-proper when the π_1-space X satisfies very restrictive conditions, such as that X has a weakly continuous duality mapping.

It was shown in [236] and in [191] that when $S: \overline{B}(0,r) \to X$ is contractive with $k < \frac{1}{2}$, then $I - S: \overline{B} \to X$ is A-proper. This follows from the fact that the radial retraction R of X onto $\overline{B}(0,r)$ is such that $\| Rx - Ry \| \leqq 2\| x - y \|$ for all $x,y \in X$ and thus S can be extended to a contraction on all of X. Consequently, Corollary 2.1 remains valid in this case under the assumption that S and C are defined only on $\overline{B}(0,r)$.

If $k \in [\frac{1}{2},1)$, then the A-properness of $I - S: \overline{B}(0,r) \to X$ (see [203]) takes place if X is reflexive and either X has a single-valued weakly continuous duality mapping or Γ_1 determined by a monotone Schauder basis and S is also weakly continuous.

In view of the discussion above, the following analog of Corollary 2.1 is valid (see [225] for somewhat more general results).

Corollary 2.2 Let X be reflexive, $S: \overline{B}(0,r) \to X$ contractive, and $C: \overline{B}(0,r) \to X$ compact. If X has either a weakly continuous duality map or X has a monotone Schauder basis and S is also weakly continuous and if $(\pi_1^\leq)$ holds on $\partial B(0,r)$, the conclusions of Corollary 2.1 hold.

We leave the proof of Corollary 2.2 to the interested reader.

It is shown in Chapter IV that if in Corollary 2.2 we replace the contractiveness assumption on $S: B \to X$ by the condition that S is nonexpansive on $\bar{B}(0,1)$ [i.e., $\| Sx - Sy \| \leq \| x - y \|$ for $x,y \in \bar{B}(0,1)$], then even when $X = l_2$ and $F = S + C$ maps $\bar{B}(0,1)$ into $\bar{B}(0,1)$, the map F may fail to have a fixed point in $\bar{B}(0,1)$. However, if one assumes that $C: \bar{B}(0,1) \to X$ is completely continuous [i.e., if $\{x_j\} \subset B(0,1)$ and $x_j \to x \in \bar{B}(0,1) \Rightarrow Cx_j \to Cx$] and X is uniformly convex, then the existence of fixed points for $F = S + C: D \subset X \to X$ has been established depending on the nature of D and the conditions satisfied by F on ∂D (see [297] for the account of various results in this area and the exhaustive references).

Although, under certain conditions on X, the theory of P_1-compact and A-proper mappings has been used (see [204,213] and Chapter IV) to obtain the existence of fixed points of mappings $F: \bar{B}(0,r) \to X$ of the form $F = S + C$, where S is nonexpansive and C completely continuous on $\bar{B}$ or even for F semicontractive in the sense of Browder [31], the constructive aspect of the theory is lost; that is, even when we know that F has a unique fixed point in B, we cannot obtain it as a strong limit of solutions $x_n \in B_n$ of approximate equations $F_n(x_n) = x_n$ for $n \in Z_+$. In view of this, we will not dwell in this section on the existence results for such mappings that are obtainable by means of the theory of A-proper mappings. An interested reader should consult the original papers. Also see Chapter IV.

In [11] Belluce and Kirk introduced the notion of a generalized contraction that lies between contractive and nonexpansive mappings and is defined to be a map $S: D \subseteq X \to X$ such that to each $x \in D$ there exists $\alpha(x) \in (0,1)$ such that $\| Sx - Sy \| \leq \alpha(x)\| x - y \|$ for all $y \in D$.

In [133] Kirk showed that the generalized contractions occur naturally among mappings that are continuously Fréchet differentiable. In fact, it was shown in [133] that if D is a bounded, open, convex set in X and $S: D \to X$ has a Fréchet derivative $S_x' \in L(X,X)$ at each x in D and the map $x \to S_x'$ is continuous, then S is a generalized contraction on D iff $\| S_x' \| < 1$ for each $x \in D$. Extending a result of [205], it was shown by Wong [307], using an argument of Kirk [132], that if $D = X$ and $S: X \to X$ is a generalized contraction, then S is P_1-compact w.r.t. $\Gamma_1 = \{X_n, P_n\}$.

In [76], Fitzpatrick established a similar result for generalized contractions defined only on closed convex subsets D of X. However, he was only able to prove the pointwise A-properness of $I - S$. As we saw in

Theorem 2.2, this suffices to establish the constructive existence of fixed points of S.

Using an argument similar to [132], the following was proved in [76].

Lemma 2.4 Let X be a reflexive π_1-space, D a closed convex set in X, and $S: D \to X$ a generalized contraction. Let $g \in X$ be such that $S(x) + g \in D$ for $x \in D$. Then $T = I - S: D \to X$ is A-proper at g.

We omit the proof of Lemma 2.4 since it will be deduced from a more general result of Webb [297] proved below.

Theorem 2.5 and Lemma 2.4 yield the following constructive fixed-point theorem proved in [76] (and earlier in [132] in a nonconstructive way).

Corollary 2.3 Let X be as in Lemma 2.4, $D \subset X$ open, convex, and bounded with $0 \in D$, $S: \overline{D} \to \overline{D}$ a generalized contraction. Then S has a unique fixed point x_0 in $\overline{D}$, and if $x_0 \notin \partial D$, then the equation $x - Sx = 0$ is uniquely and strongly a-solvable.

To prove the results of Webb [297], we need some prelimimaries. Let D be a closed convex set in X. Then the *inward set* on D at x in D is defined by

$$I_D(x) = \{z \in X : z = (1 - a)x + ay,$$
$$\text{for some } y \text{ in } D \text{ and } a \geq 1\}.$$

Note that if x is an interior point of D, then $I_D(x) = X$. Geometrically, $I_D(x)$ is the union of all rays originating at x and passing through some point y of D. Clearly, $D \subset I_D(x)$ for each $x \in D$. A map $S: D \to X$ is called *inward* if $S(x) \in I_D(x)$ for each x in D and *weakly inward* if $S(x) \in \overline{I_D(x)}$. These requirements are boundary conditions. Such maps have been studied by a number of authors, including Halpern and Bergman, Caristi, Deimling, Fitzpatrick and Petryshyn, Reich, Webb, and others (see [297]).

Remark 2.6 Before we state and prove the needed result, let us first note (see [297]) that: If X is a π_1-space, D bounded, $S: D(S) \to X$ weakly inward on D, $I - S$ A-proper at 0, and if $P_n(D) \subset D(S)$ for all large n, then the equation $Sx = x$ is feebly a-solvable.

We now prove the following extension of Lemma 2.4 contained in [297].

Lemma 2.5 Let X and D be as in Lemma 2.4. Let $S: D \to x$ be a weakly inward generalized contraction. Then $I - S$ is A-proper at 0.

Proof. Let $\{x_n \mid x_n \in D_n\}$ be a bounded sequence with $x_n - P_n S x_n \to 0$ as $n \to \infty$. Let $\varphi(z) = \lim \sup_{n \to \infty} \{\|x_n - z\|\}$; then φ is a continuous, convex function and $\varphi(z) \to \infty$ as $\|z\| \to \infty$, so φ attains its infimum over D. Let $K = \{v \in D : \varphi(v) = \inf \varphi(z), z \in D\}$; K is called the asymptotic center of $\{x_n\}$ relative to D and is a closed, bounded, convex set. For $v \in K$ we have

$$\| x_n - S(v) \| \leq \| x_n - P_n S(x_n) \| + \| P_n S(x_n)$$
$$- P_n S(v) \| + \| P_n S(v) - S(v) \|.$$

As $\| P_n \| = 1$ and S is a generalized contraction, this yields $\varphi(S(v)) \leq \alpha(v)\varphi(v)$. We claim that $\varphi(v) = \inf\{\varphi(z) : z \in \overline{I_D(v)}\}$. Indeed, since φ is continuous, if this were false, there would exist $w \in I_D(x)$ such that $\varphi(w) < \varphi(v)$. Thus $w = (1 - a)v + ay$ for some $y \in D$ and $a > 1$. By convexity of φ, this yields $\varphi(y) \leq (1/a)\varphi(w) + [1 - (1/a)]\varphi(v) < \varphi(v)$, a contradiction. As $S(v) \in \overline{I_D(x)}$ and $\alpha(v) < 1$, we now see that $\varphi(v) = 0$. Thus $x_n \to v = \varphi(v)$.   Q.E.D.

Remark 2.7 The argument used to prove Lemma 2.5 yields a simple proof of the following result:

If D is a closed convex set in a reflexive π_1-space X, $S: D \to X$ is a weakly inward generalized contraction, and if, for a bounded sequence $\{x_n\} \in D$, $x_n - S(x_n) \to 0$, then $x_n \to x$, the unique fixed point of S.

This remark generalizes some results of [132] and of [230]. Now let $S: D \to X$ be a generalized contraction and let $C: D \to X$ be compact. Following [297], we use Lemma 2.5 to obtain a result on the a-solvability of the equation $S(x) + C(x) = x$ stated as:

Theorem 2.7 Let D be a closed, convex set in a reflexive π_1-space X. Let $S: D \to X$ be a generalized contradiction, $C: D \to X$ compact, and set $S_w(x) = S(x) + w$ for x in D. Suppose that

S_w is weakly inward on D for each $w \in \overline{C(D)}$.   (*)

Then $I - S - C$ is A-proper at 0. Moreover, if D is also bounded and $P_n(D) \subset D$, the equation $S(x) + C(x) = x$ is feebly a-solvable, and strongly a-solvable if $I - S - C$ is injective.

Proof. Let $\{x_n \mid x_n \in D_n\}$ be a bounded sequence with $x_n - P_n S(x_n) - P_n C(x_n) \to 0$ as $n \to \infty$. By compactness of C, for a subsequence $\{x_j\}$, we have $C(x_j) \to w$ (say) and then $x_j - P_j S(x_j) \to w$ as $j \to \infty$. Since $I - S_w$ is A-proper at 0 by Lemma 2.5, it follows that $x_j \to x$ and $x - S(x) = w$. By continuity, $w = C(x)$, so $x - S(x) - C(x) = 0$ is feebly a-solvable, and strongly a-solvable if $I - S - C$ is injective.   Q.E.D.

Remark 2.8   Note that Krasnoselkii's condition (2.4) is much more stringent than (*) since (*) is certainly satisfied if $S + Cy \in I_D(x)$ for x,y in D. Moreover, if D has a nonempty interior, this is only a restriction for $x \in \partial D$. These comments show that Theorem 2.7 extends Theorems 2.4 and 2.5 of [230].

2.3 Fixed Points of Mappings of Strictly Semicontractive Type

Instead of considering the sum $S + C$ of the maps above, we may consider mappings formed by intertwining mappings of the foregoing type. For example, it was shown in Lemma I.2.3 that if $F: D \subset X \to X$ is a continuous *k-semicontractive mapping* [i.e., $Fx = V(x,x)$ for $x \in D$, where $V: X \times X \to X$ is a map such that for each fixed $x \in X$, $V(\cdot,x): X \to X$ is k-contractive and $V(x,\cdot): X \to X$ compact], then F is k-ball contractive and so, by Corollary I.2.1, F is P_1-compact w.r.t. Γ_1 if $k < 1$. Hence Theorem 2.5 implies the extension of Corollary 2.1, which includes the corresponding fixed-point results in [219,291].

Corollary 2.4   Let $D \subset X$ be a bounded open set with $0 \in D$. If $F: \overline{D} \to X$ is k-semicontractive with $k < 1$, which satisfies condition $(\pi_1^{\leqq})$ on ∂D, the conclusions of Theorem 2.2 hold for the equation $Fx - x = 0$.

To obtain a similar extension of Corollary 2.3, we say, following Kirk [132]:

Definition 2.2   A continuous map $F: \overline{D} \subset X \to X$ is called *strongly semicontractive relative to X* if there exists a map $U: X \times \overline{D} \to X$ such that $Fx = V(x,x)$ for $x \in \overline{D}$ and

   (i) For each $y \in \overline{D}$, $V(\cdot,y): X \to X$ is a generalized contraction.
   (ii) If $\{x_j\} \subset X$ is bounded and $\{y_j\} \subset \overline{D}$ such that $y_j \to y_0$, then $V(x_j,y_j) - V(x_j,x_0) \to 0$ in X.

The following constructive version of the fixed-point theorem in [132] has been obtained in [176] (see also [307]).

Theorem 2.8 Let X be a reflexive space with a nested scheme Γ_1, $D \subset X$ open, bounded, and convex with $0 \in D$, and $F: \overline{D} \to X$ strongly semi-contractive relative to X and satisfying $(\pi_1^{\leqq})$ on ∂D. Then the conclusions of Theorem 2.2 hold for the equation $Fx - x = 0$.

Proof. First note that the continuous map $F: \overline{D} \to X$ is also bounded. Indeed, suppose that for some $\{x_j\} \subset \overline{D}$, $\| Fx_j \| \to \infty$. We may assume that $x_j \rightharpoonup x_0$ for some $x_0 \in \overline{D}$. Then, by (ii), $\| V(x_j,x_0) - Fx_j \| \to 0$. It follows that $\{V(x_j,x_0)\}$ is an unbounded sequence. This contradicts the fact that $\overline{D}$ is bounded and $V(\cdot,x_0)$ is a generalized contraction. Next we claim that F is P_1-compact. Indeed, let $\{x_{n_j} \mid x_{n_j} \in D_{n_j}\}$ be any sequence such that $g_{n_j} \equiv \lambda x_{n_j} - P_{n_j}F(x_{n_j}) \to g$ for some g in X and any fixed $\lambda \geq 1$. By reflexivity of X, we may assume that $x_{n_j} \rightharpoonup x_0$ in X and $x_0 \in \overline{D}$ since $\overline{D}$ is weakly closed. Now, by condition (ii) and the properties of Γ_1, we see that $P_{n_j}V(x_{n_j},x_{n_j}) - P_{n_j}V(x_{n_j},x_0) \to 0$. Hence

$$\lambda x_{n_j} - P_{n_j}V(x_{n_j},x_0) = g_{n_j} + P_{n_j}V(x_{n_j},x_{n_j}) - P_{n_j}V(x_{n_j},x_0) \to g.$$

Since $V(\cdot,x_0) : X \to X$ is a generalized contraction, Lemma 2.4 (when $D = X$) or Theorem 4.1 in [76] imply that $V(\cdot,x_0)$ is P_1-compact. From this and the fact that $\lambda x_{n_j} - P_{n_j}V(x_{n_j},x_0) \to g$, we see that there exists a subsequence $\{x_{n_{j(k)}}\}$ and $x \in X$ such that $x_{n_{j(k)}} \to x$ and $\lambda x - V(x,x_0) = g$. But this subsequence also converges weakly to x_0 and hence $x = x_0$ and $\lambda x_0 - V(x_0,x_0) = \lambda x_0 - Fx_0 = g$. Hence Theorem 2.8 follows from Theorem 2.5.

Now the conclusion of Theorem 2.7 follows from Theorem 2.5.

Q.E.D.

We note that in view of Theorem 2.5, Theorem 2.8 remains valid when V in Definition 2.1 is defined only on $\overline{D}$ (i.e., $V: \overline{D} \times \overline{D} \to X$).

We complete this section with the following result proved in [81], which is related to Corollary 2.4.

Theorem 2.9 Let X be a reflexive space with scheme Γ_1 and let $F: X \to X$ be a continuous map such that $Fx = V(x,x)$ for $x \in X$, where $V: X \times X \to X$ is such that:

(i) For each $B = B(0,r) \subset X$ there exists $\alpha = \alpha(B) \in (0,1)$ and $\| V(y,x) - V(z,x) \| < \alpha \| y - z \|$ for all $x,z \in \overline{B}$ and $y \in X$.

(ii) $V(x,\cdot): X \to X$ is completely continuous for each fixed $x \in X$.

If $D \subset X$ is an open, bounded set with $0 \in D$ and F satisfies condition $(\pi_1^{\leqq})$ on ∂D, then F is P_1-compact, so the conclusions of Theorem 2.2 hold for the equation $Fx - x = 0$.

Proof. It suffices to show that $I - F$ is A-proper w.r.t. Γ_1. So let $\{x_{n_j} \mid x_{n_j} \in X_{n_j}\}$ be any bounded sequence such that $x_{n_j} - P_{n_j} F x_{n_j} \to g$ for some g in X. As usual, for convenience we replace n_j by j and also assume that $x_j \rightharpoonup x_0$ in X. Since the map $I - V(\cdot, x_0): X \to X$ is onto, we may choose $y \in X$ such that $y - V(y, x_0) = g$. We claim that $x_j \to y$. Indeed, for each $j \in Z_+$ we have the equality

$$x_j - y = x_j - P_j V(x_j, x_j) - y + P_j V(y, x_0) + P_j V(x_j, x_j)$$
$$- P_j V(y, x_0) + P_j V(y, x_j) - P_j V(y, x_j).$$

Hence

$$\| x_j - y \| \leq \| x_j - P_j F x_j - y + P_j y - P_j y + P_j V(y, x_0)\|$$
$$+ \| P_j V(y, x_j) - V(y, x_0)\| + \| P_j(V(y, x_j) - V(x_j, x_j))\|$$
$$\leq \| x_j - P_j F x_j - g \| + \| P_j y - y \|$$
$$+ \| V(y, x_j) - V(y, x_0)\| + \alpha \| y - x_j \|,$$

where $\alpha = \alpha(B(0, r))$ with $B(0, r) \supset \{x_j\}$. Now, by (i), we know that $\| V(y, x_j) - V(y, x_0) \| \to 0$. Clearly, the first two terms converge to 0. Hence $(1 - \alpha)\| x_j - y \| \to 0$, or $x_j \to y$. Since $x_j \rightharpoonup x_0$, we see that $y = x_0$ and $x_0 - F x_0 = g$. Q.E.D.

In [23], Browder obtained an existence result similar to Theorem 2.7, where X was assumed to be uniformly convex, while condition (i) was replaced by the assumption that $V(\cdot, y): \overline{D} \to X$ is nonexpansive for each $y \in \overline{D}$. Browder, Kirk, Petryshyn, Fitzpatrick, Webb, and others obtained some fixed-point theorems for intertwining maps under the assumption of various kinds on the nature of the underlying Banach space. Most of these results are included in a fixed-point theorem for 1-set contractions proved by the author in [230]. See Section 1 of Chapter IV for the proofs and further discussion of this topic.

3. CONSTRUCTIVE SURJECTIVITY THEOREMS FOR P$_\gamma$-COMPACT VECTOR FIELDS

In this section we show how the Brouwer degree theory can be used to establish constructive surjectivity theorems for various classes of P$_\gamma$-compact mappings. Our first result in this section is the following theorem for P$_1$-compact mappings, which in this generality has been proved in [236], although some of its special cases appeared earlier.

Theorem 3.1   Suppose that $F: X \to X$ is a bounded P_1-compact map which satisfies the following hypothesis:

(H1) $T = I - F$ satisfies condition $(+)$; that is, if $\{x_n\}$ is any sequence in X such that $Tx_n \to g$ for some g in X, then $\{x_n\}$ is bounded.

(H2) There is $r_0 > 0$ such that F satisfies either condition (a): F is odd on $X \backslash B(0, r_0)$ or (b): $Fx \neq \lambda x$ for $\| x \| \geq r_0$ and $\lambda \geq 1$.

Then, for each f in X, the equation

$$x - Fx = f \tag{3.1}$$

is feebly approximation solvable w.r.t. Γ_α and strongly approximation solvable if it is uniquely solvable [i.e., the Galerkin method is applicable to (3.1)].

We shall deduce Theorem 3.1 from the following more general and new result for P_γ-compact maps which prove to be useful in applications.

Theorem 3.2   Let $\gamma \in R^+$ and suppose that $F: X \to X$ is P_γ-compact and such that for some $\mu > \gamma$ the map $T_\mu = \mu I - F$ satisfies condition $(+)$. Suppose further that there exists $r_0 > 0$ such that:

(H3) If $F_n(x) = \lambda x$ for $x \in \partial B_n(0, r)$ and any $r \geq r_0$ and $n \in Z$ with $\lambda > 0$, then $\lambda \leq c$ for some $c > 0$.

(H4) F is either odd on $X \backslash B(0, r_0)$ or $Fx \neq \lambda x$ for $\| x \| \geq r_0$ and $\lambda \geq \mu$.

Then the conclusions of Theorem 3.1 hold for the equation

$$\mu x - Fx = f. \tag{3.2}$$

Proof.   Let f be an arbitrary element in X. Since $T_\mu = \mu I - F$ satisfies condition $(+)$, there exists $r \geq r_0$ and $\alpha > 0$ such that

$$\| \mu x - Fx - tf \| \geq \alpha \qquad \text{for} \quad x \in \partial B(0, r) \quad \text{and} \quad t \in [0, 1]. \tag{3.3}$$

Indeed, if this were not the case, there would exist sequences $\{x_j\} \subset X$ and $\{t_j\} \subset [0, 1]$ such that $t_j \to t_0$, $\| x_j \| \to \infty$, and $\mu x_j - Fx_j - t_j f \to 0$ as $j \to \infty$. This implies that $\mu x_j - Fx_j \to t_0 f$ with $\{x_j\}$ unbounded, in contradiction to condition $(+)$. Now we claim that there exist $n_0 \in Z_+$ and $\alpha_0 > 0$ such that

$$\| \mu x - P_n Fx - t P_n f \| \geq \alpha_0 \tag{3.4}$$
$$\text{for} \quad x \in \partial B_n(0, r), \, n \geq n_0, \quad \text{and} \quad t \in [0, 1].$$

Again, if this were not the case, there would exist sequences $\{x_{n_j} \mid x_{n_j} \in \partial B_{n_j}(0, r)\}$ and $\{t_{n_j}\} \subset [0, 1]$ such that

$$g_{n_j} \equiv \mu x_{n_j} - P_{n_j} F(x_{n_j}) - t_{n_j} P_{n_j} f \to 0 \qquad \text{as} \quad j \to \infty.$$

We may assume that $t_{n_j} \to t_0 \in [0,1]$ and note that

$$\mu x_{n_j} - P_{n_j} F x_{n_j} = g_{n_j} + t_{n_j} P_{n_j} f \to t_0 f.$$

Hence since $\mu > \gamma$ and F is P$_\gamma$-compact, there exist a subsequence $\{x_{n_{j(k)}}\}$ and $x_0 \in X$ such that $x_{n_{j(k)}} \to x_0$ and $\mu x_0 - F x_0 - t_0 f = 0$ with $x_0 \in \partial B(0,r)$, in contradiction to (3.3).

Now it follows from (3.4) and Theorem 1.1 that

$$\deg(\mu I - F_n, B_n(0,r), 0) \tag{3.5}$$
$$= \deg(\mu I - F_n - P_n f, B_n(0,r), 0) \qquad \text{for} \quad n \geq n_0.$$

To continue with the proof, we will now show that $\deg(\mu I - F_n, B_n(0,r), 0) \neq 0$ for sufficiently large $n \in Z_+$.

Suppose first that F is odd on $X \backslash B(0,r_0)$. Then F_n is odd on $X_n \backslash B_n(0,r)$ and therefore, by Theorem 1.2, $\deg(\mu I - F_n, B_n(0,r), 0) \neq 0$ for $n \geq n_0$. On the other hand, if $Fx \neq \lambda x$ for $\| x \| \geq r_0$ and $\lambda \geq \mu$, then there exists $n_1 \in Z_+$ such that

$$F_n(x) \neq \lambda x \qquad \text{for} \quad x \in \partial B_n(0,r), n \geq n_1, \quad \text{and} \quad \lambda \geq \mu. \tag{3.6}$$

Indeed, if not, we could choose sequences $\{\lambda_{n_j}\}$ with $\lambda_{n_j} \geq \mu$ and $\{x_{n_j} \mid x_{n_j} \in \partial B_{n_j}(0,r)\}$ such that $F_{n_j}(x_{n_j}) = \lambda_{n_j} x_{n_j}$ for each $j \in Z_+$. But our condition (H3) then implies that $\lambda_{n_j} \leq c$ and $\| F_{n_j}(x_{n_j}) \| \leq cr$ for all $j \in Z_+$, so we may assume that $\lambda_{n_j} \to \lambda_0 \geq \mu$. Hence $F_{n_j}(x_{n_j}) - \lambda_0 x_{n_j} = (\lambda_{n_j} - \lambda_0) F_{n_j}(x_{n_j}) \to 0$ as $j \to \infty$. Since $\lambda_0 > \gamma$ and F is P$_\gamma$-compact, it follows that there exists a subsequence $\{x_{n_{j(k)}}\}$, $x \in X$, such that $x_{n_{j(k)}} \to x$ and $Fx - \lambda_0 x = 0$ with $x \in \partial B(0,r)$, in contradiction to our condition on F in (H4) since $r \geq r_0$.

It follows from (3.6) that $\mu x - t F_n(x) \neq 0$ for $x \in \partial B_n(0,r)$, $t \in [0,1]$, and $n \geq n_2 = \max\{n_0, n_1\}$. Hence, by Theorem 1.1, $\deg(\mu I - F_n, B_n(0,r), 0) = \deg(\mu I, B_n(0,r), 0) = 1$ for $n \geq n_2$. Combining this with (3.5), we see that there exist $x_n \in B_n(0,r)$ such that $\mu x_n - F_n(x_n) - P_n f = 0$ for $n \geq n_2$. This implies that $\mu x_n - F_n(x_n) \to f$ as $n \to \infty$ and again, by the P$_\gamma$-compactness of F, there exist a subsequence $\{x_{n_j}\}$ and $x_0 \in \bar{B}(0,r)$ such that $x_{n_j} \to x_0$ and $\mu x_0 - F x_0 = f$. This proves the first part of Theorem 3.2. If x_0 is unique, then by the standard arguments, one shows that $x_n \to x_0$ in X as $n \to \infty$. \hfill Q.E.D.

Note that F in Theorem 3.2 is not assumed to be continuous or bounded. This fact will prove to be useful in applications.

Remark 3.1 Theorem 3.1 follows from Theorem 3.2 when in the latter we take $\mu = \gamma = 1$. Observe also that Theorem 3.1 is valid if the boundedness condition on F is replaced by (H3).

Remark 3.2   Note that condition ($+$) satisfied by any given map $T: X \rightarrow Y$ is equivalent to the requirement that "$T^{-1}(K)$ *be bounded in X whenever $K \subset Y$ is compact*." It is easy to show that condition ($+$), which has been used by many authors, is implied by any one of the following conditions used by various authors in their study of equations involving mappings of monotone, condensing, and A-proper type (see, e.g., [36,62,100,156, 225,275,314] and others cited there).

   Condition ($1+$): $(Tx,Kx)/\| Kx \| \rightarrow \infty$ as $\| x \| \rightarrow \infty$ (i.e., T is K-*coercive*), where K is a map of X to Y^* with $Kx \neq 0$ if $x \neq 0$.

   Condition ($2+$): $\| Tx \| \rightarrow \infty$ as $\| x \| \rightarrow \infty$ (i.e., T is *norm coercive*).

   Condition ($3+$): $\| Tx \| + (Tx,Kx)/\| Kx \| \rightarrow \infty$ as $\| x \| \rightarrow \infty$.

   Condition ($4+$): $0 \notin \overline{T(\partial B(0,r_0))}$ and $T(tx) = t^\theta T(x)$ for $\| x \| \geq r_0, t > 1$ and some $\theta > 0$.

Remark 3.3   In virtue of Remark 3.2, Theorems 3.1 and 3.2 remain valid when condition ($+$) is replaced by any one of the conditions listed in Remark 3.2 when $T = I - F$ or $T_\mu = \mu I - F$.

   Theorem 3.1 includes Corollary 1.2 in [231] when F is ball condensing and, in particular, a result in [75] when $F = S + C$ with C compact and S contractive. Moreover, in view of our discussion preceding the statement of Proposition 2.1, we see that Theorem 3.1 includes the following general result.

Corollary 3.1   Let Γ_1 be a nested scheme for (X,X), $A: X \rightarrow X$ bounded, continuous, and accretive, $S: X \rightarrow X$ ball condensing, and $C: X \rightarrow X$ compact. If $F = S + C - A$ satisfies hypotheses (H1) and (H2) of Theorem 3.1, the conclusions of the latter hold for the equation

$$x + Ax - Sx - Cx = f. \tag{3.7}$$

   A second consequence of Theorem 3.1 is the following.

Corollary 3.2   Suppose that $F: X \rightarrow X$ is P_1-compact, (H3) holds, and

(H.5) There exist a,b in R^+ and $\beta \in (0,2)$ such that

   $(Fx,w) \leq a \| x \|^\beta + b$ for all $x \in X$ and some $w \in J(x)$,

$$\tag{3.8}$$

where J is the normalized duality map. Then the conclusions of Theorem 3.1 hold.

Proof.   We will first show that (3.8) implies that $I - F$ satisfies condition ($+$). In fact, let $\{x_n\} \subset X$ be such that $x_n - Fx_n \rightarrow g$ for some $g \in X$.

Then $\| x_n - Fx_n \| \leq M$ for some $M > 0$ and all $n \in Z_+$. Therefore,

$$M \| x_n \| \geq (x_n - Fx_n, w_n) \geq \| x_n \|^2 - a\| x_n \|^\beta - b$$

for each n where $w_n \in J(x_n)$. It follows from the inequality above that $\{x_n\}$ is bounded.

To verify condition (b) of Theorem 3.1 we note that if $Fx = \lambda x$, then $\lambda\| x \|^2 \leq a\| x \|^\beta + b$, so if $\lambda \geq 1$, one has $\| x \|^2 \leq a\| x \|^\beta + b$. Clearly, those $x \in X$ that satisfy the last inequality form a bounded set. The conclusion now follows from Theorem 3.1 and Remark 3.1. Q.E.D.

Remark 3.4 Hypothesis (H5) is implied, in particular, by the condition

$$(F(x) - F(0), w) \leq 0 \qquad \text{for} \quad x \in X, w \in Jx. \tag{3.9}$$

Consequently, Theorem 3 and Corollary 2 in [219] follow as special cases of Corollary 3.2.

Proposition 3.1 Let Γ_1, A, C, and S be as in Corollary 3.1 and suppose that there exists c, $d \in R^+$ and $\beta \in (0,2)$ such that

$$((S + C)x, w) \leq c \| x \|^\beta + d \qquad \text{for each} \quad x \in X \tag{3.10}$$
$$\text{and some} \quad w \in J(x).$$

Then the conclusions of Corollary 3.1 hold for equation (3.7).

Proof. We first show that (3.10) implies that $I - F$ satisfies condition (+), where $F = S + C - A$. In fact, let $\{x_n\} \subset X$ be such that $x_n - Fx_n \to g$. Then $\| x_n - Fx_n \| \leq M$ for some $M > 0$ and all $n \in Z_+$. Therefore, using the fact that $(Ax,w) \geq (A(0),w)$ for some $w \in Jx$, we see that $M\| x_n \| \geq (x_n - Fx_n, w_n) \geq \| x \|^2 - (A(0), w_n) - c\| x \|^\beta - d$ for some $w_n \in J(x_n)$. This implies that $\tilde{M}\| x \| \geq \| x \|^2 - c\| x \|^\beta - d$ with $\tilde{M} = M + \| A(0) \|$. It follows from this inequality that $\{x_n\}$ is bounded.

To verify condition (b) of Theorem 3.1 we note that if $Fx = \lambda x$, then in view of (3.10) and the fact that $-(Ax,w) \leq -(A(0),w)$ for some $w \in J(x)$ we have for each x in X the relation $\lambda\| x \|^2 \leq c\| x \|^\beta + d + \| A(0) \| \| x \|$, so if $\lambda \geq 1$, one has $\| x \|^2 \leq c\| x \|^\beta + \| A(0) \| \| x \| + d$. Clearly, those $x \in X$ that satisfy the last inequality form a bounded set. The conclusion now follows from Corollary 3.1. Q.E.D.

The existence part of Proposition 3.1 when $C = 0$ appears in [237].

As a consequence of Theorem 3.2, we deduce the following result, which includes similar results of the author [209] when $\gamma = 0$ and in [237] when F is ball condensing.

Corollary 3.3   Suppose that $F: X \to X$ is P_γ-compact and such that F satisfies hypothesis (H3) and the following condition:

(H6) There exist sequences $\{\alpha_n\}$ and $\{\beta_n\}$ such that $\alpha_n \to \infty$, $\beta_n \to \infty$, and $\| Fx - \lambda x \| \geq \alpha_n$ when $\| x \| \geq \beta_n$ and $\lambda \geq \mu$ for any $\mu > \gamma$.

Then the conclusions of Theorem 3.2 hold.

Proof.   As before, we will first show that $\mu I - F$ satisfies condition $(+)$. Indeed, assume that $\{x_n\} \subset X$ is such that $\mu x_n - F x_n \to g$ for some g in X. Choose $n_0 \in Z_+$ such that $\| \mu x_n - F x_n \| < \alpha_{n_0}$ for all $n \in Z_+$. Then $\| x_n \| \leq \beta_{n_0}$ for all $n \geq n$, by assumption (H6).

Now we show that there exists $r_0 > 0$ such that $Fx \neq \lambda x$ for $\| x \| \geq r_0$ and $\lambda \geq \mu$. Suppose that this is not so. Then one may select $\{x_n\} \subset X$ and $\{\lambda_n\}$ with each $\lambda_n \geq \mu$ and $\| x_n \| \to \infty$ such that $F x_n - \lambda_n x_n = 0$. This clearly violates (H6).   Q.E.D.

Following Granas [105], we say that $F: X \to Y$ is *quasibounded* if there exist constants $M > 0$ and $q > 0$ such that

$$\| Fx \| \leq M \| x \| \quad \text{for all} \quad x \in X \quad \text{with} \quad \| x \| \geq q. \quad (3.11)$$

If F is quasibounded, the number $|F|$ defined by

$$|F| = \limsup_{\|x\| \to \infty} \left\{ \frac{\| Fx \|}{\| x \|} \right\} \quad (3.12)$$

is called the *quasinorm* of F. It follows that every bounded linear map F is quasibounded and its norm $\| F \|$ equals $|F|$. Furthermore, as observed by Granas, every nonlinear mapping F of X into Y that is asymptotically differentiable in the sense of Krasnoselskii [136] is quasibounded. In fact, if F is asymptotically differentiable, there exists $F_\infty \in L(X, Y)$, called the *asymptotic derivative* of F (or the *derivative* of F at ∞), such that

$$\lim_{\|x\| \to \infty} \left\{ \frac{\| F(x) - F_\infty x \|}{\| x \|} \right\} = 0. \quad (3.13)$$

It follows easily from (3.13) that F is quasibounded and that $|F| = \| F_\infty \|$. We now prove the following new result.

Proposition 3.2   Let $F: X \to X$ be P_1-compact and quasibounded. If $|F| < 1$, the conclusions of Theorem 3.1 hold.

Proof.   It follows directly from (3.12) that since F is quasibounded with $|F| < 1$, $I - F$ satisfies condition $(+)$ and the condition of (b) in (H2) of Theorem 3.1 holds for some $r_0 \geq q$. Furthermore, if $P_n Fx = \lambda x$ for $x \in X_n$ and $\| x \| \geq r_0$ with $\lambda > 0$, then $\lambda \| x \| = \| P_n Fx \| \leq \alpha M \| x \|$ (i.e., λ

$\leq \alpha M$). Hence the hypothesis (H3) of Theorem 3.2 holds. Thus Proposition 3.2 follows from Theorem 3.1 and Remark 3.1.   Q.E.D.

The existence part of the result above appears in [105] when F is compact, in [209] when F is P_0-compact, and in [237] when F is ball condensing.

4. APPROXIMATION-SOLVABILITY AND SOLVABILITY OF EQUATIONS INVOLVING MAPS OF A-PROPER TYPE

Consider the equation

$$Tx = f \qquad (x \in X, f \in Y). \tag{4.1}$$

where $T: X \to Y$ is A-proper or a uniform limit of A-proper mappings.

4.1 Approximation-Solvability

Our first abstract result in this section is the following theorem, which extends and unifies [232, Theorem 1.1] with [231, Theorem 1.3] and [225, Corollary 3.2].

Theorem 4.1   Let $\Gamma = \{X_n, V_n; E_n, W_n\}$ be admissible for (X,Y), $K: X \to Y^*$ some map, and suppose that there is a bounded and odd map $G: X \to Y$ such that:

(H1) G is A-proper w.r.t Γ and $(Gx, Kx) = \| Gx \| \| Kx \| > 0$ for $x \neq 0$.

(H2) $T_\mu = T + \mu G$ is A-proper w.r.t. Γ for each $\mu > 0$.

(H3) $Tx \neq \gamma Gx$ for all $x \in X - B(0,r)$, all $\gamma < 0$, and some $r > 0$.

(H4) T satisfies condition $(+)$ or, more generally, to each $f \in Y$ there correspond numbers $r_f \geq r$ and $\alpha_f > 0$ such that

$$\| Tx - tf \| > \alpha_f \qquad \text{for} \quad x \in \partial B(0,r_f) \quad \text{and} \quad t \in [0,1]. \tag{4.2}$$

Then

(A1) If T is A-proper w.r.t. Γ and bounded, (4.1) is feebly a-solvable w.r.t. Γ for each f in Y, and strongly if T is injective.

If there is a map $K_n: X_n \to D(W_n^*)$ such that

$$(W_n g, K_n x) = (g, Kx) \qquad \text{for all} \quad x \in X_n, g \in Y \quad \text{and each} \quad n, \tag{4.3i}$$

then the assertion in (A1) remains valid for T unbounded provided that

$$(Tx, Kx) \geq -a\| Kx \| \qquad \text{for all} \quad x \in X \quad \text{and some} \quad a \geq 0. \tag{4.3j}$$

(A2) If T is odd on $X - B(0,r)$, then the conclusions hold without (H1)–(H3) or (4.3i)–(4.3j).

Proof. Suppose that f is fixed. Then, by (H4), there are $r_f \geq r$ and $\alpha_f > 0$ such that

$$\| Tx - tf \| \geq \alpha_f \quad \text{for} \quad x \in \partial B(0,r_f) \quad \text{and} \quad t \in [0,1]. \tag{4.4}$$

(A1a) Suppose first that $T: X \to Y$ is *bounded*. Consider the homotopy

$$H(x,t) = (1 - t)Tx + tGx \quad (x \in \overline{B}(0,r), t \in [0,1]).$$

Since T is A-proper and (H2) holds, it follows that $H(\cdot,t)$ is A-proper for each $t \in [0,1]$. Moreover, $H(x,t) \neq 0$ for $x \in \partial B(0,r_f)$ and $t \in [0,1]$. If not, there would exist $x_0 \in \partial B(0,r_f)$ and $t_0 \in [0,1]$ such that $H(x_0,t_0) = (1 - t_0)T(x_0) + t_0 G(x_0) = 0$. It follows (H1) that $t_0 \neq 1$ and from (4.4) that $t_0 \neq 0$. Thus $t_0 \neq 0$. Thus $t_0 \in (0,1)$, and if we set $\mu_0 = t_0/(1 - t_0)$, we see that $Tx_0 + \mu_0 Gx_0 = 0$, in contradiction to (H3). Thus $H(x,t) \neq 0$ for $t \in [0,1]$ and $x \in \partial B(0,r_f)$ and, moreover, since T and G are bounded, H is continuous in $t \in [0,1]$, uniformly for $x \in \overline{B}(0,r_f)$. Hence, by Theorem 2.1 in Chapter IV, the generalized $\text{Deg}(H(\cdot,t),B(0,r_f),0)$ is constant in $t \in [0,1]$ and therefore $0 \notin \text{Deg}(T,B(0,r_f),0) = \text{Deg}(G,B(0,r_f),0)$ since G is odd. It follows from this and (4.4) that $0 \notin \text{Deg}(T - f,B(0,r_f),0)$. Hence, by the definition of "Deg," there exists $n_f \in Z^+$ such that the Brouwer degree $\deg(T_n - W_n f,B_n(0,r_f),0) \neq 0$ for each $n \geq n_f$. So by the Brouwer degree theory, there exists $x_n \in B_n(0,r_f)$ for each $n \geq n_f$ such that $T_n(x_n) - W_n f = 0$. This and the A-properness of T imply the existence if a subsequence $\{x_{n_j}\}$ of $\{x_n\}$ and an $x_0 \in \overline{B}(0,r_f)$ such that $x_{n_j} \to x_0$ in X and $Tx_0 = f$. The proof that the entire sequence $\{x_n\}$ coverages to x_0 if (4.1) has at most one solution for the given f in Y is obtained as follows. Suppose that $x_n \to x_0$ as $n \to \infty$. Then there exists a subsequence $\{x_{n_k}\}$, different from $\{x_{n_j}\}$, such that $\| x_{n_k} - x_0 \| \geq \epsilon$ for all large k and some $\epsilon > 0$. But $T_{n_k}(x_{n_k}) - W_{n_k} f = 0$ for all k and therefore by the A-properness of T, there exist a subsequence $\{x_{n_{k(i)}}\}$ and $x_1 \in X$ such that $x_{n_{k(i)}} \to x_1$ as $i \to \infty$ and $T(x_1) = f$ with $x_1 \neq x_0$. This contradicts the injective property of T, and thus proves the last assertion of Theorem 4.1 when T is a bounded mapping.

(A1b) Suppose now that (4.3i) and (4.3j) hold, but T is *unbounded*. Then, as in (A1), the map $H(\cdot,t)$ is still A-proper for each $t \in [0,1]$ and $H(x,t) \neq 0$ for $x \in \partial B(0,r_f)$ and $t \in [0,1]$; therefore, $\text{Deg}(H(\cdot,t),B(0,r_f),0)$ is well defined for each $t \in [0,1]$. To show that it is constant in $t \in [0,1]$, it suffices to show that there exists $n_f \in Z^+$ such that for each fixed $n \geq n_f$, $H_n(x,t) = (1 - t)T_n(x) + tG_n(x) \neq 0$ for $x \in \partial B_n(0,r_f)$ and

$t \in [0,1]$ for this then implies that $\deg(H_n(\cdot,t),B_n(0,r_f),0)$ is independent of $t \in [0,1]$ for each $n \geq n_f$. Our proof is by contradiction.

Suppose that such an n_f does not exist. Then there would exist an unbounded sequence of integers (which we identify for simplicity with the original sequence $\{n\}$), a sequence $\{t_n\} \subset (0,1)$ with $t_n \to t_0 \in [0,1]$, and $\{x_n \mid x_n \in \partial B_n(0,r_f)\}$ such that $(1 - t_n)T_n(x_n) + t_nG_n(x_n) = 0$. We distinguish among three cases:

(i) $t_0 = 0$: We note that in this case

$$T_n(x_n) = -t_n/(1 - t_n)G_n(x_n) \to 0 \qquad \text{as} \quad n \to \infty$$

since $\{G_n(x_n)\}$ is bounded. Thus since T is A-proper, there exist a subsequence, again denoted by $\{x_n\}$, and an $x \in X$ such that $x_n \to x$ in X and $Tx = 0$ with $\| x \| = r$, in contradiction to (4.4).

(ii) $t_0 = 1$: Since $t_nG_n(x_n) = (t_n - 1)T_n(x_n)$ for each n, it follows from this, (H1), and (4.3i) that $t_n\| G(x_n) \| \| Kx_n \| = (t_n - 1)(Tx_n,Kx_n)$. Since $(t_n - 1) < 0$ and $(Tx_n,Kx_n) \geq -a\| Kx_n \|$ by (4.3i)–(4.3j), it follows that $(1 - t_n)(Tx_n,Kx_n) \geq -(1 - t_n)\| Kx_n \|$ by (4.3j). Therefore, for each n we have

$$t_n\| Gx_n \| \| Kx_n \| \leq a(1 - t_n)\| Kx_n \|.$$

Hence $\| Gx_n \| = a(1 - t_n)/t_n$. Since $\{\| Gx_n \|\}$ is bounded, there exists a subsequence, again denoted by $\{\| Gx_n \|\}$, such that $\| Gx_n \| \to c_1$ for some $c_1 \geq 0$. But $a(1 - t_n)/t_n \to 0$ as $n \to \infty$ and thus $c_1 = 0$. This implies that $W_nG(x_n) = G_n(x_n) \to 0$ as $n \to \infty$ if (4.3i)–(4.3j) hold. This and the A-properness of G imply the existence of a subsequence, again denoted by $\{x_n\}$, and an x in X such that $x_n \to x$ in X as $n \to \infty$ and $Gx = 0$ with $\| x \| = r$ in contradiction to (H1).

(iii) $0 < t_0 < 1$: Since $T_n(x_n) + (t_0/(1 - t_0))G_n(x_n) = (t_0/(1 - t_0) - t_n/(1 - t_n))G_n(x_n) \to 0$ as $n \to \infty$, (H2) implies the existence of $\{x_{n_j}\}$ and x such that $x_{n_j} \to x$ and $Tx + (t_0/(1 - t_0))Gx = 0$ with $x \in \partial B$, in contradiction to (H3).

The discussion above shows that $\mathrm{Deg}(H(\cdot,t),B(0,r_f),0)$ is also constant in $t \in [0,1]$ when (4.3i) and (4.3j) hold. This proves the validity of our second assertion in (A1).

(A2) Now suppose that T is odd. Since T satisfies (4.2), it follows that $H(x,t) = Tx - tf$ is an admissible A-proper homotopy and $H(x,t) \neq 0$ for $x \in B(0,r_f)$ and $t \in [0,1]$. Hence, by Theorem 2.1 in Chapter IV, $0 \notin \mathrm{Deg}(T - f,B(0,r_f),0) = \mathrm{Deg}(T,B(0,r_f),0)$. This and the definition of "Deg" imply the existence of $n_f \in Z^+$ such that $\deg(T_n - W_nf,B_n(0,r_f),0) \neq 0$ for each $n \geq n_f$. Hence, for each $n \geq n_f$, there exists $x_n \in \bar{B}_n(0,r_f)$ such that $T_n(x_n) - W_nf = 0$. Since $T_n(x_n) - W_nf \to 0$ as $n \to \infty$ and $\{x_n\}$ is bounded, the A-properness of T implies the existence of a subsequence $\{x_{n_j}\}$ and an element $x \in \bar{B}(0,r_f)$ such that $x_{n_j} \to x_0$ in X as $j \to$

∞ and $Tx = f$. This proves the validity of (A2) without using the condition that there exist operators $K: X \to Y^*$ and $G: X \to Y$ which satisfy hypotheses (H1)–(H3) or (4.3i)–(4.3j).   Q.E.D.

The following corollary will prove to be useful in what follows.

Corollary 4.1   Let $T: X \to Y$ be A-proper w.r.t. Γ and suppose that T satisfies $(+)$.

(a1) If T is also odd on $X - B(0,r)$ for some $r \geq 0$, then Eq. (4.1) is feebly a-solvable w.r.t. Γ for each $f \in Y$, and strongly a-solvable if (4.1) is uniquely solvable.

(a2) The same conclusion holds for Eq. (4.1) when T is not odd provided that T is bounded and there exists a bounded map $G: X \to Y$ such that

(h2) G is odd, A-proper, $(Gx,Kx) = \| Gx \| \| Kx \| > 0$ for $x \neq 0$, and $T + \mu G$ is A-proper for each $\mu > 0$.

(h3) $Tx \neq \gamma Gx$ for all $x \in X - B(0,r)$, all $\gamma < 0$, and some $r \geq 0$.

Remark 4.1   The assertions of Corollary 4.1 remain valid if instead of condition $(+)$ one assumes that T satisfies any one of conditions $(1+)$, $(2+)$, $(3+)$, or $(4+)$ of Remark 3.2.

It will be shown in Section 4.3 that Theorem 4.1 includes a number of special cases depending on the choice of Y and K.

4.2   Solvability of Equations Involving Limits of A-Proper Maps

In this section we use Theorem 4.1 to deduce the following general surjectivity theorem for the map $T: X \to Y$, which is *not* A-proper but a uniform limit of a special sequence of A-proper maps. Theorem 4.2 below is an extension of [211, Theorem 2.4], [220, Theorem 1.3], and as we shall see in Section 4.3, a number of surjectivity theorems for maps of monotone type, 1-ball-contractive vector field, and others.

Theorem 4.2   Let $\Gamma,T,G: X \to Y$ and $K: X \to Y^*$ be such that hypotheses (H1)–(H4) of Theorem 4.1 hold. Then

(A3) If T is bounded and satisfies condition $(+ +)$ (i.e., if $\{x_j\} \subset X$ is bounded and $Tx_j \to g$ for some g in Y, then there is an x in X such that $Tx = g$), then $T(X) = Y$. If we also assume that (4.3i) holds and $(Tx,Kx) \geq -a\| Kx \|$ for $x \in X$, then $T(X) = Y$ for T not necessarily bounded.

(A4) If T is odd on $X - B(0,r)$, then $T(X) = Y$ without the hypotheses (H1)–(H3) and the condition that T is bounded.

Proof. (A3) It follows from (H2) that $T_\lambda = T + \lambda G$ is A-proper w.r.t. Γ for each fixed $\lambda > 0$, while (H4) and the boundedness of G imply the existence of a number $\lambda_f > 0$ such that

$$\| T_\lambda x - tf \| \geq 2^{-1}\alpha_f \quad \text{for} \quad x \in \partial B(0,r_f),\ t \in [0,1],\ \lambda \in (0,\lambda_f).$$

$$(4.2')$$

Moreover, (H2) implies the A-properness of $T_\lambda + \mu G$ for each $\mu > 0$ and (H3) shows that $T_\lambda(x) \neq \gamma Gx$ for all $x \in X - B(0,r)$ and $\gamma < 0$. Since, obviously, T_λ is bounded when T is bounded and $(T_\lambda x, Kx) \geq -a\| Kx \|$ if $(Tx,Kx) \geq -a\| Kx \|$ for $x \in X$, it follows that T_λ satisfies all the conditions of Theorem 4.1 for each fixed $\lambda \in (0,\lambda_f)$. Hence applying Theorem 4.1 to the equation $T_{\lambda_k}(x) = Tx + \lambda_k Gx = f$, where $\lambda_k \in (0,\lambda_f)$ is such that $\lambda_k \to 0^+$ as $k \to \infty$, in each of the two cases (i.e., T bounded or T unbounded) we find an element $x_k \in B(0,r_f)$ such that $T(x_k) + \lambda_k G(x_k)$ $= f$ for each $k \in Z_+$. Since $\lambda_k \to 0^+$ as $k \to \infty$ and $\{Gx_k\}$ is bounded, it follows that $T(x_k) = f - \lambda_k G(x_k) \to f$ in Y as $k \to \infty$. This and the condition $(++)$ imply the existence of an $x \in X$ such that $Tx = f$. Since f in Y was fixed but arbitrary, it follows that $T(X) = Y$.

(A4) Since T and G are odd, it follows that T_λ is odd and, in view of (4.2'), $T_\lambda(x) \neq tf$ for all $x \in B(0,r_f)$, $t \in [0,1]$, and $\lambda \in (0,\lambda_f)$. Hence, by Theorem 2.1 in Chapter IV, $\text{Deg}(T_\lambda, B(0,r_f),0) \neq \{0\}$. Thus for each λ_k $\in (0,\lambda_k)$ with $\lambda_k \to 0^+$, there exists $x_k \in \overline{B}(0,r_f)$ such that $T(x_k) + \lambda_k G(x_k)$ $= f$. Since G is bounded and $\lambda_k \to 0$, we get $T(x_k) = f - \lambda_k G(x_k) \to f$ as $k \to \infty$. In view of this, condition $(++)$ implies the existence of an x $\in X$ such that $Tx = f$. Q.E.D.

Remark 4.2 It is interesting to note that when $T: X \to Y$ is bounded, then Theorems 4.1 and 4.2 provide the solvability results for (4.1) without condition (4.3i), which requires Γ to be such that $(W_n g, K_n x) = (g, Kx)$ for x in X_n and $g \in Y$. If $Y \neq X^*$, the latter condition essentially involves K in the construction of Γ and in some cases (see [172,225]) this imposes a considerable restriction on T for it to be A-proper, or a uniform limit of A-proper maps, with respect to such schemes.

4.3 Special Cases

To obtain some interesting special cases of Theorem 4.1 and 4.2 and to indicate their generality and unifying property, after some historical comments on accretive maps, we establish here some of their basic properties in the form due to Webb [295], which he proved by direct and simple arguments.

It is by now well known (e.g., [36,161]) that the study of accretive operators is intimately connected with the theory of evolution differential equations in Banach spaces; indeed, they are the infinitesimal generators of nonlinear contraction semigroups. Strongly accretive operators have been shown to be A-proper in the past by two types of arguments. The first and direct argument which was used by the author [222] required that the space X possess a weakly continuous duality mapping, which hypothesis is known to be very restrictive outside Hilbert space. The second one, the fact that strongly accretive operators are surjective [63], was used by the writer to prove that a continuous strongly accretive map is A-proper (e.g., see Theorem 3.1G in [225]).

In [295], Webb used some results of [247] to give a direct proof of the fact that even a demicontinuous strongly accretive mapping is A-proper if X has a uniformly convex dual X^* without the assumption that the duality map is weakly continuous, and the surjectivity result is part of the conclusion, not the hypothesis. Following Kato [128], we have

Definition 4.1 Let X be real. A mapping $T: D(T) \to X$ is said to be *accretive* if for all x,y in $D(T)$ and all $\lambda > 0$,

$$\| \lambda x + Tx - (\lambda y + Ty) \| \geq \lambda \| x - y \|, \tag{4.5}$$

or equivalently, assuming that (f,x) is the value of $f \in X^*$ at $x \in X$,

$$(Tx - Ty, w) \geq 0 \qquad \text{for all} \quad x,y \in D(T) \quad \text{and some} \quad w \in J(x - y), \tag{4.5}'$$

where $J: X \to X^*$ is the normalized duality mapping defined uniquely by

$$Jx = \{w \mid w \in X^* \qquad \text{such that} \quad (w,x) = \| x \| \| x \|$$
$$\text{and} \quad \| w \| = \| x \| \}.$$

It is known [128] that J is single valued and uniformly continuous on bounded subsets of X if X^* is uniformly convex. It was shown by the author that even for general X^*, a Banach space X is strictly convex if and only if the multivalued $J: X \to X^*$ is strictly monotone [i.e., $(f - g, x - y) > 0$ for $f \in Jx$, $g \in Jy$]. For other properties of J, see [222,244] and [36,161].

Definition 4.1′ We say that $T: D(T) \to X$ is *strongly accretive* (i.e., λ-accretive) if there exists a constant $\lambda > 0$ such that

$$(Tx - Ty, J(x - y)) \geq \lambda \| x - y \|^2 \qquad \text{for all} \quad x,y \in D(T). \tag{4.6}$$

Theorem 4.3 Let $\Gamma_1 = [X_n, P_n\}$ a nested projectionally complete scheme for X with $\| P_n \| = 1$ for each n (i.e., X is a π_1-space), let X^* be uniformly convex and let $T: X \to X$ be demicontinuous and strongly accretive. Then T is A-proper w.r.t. Γ_1 and $Tx = f$ is uniquely approximation solvable for each $f \in X$.

To prove Theorem 4.3 of Web [295] we first need the following.

Proposition 4.1 Let X be a separable space with X^* uniformly convex and let $\{x_n\}$ be a bounded sequence in X. Then there exist a subsequence $\{x_k\}$ (say) and a point $v \in X$ such that $\{J(x_k - v)\}$ converges weakly to 0 in X^*.

Proof. By Lemma 1.1 of Reich [247] there is a subsequence $\{x_k\}$ such that $\varphi(\zeta) = \lim_k \| x_k - \zeta \|$ exist for all $\zeta \in X$. Since φ is continuous and convex and since $\varphi(\zeta) \to \infty$ as $\| \zeta \| \to \infty$, φ attains its infimum over X at some v. As $J(x)$ is a subdifferential of the convex function $\frac{1}{2}\| x \|^2$([36]), we have $(y - x; J(x)) \leq \frac{1}{2}\| y \|^2 - \frac{1}{2}\| x \|^2$ for all x,y in X. Thus, for all $\zeta \in X$,

$$(\zeta - v, J(x_k - \zeta)) \leq \tfrac{1}{2}\| x_k - v \|^2 - \tfrac{1}{2}\| x_k - \zeta \|^2.$$

For a fixed $t > 0$ and $\xi \in X$, let $z_t = v + t\xi$. Then $\lim \sup_{k\to\infty}(t\zeta, J(x_k - z_t)) \leq 0$, since $\varphi(v)$ is the infimum. Cancel $t > 0$ and then let $t \to 0$. As J is uniformly continuous on bounded sets, $\lim \sup_{k\to\infty}(\zeta, J(x_k - v)) \leq 0$. Replacing ζ by $-\zeta$ proves that $\{J(x_k - v)\}$ converges weakly to 0.
                                                                                Q.E.D.

Proof of Theorem 4.3. Let $\{x_j \mid x_j \in X_j\}$ be any bounded sequence with $P_j Tx_j \to g$ for some g in X. Let $\lambda > 0$ be such that $T - \lambda T$ is accretive [i.e., (4.5) holds]. Since $P_j^* Jx = Jx$ for each $x \in X_j$ [38], we have for all $y \in X_j$,

$$(P_j Tx_j - P_j Ty, J(x_j - y)) = (P_j Tx_j - Ty, J(x_j - y) \geq \lambda\| x_j - y \|^2.$$

Thus

$$\lim_{j\to\infty} \sup(g - Ty, J(x_j - y)) \geq \lim_{j\to\infty} \sup \lambda\| x_j - y \|^2. \qquad (4.7)$$

Let $\{x_k\}$ and $v \in X$ be the subsequence and a point given by Proposition 4.1. Since $P_m v \to v$ as $m \to \infty$ and T is demicontinum, $\| g - T(P_m v) \| \leq M$ for some $M > 0$. Given $\epsilon > 0$, fix m so that $\| P_m v - v \| < \epsilon$ and

$$\| J(x_k - v) - J(x_k - P_m v) \| < \lambda\epsilon^2/M, \qquad \text{uniformly in} \quad k.$$

Then

$$(g - T(P_m v), J(x_k - P_m v)) \leq (g - T(P_m v), J(x_k - v)) + \lambda \epsilon^2.$$

Take $y \in P_n v$ in (4.7) and use the fact that $\{J(x_k - v)\} \rightharpoonup 0$ weakly to obtain

$$\lim_{k \to \infty} \sup \| x_k - P_m v \|^2 \leq \epsilon^2.$$

As $\| x_k - v \| \leq \| x_k - P_m v \| + \| P_m v - v \|$, this yields

$$\lim \sup \| x_k - v \| \leq 2\epsilon.$$

Thus $x_k \to v$. As T is demicontinuous, $\{Tx_k\}$ converges weakly to Tv. For every u in $\cup_n X_n$ [i.e., u lies in some $X_m (m \geq k)$]

$$(g, Ju) + \lim_k (P_k Tx_k, Ju) = \lim_k (Tx_k, Ju) = (Tv, Ju).$$

Also, since J is uniformly continuous and $J(X) = X^*$ (see [126]), the set of all such Ju is dense in X^*. Thus $Tv = g$.

To prove that the equation (i), $Tx = f$, is uniquely approximation solvable w.r.t. $\Gamma_1 = \{X_n, P_n\}$, we note that for any $f \in X$ and $x \in X$,

$$(P_n Tx - P_n f, Jx) = (Tx - f, Jx) \geq \lambda \| x \|^2 + (T(0) - f, Jx) \qquad (4.8)$$
$$\geq \lambda \| x \|^2 - (\| T(0) \| + \| f \|) \| x \| \geq 0$$

for $\| x \| = r$, r sufficiently large. It follows from (4.8) that there exists $x_n \in X_n \cap B(0, r)$ with $P_n T(x_n) = P_n f$. The A-properness of T concludes the argument.   Q.E.D.

Corollary 4.2 Under the same hypotheses on X, if T is accretive and demicontinuous, the $\mu I + T$ is surjective for each $\mu > 0$ (i.e., T is m-accretive)

Corollary 4.2 was proved in [36, Theorem 9.14] for spaces not (π_1), using the differential equation approach.

It was noted in [296] that if in Theorem 4.3 one deletes "strongly," the A-properness property is lost, but if X is also uniformly convex, the accretive map has both the closedness property and is pseudo-A proper in the sense of the writer [233]. We prove first the following result due to Webb [296].

Theorem 4.4 Let X a separable, uniformly convex Banach space with uniformly convex dual X^*. Let $T: D(T) \to X$ be demicontinuous and m-accretive and suppose that $\{x_n\}$ is a bounded sequence such that $Tx_n \to$

f for some f in X. Then there exists $v \in D(T)$ such that $Tv = f$ [i.e., T satisfies $(++)$].

To prove Theorem 4.4 and others, we need the following extension of Proposition 4.1, which in turn depends on some extended results of Reich [247].

Proposition 4.2 Let X be as in Theorem 4.4, and let $\{x_n\} \subset X$ be bounded. Then there exists a subsequence $\{x_k\}$ (say) such that

$$\varphi(z) = \lim_k \| x_k - z \|$$

exists for all $z \in X$. Moreover, there is a unique $v \in X$ such that $J(x_k - v) \to 0$ in X^*. In fact, v is the unique point in X at which φ attains its infinum.

Proof. The existence of $\varphi(z)$ is Lemma 1.1 of Reich [247]. From Proposition 4.1 there exists v in X that minimizes φ over X and moreover, $J(x_k - v) \to 0$. As X is uniformly convex, the minimizer v is unique (see, e.g., [69]). It remains to prove that if $J(x_k - w) \to 0$ for some w in X, then $w = v$. As $J(x)$ is the subdifferential of $\frac{1}{2} \| x \|^2$, we have

$$(z - v, J(x_k - Z)) \leq \tfrac{1}{2}\| x_k - v \|^2 - \tfrac{1}{2}\| x_k - z \|^2 \tag{4.9}$$
$$\text{for each } k \text{ and all } z \in X.$$

We observe, for later reference, that this implies that

$$\lim_{k \to \infty} (z - v, J(x_k - z)) < 0 \qquad \text{for all} \quad z \in X. \tag{4.10}$$

Now suppose that $J(x_k - w) \to 0$. Taking $z = w$ in (4.9) and passing to the limit gives

$$\lim_{k \to \infty} \| x_k - w \|^2 \leq \lim_{k \to \infty} \| x_k - v \|^2,$$

and as v is the unique minimizer, this shows that $w = v$. Q.E.D.

Proof of Theorem 4.4. Let $\{x_k\}$ and v be as given in Proposition 4.2. As T is accretive, for each k and all $z \in D(T)$,

$$(Tx - Tz, J(x - z)) \geq 0.$$

Thus

$$\liminf_{k \to \infty} (f - Tz, J(x_k - z)) \geq 0 \qquad \text{for all} \quad z \in D(T).$$

By (4.10) in the proof of the proposition,

$$\liminf_{k \to \infty} (v - z, J(x_k - z)) \geq 0 \qquad \text{for all} \quad z \in X,$$

so that

$$\liminf_{k \to \infty} (v + f - (z + Tz), J(x_k - z)) \geq 0 \qquad \text{for all} \quad z \in D(T).$$

As T is m-accretetive, given $\xi \in X$ and $t \geq 0$, there exists $z_t \in D(T)$ such that

$$z_t + Tz_t = v + f - t\xi.$$

Thus for $t > 0$, $\liminf (t\xi, J(x_k - z_t)) \geq 0$. Cancel $t > 0$ and then let $t \to 0$. Since J is uniformly continuous on bounded sets, $\| J(x_k - z_t) - J(x_k - z_0) \| \to 0$, uniformly in k. Therefore, $\liminf_{k \to \infty} (\xi, J(x_k - z_0)) \geq 0$. As ξ is arbitrary, replacing ξ by $-\xi$ proves that $J(x_k - z_0) \to 0$. By Proposition 4.2, $z_0 = v$, so that $Tv = f$.   Q.E.D.

Corollary 4.3 Under the conditions Theorem 4.4, the image of a closed ball under m-accretive map $T: X \to X$ is closed.

The corollary was first proved by Browder [23] for nonseparable X but by a complicated argument.

The author of [296] also proved that under conditions of Theorem 4.4 with X also a π_1-space, a demicontinuous accretive map $T: D(T) \to X$ is also pseudo-A proper in the sense of the writer [233], where T is said to be *pseudo-A proper* w.r.t. Γ_1 if when $\{x_k \mid x_k \in X_k\}$ is any bounded sequence such that $P_k T(x_k) \to g$ for some g in X, then there exists a vector $x \in D(T)$ such that $Tx = g$.

It is not hard to show that a slight variant of the argument used in [296] to prove the latter assertion can also be used to show that

Proposition 4.3 Under the hypotheses of Theorem 4.4 with X also a π_1-space, a demicontinuous and accretive mapping $T: X \to X$ is weakly A-proper w.r.t. Γ_1.

We will now deduce some special cases of Theorems 4.1 and 4.2.

A. Let X be a separable and reflexive Banach space. In view of the results of Asplund and Kadec, we may and will assume, without loss of generality, that the norm in X is such that X and X^* are locally uniformly convex. Now, if in Theorem 4.2 we set $Y = X^*$, $E_n = X_n^*$, and $W_n = V_n^*$, the simplest choice of the mappings $K: X \to X$, $K_n: X_n \to X_n$, and

$G: X \to X^*$ for which (H1) holds and the scheme $\Gamma_I = \{X_n, V_n; X_n^*, V_n^*\}$ satisfies condition (4.3i) is to take $K = I$, $K_n = I_n$, and $G = J$, where J is the normalized duality map of X into X^*.

In this case Theorem 4.2 reduces to the new surjectivity result in [211].

Theorem 4.5 Let $T: X \to X^*$ satisfy conditions $(+)$ and $(++)$ and suppose that $T + \mu J$ is A-proper w.r.t. Γ_I for each $\mu > 0$.

(A5) If T is odd on $X - B(0,r)$ for some $r \geq 0$, then $T(X) = X^*$.

(A6) $T(X) = X^*$ also when T is not odd, provided one also assumes that

(H3A) $Tx \neq \gamma Jx$ for all $x \in X - B(0,r)$, all $\gamma < 0$, and some $r \geq 0$, and (4.11) T is either bounded or $(Tx,x) \geq -a\| x \|$ for all x and some $a > 0$.

To state an important corollary of Theorem 4.5, we recall:

Definition 4.2 $T: X \to X^*$ is said to be

(1) *Monotone* if $(Tx - Ty, x - y) \geq 0$ for all x,y in X.

(2) *Pseudomonotone* if $x_j \rightharpoonup x$ in X and $\lim \sup(Tx_j, x_j - x) \leq 0$ imply that $(Tx, x - v) \leq \lim \inf(Tx_j, x_j - v)$ for all v in X.

(3) *Semimonotone* if there is $V: X \times X \to X^*$ such that $T(x) = V(x,x)$ for $x \in V$, $V(x,\cdot)$ is monotone and hemicontinuous and $V(\cdot,x)$ is completely continuous for each fixed $x \in X$.

(4) *Quasimonotone* if $x_j \rightharpoonup x$ in X implies that $\lim \sup(Tx_j, x_j - x) \geq 0$.

Monotone maps were introduced independently by Vainberg-Kachurowski and Zarantonello, and were studied further by Minty, Browder, Kachurowskii, Rockafellar, Brezis, Fitzpatrick, and others; the applications of the monotone mapping theory to elliptic PDEs were initiated by Browder (see [17,21,29,33,39,67] for references). The study of a special class of pseudomonotone maps, the so-called mappings of the calculus of variations type, were initiated (with application to PDEs) by Leray and Lions [152], of the pseudomonotone maps by Brezis [16], and those of semimonotone maps by Browder [25,26]. Hess [113] and Calvert–Webb [44] introduced the notion of a quasimonotone map. Extending a result of [36], it was shown in [44] that if $T: X \to X^*$ is demicontinuous and quasimonotone, then $T_\mu = T + \mu J$ satisfies condition (S_+) of Browder [35] for each $\mu > 0$ [i.e., if $x_j \rightharpoonup x$ in X and $\lim \sup(T_\mu x_j, x_j - x) \leq 0$, then $x_j \to x$ in X]. It follows trivially from (1) that if T is monotone, then T is quasimonotone; moreover, it was shown in [15] that if T is either semimonotone or locally bounded and pseudomonotone, T is demicontinuous and quasimonotone. Note that if A is monotone and C compact, then $T = A + C$ is quasimonotone, but it is neither pseudomonotone nor semimonotone if C is not completely continuous.

In view of these remarks and the fact, proved in [231] (see also [36]) that if T is demicontinuous, quasimonotone, and *semibounded* (i.e., $\{Tx_j\}$ is bounded in X^* whenever $\{x_j\} \subset X$ and $\{(Tx_j, x_j)\}$ are bounded), then $T_\mu = T + \mu J$ is A-proper w.r.t. Γ_I for each $\mu > 0$, we deduce from Theorem 4.5 the following new existence theorem of Calvert–Webb [44] for T coercive and of Fitzpatrick [75] and Petryshyn [232] for T subcoercive.

Proposition 4.4 Suppose that $T: X \to X^*$ satisfies conditions ($+$) and ($++$), T is demicontinuous and quasimonotone, and T is either bounded or $(Tx,x) \geq -a\| x \|$ for $x \in X$ with T semibounded. Then $T(X) = X^*$ provided that

(H3) $Tx \neq \gamma Jx$ for all $\| x \| \geq r$, all $\gamma < 0$, and some $r > 0$.

It is known (see [24–26]) that if T is demicontinuous and either monotone, semimonotone, or pseudomonotone, then T satisfies condition ($++$); moreover, a monotone map is semibounded and demicontinuous if it is hemicontinuous (see [127]). Hence, in view of this and the discussion preceding Proposition 4.4, we see that the latter implies the following corollary, which extends and unifies many earlier results.

Corollary 4.4 Let $T: X \to X^*$ be demicontinuous and satisfy condition ($+$). Then $T(X) = X^*$ if T also satisfies one of the conditions:

(a) T is monotone.
(b) T is semimonotone, semibounded, and satisfies (H3A).
(c) T is pseudomonotone, satisfies (H3A), and is either bounded or $(Tx,x) \geq -a\| x \|$ for $x \in X$ with T semibounded.

Remark 4.3 Note that just as in the fixed-point theory, condition (H3A) is the weakest Leray–Schauder boundary condition that must be imposed on T, when T is not monotone, for T to be surjective. As has already been noted, (H3A) holds if, for example, there is an $r > 0$ such that $(Tx,x) \geq 0$ or even $\| Tx \| + (Tx,x)/\| x \| \geq 0$ for $\| x \| \geq r$. When T is monotone we may always assume that $(Tx,x) \geq 0$ for all $x \in X$ since otherwise we can replace T by $T' = T - T(0)$.

Note also that by Remark 3.2 ($3+$), Proposition 4.4, and Corollary 4.4 [for (b) or (c)] remain valid when conditions ($+$) and (H3A) are replaced by the single assumption

$$\| Tx \| + (Tx,x)/\| x \| \to \infty \qquad \text{as} \quad \| x \| \to \infty.$$

which was used in [25,75,154,174].

Remark 4.4 In case T is coercive, Corollary 4.4(a) was proved independently by Browder [29] and Minty [177]. In its present form it was proved by Rockafellar [252], although under some additional conditions on X it was proved earlier by Browder [26]. In case T is coercive, Corollary 4.4(b) was proved by Browder [26] and Corollary 4.4(c) by Brezis [16] (see also Leray-Lions [152]) when T is bounded. The existence results of Corollary 4.4(a) and (b) sharpen the existence theorems for semimonotone and pseudomonotone mappings under subcoercive assumptions of Browder, Browder-Hess, DeFigueiredo-Gupta, Fitzpatrick, and Petryshyn (see [211] for exact references). It should be added that except for the writer, none of the authors above assume that X is separable.

B.1. To deduce the surjectivity results from Theorem 4.2 for mappings $T: X \to X$ when X is a separable Banach space with a projectionally complete scheme $P_1 = \{X_n, V_n; X_n, P_n\}$ with $\| P_n \| = 1$, we set $Y = X$, $E_n = X_n$, and $W_n = P_n$ and note that if, for example, we choose $G = I$, $K = J$, and $K_n = P_n^* J$, then Theorem 4.2 yields the following new

Theorem 4.6 Let $T: X \to X$ satisfy conditions $(+)$ and $(++)$ and suppose that $T + \mu I$ is A-proper w.r.t. Γ_1 for each $\mu > 0$.

(A7) If T is odd on $X - B(0,r)$ for some $r \geq 0$, then $T(X) = X$.
(A8) $T(X) = X$ also when T is not odd provided that one assumes that (H3B) $Tx \neq \gamma x$ for $x \in X - B(0,r)$ and $\gamma < 0$ and

Either T is bounded or $(Tx,Jx) \geq -a\| x \|$ for $x \in X$. (4.11)

Remark 4.5 It is known that if $F: X \to X$ is 1-ball contractive and $T = I - F$, then T is bounded and $T + \mu I$ is A-proper w.r.t. Γ_1 for each $\mu > 0$. Moreover, the condition $Fx \neq \lambda x$ for $\| x \| \geq r$ and $\lambda > 1$, used in [237], is equivalent to (H3B) when $T = I - F$. Hence Theorem 1.1 in [237] (when F is 1-ball contractive) is a special case of Theorem 4.6. Let us add that condition $(++)$ can be dropped if F is ball condensing.

It follows from the result of Martin [161] that if $T: X \to X$ is continuous and accretive [i.e., $(Tx - Ty, w) \geq 0$ for all $x,y \in X$ and some $w \in J(x - y)$], then $T + \mu I$ is A-proper w.r.t. Γ_1. Consequently, Theorem 4.6

implies the validity of the following result from [23], preceding a related result in [296].

Corollary 4.5 Let $T: X \to X$ be continuous, accretive, and satisfy condition $(+)$, Then $T(X)$ is dense in X and $T(X) = X$ if T also satisfies condition $(++)$.

Further, let us state an important corollary of Theorem 4.1 when $Y = X$. Extending a lemma of Toland and Webb [275], it was shown in [220], and independently in [171], that if $F: X \to X$ is ball condensing and $T: X \to X$ accretive and continuous, then $T = I + A - F$ is A-proper w.r.t. Γ_1. Since the map $T + \mu I$ is also A-proper for each $\mu > 0$, an immediate consequence of Theorem 4.1 is the following constructive result, whose special cases appear in [171,175].

Corollary 4.6 Let $F: X \to X$ be ball condensing, $A: X \to X$ accretive and continuous, and suppose that $T = I + A - F$ satisfies condition $(+)$. If there is $r > 0$ such that either T is odd on $X - B(0,r)$, or $(F - A)(x) \neq \lambda x$ for $x \in X - B(0,r)$ and $\lambda > 1$, then the equation $x + Ax - Fx = f$ is feebly a-solvable w.r.t. Γ_1 for each f in X, and strongly if it is uniquely solvable for a given f.

Remark 4.6 Suppose that there are constants $a_0 > 0$, a_1, and $\beta \in (0,2)$ such that

$$(Fx,w) \leq a_0\| x \|^\beta + a_1 \qquad \text{for all} \quad x \in X \quad \text{and some} \quad w \in J(x).$$

$$(4.12)$$

Then $T = I + A - F$ satisfies condition $(+)$ and there is $r > 0$ such that $(F - A)(x) \neq \lambda x$ for $\| x \| \geq r$ and F satisfying (4.12).

To see this, suppose that $Tx_j \to g$ in X. Then $\| Tx_j \| \leq M$ for all j and for some $M > 0$. Let $\tilde{T} = I + A$ and note that

$$(\tilde{T}x_j,w_j) = \| x \|^2 + (Ax_j - A(0) + A(0), w_j) \geq \| x_j \|^2 - (A(0), w_j)$$

$$\geq \| x_j \|^2 - c_1\| x_j \|, \qquad \text{where} \quad c_1 = \| A(0) \|.$$

Hence it follows from this and (4.12) that

$$M \| x_j \| \geq (\tilde{T}x_j - Fx_j, w_j) = (\tilde{T}x_j,w_j) - (Fx_j,w_j)$$

$$\geq \| x_j \|^2 - a_0\| x_j \|^\beta - c_1\| x_j \| - a_1 \qquad \text{for each} \quad j \in Z^+.$$

This implies that for each $j \in Z^+$ we have

$$M \geq \| x_j \| - a_0 \| x_j \|^{\beta - 1} - c_1 - a_1 / \| x_j \| \quad (\beta \in (0,2)).$$

It obviously follows from the inequality above that $\{x_j\}$ is bounded.

To verify the second hypothesis mentioned above, we note that if $(F - A)(x) = \lambda x$, then $(A + \lambda)(x) = Fx$, so for some $w \in J(x)$ we have

$$(Ax, w) + \| x \|^2 = (Fx, w). \tag{4.13}$$

Since $(Ax, w) \geq -c_1 \| x \|$ for $x \in X$, (4.12) and (4.13) imply that

$$\lambda \| x \|^2 \leq a_0 \| x \|^\beta + a_1 + c_1 \| x \|.$$

So if $\lambda \geq 1$, one has $\| x \|^2 \leq a_0 \| x \|^\beta + a_1 + c_1 \| x \|$. Clearly, those $x \in X$ which satisfy the preceding inequality form a bounded set. Hence (H3B) of Theorem 4.6 holds and thus Corollary 4.6 remains valid if condition $(+)$ is replaced by (4.12).

Finally, we remark that the existence result analogous to Corollary 4.6 when F is set condensing and satisfies condition (4.12) was proved by Schöneberg [260] using a rather complicated degree theory for semicondensing mappings. When $A = I$ we get the result of [237].

Finally, in view of Corollary 4.2 and Theorem 4.4 of Webb, we have the following stronger version of Corollary 4.5, which properly extends Theorem 3 of Webb [296].

Corollary 4.7 If X and X^* are uniformly convex Banach spaces and if $T: X \to X$ is accretive, demicontinuous, and satisfies condition $(+)$, then $T(X) = X$.

4.4 Error Estimate for the General Projection Method

Let X and Y be real Banach spaces with the projectionally complete scheme $\Gamma_p = \{X_n, P_n; Y_n, Q_n\}$, D an open set in X, $T: \overline{D} \subset X \to Y$ a Fréchet-differentiable A-proper (w.r.t. Γ_p) mapping, $D_n = D \cap X_n$, and $T_n = Q_n T|_{\overline{D}_n}: D_n \subset X_n \to Y_n$ for each $n \in Z^+$. In this section we prove a theorem stated in [225] which provides the estimate for the error $\| x_0 - x_n \|$ when $x_0 \in D$ is an isolated solution of the equation

$$T(x) = f \quad (x \in \overline{D}, f \in Y) \tag{4.14}$$

with an A-proper F-derivative T'_{x_0} at $x_0 \in D$ and $x_n \in D_n$ is a solution of the approximate equation

$$T_n(x) = Q_n f \quad (x_n \in \overline{D}_n, Q_n f \in Y_n), \quad n = 1, 2, 3, \ldots \tag{4.15}$$

such that $x_n \to x_0$ in X as $n \to \infty$ [i.e., equation (4.14) is uniquely approximation solvable w.r.t. Γ_p for a given f in Y].

To state and prove our theorem, we first recall that $T: D \subset X \to Y$ is said to be *F-differentiable* at $x \in D$ if there exists an operator $T'_x \in L(X,Y)$, called the *F-derivative of T at x*, such that for each $y \in D$ one has

$$Tx - Ty = T'_x(x - y) \tag{4.16}$$

$$+ R(x,y) \quad \text{with} \quad \frac{\| R(x,y) \|}{\| x - y \|} \to 0 \quad \text{as} \quad \| x - y \| \to 0.$$

It is known (see [136]) and easy to prove that if $C: D \subset X \to X$ is compact and *F*-differentiable at $x \in D$, the C'_x is also compact. It is also known that if C is *k*-set or *k*-ball contractive and *F*-differentiable at $x \in D$, then $C'_x \in L(X,X)$ is also *k*-set or *k*-ball contractive (see [195]. However, the following example in [62] shows that the *F*-derivative of an A-proper map need not, in general, be A-proper.

Example 4.1 Let $X = (c)_0$ with basis $\{\varphi_i\}$ and let $\Gamma_1 = \{X_n, P_n\}$ be a projectionally complete scheme determined by $\{\varphi_i\}$. Define an operator $T: X \to X$ by

$$T(\sum_{i \geq 1} x_i \varphi_i) = \tfrac{1}{3} \sum_{i \geq 1} x_i^3 \varphi_i \quad \text{for} \quad x \in X. \tag{4.17}$$

Then one can easily check that $T: X \to X$ is A-proper w.r.t. Γ_1, but its *F*-derivative T'_{x_0} at $x_0 = \sum_{i \geq 1} i^{-1/2} \varphi_i$ is *not* A-proper w.r.t. Γ_1.

However, it was shown by Petryshyn and Tucker [236] that if X is reflexive, $F: X \to X$ is *F*-differentiable at each $x \in X$, and $F': X \to L(X,X)$ is completely continuous [i.e., $F'_{x_j} \to F_x$ in $L(X,X)$ if $x_j \rightharpoonup x$ in X], then F'_x is P-compact at each $x \in X$ if and only if F is P-compact. Using the technique of [236], it was shown by Fitzpatrick in [76] that an analogous result holds for an *F*-differentiable A-proper mapping $T: X \to Y$ with X reflexive under the following slightly weaker condition:

$$T'_{y_n}(x - z_n) - T'_x(x - z_n) \to 0 \quad \text{when} \quad z_n \rightharpoonup x \quad \text{and} \quad y_n \rightharpoonup x \quad \text{in} \quad X.$$

In [76], Fitzpatrick developed the theory of *F*-differentiable A-proper mappings. First, it was shown that

Lemma 4.1 If $T \in L(X,Y)$ is A-proper and injective, then $\text{Deg}(T,B(0,r),0) = \text{Deg}(T,B(x_0,r), T(x_0))$ for any x_0 in X and $r > 0$.

Then, in addition to other results, he extended the Leray-Schauder formula (see [136]) to A-proper mappings by proving

Lemma 4.2 Let $D \subset X$ be open and bounded and let $T: \overline{D} \subset X \to Y$ be A-proper w.r.t. Γ_p with F-derivative T'_{x_0} at $x_0 \in D$ such that T'_{x_0} is A-proper w.r.t. Γ_p and injective. Then x_0 is an isolated solution of $Tx = T(x_0)(x \in D)$ and there exists a neighborhood $\overline{B}(x_0,r) \in D$ such that

$$\mathrm{Deg}(T,B(x_0,r),T(x_0)) = \mathrm{Deg}(T'_{x_0},B(x_0,r),T'_{x_0}(x_0)).$$

Combining Lemmas 4.1 and 4.2, we obtain the following generalization of a classical result which is quite useful.

Corollary 4.8 Under condition of Lemma 4.2,

$$\mathrm{Deg}(T,B(x_0,r),T(x_0)) = \mathrm{Deg}(T'_{x_0},B(0,r),0).$$

These and other differentiability results from [76] are proved in Chapter IV.

In what follows we say that $T: \overline{D} \subset X \to Y$ is *continuously* F-differentiable at $x_0 \in D$ if T is F-differentiable in a neighborhood of x_0 in D and $\| T'_x - T'_{x_0} \| \to 0$ as $x \to x_0$ in X. In view of (4.16), this implies that

$$\| R(x,y) \| / \| x - y \| \to 0 \quad \text{as} \quad x \to x_0 \text{ and } y \to y_0 \text{ in } X. \quad (4.18)$$

We are now in position to prove the main result of this section, which, among others, extends [136, Theorem 3.2 in Chapter III] dealing with the speed of convergence of the Galerkin method when the latter is applied to equation (4.14) when $Y = X$, $f = 0$, and $T = I - C$ with $C: X \to X$ compact.

Theorem 4.7 Let $D \subset X$ be open and $T: \overline{D} \to Y$ A-proper. Let $x_0 \in D$ be a solution of equation (4.14) and suppose that T is F-differentiable at x_0 with T'_{x_0} injective and A-proper. Then equation (4.14) is strongly approximation solvable in $B(x_0,r) \subset D$ for some $r > 0$; that is, there exist $r > 0$ and $M_0 \geq 1$ such that equation (4.14) has x_0 as its only solution in $\overline{B}(x_0,r) \subset D$ and the approximate equation (4.15) has a solution $x_n \in B_n(x_0,r)$ for each $n \geq n_0$ such that $x_n \to x_0$ in X as $n \to \infty$. Moreover, if we set $\gamma = \| (T'_{x_0})^{-1} \|$, then for any $\xi \in (0,\gamma)$ there exists an $n_1 \, (\geq n_0)$ such that

$$\| x_n - x_0 \| \leq (\gamma - \xi)^{-1} \| Tx_n - f \| \quad \text{for} \quad n \geq n_1. \quad (4.19)$$

If we also assume that T is continuously F-differentiable at x_0, there exists $n_2 \, (\geq n_1)$ such that equation (4.15) is uniquely solvable for each $n \geq n_2$. Moreover, there is a constant M that depends on $\| T'_{x_0} \|$, γ, and β

$= \sup_n \| Q_n \|$ such that

$$\| x_n - x_0 \| \le M \| x_0 - P_n x_0 \| \qquad \text{for} \quad n \ge n_2. \tag{4.20}$$

Proof. Since, by hypothesis, the A-proper mapping $T: \overline{D} \subset X \to Y$ has an injective A-proper F-derivative T'_{x_0} at the solution $x_0 \in D$ of equation (4.14), it follows from Lemma 4.2 and Corollary 4.8 that x_0 is an isolated solution and there exist $\overline{B}(x_0, r) \in D$ such that

$$0 \notin \mathrm{Deg}(T, B(x_0, r), f) = \mathrm{Deg}(T'_{x_0}, B(0, r), 0)$$

because $T'_{x_0} \in L(X, Y)$ is A-proper and injective. Hence, by the definition of the generalized degree, Deg (see Chapter IV), there exists $n_0 \ge 1$ such that the Brouwer degree $\deg(T_n, B_n(x_0, r), Q_n f) \ne 0$ for $n \ge n_0$. Hence there exists $x_n \in \overline{B}_n(x_0, r)$ such that $T_n(x_n) \to Q_n f$ and so, by the A-properness of T and the property of $B(x_0, r)$, $x_n \to x_0$ in X as $n \to \infty$. In view of this, relation (4.16), our condition on T'_{x_0}, and the equality

$$T'_{x_0}(x_0 - x_n) = f - T x_n - R(x_0, x_0 - x_n) \tag{4.21}$$

we see that $\| T'_{x_0}(x) \| \ge \gamma \| x \|$ for all $x \in X$ and thus for any $\epsilon \in (0, \gamma)$ there exists $n_1 (\ge n_0)$ such that $\| R(x_0, x_0 - x_n) \| \le \epsilon \| x_0 - x_n \|$ by (4.16). This and (4.21) imply the validity of the error estimate (4.19).

To prove the second part of Theorem 4.7, note that if $\{z_n\} \subset B(x_0, r)$ is such that $z_n \to x_0$ in X as $n \to \infty$, then since T is assumed to be continuously F-differentiable at x_0, there exist a constant $c > 0$ and an integer $n_2 (\ge n_1)$ such that

$$\| Q_n T'_{z_n}(x) \| \ge c \| x \| \qquad \text{for all} \quad x \in X_n \quad \text{and all} \quad n \ge n_0. \tag{4.22}$$

From (4.22) and the fact that

$$\| R(P_n x_0, P_n x_0 - x_n) \| / \| P_n x_0 - x_n \| \to 0 \qquad \text{as} \quad n \to \infty,$$

it follows that equation (4.15) is uniquely solvable in $B(x_0, r)$ for each $n \ge n_2$. Setting $z_n = P_n x_0$ and using the fact that $Q_n T x_n = Q_n f$ with $T x_0 = f$ and the equality

$$\begin{aligned} Q_n T'_{z_n}(x_n - z_n) &= Q_n T x_n - Q_n T_{z_n} - Q_n R(z_n, x_n - z_n) \\ &= Q_n \{ T x_0 - T_{z_n} - R(z_n, x_n - z_n) \}, \end{aligned}$$

we see that to any given $\epsilon \in (0, c)$ there corresponds $n_3 (\ge n_2)$ such that

$$(c - \epsilon) \| x_n - z_n \| \le \beta \| T x_0 - T_{z_n} \|$$
$$\text{for} \quad n \ge n_2 \quad \text{where} \quad \| Q_n \| \le \beta \quad \text{for} \quad n \in Z^+.$$

On the other hand, $\| T x_0 - T_{z_n} \| \le \| T'_{x_0} \| \| z_n - x_0 \| + \epsilon_1 \| z_n - x_0 \|$

for $n \geq n_1$ $(\geq n_3)$ and any given $\epsilon_1 > 0$. Combining the inequalities above, we obtain the error estimate (4.20).   Q.E.D.

Remark 4.7 Theorem 4.7 is applicable, in particular, to the case when $Y = X$, $f = 0$, and $T = I - F$, where $F: \overline{D} \subset X \to X$ is P_1-compact and, in particular, ball condensing or compact. Thus Theorem 3.2 in [136, Chapter III], as well as Theorem 7 in [203] for $\gamma = 1$ (without the restriction assumption (23) used in [203]) are deducible as corollaries of Theorem 4.7.

4.5 Nonlinear Friedrichs Extension with Error Estimate for the Galerkin Method

Let H be a Hilbert space and let $P: D_P: H \to H$ be a densely defined nonlinear, unbounded, and K-strongly stable operator studied by the author in [217]. The purpose of this paper is to use the theory of the solvable Friedrichs type of extension from [217] (see Section 4.6) to prove the convergence in a generalized energy space H_0 of the Galerkin approximants $x_n \in H_n \subset D_P$ without the restrictive boundedness assumption imposed in [217], provide the error estimate in the H_0-norm, which in some sense is the best possible, and apply these abstract results to the constructive solvability of the nonlinear partial differential equation [86], which is important in the theory of stationary flows and other fields.

Preliminaries

In this section we provide readers with some definitions and results with which they may not be familiar. Let $(\cdot,\cdot)$ and $\| \cdot \|$ denote the inner product and norm in H, respectively. A linear operator T defined on dense domain $D_T \subset H$ is called K-*positive definite* (K-p.d.) if there exists a closable operator K with $D_K \supseteq D_T$ mapping D_T onto a dense subspace $K(D_T)$ of H and two constants $\alpha > 0$ and $\beta > 0$ such that for all $u \in D_T$,

$$(Tu,Ku) \geq \alpha\| u \|^2, \qquad \| Ku \|^2 \leq \beta(Tu,Ku). \qquad (\#)$$

T is called K-*symmetric* if $(Tu,Kv) = (Ku,Tv)$ for all $u,v \in D_T$. We note that when H is real, K-p.d. operators need not be K-symmetric, but when H is complex, each K-p.d. operator T is K-symmetric since $(Tu,Kv) = (Ku,Tv)$ by polarization.

Suppose that T is K-p.d. (and K-symmetric if H is real) and let H_0 denote the completion of D_T with respect to

$$[u,v] = (Tu,Kv), \qquad | u |_0 = [u,u]^{1/2} \qquad \text{for all} \quad u,v \in D_T.$$

Then it is known [216] that H_0 can be regarded as a subset of H, K can be extended to a bounded operator K_0 of H_0 into H, and T has a unique closed K_0-p.d. and K_0-symmetric extension T_0, called the *generalized Friedrichs extension of* T such that $T_0 \supseteq T$ and T_0 has a bounded inverse T_0^{-1} defined on all of H and T_0 satisfies (#). When $K = I$, T_0^{-1} is the Friedrichs self-adjoint extension. For the exhaustive survey of results in this area of this and other authors and for applications of the linear theory of K-p.d. operators to PDEs, see the monograph of Filippov [73].

In [217] the author extended his procedure of [216] to yield a nonlinear version of the theory of Friedrichs extension of densely defined nonlinear and unbounded mapping $P: D_P = D_T \subset H \to H$, which satisfies the following two conditions:

$$| (Pu - Pv, K(u - v)) | \geq \eta | u - v |_0^2 \qquad (4.23)$$
$$\text{for all} \quad u,v \in D_P \quad \text{and some} \quad \eta > 0.$$

$$((P(u_n) - P(u_m), Kh) \to 0 \qquad \text{as} \quad n,m \to \infty \quad \text{and} \quad h \in D_P \qquad (4.24)$$

whenever $\{u_n\} \subset D_T$ is a Cauchy sequence in H_0. Under these conditions the extension P_0 is bijective, unique, H_0-demiclosed [if $\{u_n\} \subset D_P$ is such that $u_n \to u$ in H_0 and $P(u_n) \to g$ in H, then $u \in D_P$ and $P(u) = g$] and structurally given by $P_0 = T_0 W_0$, where W_0 is an extension of $W \equiv T_0^{-1}P$ in H_0 such that $W \subset W_0 \subset \overline{W}$ with $\overline{W}$ a weak closure in H_0 and $R_{W_0} = D_{T_0}$. It follows from (4.24) that $\overline{W}: H_0 \to H_0$ is demicontinuous and from (4.23) that

$$\eta | u - v |_0^2 = | [\overline{W}u - \overline{W}v, u - v] | \qquad \text{for all} \quad u,v \in H_0. \qquad (4.25)$$

We refer to $P: D_P \subset H \to H$ satisfying (4.23) as *K-strongly stable* on D_P and to $\overline{W}: H_0 \to H_0$ satisfying (4.25) as *strongly stable* on H_0. It was noted in [314] that when one treats elliptic equations with the aid of the unbounded Friedrichs extension, one works in terms of "unbounded" operators. In certain fields (e.g., quantum theory and others) the use of unbounded operators is indispensable.

Suppose now that H is separable. Then H_0 is also separable, so we can construct a sequence $\{H_n\}$ of finite-dimensional subspaces of D_P such that $\text{dist}_{H_0}(u,H_n) \to 0$ as $n \to \infty$ for each u in H_0. We will also assume that $\{H_n\}$ is monotonic. Utilizing the structure of P_0, the author proved in [217] (see Theorems 4.8 and 4.9) that if P satisfies conditions (4.23)–(4.24), the equation

$$Pu = f \qquad (u \in D_P, f \in H) \qquad (4.26)$$

has a unique *strong solution* $u_0 \in D_{P_0} \subset H_0$ for any given f in H such that

$$P_0(u_0) = f; \tag{4.27}$$

moreover, if $\overline{W}$ is bounded, the strong solution u_0 was constructed as the unique limit in H_0 of the Galerkin approximants $u_n \in H_n$ determined by

$$(P(u_n), Kv) = (f, Kv) \qquad \text{for all} \quad v \in H_n, \quad n = 1, 2, \ldots. \tag{4.28}$$

Main New Results

Using the structure of the solvable Friedrichs extension $P_0 = T_0 W_0$, in this section we prove the convergence of the Galerkin approximants $u_n \to u_0$ in H_0 without the restrictive assumption that $\overline{W}$ is bounded in H_0 and we also provide the error estimate for $|u_0 - u_n|_0$, which appears to be the best possible even when $K = T = I$ and $P: H \to H$. To state our first new result we need the following notion. The operator P is H_0-*locally Lipschitz continuous* if to each u in D_P there exist a neighborhood $V(u)$ in H_0 and a number $l \equiv l(u) > 0$ such that

$$\begin{aligned}
|(Pu - Pv, Kh)| &\le l\,|u - v|_0\,|h|_0 \\
&\text{for all} \quad u, v \in D_P \cap V(u) \quad \text{and} \quad h \quad \text{in} \quad D_P.
\end{aligned} \tag{4.29}$$

Theorem 4.8 Let H be a Hilbert space and let $P: D_P \equiv D_T \subset H \to H$ be a nonlinear mapping satisfying conditions (4.23) and (4.24). Then:

(a) For each $f \in H$, equation (4.26) has a unique strong solution $u \in D_{P_0} \subset H_0$ satisfying (4.27), where P_0 is the generalized Friedrichs extension of P.

(b) If H is also separable, the Galerkin equation (4.28) has a unique solution $u_n \in H_n$ for each $n \ge 1$ such that $|u - u_n|_0 \to 0$ as $n \to \infty$.

(c) If P is also H_0-locally Lipschitz continuous on D_P, the speed of convergence is characterized by the estimate

$$|u - u_n|_0 \le \frac{l}{\eta}\,\text{dist}_{H_0}(u, H_n) \qquad \forall\, n \ge n_0. \tag{4.30}$$

Proof. The validity of assertion (a) follows from the discussion preceding the statement of the theorem. To prove assertion (b), we note that because the generalized Friedrichs extension P_0 is given structurally by $P_0 = T_0 W_0$, we see that the solvability of equations (4.28) is equivalent to the solvability of $[W_0(u_n), v] = [T_0^{-1} f, v]$ or of equations

$$[\overline{W}(u_n), v] = [f_0, v] \qquad \text{for all} \quad v \text{ in } H_n \text{ with} \tag{4.31}$$

$$f_0 = T_0^{-1} f \in H_0.$$

It follows from (4.23) and (4.24), which is also satisfied by $P_0 = T_0 W_0$, that $\overline{W}: H_0 \to H_0$ is demicontinuous and strongly stable, that is,

$$\eta| u - u_n |_0^2 = | [\overline{W}u - \overline{W}u_n, u - u_n] |. \tag{4.32}$$

This and Proposition 1.4 in [51] imply that the equation

$$\overline{W}(u) = f_0 \qquad (u \in H_0, f_0 = T_0^{-1}f \in H_0) \tag{4.33}$$

is uniquely approximation solvable [i.e., equation (4.31) has a unique solution $u_n \in H_n$ for each $n \geq 1$ and each $f_0 \in H_0$ such that $u_n \to u_0$ strongly for some $u_0 \in H_0$ and $\overline{W}(y_0) = f_0$]. Since $f_0 = T_0^{-1}f$, we see that $\overline{W}(u_0) \in D_{T_0} = R_{W_0}$, $u_0 \in D_{P_0}$, and $P_0(u_0) = f$. Because (4.26) has a unique strong solution satisfying (4.27), it follows that $u_0 = u$.

To prove assertion (c) we note that since (4.27) (with $u_0 = u$) holds if and only if

$$(P_0(u), K_0 v) = (f, K_0 v) \qquad \text{for all} \quad v \in H_0,$$

and

$$(P_0(u_n), Kv) = (f, Kv) \qquad \text{for all} \quad v \in H_n, \quad n = 1, 2, \ldots,$$

it follows that

$$(P_0(u) - P_0(u_n), Kv) = 0 \qquad \text{for all} \quad v \in H_n. \tag{4.34}$$

In view of the equality (4.34) and the inquality (4.23), for all $v \in H_n$ we have

$$\begin{aligned} \eta| u - u_n |_0^2 &\leq | (P_0(u) - P_0(u_n), K_0(u - u_n)) | \\ &= | (P_0(u) - P(u_n), Ku) - (P_0(u) - P_0(u_n), Ku_n) | \\ &= | (P_0(u) - P_0(u_n), K(u - v)) |. \end{aligned}$$

Setting $h = u - v$, the inequality above shows that

$$\eta| u - u_n |_0^2 \leq | (P_0(u) - P_0(u_n), Kh) |, \qquad h = u - v. \tag{4.35}$$

Since $| u - u_n |_0 \to 0$ as $n \to \infty$ by assertion (b) and P_0 is locally Lipschitz continuous, it follows from this and (4.35) that there exist a neighborhood $V(u)$ of u in H_0, a number $l \equiv l(u) > 0$, and $n_0 \in Z^+$ such that $\{u_n\} \subset V(u) \cap D_P$ for all $n \geq n_0$, and thus for all v in H_n we have

$$| (P_0(u) - P_0(u_n), Kh) | \leq l | u - u_n | | h | = l | u - u_n | | u - v |$$

since $h = u - v$. Combining the last inquality with (4.35) and canceling $| u - u_n |$ on both sides, we have the looked-for inequality (4.30), that is,

$$| u - u_n |_0 \leq \frac{l}{\eta} \, \text{dist}_{H_0}(u, H_n), \qquad \forall \, n \geq n_0;$$

where l is a small local Lipschitz constant that depends on u but not on the subspaces $\{H_n\}$. Q.E.D.

Corollary 4.9 If $P: H \to H$ is demicontinuous and strongly stable, that is,

$$\gamma \| u - v \|^2 \leq | (Pu - Pv, u - v) | \tag{4.36}$$
$$\text{for all} \quad u, v \text{ in } H \quad \text{and some} \quad \gamma > 0,$$

then the Galerkin equation (4.28) (with $K = I$) has a unique solution $u_n \in H_n \subset H$ for each $n \geq 1$ such that $\| u - u_n \| \to 0$ as $n \to \infty$.

If P is also locally Lipschitz continuous on H, the speed of convergence is characterized by

$$\| u - u_n \| \leq \frac{l}{\gamma} \operatorname{dist}_H(u, H_n) \quad \text{for all} \quad n \geq u_0. \tag{4.37}$$

Proof. Corollary 4.9 is an immediate consequence of Theorem 4.8 when in the latter we put $K = T = I$ with I an identity on $H_0 \equiv H$. Q.E.D.

Remark 4.8 To the best of my knowledge there are no sharp estimates of the form (4.30) when the Galerkin method is applied to equations (4.36) involving unbounded densely defined abstract nonlinear operators. The estimate of [51, Theorem 5] is not as sharp as (4.30); and moreover, it was obtained there under somewhat restrictive conditions:

$$| [\overline{W}u - \overline{W}v, u - v] | \leq C(M) | u - v |_0 \quad \text{for all} \quad u, v \text{ in } H \tag{4.38}$$

when $| u |_0, | v |_0 \leq M$ for any $M \in R^+$ and some constant $C(M) > 0$.

Remark 4.9 Since monotonicity of $\{H_n\}$ was used only to prove that equation (4.28) is uniquely approximation solvable in H_0 when $\overline{W}: H_0 \to H_0$ is strongly stable and demicontinuous, it is known and easy to show that when $\overline{W}$ is also bounded (or continuous) in H_0, then equation (4.33) is uniquely approximation solvable in H_0 *without* the assumption that $\{H_n\}$ is monotonic. Thus in this case Theorem 4.8 is *valid without* the monotonicity condition on $\{H_n\}$. Similar remarks apply to Corollary 4.9, where $W = P$ and $H_0 = H$.

Remark 4.10 The whole point in our discussion of the constructive approach to the solvability of equation (4.26) was to obtain the error estimate (4.30). When Theorem 4.1 is applied to ODEs and PDEs, then with a judicious choice of finite-dimensional spaces H_n (e.g., finite element spaces) in suitable Sobolev spaces, one can apply the interpolation and approximation function theory results directly to estimate

$$\text{dist}_{H_0}(u, H_n) = \inf_{v \in H_n} |u - v|_0.$$

This together with (4.30) gives the upper bounds for the Galerkin error $|u - u_n|_0$ even for equations involving unbounded nonlinear densely defined operators. A similar remark applies to Corollary 4.9, when $P: H \to H$ is a nonlinear operator defined on all of H but not necessarily bounded or continuous.

It should be noted that Theorem 4.8 and Corollary 4.9 contain a number of special cases, depending on the choice of operators T and K and the smoothness of P.

Here we restrict ourself to the following consequence of Theorem 4.8, which is useful when Galerkin's method is applied to the solvability of boundary value problems for nonlinear or semilinear ordinary and elliptic differential equations when we seek a weak or generalized solution in a suitable Sobolev space and where the corresponding operator is not necessarily strongly monotone and Lipschitz continuous on bounded sets, as was assumed in [284] and others.

When in Theorem 4.8 we let $D_T = H$, $T = K = I$, assume that $P: H \to H$ is either continuous, or bounded and demicontinuous, and *strongly stable*; that is, for all $u, v \in H$ and some $\eta > 0$,

$$\eta \| u - v \|^2 \leq | (Pu - Pv, u - v) |, \tag{4.39}$$

and $\{H_n\} \subset H$ is such that $\text{dist}_H(u, H_n) \to 0$ as $n \to \infty$ for each u in H with $\{H_n\}$ *not* necessarily monotonic, then we have the following special case of Corollary 4.9.

Corollary 4.10   If $P: H \to H$ is strongly stable and continuous or bounded and demicontinuous, then the Galerkin equation (4.28) (with $K = I$) has a unique solution $u_n \in H_n$ for each $n \geq 1$ such that $u_n \to u$ in H as $n \to \infty$ and u is the unique solution of equation (4.26) (with $D_P = H$). If P is also locally Lipschitz continuous, then the speed of convergence is characterized by the estimate

$$\| u - u_n \| \leq \frac{l}{\eta} \, \text{dist}_H(u, H_n). \tag{4.40}$$

Remark 4.11   The existence part of Corollary 4.9 was first proven in [312] for $P: H \to H$ bounded and extended in [27] without this condition, while the proof of the convergence of the Galerkin method without error estimate was given in [204] for P continuous. For more details concerning this class of maps, see [51,73,99].

Remark 4.12 Note that when $B: H \to H$ is strongly monotone (with constant $b \geq 0$) and $C: H \to H$ is Lipschitz continuous (with constant $c > 0$), then $P \equiv B + C$ is strongly stable if $b - c > 0$ since for all u,v in H we have

$$| (Pu - Pv, u - v) | \geq (Bu - Bv, u - v) - | (Cu - Cv, u - v) |$$
$$\geq (b - c)\| u - v \|^2,$$

and thus Corollary 4.10 applies to the latter class of mappings.

Remark 4.13 Corollaries 4.9 and 4.10 remain valid when $P: X \to X^*$, where X is a separable reflexive Banach space, X^* its dual, and the pair (X,X^*) has a projectionally complete scheme.

It was shown in [86] that the problem associated with the study of stationary flow downward along a plain wall reduces to determining a solution u on G of the diffusion equation:

$$Pu \equiv -\Delta u + g(x) \frac{\partial u}{\partial y} + f(u,y) = h(x,y) \qquad [u \in L_2(G)] \qquad (4.41)$$

$$u(x,0) = u(0,y) = \frac{\partial u}{\partial y}(x,b) = \frac{\partial u}{\partial x}(a,b) = 0, \qquad (4.42)$$

where $G \equiv \{(x,y) \mid 0 \leq x \leq a, 0 \leq y \leq b\}$. In this problem $g(x)$ describes the velocity of the fluid and f the possible chemical reaction of the components in the liquid.

To apply our Theorem 4.8, set $Tu \equiv -\Delta u$ with $D(T) = D(P) \equiv \{u \in C^2(G) \mid u$ satisfies (4.42)$\}$ and let H_0 be the completion of $D(T)$ with respect to $[u,v] = (-\Delta u,v) = (Tu,v)$ with $| \cdot |_0 = [\cdot]$, where $(\cdot,\cdot)$ and $\| \cdot \| = (\cdot)^{1/2}$ is the metric of $L^2(G)$. Assume that functions g and f satisfy the following conditions:

(a) $g: [0,a] \to R^+$ is continuous and nonnegative
(b) $f: R \times [0,b] \to R$ is continuous and increasing in its first variable u, $f(0,y) = 0$, and $f(u,y)$ is H_0-locally Lipschitzian in u; that is, to each u in $D(T)$ there exist a neighborhood $V(u)$ in H_0 and $l \equiv l(u)$ such that $\| f(u;\cdot) - f(v,\cdot) \| \leq l\| u - v \|_0$ for $u,v \in V(u) \cap D(T)$.

Let $\{H_n\} \subset D(P)$ be a sequence of finite-dimensional spaces such that $\text{dist}_{H_0}(v,H_n) \to 0$ as $n \to 0$ for each v in H_0, and let $u_n \in H_n$ be a Galerkin solution determined by

$$(Pu_n, v) = (f, v) \qquad \forall \, v \in H_n. \tag{4.43}$$

Proposition 4.5 Let P be defined by (4.41)–(4.42) and assume that conditions (a) and (b) are satisfied. Then for each h in $L_2(G)$ there exists a unique strong solution u_0 of (4.41)–(4.42) in $D(P_0) \subset H_0$, where P_0 is the generalized Friedrichs type extension of P, the Galerkin equation (4.43) has a unique solution $u_n \in H_n$ for each $n \geq 1$ and the speed of convergence of $|u_0 - u_n|_0$ is characterized by

$$|u_0 - u_n\| \leq (q + l) \, \mathrm{dist}_{H_0}(u_0, H_n), \tag{4.44}$$

where $q = \max g(x)$ for $x \in [0, a]$.

Proof. The assertion of Proposition 4.5 follows immediately from Theorem 4.8. Indeed, it is not hard to see that for u, v in $D(P)$, we have

$$(Pu - Pv, u - v) = |u - v|_0^2$$

$$+ f(u, y) - f(v, y)(u - v) + \left(g(x) \frac{\partial(u - v)}{\partial y}, u - v \right) \geq |u - v|_0^2,$$

since by assumption (b) the second term is not negative, while one easily shows that

$$\left(g(x) \frac{\partial(u - v)}{\partial y}, u - v \right) = \int_0^a \tfrac{1}{2} g(x) [u(x, b) - v(x, b)]^2 \, dx \geq 0.$$

On the other hand, one easily shows that $\| Nu - Nv \| \leq (q + l) | u - v |_0$ for all $u, v \in V(u_0) \cap D(P)$. Thus Proposition 4.5 follows from Theorem 4.7. Proposition 4.5 improves the results of [51, 86].   Q.E.D.

4.6   Friedrichs Extension of Nonlinear Operators

For the sake of completeness we obtain in this section the nonlinear version of Theorem II.2.2 by constructing the Friedrichs extension for a densely defined nonlinear operator P that we used in Section 4.5.

Let P be a nonlinear operator transforming a dense domain $D_P \subset H$ into H and let T be a linear K-p.d. operator defined on $D_T = D_P$. In analogy to the concepts introduced by Browder, Minty, and others, we say that P is H_0-demicontinuous if $\{u_n\} \subset D_P$, $u \in D_P$, and $u \to u$ strongly in H_0 imply $Pu_n \to Pu$ weakly in H; P is H_0-locally bounded if Pu_n is bounded in H whenever $\{u_n\} \subset D_P$ is a Cauchy sequence in H_0; P is H_0-demiclosed if $\{u_n\} \subset D_P$, $u_n \to u$ strongly in H_0, and $Pu_n \to g$ weakly in H imply $u \in D_P$ and $Pu = g$; P is strongly H_0-monotonic on D_P if for all u and v in D_P

$$\text{Re}(Pu - Pv, K(u - v)) \geq \gamma |u - v|^2, \qquad \gamma > 0. \tag{4.45}$$

Evidently, if $K = T = I$ on D_P, our definitions are identical with those considered in [27–30], [119,127,178], and [312].

Theorem 4.9 Let T be K-p.d. and P be a nonlinear mapping of $D_P = D_T$ into H. If for some positive constant $\eta > 0$,

$$|(Pu - Pv, K(u - v))| \geq \eta |u - v|^2, \qquad \forall u,v \in D_P \tag{4.46}$$

and

$$|(Pu_n - Pu_m, K_0 h)| \to 0 \quad (n,m \to \infty), \qquad h \in H_0, \tag{4.47}$$

whenever $\{u_n\} \in D_P$ is a Cauchy sequence in H_0, then P has an extension P_0 such that $P_0 \supset P$, P_0 is a one-to-one mapping of D_{P_0} onto H, P_0 is given by

$$P_0 = T_0 W_0, \tag{4.48}$$

where W_0 is a certain extension of $T_0^{-1} P$ in H_0, and P_0 is H_0-demiclosed. Furthermore, P_0 is unique.

Proof. Let T_0 be the K_0-p.d. extension of T constructed by Theorem II.2.1, and let W be an operator in H_0 with domain $D_W = D_p \subset H_0$ and range $R_W \subset H_0$ defined by $W \equiv T_0^{-1} P$. Note that in view of (4.46), (4.47), and the definition of W,

$$[Wu_n - Wu_m, h] \to 0 \quad (n,m \to \infty), \qquad h \in H_0, \tag{4.49}$$

whenever $|u_n - u_m| \to 0$ $(n,m \to \infty)$ with $u_n \in D_P$; that is, W maps every strongly Cauchy sequence $\{u_n\}$ $(\subset D_P)$ in H_0 into a weakly Cauchy sequence $\{Wu_n\}$ in H_0.

Let us now extend W by weak closure to $\hat{W}$ mapping H_0 into H_0 as follows: If $u \in D_W$, we put $\hat{W}u = Wu$; if $u \in \overline{D}_W = H_0$, there is a sequence $\{u_n\}$ in D_W such that $u_n \to u$ strongly in H_0, and consequently, $\{Wu_n\}$ is a weakly convergent sequence in H_0. Since H_0 is weakly complete, there is a unique element u^* in H_0 such that $u^* = \text{weak } \lim_n(Wu_n)$. Note that any two sequences $\{u_n'\}$ and $\{u_n''\}$ in D_P with the same limit u in H_0 must have weak $\lim_n(Wu_n') = \text{weak } \lim_n(Wu_n'')$ since otherwise the sequence of Wu's would have no limit. Thus u^* depends only on u. We may therefore take $\hat{W}u = u^* = \text{weak } \lim_n(Wu_n)$. (No contradiction with the previous definition of $\hat{W}$ on D_W is possible for if $u \in D_W$, we may take $u_n = u$ for each n.)

Thus it follows from the construction of $\hat{W}$ that it is a demicontinuous mapping of H_0 into H_0. Furthermore, $\hat{W}$ is such that for all u and v in H_0,

$$| [\hat{W}u - \hat{W}v, u - v] | \geq \eta | u - v |^2. \tag{4.50}$$

To see this, let u and v be any elements in H_0 and $\{u_n\}$ and $\{v_n\}$ be sequences in D_W so that $| u_n - u | \to 0$ and $| v_n - v | \to 0$, as $n \to \infty$. Then, by demicontinuity of $\hat{W}$ in H_0, $\{Wu_n - Wv_n\} \rightharpoonup \hat{W}u - \hat{W}v$ weakly in H_0. Hence the passage to the limit in the inequality

$$| [\hat{W}u_n - \hat{W}v_n, u_n - v_n] | \geq \eta | u_n - v_n |^2$$

[which, in view of (4.46), is valid for all elements in D_W] yields the validity of (4.50) for all u and v in H_0.

Since $\hat{W}$ is a demicontinuous mapping of H_0 into H_0 satisfying the inequality (4.50), Browder's theorem [30] implies that $\hat{W}$ maps H_0 onto H_0 and has a continuous inverse defined on $H_0 = R_{\hat{W}_0}$.

Thus we may consider a mapping W_0 in H_0 such that $W \subset W_0 \subset \hat{W}$ with $R_{W_0} = D_{T_0}$. If we now define P_0 on $D_{P_0} = D_{W_0}$ by $P_0 \equiv T_0 W_0$, it is easy to see that $P_0 \supset P$ and that P_0 is a one-to-one mapping of D_{P_0} onto H. Indeed, for $u \in D_P$ we have $W_0 u = Wu = T_0^{-1} Pu$ and hence $P_0 u = T_0 W_0 u = Pu$ (i.e., $P_0 \supset P$); furthermore, since $R_{W_0} = D_{T_0}$ and T_0 maps R_{W_0} onto H, P_0 maps D_{P_0} onto H; finally, if $P_0 u = f$ and $P_0 v = f$, the definition of P_0 and (4.50) imply that

$$0 = | (P_0 u - P_0 v, K_0(u - v)) |$$

$$= | [W_0 u - W_0 v, u - v] | \geq \eta | u - v |^2,$$

from which we derive the equality $u = v$.

To prove the other assertion of Theorem 4.9 note that if $\{u_n\} \subset D_P$ with $u_n \to u_0$ strongly in H_0 and $P_0 u_n \to f$ weakly in H, then by demicontinuity of $\hat{W}$ in H_0, the continuity of T_0^{-1} in H, and the structure of P_0 we find that

$$\hat{W}u_n \rightharpoonup \hat{W}u_0$$

weakly in H_0 and

$$T_0^{-1} P_0 u_n = W_0 u_n \rightharpoonup T_0^{-1} f$$

weakly in H; that is, $[\hat{W}u_n, h] \to [\hat{W}u_0, h]$ for every h in H_0 and $(P_0 u_n, z) \to (f, z)$ for every z in H and, in particular, for every $z = Kh$ with $h \in D_P$. Since $\hat{W} = W_0 = W$ on D_P and $P_0 = T_0 W_0$ we find that $[\hat{W}u_0, h] = [T_0^{-1} f, h]$ for every $h \in D_P$. Since D_P is dense in H_0, $\hat{W}u_0 = T_0^{-1} f$. Hence $\hat{W}u_0 \in D_{T_0}$, that is,

$$u_0 \in D_{W_0} = D_{P_0} \quad \text{and} \quad P_0 u_0 = T_0 W_0 u_0 = f;$$

hence P_0 is H_0-demiclosed.

Finally, to prove the uniqueness of P_0 note first that (P_0u, K_0v) is continuous in u on H_0 for each fixed v in H_0. This follows from the demicontinuity of $\hat{W}$ and the equation $(P_0u, K_0v) = [\hat{W}u, v]$ which holds for each u in D_P and v in H_0. Since the latter equation would be valid for any P_0 satisfying the conditions of our Theorem 4.9, it is easy to verify that these conditions determine P_0 uniquely.   Q.E.D.

Corollary 4.11   If T is K-p.d. and $P = T + S$ is such that $D_S \supseteq D_T$,

$$| (Pu - Pv, K(u - v)) | \geq \eta_1 | u - v |^2, \qquad \eta_1 > 0, \quad u,v \in D_T \tag{4.51}$$

and

$$(Su_n - Su_m, K_0h) \to 0 \quad (n,m \to \infty), \qquad h \in H_0, \tag{4.52}$$

wherever $\{u_n \mid u_n \in D_P\}$ is a Cauchy sequence in H_0, then

$$P_0 = T_0(I + R_0), \qquad R = T_0^{-1}S. \tag{4.53}$$

Proof. The conditions (4.51) and (4.52) imply that $P = T + S$ satisfies (4.46) and (4.47) with $\eta = \eta_1$. Hence, by Theorem 4.9, P has a solvable extension $P_0 = T_0 W_0$, where $W_0 \supseteq W = T_0^{-1}P$ is the restriction of $\hat{W}$ such that $R_{W_0} = D_{T_0}$. Since

$$W = T_0^{-1}(T + S) = T_0^{-1}T + T^{-1}S = I + R$$

on D_T and, by (4.52), the operator $R = T_0^{-1}S$ (defined on $D_T \subset H_0$) has a demicontinuous extension $-\hat{R} = \hat{W} - I$ with $-R_0 = I - W_0$. This implies the validity of (4.53).

In applications, as for example in elasticoplasticity, it often happens that instead of (4.47) it is easier to verify a stronger condition for which the assertions of Theorem 4.9 remain valid. In fact, the following corollary is an immediate consequence of Theorem 4.9.

Corollary 4.12   Let T be K-p.d. and P be a nonlinear mapping of $D_P = D_T$ into H such that

$$| (Pu - Pv, K(u - v)) | \geq \eta | u - v |^2,$$
$$\eta > 0, \quad u,v \in D_P, \tag{4.54}$$
$$| (Pu - Pv, K_0h) | \leq \theta | u - v | | h |,$$
$$\theta > 0, \quad u,v \in D_P, \quad h \in D_{T_0}; \tag{4.55}$$

then P has an H_0-demiclosed extension P_0 such that $P_0 \supseteq P$, P_0 is a one-to-one mapping of D_{P_0} onto H, and

$$P_0 = T_0 W_0, \tag{4.56}$$

where W_0 is a certain extension of $W = T_0^{-1} P$ in H_0.

Remark 4.14 Let us remark that in view of our stronger condition (4.55) the operator $W = T_0^{-1} P$ actually satisfies the Lipschitizian condition on the subset D_T of H_0. Indeed, if u and v are arbitrary elements of D_T and $h = Wu - Wv$, then by (4.55),

$$| h |^2 = [Wu - Wv, h] = (Pu - Pv, Kh) \le \theta | u - v | | h |,$$

and consequently, W satisfies the Lipschitz condition

$$| Wu - Wv | \le \theta | u - v |.$$

Hence there exists a unique Lipschitzian extension $\hat{W}$ of W to all of H_0 such that $\hat{W}u = Wu$ for $u \in D_T$ and

$$| \hat{W}u - \hat{W}v | \le \theta | u - v | \quad \text{and} \quad | [\hat{W}u - \hat{W}v, u - v] | \ge \eta | u - v |^2$$

for all $u, v \in H_0$. In this case we can apply the result of Zarantonello [23] to show that $\hat{W}$ maps H_0 onto H_0 and thus use the mapping $\hat{W}$ in our construction of P_0. This we will do in the next two corollaries.

Let us also remark that in this case it is not necessary for the restrictive condition (4.55) to hold for all $h \in D_{T_0}$. Indeed, it follows from the proof of the Lipschitzian property of W that it is sufficient for (4.55) to hold only for all $h \in D_{T_0}$ of the form $h = T_0^{-1}(Wu - Wv)$ with $u, v \in D_{T_0}$.

The following two corollaries determine the equality $D_{P_0} = D_{T_0}$, which is useful in the study of the regularity of solutions of (4.26).

Corollary 4.13 If T is K-p.d. and $P = T + S$ is such that

$$| (Pu - Pv, K(u - v)) | \ge \eta_1 | u - v |^2, \qquad \eta_1 > 0, \quad u, v \in D_P \tag{4.57}$$

$$\| Su - Sv \| \le \theta_1 | u - v |, \qquad \theta_1 > 0, \quad u, v \in D_P, \tag{4.58}$$

then $D_{P_0} = D_{T_0}$ and

$$P_0 = T_0 + S_0, \tag{4.59}$$

where S_0 is an extension of S in H_0.

Proof. It is easy to prove that in view of (4.58), $P = T + S$ also satisfies the condition (4.55) with $\theta = 1 + \theta_1 \sqrt{\alpha_2}$. Hence $P_0 = T_0(I + N_0)$, where N_0 is the restriction of $\bar{N} = (T_0^{-1})^\smile = T_0^{-1}\bar{S}$ with $\bar{S}$ being extension of S to H_0 [which, in view of (4.58), certainly exists]. Now $\bar{W}u \in D_{T_0}$ if and only if $u \in D_{T_0}$. This follows from the fact that $\bar{W} = I + T_0^{-1}\bar{S}$ and $T_0^{-1}\bar{S}u \in D_{T_0}$ for all u in H_0. Thus $D_{W_0} = D_{T_0}$; hence $D_{P_0} = D_{T_0}$ and $P_0 = T_0 + S_0$, where we have put $S_0 = T_0 N_0$.

$$\text{Q.E.D.}$$

Corollary 4.14 Let T be K-p.d. and K be closed in $D_K = D_T$. If P satisfies the conditions of Corollary 4.12 (or even the weaker conditions of Theorem 4.9), then $P_0 = P$ (i.e., P is a one-to-one mapping of D_P onto H).

Proof. In view of our additional hypothesis on K, Theorem 2 in [216] implies that $T_0 = T$, $K_0 = K$, and $H_0 = D_T$. Hence $\bar{W} = W_0 = W$ and, by Corollary 4.12 (or by Theorem 4.9), $P_0 = P$.   Q.E.D.

Remark 4.15 If $P = L$, where L is a linear mapping of $D_L (= D_T)$ into H, the conditions and the assertions of Corollaries 4.12, 4.13, and 4.14 reduce to the corresponding conditions and assertions of Theorem 3, Corollary 4, and Theorem 4 in [216], respectively. The assertion of Corollary 4.11 with the stronger condition $| (Su - Sv, K_0 h) | \leq \theta_2 | u - v | | h |$ reduces to Corollary 3 in [216].

The following theorem and corollary establish a two-way connection between the range and the H_0-demicontinuity of an H_0-locally bounded operator satisfying the condition (4.46).

Theorem 4.10 Let T be K-p.d., K be closed with $D_K = D_T$, and P satisfy the inequality (4.46). If there is a constant $M > 0$ such that for every Cauchy sequence $\{u_n\}$ in H_0 and every $h \in H_0$,

$$| (Pu_n, K_0 h) | \leq M| h |, \tag{4.60}$$

then P maps D_P onto H if and only if P is H_0-demicontinuous.

Proof. (Necessity). Let us first note that, in view of our conditions on K, Theorem 2 in [216] implies that $T_0 = T$, $K_0 = k$, and $H_0 = D_T$. Let W be the operator in H_0 defined by $W \equiv T^{-1}P$.

If we assume that P maps $D_P (= H_0)$ onto H, then W maps H_0 onto H_0 since T^{-1} maps H onto H_0. Let $\{u_n\}$ be a Cauchy sequence in H_0. Since H_0 is complete, there is $u_0 \in H_0$ such that $u_n \to u_0$ strongly in H_0 and, in view of (4.60), $| [Wu_n, h] | \leq M| h |$ for every $h \in H_0$. Hence $\{Wu_n\}$

is itself a bounded sequence in H_0. Since W maps every Cauchy sequence $\{u_n\}$ in H_0 into a bounded sequence $\{Wu_n\}$ in H_0 and the latter is weakly precompact in H_0, it suffices to show that there is a subsequence of $\{Wu_n\}$ converging weakly to Wu_0 in H_0. Now let $\{Wu_{n_k}\}$ be a subsequence of $\{Wu_n\}$ that converges weakly in H_0 to some element, say p, in H_0. Hence, in view of (4.46) and the fact that $D_P = H_0$, for every v in H_0 we have the inequality

$$|\,[Wu_{n_k} - Wv, u_{n_k} - v]\,| \geq \eta\,|\,u_{n_k} - v\,|^2. \tag{4.61}$$

Passing to the limit in (4.61) as $n_k \to \infty$, we get the inequality

$$|\,[p - Wv, u_0 - v]\,| \geq \eta\,|\,u_0 - v\,|^2 \tag{4.62}$$

valid for each v in H_0. Applying the Schwarz inequality to (4.62), we get

$$\eta\,|\,u_n - v\,|^2 \leq |\,p - Wv\,|\,|\,u_0 - v\,|.$$

This shows that for each v in H_0, we have the inequality $\eta\,|\,u_0 - v\,| \leq |\,p - Wv\,|$. Since $R_W = H_0$, there exists a $y \in D_W = H_0$ such that $p = Wy$ and $\eta\,|\,u_0 - v\,| \leq |\,Wy - Wv\,|$ for each $v \in H_0$. If we take $v = y$, the last inequality implies that $u_0 = y$ and $p = Wu_0$. Thus $Wu_n \to Wu_0$ weakly in H_0, whenever $u_n \to u_0$ strongly in H_0. This and the definition of W and (4.43) imply that $Pu_n \to Pu_0$ weakly in H (i.e., P is H_0-demicontinuous).

(Sufficiency). Suppose that P is H_0-demicontinuous. Then for every $z \in H$, $(Pu_n,z) \to (Pu_0,z)$. Since K has a bounded inverse defined on all of H, for every $z \in H$ there is a unique $h \in D_K = H_0$ such that $z = Kh$. Defining W by $W \equiv T^{-1}P$ we find that W maps H_0 into H_0 and that

$$[Wu_n,h] = (Pu_n,z) \to (Pu_0,z) = [Wu_0,h]$$

for every h in H_0 whenever $u_n \to u_0$ strongly in H_0. Hence W is a demicontinuous mapping of H_0 into H_0 such that

$$|\,[Wu - Wv, u - v]\,| \geq \eta\,|\,u - v\,|^2$$

for all u and v in H_0. Thus, by Browder's theorem [30], W maps H_0 onto $H_0 = D_T$. Since T maps D_T onto H, this implies that $TW = P$ maps D_p onto H and completes the proof of Theorem 4.10. Q.E.D.

Corollary 4.15 If P is a locally bounded mapping of H into H such that

$$|\,(Pu - Pv, u - v)\,| \geq c\|\,u - v\,\|^2, \qquad \forall\, u,v \in H, \tag{4.63}$$

then P is onto H if and only if P is demicontinuous.

Proof. Corollary 4.15 is a special case of Theorem 4.10 if in it we take $T = K = I$.   Q.E.D.

Strongly H_0-monotonic operators Let us observe in passing that condition (4.46) of Theorem 4.9 or (4.63) of Corollary 4.15 is obviously satisfied when the nonlinear operator P is strongly H_0-monotonic, that is, if there is a constant $\gamma > 0$ such that

$$\text{Re}(Pu - Pv, K(u - v)) \geq \gamma |u - v|^2, \qquad u,v \in D_P = D_T. \quad (4.64)$$

Sometimes, in applications, this is the condition that is easier to verify. Hence the theorems and corollaries proved above are valid for strongly H_0-monotonic operators with the corresponding additional conditions. Similarly, instead of (4.55), it is sufficient to assume a slightly weaker condition

$$|\text{Re}(Pu - Pv, K_0(u - v))| \leq \beta |u - v||h|, \qquad (4.64)'$$
$$\beta > 0, \quad u,v \in D_P, \quad h \in D_{T_0},$$

valid for all h in D_{T_0} of the form $h = T_0^{-1}(Pu - Pv)$ with $u,v \in D_f$.

Thus it appears to be useful to have some easily verifiable tests for the H_0-monotonicity of an operator. To this end the following lemma appears to be convenient (see also Minty [178]).

Lemma 4.3 If P has the property that for any $x,z \in D_p$ and real t there is a constant $\gamma > 0$ so that

$$\left[\frac{d}{dt}\text{Re}(Kh, P(z + th))\right]_{t=0} \geq \gamma |h|^2, \qquad h = x - z, \qquad (4.65)$$

then P is strongly H_0-monotonic on D_P.

Proof. Let x and y be any elements in D_P and $u = x - y$; let $f(s)$ be the real-valued function defined for $0 \leq s \leq 1$ by $f(s) = \text{Re}(Ku, P(y + su))$. In view of our conditions, it is not hard to see that $f(s)$ is differentiable on $(0,1)$, and hence by the mean-value theorem there is a ξ such that

$$f(1) - f(0) = \text{Re}(K(x - y), Px - Py) = f'(\xi)$$
$$= \left[\frac{d}{ds}\text{Re}(Ku, P(y + su))\right]_{s=\xi}, \qquad 0 < \xi < 1.$$

That is, letting $z = y + \xi u$, $t = \Delta s/(1 - \xi)$, and noting that $h = (1 -$

$\xi)u = (1 - \xi)(x - y)$, we get

$$f'(\xi) = \lim_{\Delta s \to 0} \frac{\operatorname{Re}(Ku, P(z + \Delta su) - Pz)}{\Delta s}$$

$$= \lim_{t \to 0}(1 - \xi)^{-2} \frac{\operatorname{Re}(Ku, P(z + \Delta su) - Pz)}{t} \tag{4.66}$$

$$= (1 - \xi)^{-2}\left[\frac{d}{dt}\operatorname{Re}(Kh, P(z + th))\right]_{t=0}.$$

On the other hand, since z and x belong to D_P and

$$h = x - z = (1 - \xi)(x - y),$$

(4.66) and our assumption (4.65) imply that

$$\operatorname{Re}(K(x - y), Px - Py) \geq \frac{\gamma}{(1 - \xi)^2}|h|^2 = \gamma|x - y|^2$$

for any x and y in D_P. This shows that P is strongly H_0-monotonic.

Q.E.D.

5. SOLVABILITY OF EQUATIONS INVOLVING SEMILINEAR WEAKLY A-PROPER MAPPINGS*

In this section we establish existence theorems for abstract and differential semilinear equations with linear part L having a nontrivial kernel $N(L)$, and the nonlinear part N is such that the operator $L + N$ is weakly A-proper (see Definition 5.1 below). Some of the results presented here appeared in [212] and in [226].

It should be pointed out that the class of weakly A-proper mappings is very general, and thus as we indicate below, our existence results will include as special cases many results of other authors obtained by them for narrower classes of mappings.

Let X and Y be real Banach spaces. We are concerned with the problem of the existence of a solution of

$$Lx + Nx = f \qquad (x \in X) \tag{5.A}$$

for a given f in Y, where $L \in L(X, Y)$ is Fredholm of index zero and N: $X \to Y$ is a nonlinear map of a particular type such that $N(x) = o(\| x \|)$ as $\| x \| \to \infty$. It is known that many boundary value problems for ordinary

---

* From *Nonlinear Equations in Abstract Spaces* (V. Lakshmikantham, ed.), Academic Press, Orlando, Florida (1978).

and partial differential equations can be reformulated as abstract operator equations of the type above if X and Y are chosen judiciously.

If either N or the generalized inverse of L is compact, one can treat such problems by using the Leray–Schauder or Mawhin degree theory or the Schauder fixed-point theorem (see [98,163]). When compactness is not available, which is often the case with differential operators when N depends on the highest-order derivative or the underlying domain in R^n is unbounded, the recent surjectivity theorems for monotone-type operators or the degree theory for condensing vector fields proved to be useful in treating certain classes of semilinear equations (see [18,77,98,188]). In this section we show how the theory of mappings of A-proper type can be used successfully in treating the existence problems above under conditions that are weaker than those of other authors.

The purpose of this is the following. In Sections 5.1 and 5.2 we present some abstract existence results for (5.A) when $L + N$ is *weakly A-proper*, and in Section 5.3 we apply them to the solvability of certain differential equations. Using the Brouwer degree theory and some properties of A-proper mappings, it is first recalled that if L is A-proper and injective, (5.A) is solvable for each f in Y. It is then shown in Theorem 5.1 that if $N(L) \neq \{0\}$, one can always construct a map $C \in L(X,Y)$ such that $L + \lambda C$ is A-proper and injective for each $\lambda \neq 0$, while $L + \lambda C + N$ remains weakly A-proper, and for each $k \in Z_+$ the equation

$$Lx_k + \lambda_k Cx_k + Nx_k = f \qquad (\lambda_k \to 0 \quad \text{as} \quad k \to \infty) \qquad (5.B)$$

has a solution $x_k \in X$ for each f in Y. Thus if $\{x_k\}$ is bounded for some f in Y, then (5.A) is a solvable provided that $T(B)$ is closed whenever B is a closed ball in the space X. In Theorem 5.2 the boundedness of $\{x_k\}$ is established under various asymptotic positivity conditions which have their origin in the existence results of Landesman and Lazer [148] and which have since been extended in various directions by a number of authors (see [18,47,71,77,85,114,186,188,227,234] and others cited there).

It turns out that the class of weakly A-proper mappings $T = L + N$ is not only well suited for the problem at hand but it is also large enough so as to allow us to extend and unify the abstract results of [71] (including some in [92,186]) with some of [77] (and [18] if the linear part is bounded), which were obtained by these authors by using various types of arguments. Indeed, it is shown in Section 5.2 that if $L \in L(X,Y)$ is A-proper and $N: X \to Y$ is either compact, weakly continuous, or of type (KM), then $T = L + N$ is weakly A-proper and $T(B)$ is closed provided that if N is of type (KM) some additional conditions hold. In view of this, Corollary 5.1 extends Theorems 1, 2, and 3 of DeFigueiredo [71] and some results in [92,186], while Theorems 5.1 and 5.2 extend some results of

Brezis and Nirenberg [18] to weakly A-proper maps. It is also shown that if $L \in L(X,Y)$ is Fredholm of index 0 and $N(L + C)^{-1}$ is ball condensing, one can construct a special scheme $\hat{\Gamma}$ for (X,Y) so that L and $L + N$ are A-proper w.r.t. $\hat{\Gamma}$, and thus Propositions 2.4 and 2.5 of Fitzpatrick [77] also follow from Theorem 5.2. Now the author of [77] uses the degree theory for condensing vector fields to study (5.A), and therefore his results are also valid for set-condensing maps $F = N(L + C)^{-1}$.

In case (5.A) is a differential boundary value problem our conditions on $T = L + N$ allow N to depend on the highest-order terms without the condition (used in [77]) that N is Lipschitzian with respect to them. Consequently, our existence results allow application to broader classes of differential boundary value problems, including those whose underlying domain in R^n is unbounded (see Problem 5.2).

In Section 5.3 we show what type of boundary value problems for ODEs and PDEs can be treated by our existence results. Problems 5.1 and 5.4 in this section were first studied in [77]. We include them here to indicate the difference in assumptions for the respective abstract existence results to be applicable.

5.1   Perturbation Method in the Solvability of (5.A)

Let X and Y be Banach spaces with an admissible scheme $\Gamma = \{X_n, V_n, E_n, W_n\}$. It will be seen below that the following class of maps is not only well suited for the study of the existence of solutions of semilinear equations of the form (5.A) but is also large enough so as to allow us to extend and unify the abstract existence results of [71,92,186] with some of those in [77] obtained by these authors for special classes of maps satisfying different conditions and using various types of arguments. We also indicate the extension of some results in [17,18] when the linear part is bounded.

Definition 5.1 $T: X \to Y$ is said to be *weakly A-proper (respectively, A-proper)* w.r.t. Γ if $T_n: X_n \to E_n$ is continuous for each n and if $\{x_{n_j} \mid x_{n_j} \in X_{n_j}\}$ is any bounded sequence such that $\| T_{n_j}(x_{n_j}) - W_{n_j}g \| \to$ 0 as $j \to \infty$ for some g in Y, with $T_n = W_n T|_{X_n}$, then there exist a subsequence $\{x_{n_{j(k)}}\}$ and $x \in X$ such that $x_{n_{j(k)}} \rightharpoonup x \in X$ (respectively, $x_{k_{j(k)}} \to$ x in X) and $Tx = g$.

The generality of the class of weakly A-proper maps is illustrated by the following examples (others and the definitions of some of the notions used here will be given later).

Example 5.1 All A-proper maps. This class includes compact and ball-condensing vector fields, maps of type (S) and (KS) as well as of strongly K-monotone type (see [36,210,229]). Extending a recent result in [275], it was shown independently in [171] and in [220] that the class of A-proper maps also includes the maps $T: X \to X$ of the form $T = I - F + N + C$, where F is ball condensing, N accretive and continuous, and C is compact.

The interesting feature of the latter class is that neither the theory of accretive nor condensing-type mappings is applicable to equations involving the class of maps described above.

Example 5.2 Semibounded mappings $T: X \to X^*$ of type (M) when X is reflexive and, in particular, pseudomonotone, semimonotone, monotone, and weakly continuous maps (see Lemma 5.2 below).

The following example (see Section 5.2) indicates the type of semilinear mappings that are weakly A-proper.

Example 5.3 Suppose that X is reflexive, $L \in L(X,Y)$ is A-proper, N: $X \to Y$ is of type (KM), and $C: X \to Y$ is completely continuous. Then $T = L + N + C$ is weakly A-proper provided that $\phi(x) = (Lx,Kx)$ is weakly lower semicontinuous, where $K: X \to Y^*$ satisfies suitable conditions.

To state our first result we recall that a map $A \in L(X,Y)$ is called Fredholm if it has a finite-dimensional null space $N(A)$ of dimension d and a closed range $R(A)$ of finite codimension $d^* = \dim N(A^*)$. The index of A is $\text{ind}(A) = d - d^*$. We say that a nonlinear map $N: X \to Y$ is *asymptotically zero* if $\| Nx \|/\| x \| \to 0$ as $\| x \| \to \infty$ or, equivalently, $N(x) = o(\| x \|)$ as $\| x \| \to \infty$.

The proof of Theorem 5.1 below is based on the following lemma, which has an independent interest.

Lemma 5.1 Suppose $A \in L(X,Y)$ is A-proper w.r.t. Γ and injective. If $N: X \to Y$ is asymptotically zero and $T = A + N$ is weakly A-proper w.r.t. Γ, then the equation

$$Ax + Nx = f \tag{5.1}$$

has a solution for each f in Y.

Proof. Since the proof of Lemma 5.1 is very similar to the proof of Theorem 2 in [232], we give only the outline here. Now to each $f \in Y$ there are $r_f > 0$ and $n_f \in N$ such that for all $t \in [0,1]$ and all $n \geq n_f$, we have

$$W_n A(x) + (1 - t)W_n(Nx - f) \neq 0 \tag{5.2}$$
$$\text{for} \quad x \in X_n \quad \text{with} \quad \| x \| = r_f.$$

Indeed, if (5.2) were not true for some $f \in Y$, there would exist sequences $\{x_{n_j} \mid x_{n_j} \in X_{n_j}\}$ and $t_{n_j} \in [0,1]$ such that

$$\| x_{n_j} \| \to \infty \quad \text{as} \quad j \to \infty \quad \text{and} \quad W_{n_j}A(x_{n_j})$$
$$+ (1 - t_{n_j})W_{n_j}(Nx_{n_j} - f) = 0$$

for each j. Since $\{W_n\}$ is uniformly bounded and $\| x_{n_j} \| \to \infty$, it follows from the last equation and our condition on N that $A_{n_j}(z_{n_j}) = (t_{n_j} - 1)W_{n_j}(Nx_{n_j} - f)/\| x_{n_j} \| \to 0$ as $j \to \infty$, where $z_{n_j} = x_{n_j}/\| x_{n_j} \|$. Since $\{z_{n_j} \mid z_{n_j} \in X_{n_j}\}$ is bounded and $A_{n_j}(z_{n_j}) \to 0$, the A-properness of A implies the existence of a subsequence $\{z_{n_{j(k)}}\}$ and $z \in X$ such that $z_{n_{j(k)}} \to z$ and $Az = 0$ with $\| z \| = 1$, in contradiction to the condition that A is injective. Thus (5.2) holds for each $n \geq n_f$ and therefore, by the finite-dimensional degree theory, there exists $x_n \in X_n \cap B(0,r_f)$ such that $T_n(x_n) - W_n f = 0$ for each $n \geq n_f$. In view of this and the weak A-properness of $T = A + N$, there exists a subsequence $\{x_{n_j}\}$ and $x \in X$ such that $x_{n_j} \to x$ and $Tx = f$.   Q.E.D.

It is known that even when $T = A + N$ is A-proper, equation (5.1) need not have a solution for each f in Y if $N(A) \neq \{0\}$. Extending a result of [118], it was shown in [225] that if $\text{ind}(A) = 0$, T is A-proper, and $R(N) \subseteq R(A)$, (5.1) is solvable if and only if $f \in R(A)$. However, it was shown in [56] that the condition "$R(N) \subseteq R(A)$" is very restrictive at least when applied to partial differential equations with sufficiently smooth coefficients.

Employing a perturbation method used in similar situations by other authors (e.g., [18,71,114,212]) and Lemma 5.1, we first prove general existence Theorem 5.1 for (5.A) with $N(L) \neq \{0\}$ and then use it as a basis for obtaining existence results for the general class of weakly A-proper maps $T = L + N$ with N satisfying certain asymptotic positivity conditions. In what follows B will always denote an arbitrary closed ball in X.

Theorem 5.1 Suppose that $L \in L(X,Y)$ is A-proper w.r.t. Γ with $\text{ind}(L) = 0$ and $N: X \to Y$ is a mapping such that $T = L + N$ is weakly A-proper w.r.t. Γ, $N(x) = o(\| x \|)$ as $\| x \| \to \infty$ and $T(B)$ is closed in Y for each B

$\subset X$. Then we can construct a compact map $C \in L(X,Y)$ such that

$$Lx + Nx = f \qquad (x \in X, f \in Y) \qquad\qquad (5.3)$$

is solvable for a given f in Y provided that

(A) Either the set $S_k^+ \equiv \{x_k \in X \mid Lx_k + \lambda_k Cx_k + Nx_k - f = 0, \lambda_k > 0$ and $\lambda_k \to 0$ as $k \to \infty\}$ or the set $S_k^- \equiv \{x_k \in X \mid Lx_k + \lambda_k Cx_k + Nx_k - f = 0, \lambda_k < 0$ and $\lambda_k \to 0$ as $k \to \infty\}$ is bounded by a constant independent of k.

Proof. It was shown in Theorem II.3.1 that since $L \in L(X,Y)$ is A-proper w.r.t. Γ, it follows that L is Fredholm with $\text{ind}(L) \geq 0$, so by our hypotheses, $d = d^*$. Hence there exists a closed subspace X_1 of X and subspace Y_2 of Y with $\dim Y_2 = d$ such that $X = N(L) \oplus X_1$, $L(X_1) = R(L)$, L is injective on X_1, and $Y = Y_2 \oplus R(L)$. Let M be an isomorphism of $N(L)$ onto Y_2, let P be the linear projection of X onto $N(L)$, and let L_λ be a linear mapping of X into Y defined by $L_\lambda = L + \lambda C$, where $C = MP$ and $\lambda \neq 0$. Since L_λ is A-proper and C is compact, L_λ is A-proper w.r.t. Γ for each $\lambda \neq 0$. Moreover, L_λ is injective for each $\lambda \neq 0$. Indeed, if $L_\lambda(x) = Lx + \lambda Cx = 0$, then $Lx = -\lambda Cx$ and since $\lambda \neq 0$ and $R(L) \cap Y_2 = \{0\}$, it follows that $Lx = 0$ and $Cx = 0$. Hence $MPx = 0$, so $Px = 0$ since M is injective. Consequently, $x \in X_1$ with $Lx = 0$. Thus $x = 0$ since L is injective on X_1; that is, L_λ is injective for $\lambda \neq 0$ and, in fact, bijective by Theorem II.1.1.

Furthermore, the map $T_\lambda = L_\lambda + N$ is weakly A-proper for each fixed $\lambda \neq 0$. Indeed, let $\{x_{n_j} \mid x_{n_j} \in X_{n_j}\}$ be any bounded sequence such that $g_{n_j} \equiv W_{n_j} T_\lambda(x_{n_j}) - W_{n_j}(g) \to 0$ as $j \to \infty$ for some g in Y. Since $\{x_{n_j}\}$ is bounded and C is compact, we may assume, without loss of generality, that $C(x_{n_j}) \to z$ in Y for some z in Y. This and the uniform boundedness of $\{W_n\}$ imply that

$$W_{n_j} L(x_{n_j}) + W_{n_j} N(x_{n_j}) - W_{n_j}(g - \lambda z) = g_{n_j} - \lambda W_{n_j}(Cx_{n_j} - z) \to 0$$

as $j \to \infty$. Hence, by the weak A-properness of $L + N$, there exist a subsequence $\{x_{n_{j(k)}}\}$ and $x \in X$ such that $x_{n_{j(k)}} \rightharpoonup x$ as $k \to \infty$ and $Lx + Nx = g - \lambda z$. But since $C \in L(X,Y)$ is completely continuous, it follows that $C(x_{n_{j(k)}}) \to Cx = z$. Hence $T_\lambda x = g$; that is, T_λ is weakly A-proper w.r.t. Γ for each $\lambda \neq 0$.

Now let $\{\lambda_k\}$ be a sequence such that either $\lambda_k \to 0^+$ or $\lambda_k \to 0^-$ as $k \to \infty$ (i.e., $\lambda_k > 0$ or $\lambda_k < 0$ and $\lambda_k \to 0$ as $k \to \infty$) and observe that $T_k \equiv L + \lambda_k C + N$ satisfies the conditions of Lemma 5.1 for each k. Hence, in either case, for each f in Y and each k there exists a vector $x_k \in X$

such that

$$Lx_k + \lambda_k Cx_k + Nx_k = f \qquad \text{for each } k. \tag{5.4}$$

Now if f is such that condition (A) holds, then in either case there exists a constant $r_f > 0$ such that $\| x_k \| \le r_f$ for all k. In view of this and the convergence of $\{\lambda_k\}$ to 0, it follows that $\{x_k\} \subset \bar{B}(0, r_f)$ and $Lx_k + Nx_k \to f$ as $k \to \infty$. Since by assumption $T(\bar{B}(0, r_f))$ is closed, it follows that in either case there exists $x_0 \in \bar{B}(0, r_f)$ such that $Lx_0 + Nx_0 = f$.

Q.E.D.

We will now show that the boundedness condition (A) is satisfied if in addition to the growth condition $N(x) = o(\| x \|)$ as $\| x \| \to \infty$ we assume also that N satisfies certain asymptotic positivity conditions. Such conditions have their origin in the existence result of Landesman and Lazer [148] for PDEs, and have since been considered by a number of authors, including Nečaš [186], Fučik [90], Fučik et al. [92], DeFigueiredo [71], Fitzpatrick [77], Nirenberg [188], Brezis and Nirenberg [18], Petryshyn [212,227] and many others (for further references, see [234]).

It should be added that we mention explicitly only those authors to whose work a direct reference is made here. However, the problem of the solvability of semilinear abstract and differential equations, especially by means of the Lyapunov-Schmidt method, has been studied by many authors, including Cesari, Mahwin, Cronin, Nirenberg, Kannan, Schechter, Hale, Schur, Berger, Ahmad, Hess, Gustafson, Sather, Osborne, Lazer, Ambrozetti, Mancini, Prodi, Gupta, Leach, Dancer, Williams, and others. For an excellent survy of the results of these and other authors, see the monographs [48,98,234].

As the first consequence of Theorem 5.1 we deduce the following result, which, as will be shown below, extends abstract results of [71,92,186] and some results of [77]. For bounded linear part we also extend some results in [18].

Theorem 5.2 Suppose that Y is a Hilbert space. $L \in L(X, Y)$ is A-proper w.r.t. Γ with $\text{ind}(L) = 0$, and $N: X \to Y$ is such that $T = L + N$ is weakly A-proper w.r.t. Γ and $T(B)$ is closed in Y for each $B \subset X$. Let M and P be as defined in the proof of Theorem 5.1.

Then, for a given f in Y, equation (5.3) is solvable provided that one of the following three conditions holds:

B (1) $N(x) = o(\| x \|)$ as $\| x \| \to \infty$, and
  (2) Either $B^+: \overline{\lim}(Nx_k, My) > (f, My)$ or $B^-: \underline{\lim}(Nx_k, My) < (f, My)$ whenever $\{x_k\} \subset X$ is such that $\| x_k \| \to \infty$ and $x_k / \| x_k \| \to y \in N(L)$.

C (1) $\| Nx \| \le a\| x \|^\alpha + b$ for all $x \in X$, some $\alpha \in [0,1)$ and $a > 0$, $b \ge 0$, and

(2) Either C^+: $\overline{\lim} \, (N(t_k\bar{y}_k + t_k^\alpha z_k), My) > (f,My)$ or C^-: $\underline{\lim} \, (N(t_k\bar{y}_k + t_k^\alpha z_k), My) < (f,My)$ whenever $y \in N(L) \cap \partial B(0,1)$ and sequences $\{t_k\} \subset R^+$, $\{\bar{y}_k\} \subset N(L)$ and $\{z_k\} \subset X_1$ are such that $\bar{y}_k \to y$ in X, $t_k \to \infty$ and $\| z_k \| \le c$ for all k and some $c > 0$.

D (1) Condition $C(1)$ holds, and

(2) Either D^+: $\overline{\lim} \, (Nx_k, My_k) > (f,My)$ or D^-: $\underline{\lim}(Nx_k M_k) < (f,My)$ whenever $\{x_k\} \subset X$ is such that $\| Px_k \| \to \infty$ and $y_k = Px_k/\| Px_k \| \to y \in N(L)$.

Proof. It follows from C(1) that $N(x) = o(\| x \|)$ as $\| x \| \to \infty$. Thus to deduce Theorem 5.2 from Theorem 5.1, it suffices to show that the set S_k^+ is bounded if either B^+, $C^+ - C(1)$, or $D^+ - C(1)$ holds, while S_k^- is bounded if either B^-, $C^- - C(1)$, or $D^- - C(1)$ holds (i.e., S_k^+ or S_k^- is bounded by a constant independent of k).

Let $\{\lambda_k\}$ be such that $\lambda_k \to 0^+$ as $k \to \infty$ and let $\{x_k\} \subset X$ be any sequence such that (5.4) holds for each k. We claim that B^+, $C^+ - C(1)$, or $D^+ - C(1)$ implies the boundedness of $\{x_k\}$. If this were not the case, then since $\| N(x_k) \|/\| x_k \| \to 0$ as $t_k \equiv \| x_k \| \to \infty$ and $\{u_k\} \equiv \{x_k/\| x_k \|\}$ is bounded, it follows from this and (5.4) that

$$L(u_k) = f/\| x_k \| - \lambda_k C(u_k) - N(x_k)/\| x_k \| \to 0 \qquad \text{as} \quad k \to \infty. \tag{5.5}$$

Hence, since L is proper, there exists a subsequence that we again denote by $\{u_k\}$ such that $u_k \to y$ in X, $\| y \| = 1$, and $L(y) = 0$. Now taking the inner product of (5.4) with $M(y)$ and noting that $Y = N(L^*) \oplus R(L)$, M: $N(L) \to M(L^*)$, and $C = MP$, we see that $(Lx_k,My) = 0$, $(Cu_k,My) \to (Cy,My) = (My,My) > 0$ as $k \to \infty$, $\lambda_k(Cx_k,My) + (Nx_k - f,My) = 0$ for each k, and $\lambda_k\| x_k \|(Cu_k,My) > 0$ for all k sufficiently large. This implies that

$$(Nx_k - f,My) < 0 \qquad \text{for all large} \quad k. \tag{5.6}$$

Suppose first that B^+ holds. In view of this, (5.6) implies that $\overline{\lim}(Nx_k,My) \le (f,My)$, in contradiction to B^+. Thus the set S_k^+ is bounded if B^+ holds. Suppose now that $C^+ - C(1)$ holds. Since $X = N(L) \oplus X_1$, we can write $x_k = v_k + w_k$ with $v_k = Px_k \in N(L)$ and $w_k = (I-P) x_k \in X_1$ for each k. Let Q be the orthogonal projection of Y onto $R(L)$. Then since $Y = N(L^*) \oplus R(L)$ and C: $Y \to N(L^*)$, applying Q to (5.4), we get the equality $Lx_k + Q(Nx_k - f) = 0$ or $Lw_k = Q(f - Nx_k)$. Since $L = L|_{X_1}$: $X_1 \to R(L)$ has a bounded inverse, it follows from the last equality and C(1) that

$$\| w_k \| \leq a_1 \| x_k \|^\alpha + a_2 \qquad \text{for some} \quad a_1 > 0 \quad \text{and} \quad a_2 > 0. \qquad (5.7)$$

Setting $u_k = x_k/\| x_k \| = \bar{y}_k + \bar{w}_k$, we see that $\| u_k \| = 1$, $\bar{y}_k = v_k/\| x_k \| \in N(L)$, $\bar{w}_k = w_k/\| x_k \| \in X_1$, $\bar{w}_k \to 0$ in X, $\bar{y}_k \to y \in N(L)$ with $\| y \| = 1$, and $x_k = t_k \bar{y}_k + w_k$ with $w_k = t_k^\alpha z_k$, where $\{z_k\} \subset X_1$ is bounded in view of (5.7). Thus (5.6) and condition C^+ imply that

$$\overline{\lim}(Nx_k,My) = \overline{\lim}(N(t_k \bar{y}_k + t_k^\alpha z_k),My) \leq (f,My),$$

in contradiction to C^+. Thus the set S_k^+ is also bounded if $C^+ - C(1)$ holds.

Suppose now that $D^+ - C(1)$ holds. Since $u_k = x_k/\| x_k \| = \bar{y}_k + \bar{w}_k$ with $\bar{w}_k \to 0$ and $\bar{y}_k \to y \in N(L)$ with $\| y \| = 1$, it follows that $\| v_k \|/\| x_k \| = \| Px_k \|/\| x_k \| \to 1$ as $k \to \infty$. Hence $\| Px_k \| \to \infty$ and $y_k = Px_k/\| Px_k \| = v_k/\| v_k \| \to y$ as $k \to \infty$. Now taking the inner product of (5.4) with My_k and noting that $(Lx_k,My_k) = 0$, we get the equality.

$$\lambda_k(Cx_k,My_k) + (Nx_k - f,My_k) = 0 \text{ for each } k.$$

Since $u_n = x_k/\| x_k \| \to y$ and $y_k \to y$ we see that $(Cu_k,My_k) \to (Cy,My) = (My,My) > 0$ as $k \to \infty$. This and the fact that $\lambda_k > 0$ imply that $\lambda_k(Cx_k,My_k) = \lambda_k \| x_k \| (Cu_k,My_k) > 0$ for all k sufficiently large. In view of this, the last equality implies that $(Nx_k - f,My_k) < 0$ for all large k and therefore $\overline{\lim}(Nx_k,My_k) \leq (f,My)$, in contradiction to D^+. Thus the set S_k^+ is also bounded if $D^+ - C(1)$ holds.

In a similar way one shows that the set S_k^- is bounded if either B^-, $C^- - C(1)$, or $D^- - C(1)$ holds. Indeed, if not, then by the same argument we would be led to one of the equations $\lambda_k(Cx_k,My) + (Nx_k - f,My) = 0$ or $\lambda_k(Cx_k,My_k) + (Nx_k - f,My_k) = 0 \,\forall\, k$ with $\lambda_k(Cx_k,My) = \lambda_k \| x_k \| (Cu_k,My) < 0$ and $\lambda_k(Cx_k,My_k) = \lambda_k \| x_k \| (Cu_k,My_k) < 0$ for all sufficiently large k since $\lambda_k < 0$ and $0 < (My,My) = \lim(Cu_k,My) = \lim(Cu_k,My_k)$. Thus $(Nx_k - f,My) \geq 0$ in the first case and $(Nx_k - f,My_k) \geq 0$ in the second for all large k. Thus, as before, we obtain the contradiction in each case. Hence S_k^- is bounded.   Q.E.D.

Note 5.1 It should be emphasized that whenever we say that S_k^+ or S_k^- is bounded, we always mean that S_k^+ or S_k^- is bounded by some constant independent of k.

Remark 5.0 It is not hard to show that the asymptotic positivity condition B^+ is equivalent to the hypothesis:

(B_1^+) $\underline{\lim}(Nx_k,My) > (f,My)$ whenever $\{x_k\} \subset X$ is such that $\| x_k \| \to$

∞ and $x_k/\|x_k\| \to y \in N(L)$, while the condition B^- is equivalent to the hypotheses:

(B_1^-) $\overline{\lim}(Nx_k,My) < (f,My)$ whenever $\{x_k\} \subset X$ is such that $\|x_k\| \to \infty$ and $x_k/\|x_k\| \to y \in N(L)$.

The same can be said about the other conditions.

Remark 5.1 If $N: X \to Y$ is also continuous and $T = L + N$ is A-proper w.r.t. Γ, then the conclusions of Theorems 5.1 and 5.2 hold without the assumption that $T(B)$ is closed since, as has been shown earlier by Theorem II.1.1, every continuous A-proper map is proper and, in particular, $T(B)$ is closed in Y for each $B \subset X$.

Remark 5.2 Assuming that $X \subset Y$ and $Y = N(L) \oplus R(L)$ the asymptotic growth and positivity conditions $B(1)$–$B(2)$ (in the form stated in Remark 5.0 with $M = I$) were used in [77] for a special case of maps to be discussed in Section 5.2, while conditions $C(1)$–$C(2)$ were used earlier in [71] as extensions of the hypothesis used in [92,186] for $X = Y$. We add that when $N: X \to Y$ satisfies condition $C(1)$, it is not difficult to show that the positivity conditions imposed in [92, Theorem 2.3] and in [77, Proposition 2.5] imply $D(2)$ of Theorem 5.2.

Remark 5.3 In case $Y = X = H$, where H is a separable Hilbert space and thus H always has a projectionally complete scheme $\{X_n,P_n\}$, Theorem 5.2 takes a particularly simple form when one assumes that $N(L) = N(L^*)$ since in that case $M = I$ and the perturbation equation (5.4) takes the form

$$Lx_k + \lambda_k Px_k + Nx_k = f, \tag{5.8}$$

where P is the orthogonal projection of H onto $N(L)$. It will be shown in Section 5.2 that in this case Theorem 5.2 extends a number of results obtained earlier by other authors for special classes of weakly A-proper mappings.

We conclude this section with the indication of how one can use Theorem 5.2 (when $Y = X = H$) to obtain existence results for the class of weakly A-proper mappings under the asymptotic growth and positivity conditions used by Brezis and Nirenberg [18] for a different but related class of maps which we indicate at the end of this section (see also Section 5.2). We may add that our proofs for the existence of solutions are somewhat simpler than those given in [17,18] for characterization of ranges of semilinear operators. To state further results, we first recall that if $L \in L(H,H)$ is a Fredholm map with $N(L) = N(L^*)$, then there exists a constant $\alpha_0 > 0$ such that $\| Lx \| \geq \alpha_0 \| w \|$, where $x = v + w$ with $v \in N(L)$ and $w \in R(L)$. Hence $| (Lx,x) | = | (Lx,w) | \leq \| Lx \| \| w \| \leq (1/\alpha_0) \| Lx \|^2$, that is, $(Lx,x) \geq -(1/\alpha_0) \| Lx \|^2$ for all x in H. In what follows we shall take $\alpha > 0$ to be the largest $\alpha_0 > 0$ such that $(Lx,x) \geq -(1/\alpha) \| Lx \|^2$ for all x in H.

Corollary 5.1   Suppose that $L \in L(H,H)$ is A-proper with $N(L) = N(L^*)$ and $T = L + N$ is weakly A-proper. Then, for a given $f \in H$, equation (5.3), $Lx + Nx = f$, is solvable provided that the following additional conditions hold:

$$N(x) = o(\| x \|) \quad \text{as} \quad \| x \| \to \infty \quad \text{and} \quad T(B) \text{ is closed for each } B \subset H.$$
$$(5.9)$$

$$(Nx - Ny,x) \geq \gamma \| Nx \|^2 - C(y) \ \forall \ x,y \in H \quad \text{and some} \\ \text{positive} \quad \gamma < \alpha, \quad \text{where} \quad C(y) \quad \text{is independent of} \quad x. \quad (5.10)$$

$$\liminf_{t \to \infty}(N(tx),x) > (f,x) \quad \forall \ x \in N(L), \qquad \| x \| = 1. \quad (5.11)$$

Proof.   In view of Theorem 5.1(A), to prove Corollary 5.1 it suffices to show that the set S_k^+ is bounded. If not, there would exist sequences $\{\lambda_k\} \subset R^+$ and $\{x_k\} \subset X$ with $\lambda_k \to 0^+$ as $k \to \infty$ such that x_k satisfies (5.8) for each k and $\| x_k \| \to \infty$ as $k \to \infty$. In view of this and the first part of (5.9) we may assume that $y_k \equiv x_k/\| x_k \| \to y \in N(L)$ with $\| y \| = 1$.

Taking the inner product of (5.8) with x_k, we get

$$(Lx_k,x_k) + \lambda_k(Px_k,x_k) + (Nx_k,x_k) = (f,x_k). \quad (5.12)$$

Since $\lambda_k(Px_k,x_k) = \lambda_k \| Px_k \|^2 > 0$, it follows from (5.12) that

$$(Lx_k,x_k) + (Nx_k,x_k) \leq (f,x_k) \quad \text{for each} \quad k. \quad (5.13)$$

Now it follows from (5.10) that for each fixed $t > 0$ and each k,

$$(Nx_k,x_k) \geq \frac{1}{\gamma} \| Nx_k \|^2 + (N(ty,x_k) - C(ty). \quad (5.14)$$

On the other hand, applying $Q = I - P$ to (5.8), we get $Lx_k = Q(f - Nx_k)$ and therefore

$$\| Lx_k \|^2 \leq \| f - Nx_k \|^2 = \| f \|^2 - 2(Nx_k,f) + \| Nx_k \|^2.$$

Hence

$$(Lx_k,x_k) \geq -\frac{1}{\alpha} \| Lx_k \|^2 \geq -\frac{1}{\alpha} \| Nx_k \|^2 + \frac{2}{\alpha}(Nx_k,f) - \frac{1}{\alpha} \| f \|^2.$$
$$(5.15)$$

In view of (5.14)–(5.15) and the assumption that $\gamma < \alpha$, it follows from (5.13) that

$$(N(ty),x_k) - C(ty) + \frac{2}{\alpha}(Nx_k,f)$$

$$-\frac{1}{\alpha}\|f\|^2 \leq (f,x_k) \qquad \text{for each} \quad k.$$

(5.16)

It follows from (5.16) that

$$(N(ty),y_k) \leq (f,y_k) + \frac{C(ty)}{\|x_k\|} + \frac{\alpha^{-1}\|f\|^2}{\|x_k\|} + \frac{2\alpha^{-1}\|f\|\|Nx_k\|}{\|x_k\|}.$$

Since $y_k \to y$ and $\|Nx_k\|/\|x_k\| \to 0$ as $k \to \infty$, taking the limit superior in the last inequality as $k \to \infty$ leads to $(N(ty),y) \leq (f,y)$ for each $t > 0$ and $y \in N(L)$ with $\|y\| = 1$. Hence lim inf$_{t\to\infty}(N(ty),y) \leq (f,y)$ with $y \in N(L)$ and $\|y\| = 1$, in contradiction to (5.11). Thus S_k^+ is bounded.

Q.E.D.

Since $T + F$ remains weakly A-proper when F is completely continuous (i.e., $Fu_j \to Fu$ whenever $u_j \to u$ in H), the same arguments used to prove Corollary 5.1 also yield the validity of the following practically useful result.

Corollary 5.2 Suppose that L and N satisfy conditions of Corollary 5.1 except for (5.9), which is replaced by

$$Nx = o(\|x\|) \quad \text{as} \quad \|x\| \to \infty \quad \text{and} \quad T = L + N \quad \text{is demiclosed}$$

(5.9')

(i.e., if $u_j \to u$ in H and $Tu_j \to h$, then $Tu = h$).

If F is completely continuous, $R(F)$ is bounded (a weaker condition suffices) and $\underline{\lim}(Fx_k,x_k)/\|x_k\| = \eta \geq 0$ whenever $\|x_k\| \to \infty$ as $k \to \infty$, then the equation $Lx + Nx + Fx = f$ is solvable.

As our final result in this section we have

Corollary 5.3 Suppose that $L \in L(H,H)$ is A-proper with $N(L) = N(L^*)$ and $T = L + N$ is weakly A-proper. Then, for a given f in H, equation (5.3) is solvable provided that the following additional conditions hold:

$$\|Nx\|/\|x\|^{1/2} \to 0 \quad \text{as} \quad \|x\| \to \infty \quad \text{and}$$

$$T(B) \quad \text{is closed for each} \quad B \subset H.$$

(5.17)

$$\varliminf(Nx_k,u_k) > (f,y) \quad \text{whenever} \quad \{x_k\} \subset X \quad \text{is such}$$
$$\text{that} \quad \| x_k \| \to \infty \quad \text{and} \quad u_k \equiv x_k/\| x_k \| \to y \in N(L) \qquad (5.18)$$
$$\text{with} \quad \| y \| = 1.$$

Proof. By Theorem 5.1(A), it suffices to show that the set S_k^+ is bounded if (5.17) and (5.18) hold. If S_k^+ were not bounded, we could find $\{x_k\}$ satisfying (5.8) such that $\| x_k \| \to \infty$ and $u_k \equiv x_k/\| x_k \| \to y \in N(L)$ with $\| y \| = 1$. Taking the inner product of (5.8) with x_k and noting that $\lambda_k(Px_k,x_k) > 0$, we get (5.13). Since $\| Lx_k \|^2 \leq \| f - Nx_k \|^2 \leq 2(\| f \|^2 + \| Nx_k \|^2)$, we see that

$$(Lx_k,x_k) \geq -\frac{1}{\alpha} \| Lx_k \|^2 \geq -\frac{2}{\alpha} (\| f \|^2 + \| Nx_k \|^2) \qquad \text{for all} \quad k.$$

It follows from this and (5.13) that

$$(Nx_k,x_k) \leq (f,x_k) + \frac{2}{\alpha} (\| f \|^2 + \| Nx_k \|^2).$$

Dividing the last inequality by $\| x_k \|$, we get the relation

$$(Nx_k,u_k) \leq (f,u_k) + \frac{2}{\alpha} \left(\frac{\| f \|^2}{\| x_k \|} + \frac{\| Nx_k \|^2}{\| x_k \|} \right).$$

Since $u_k \to y \in N(L)$, it follows from the last inequality and the first part of (5.17) that $\varliminf(Nx_k,u_k) \leq (f,y)$, in contradiction to (5.18).   Q.E.D.

Remark 5.4 If $(Lx,x) \geq 0$ for $x \in H$, then instead of the condition that $Nx = o(\| x \|^{1/2})$ as $\| x \| \to \infty$, it suffices to assume that $Nx = o(\| x \|)$ as $\| x \| \to \infty$.

Remark 5.5 Corollary 5.1 is related to [18, Corollary II.7], where it is assumed that $L: D(L) \subset H \to H$ is unbounded but $(\lambda P + L)^{-1}: H \to H$ is compact and $N: H \to H$ is monotone. The same can be said about Corollary 5.3 and the corollary of [17, Theorem 13] attributed in [17] to [18]. See [18] for the discussion of the condition (5.10) and its usefulness in applications.

Remark 5.6 It was shown in [18] that if $N: H \to H$ is any mapping, then one can define a "recession function" of N for any $y \in H$ by

$$J_N(y) = \lim_{\substack{t \to \infty \\ v \to y}} \inf(N(tv),v) \qquad (5.19)$$

and the function $J_N(y)$ thus defined is lower semicontinuous from H to $[-\infty,\infty]$ with $J_N(\lambda y) = \lambda J_N(y)$ for each $\lambda > 0$ and $y \in H$; moreover, if

N is monotone, hemicontinuous, and $\| Nx \|/\| x \| \to 0$ as $\| x \| \to \infty$, then $J_N(y) = \lim \inf_{t\to\infty}(N(ty),y)$.

It follows from this that if in Corollary 5.3 the map N is assumed to be monotone, then since $(Nx_k,y_k) = (N(t_ku_k,u_k)$ for each k with $t_k \to \infty$ and $u_k \to y$, it follows from the definition of $J_N(y)$ that condition (5.18) is equivalent to

$$J_N(y) > (f,y), \qquad y \in N(L), \qquad \| y \| = 1. \qquad (5.20)$$

Moreover, since $|(Nx_k,u_k) - (Nx_k,y)| \to 0$ whenever $R(N)$ is bounded, it follows that in this case condition (5.20) is equivalent to the hypothesis (B_1^+) of Remark 5.0; that is, $\underline{\lim}(Nx_k,y) > (f,y)$ whenever $\{x_k\} \subset X$ is such that $\| x_k \| \to \infty$ and $u_k \equiv x_k/\| x_k \| \to y \in N(L)$. Indeed, since $0 = \underline{\lim}\{(Nx_k,u_k) - (Nx_k,y)\} \le \underline{\lim}(Nx_k,u_k) - \underline{\lim}(Nx_k,y)$, we see that $\underline{\lim}(Nx_k,y) \le \underline{\lim}(Nx_k,u_k) = J_N(y)$ [i.e., B_1^+ implies (5.20)]. On the other hand, since $0 = \underline{\lim}\{(Nx_k,y) - (Nx_k,u_k)\} \le \underline{\lim}(Nx_k,y) - \underline{\lim}(Nx_k,u_k)$, we get that $J_N(y) = \underline{\lim}(Nx_k,u_k) \le \underline{\lim}(Nx_k,y)$ [i.e., (5.20) implies B_1^+].

5.2 Special Cases of Weakly A-Proper Maps

In this section we discuss some special classes of maps $N: X \to Y$ for which $T = L + N$ is weakly A-proper and $T(B)$ is closed in Y. As special cases we deduce from Theorems 5.1 and 5.2 the existence results of other authors mentioned in the introduction to Section 5.1.

We begin with the following notion, which is due to Brezis [16] when $Y = X^*$. In what follows we say that $T: X \to Y$ is of *type* (KM) provided that the following conditions hold:

(i) If $x_j \to x$ in X, $Tx_j \to g$ in Y, and $\overline{\lim}(Tx_j,Kx_j) \le (g,Kx)$, then $Tx = g$, where K is a suitable map of X into Y^*.

(ii) T is continuous from finite-dimensional subspaces of X into Y equipped with weak* topology.

We shall also say that T is *semibounded* if $\{Tx_j\}$ is bounded whenever $\{x_j\}$ and $\{(Tx_j,Kx_j)\}$ are bounded.

Lemma 5.2 Suppose that X is reflexive, $\Gamma = \{X_n,V_n;E_n,W_n\}$ is admissible for (X,Y), $K: X \to Y^*$ continuous with $R(K)$ dense in Y^*, and $K_n: X_n \to D(W_n^*)$ is such that $\{K_n(z_n)\}$ is uniformly bounded whenever $\{z_n \mid z_n \in X_n\}$ is bounded and for each n we have

$$(W_ng,K_nx) = (g,Kx) \qquad \text{for all} \quad x \in X_n \quad \text{and} \quad g \in Y. \qquad (5.21)$$

(a) If $T: X \to Y$ is weakly continuous, then T is weakly A-proper w.r.t. Γ and $T(B)$ is closed for each closed ball $B \subset X$.

(b) If $T: X \to Y$ is semibounded and of type (KM), then T is weakly A-proper w.r.t. Γ and $T(B)$ is closed provided that K is also weakly continuous.

We omit the proof of Lemma 5.2 since the proof of (a) is similar to the proof of Proposition 2 in [233], while the proof of (b) is essentially the same as the proof of Propositions 11 and 12 in [233].

Remark 5.7 The conditions on K in (b) of Lemma 5.2 are certainly satisfied if $K \in L(X,Y^*)$ and $R(K)$ is dense in Y^*. In applications this is often the case, as we shall see in Problem 5.1 of Section 5.3.

In view of Lemma 5.2, Theorems 5.1 and 5.2 yield the following:

Corollary 5.4 Suppose that X, Γ, K, and K_n are as in Lemma 5.2, $L \in L(X,Y)$ is A-proper w.r.t. Γ with $\text{ind}(L) = 0$, $N: X \to Y$ is such that $N(x) = o(\| x \|)$ as $\| x \| \to \infty$, and one of the following three conditions holds:

(H1) N is compact.
(H2) N is weakly continuous.
(H3) N is semibounded and of type (KM) and $\phi(x) = (Lx,Kx)$ is weakly lower semicontinuous [i.e., $x_j \rightharpoonup x \Rightarrow \phi(x) \le \underline{\lim} \, \phi(x_j)$].

Then (5.3), $Lx + Nx = f$ has a solution for a given f in Y provided that condition (A) of Theorem 5.1 holds or Y is a Hilbert space and condition (B), (C), or (D) of Theorem 5.2 holds.

Proof. (H1) Since $N: X \to Y$ is compact, $T = L + N$ is a continuous A-proper mapping. Hence in this case the conclusion of Corollary 5.4 follows from Theorem 5.1 and Remark 5.1 if (A) holds or from Theorem 5.2 and Remark 5.1 if Y is a Hilbert space and (B), (C), or (D) holds.

(H2) Since $T = L + N: X \to Y$ is weakly continuous, it follows from Lemma 5.2(a) that T is weakly A-proper w.r.t. Γ and $T(B)$ is closed in Y for each $B \subset X$. Thus Corollary 5.2 (H2) follows from Theorem 5.1 if (A) holds and from Theorem 5.2 if Y is a Hilbert space and (B), (C), or (D) holds.

(H3) In view of Lemma 5.2(b), to establish the assertion of Corollary 5.2 when (H3) holds, it suffices to show that $T = L + N$ is semibounded and of type (KM) and then use Theorem 5.1 or 5.2. The fact that T is semibounded follows easily from the boundedness of L and semiboundedness of N. To show that T is of type (KM), we let $x_j \rightharpoonup x$ in X, $Tx_j \rightharpoonup h$ in Y, and $\overline{\lim}(Tx_j,Kx_j) \le (h,Kx)$. Since $Lx_j \rightharpoonup Lx$ in Y, it follows that

$Nx_j \rightharpoonup h - Lx$ and thus, by our condition on ϕ, we have

$$\overline{\lim}(Nx_j, Kx_j) = \overline{\lim}\{(Tx_j, Kx_j) - (Lx_j, Kx_j)\}$$
$$\leq \overline{\lim}(Tx_j, Kx_j) - \underline{\lim}(Lx_j, Kx_j) \leq (h - Lx, Kx).$$

Since N is of type (KM), it follows that $Nx = h - Lx$ or $Tx = h$ [i.e., T is of type (KM)].   Q.E.D.

Remark 5.8 If $K \in L(X, Y^*)$ and $A, C \in L(X, Y)$ are such that $(Ax, Kx) \geq 0$ for $x \in X$ and C is compact, then $\phi(x) = (Lx, Kx)$ is weakly lower semicontinuous, where $L = A + C$ or $L = A - C$.

Special case If we set $Y = X^*$, where X is a separable reflexive space, the natural choice for Γ, K, and K_n that satisfy the condition of Lemma 5.2 is $\Gamma = \Gamma_I = \{X_n, V_n; X_n^*, V_n^*\}$, $K = I$, and $K_n = I_n$, with I and I_n denoting the identities in X and X_n, respectively. In this case, the map N in (H3) is of type (M) in the sense of Brezis [16]. This class includes all the hemicontinuous monotone mappings and the class of pseudomonotone maps introduced in [16]. Hence Corollary 5.4 is valid for these maps $N: X \to X^*$.

In case $Y = X = H$, where H is a separable Hilbert space, Corollary 5.4 extends essentially Theorems 1, 2, and 3 of [71] as well as the corresponding earlier results of [92,186].

Remark 5.9 The following observations will prove to be useful in applications of the results above to semilinear differential equations.

(i) Suppose that H_0 and H are Hilbert spaces such that $H_0 \subseteq H$, H_0 is dense in H, and the embedding of H_0 into H is compact. If $L \in L(H_0, H_0)$ and if there exist constants $c_0 > 0$ and c_1 such that

$$(Lu, u)_0 \geq c_0 \| u \|_0^2 - c_1 \| u \|^2 \qquad \text{for all} \quad u \in H_0, \tag{5.22}$$

then L is A-proper w.r.t. any projectionally complete scheme $\Gamma_0 = \{X_n, P_n\}$ for (H_0, H_0), $\text{ind}(L) = 0$, and the functional $\phi(u) = (Lu, u)_0$ is weakly lower semicontinuous (see [225]). It is known (see [87]) that in view of Gärding's inequality, the generalized Dirichlet forms on a Sobolev space $\overset{\circ}{W}_2^m(Q)$ corresponding to strongly elliptic operators of order $2m$ give rise to mappings $L \in L(\overset{\circ}{W}_2^m, \overset{\circ}{W}_2^m)$ satisfying (5.22).

(ii) Let X be a real Banach space and let Q be any bounded set in X. Then the *ball measure of noncompactness* of Q, $\chi(Q)$, is defined to be $\chi(Q) = \inf\{r > 0 \mid Q \text{ can be covered by a finite number of balls with radii}$

$\leq r\}$. A continuous mapping $F: X \to X$ is said to be a k-ball *contraction* (respectively, *ball condensing*) if $\chi(F(Q)) \leq k\chi(Q)$ for all bounded sets $Q \subset X$ and some $k \geq 0$ [respectively, $\chi(F(Q)) < \chi(Q)$ whenever $\chi(Q) \neq 0$]. For the survey of the theory of these classes of mappings, see [253]. It is known that if $F: X \to X$ is k-ball contractive with $k \in [0,1)$, then $L = I \pm F$ is A-proper w.r.t. $\Gamma_1 = \{X_n, P_n\}$. Moreover, if $X = H$ and $F \in L(H,H)$, then F can be represented as $F = A + C$ with C compact and $\| A \| \leq (k + 1)/2$ (see [289]). It follows from this and Remark 5.8 that L is A-proper w.r.t. Γ_1, $\text{ind}(L) = 0$, and $\phi(u) = (Lu,u)$ is weakly lower semicontinuous on H.

To deduce some existence results in [77] from Theorem 5.2, we need

Lemma 5.3 Suppose that $\Gamma_\alpha = \{X_n, P_n\}$ is projectionally complete for (X,X). If $L \in L(X,Y)$ is Fredholm with $\text{ind}(L) = 0$, $Y_n = (L + C)(X_n) \subset Y$ for each n, and Q_n is a projection of Y onto Y_n, where $C \in L(X,Y)$ is the compact map constructed in the proof of Theorem 5.1, then L is A-proper w.r.t. the projectionally complete scheme $\Gamma_c = \{X_n, P_n; Y_n; Q_n\}$ for (X,Y).

Proof. It is not hard to show that since $L + C$ is a linear homeomorphism of X onto Y, the scheme Γ_c is projectionally complete for (X,Y). To show that L is A-proper w.r.t. Γ_c, let $\{x_{n_j} \mid x_{n_j} \in X_{n_j}\}$ be any bounded sequence such that $Q_{n_j}L(x_{n_j}) - Q_{n_j}g \to 0$ for some g in Y [i.e., $Q_{n_j}L(x_{n_j}) \to g$ as $j \to \infty$ because $Q_{n_j}g \to g$ in Y as $j \to \infty$]. Since $\{x_{n_j}\}$ is bounded, C compact, and $Q_n h \to h$ for each h in Y, we may assume that $C(x_{n_j}) \to g_1$ for some g_1 in Y, so $Q_{n_j}C(x_{n_j}) \to g_1$ in Y. Therefore, $y_{n_j} \equiv Q_{n_j}(L + C)(x_{n_j}) \to g + g_1 \equiv y$ in Y as $j \to \infty$. But $Q_{n_j}(L + C)(x_{n_j}) = (L + C)(x_{n_j})$ for each j, and therefore $x_{n_j} = (L + C)^{-1}(y_{n_j}) \to (L + C)^{-1}y \equiv x$ in X. Hence $(L + C)x = g + g_1$ with $g_1 = \lim_j C(x_{n_j}) = C(x)$ (i.e., $Lx = g$ and thus L is A-proper w.r.t. Γ_c).   Q.E.D.

Corollary 5.5 Suppose that $L \in L(X,Y)$ is Fredholm with $\text{ind}(L) = 0$ and Γ_c is the scheme constructed above with $\| Q_n \| = 1$ for each n. Suppose that $N: X \to Y$ is such that $N(L + C)^{-1}: Y \to Y$ is ball-condensing.

If, for a given $f \in Y$, condition (A) holds or Y is a Hilbert space and either $B(1)$–$B(2)$ or $D(1)$–$D(2)$ of Theorem 5.2 holds, then Eq. (5.3), $Lx + Nx = f$, is solvable.

Proof. In view of Lemma 5.3 and Remark 5.1, Corollary 5.5 will follow from Theorem 5.1 when (A) holds and from Theorem 5.2 when either $B(1)$–$B(2)$ or $D(1)$–$D(2)$ holds if we show that $T = L + N$ is A-proper w.r.t. Γ_c. So let $\{x_{n_j} \mid x_{n_j} \in X_{n_j}\}$ be any bounded sequence such that

$Q_{n_j}L(x_{n_j}) + Q_{n_j}N(x_{n_j}) \to g$ for some g in Y. If we set $y_{n_j} = (L + C)(x_{n_j})$ for each j, then $\{y_{n_j} \mid y_{n_j} \in Y_{n_j}\}$ is bounded in Y and $y_{n_j} + Q_{n_j}A(y_{n_j}) \to g$ in Y as $j \to \infty$, where $A: Y \to Y$ is a ball-condensing mapping given by $A = (N - C)(L + C)^{-1}$. Since $I + A$ is A-proper w.r.t. $\{Y_n, Q_n\}$, there exist a subsequence $\{y_{n_{j(k)}}\}$ and $y \in Y$ such that $y_{n_{j(k)}} \to y$ in Y as $k \to \infty$ and $y + Ay = g$. Hence $x_{n_{j(k)}} = (L + C)^{-1}y_{n_{j(k)}} \to x \equiv (L + C)^{-1}y$ in X as $k \to \infty$, $(L + C)x = y$ and $y + Ay = (L + C)x + (N - C)x = g$ (i.e., $Lx + Nx = g$). Q.E.D.

The second part of Corollary 5.5 includes [77, Propositions 2.4 and 2.5], where it is also assumed that $X \subset Y$, $L: X \to Y$ is such that $Y = N(L) \oplus R(L)$. It should be added that the author of [77] uses the degree theory for condensing vector fields to study the solvability of (5.3), and therefore his results are also valid for set-condensing maps and for the space X, which need not have a projectionally complete scheme. However, the arguments of [77] cannot be used to study the solvability of equation (5.3) for the more general class of weakly A-proper maps treated here.

5.3 Some Applications to Differential Equations

In this section we deduce some consequences of the results of Sections 5.1 and 5.2 for semilinear ordinary and partial differential equations. Some of the problems we treat here show that they can not be put into the framework to which the theories of compact and condensing operators or the theory of monotone operators are applicable. However, as we shall see, the results for the more general class of A-proper and weakly A-proper maps are applicable.

Problem 5.1 treats a boundary value problem for an ODE of second order whose nonlinear part N depends also on u'' and satisfies certain "monotonicity" conditions which preclude the applicability of the condensing-type mapping results obtained in [77], where the same problem is treated under Lipschitz-type conditions on N.

Problem 5.2 treats the case where unbounded discontinuous operators are involved. The abstract results below are modeled on situations that arise when one attempts to treat semilinear elliptic problems on *unbounded* domains in which case the resolvent operators are not compact and, in general, the nonlinear part N is also noncompact.

Problem 5.3 deals with the existence of weak solutions for elliptic semilinear equations of order $2m$ treated by other authors earlier under somewhat different conditions and by different methods. We include it here so as to illustrate the applicability of the results of Section 5.1, which we believe provides a somewhat simpler proof for the existence of solutions (see [77,148] and others).

Problem 5.4 treats a semilinear elliptic second-order equation with Neumann boundary condition studies in [77] under more restrictive conditions on the nonlinear part. The results in [77] cannot be applied to the problem treated here.

Problem 5.1 To compare our existence theorem (Theorem 5.3 below) with that of [77, Theorem 3.3], we first study the solvability of the boundary value problem:

$$\begin{cases} -u''(t) - g(t,u'(t),u''(t)) + f(t,u(t),u'(t)) = 0 \\ u(0) = u(1), \qquad u'(0) = u'(1) \end{cases} \tag{5.23}$$

treated in [77] under rather restrictive condition on g. To state the hypotheses under which (5.23) has a solution in $W_2^2([0,1])$, we let $Y = L_2([0,1])$ and $X = \{u \mid u \in W_2^2([0,1]), u(0) = u(1), u'(0) = u'(1)\} \subset Y$, where Y and X are separable real Hilbert spaces with the respective norms $\| \cdot \|$ and $\| \cdot \|_2 = (\| u \|^2 + \| u' \|^2 + \| u'' \|^2)^{1/2}$ and the corresponding inner products $(\cdot,\cdot)$ and $(\cdot,\cdot)_2$. Defining $L: X \to Y$ by $L(u)(t) = -u''(t)$, for $u \in X$ and $t \in [0,1]$, it is easy to show that $L \in L(X,Y)$, $\mathrm{ind}(L) = 0$, $N(L)$ consists of constant functions, $Y = N(L) \oplus R(L)$ and $L + I: X \to Y$ is a linear homeomorphism, where I denotes the inclusion map of X into Y which is compact by the Sobolev embedding theorem (see [1]).

Let $\{X_n\} \subset X$ be a sequence of finite-dimensional spaces, such that $\mathrm{dist}(u,X_n) = \inf\{\| u - v \|_2 : v \in X_n\} \to 0$ as $n \to \infty$ for each u in X. Let K be the homeomorphism of X onto Y given by $K = L + I$, and for each n, let $Y_n = K(X_n) \subset Y$. Then if $P_n: X \to X_n$ and $Q_n: Y \to Y$ denote orthogonal projections, the scheme $\Gamma_0 = \{X_n,P_n;Y_n,Q_n\}$ is projectionally complete for (X,Y) and, since

$$(Lu,Ku) \geq \| u \|_2^2 - \| u \|^2 \qquad \text{for} \quad u \in X, \tag{5.24}$$

the map $L: X \to Y$ is A-proper w.r.t. Γ_0 with $\mathrm{ind}(L) = 0$. We impose the following conditions on the functions g and f:

(a1) $g: [0,1] \times R^2 \to R$ is continuous and there are constants $p > 0$ and $c_0 \in (0,1)$ such that $| g(t,s,r) | \leq p$ for all $t \in [0,1]$ and $r,s \in \mathbb{R}$ and $[g(t,s,r_1) - g(t,s,r_2)](r_1 - r_2) \geq -c_0 | r_1 - r_2 |^2$ for all $t \in [0,1]$ and $s,r_1,r_2 \in \mathbb{R}$.

(b1) $f: [0,1] \times \mathbb{R}^2 \to \mathbb{R}$ is continuous and there exist $a(x) \in Y$, $\beta \in (0,1)$ and $\gamma > 0$ such that $| f(t,s,r) | \leq a(t) + \gamma(| s |^\beta + | r |^\beta)$ for $t \in [0,1]$ and $r,s \in \mathbb{R}$.

We define the operators $G,C: X \to Y$ by $G(u)(t) = -g(t,u'(t),u''(t))$ and $C(u)(t) = f(t,u(t),u'(t))$ for $t \in [0,1]$ and $u \in X$. First note that since X is compactly embedded into $C^1([0,1])$, the conditions on f in (b1) imply that C is compact, as a map from X to Y, and

$$\| C(u) \| \le \| a \| + \gamma_1 \| u \|_2^\beta \qquad \text{for} \quad u \in X. \tag{5.25}$$

Now it is not hard to show that, in virtue of (a1), $G: X \to Y$ is continuous, $\| Gu \| \le p$ for $u \in X$, and $(Gu - Gv, Ku - Kv) \ge -c_0\| u'' - v'' \|^2 - 2p\| u - v \| + (g(t,u',v'') - g(t,v',v'')), u'' - v'')$ for $u,v \in X$. It follows from this and (5.24) that for all $u,v \in X$ we have

$$((L + G)u - (L + G)v, Ku - Kv) \ge (1 - c_0) \| u - v \|_2^2$$
$$- \| u - v \|^2 - 2p\| u - v \| + (g(t,u',v'') - g(t,v',v'')), u'' - v'').$$

This and the results in [225] imply:

Lemma 5.4 If g and f satisfy conditions (a1) and (b1), respectively, then the operator $T = L + G + C: X \to Y$ is continuous and A-proper w.r.t. Γ_0 and $Nu = o(\| u \|_2)$ as $\| u \|_2 \to \infty$ with $N = G + C$.

Our first result in this section, Theorem 5.3 below, improves [77, Theorem 3.3] and the corresponding results in [122] in that our nonlinearity g is not assumed to satisfy the Lipschitz condition of [77], while f is not assumed to satisfy the growth conditions of [122] (no u'' is permitted in the nonlinear part in [122]).

Theorem 5.3 Suppose that (a1) and (b1) hold and that there exist $N_0 > 0$ and $\delta > p$ such that either (H1) or (H1') holds, where

(H1) For $t \in [0,1]$ and $r \in \mathbb{R}$ we have $f(t,r,s) \ge \delta$ when $s \ge N_0$ and $f(t,r,s) \le -\delta$ when $s \le -N_0$.
(H1') For $t \in [0,1]$ and $r \in \mathbb{R}$ we have $f(t,s,r) \le -\delta$ when $s \ge N_0$ and $f(t,r,s) > \delta$ when $s \le -N_0$.

Then the boundary value problem (5.23) has a solution in $W_2^2([0,1])$.

Proof. By virtue of Lemma 5.4 and Theorem 5.2 (with Remarks 5.0 and 5.1), to prove Theorem 5.3, it suffices to show that (H1) implies B_1^+, while (H1') implies B_1^-. Thus suppose first that (H1) holds. We have to show that if $\{u_n\} \subset X$ is a sequence such that $\| u_n\|_2 \to \infty$ and $\{u_n/\| u_n \|_2\}$ converges in X to the function whose constant value is $+1$ (respectively,

-1), then

$$\underline{\lim} \int_0^1 g(t,u_n'(t),u_n''(t)) + f(t,u_n(t),u_n'(t)) \, dt > 0$$

(respectively, $\overline{\lim} \int_0^1 \{g(t,u_n',u_n'') + f(t,u_n,u_n')\} \, dt < 0$).

Suppose first that $y_n \equiv u_n/\| u_n \|_2 \to +1$ in X as $n \to \infty$. Then, since X is compactly embedded into $C([0,1])$, $y_n(t) \to 1$ uniformly on $[0,1]$. Hence there exists $n_0 \geq 1$ such that $u_n(t) = \| u_n \| y_n(t) \geq N_0$ if $n_0 \geq N_0$ and $t \in [0,1]$ and therefore, by (H1),

$$g(t,u_n',u_n'') + f(t,u_n,u_n') \geq f(t,u_n(t),u_n'(t)) - | g(t,u_n',u_n'') |$$
$$\geq \delta - p > 0$$

for $t \in [0,1]$ and $n \geq n_0$. This implies the inequality above. In a similar way, one shows that (H1′) implies B_1^- of Remark 5.0.   Q.E.D.

Remark 5.10 In [77] the solvability of (5.23) was studied by means of the degree theory of condensing vector fields under the more restrictive and essential assumption that

$$| g(t,s,r_1) - g(t,s,r_2) | \leq c_0 | r_1 - r_2 | \quad \text{for some} \quad c_0 \in (0,1) \qquad (5.26)$$

and all $t \in [0,1]$ and $s,r_1,r_2 \in \mathbb{R}$. The author of [77] also assumes that $g(t,s,r)$ is continuous in s uniformly for $(t,r) \in [0,1] \times \mathbb{R}$.

Problem 5.2 The abstract results below are modeled on situations that arise when one attempts to treat semilinear elliptic problems on *unbounded* domains (see [275] and others cited there, where the global bifurcation phenomenon is treated for such operators). The interesting feature of this problem is that the resolvents of the corresponding elliptic operators are no longer compact, the Sobolev embedding theorems in general are no longer valid, and unless very strict conditions are imposed, the nonlinear part is not compact. As a second example we consider the solvability of the equation

$$Au - \lambda u + Nu = f \quad [u \in D(A), f \in H, \lambda \in \mathbb{R}], \qquad (5.27)$$

where H is a real separable Hilbert space and A, N, and λ are assumed to satisfy the following hypotheses:

(a2) A is a densely defined, positive definite, self-adjoint operator whose essential spectrum $\sigma_e(A)$ is bounded below; that is, there is a number $\gamma > 0$ such that for each $\epsilon > 0$, $\sigma(A) \cap (-\infty, \gamma - \epsilon)$ consists of a nonempty set of isolated eigenvalues, each of finite multiplicity, with

$$\lambda_0 < \lambda_1 < \lambda_2, \ldots, < \gamma.$$

(b2) The number λ is an eigenvalue of A such that $| \lambda | \gamma^{-1} < 1$.

(c2) $N: D(A^{1/2}) \to H$ is a map such that $N = N_1 + N_2$, where $N_2: H_0 \to H$ is compact, $N_1: H_0 \to H$ is continuous and monotone [i.e., $(N_1 u - N_1 v, u - v) \geq 0 \ \forall \ u,v \in H_0$], and $\| Nu \|/\| u \|_0 \to 0$ as $\| u \|_0 \to \infty$, where H_0 is the completion of $D(A)$ in the metric $[u,v] = (Au,v)$ and $\| u \|_0 = (Au,u)^{1/2}$ for all $u,v \in D(A)$.

It is known that H_0 is continuously embedded into $H, H_0 = D(A^{1/2})$, $\| u \|_0 = \| A^{1/2} u \|$ for all u in H_0, A has an inverse A^{-1} defined on H, $A^{-1}: H \to H$ is bounded, self-adjoint, and positive, its square root $A^{-1/2}: H \to H$ is also self-adjoint, positive, and bounded, and the square root $A^{1/2}$, considered as a mapping from H_0 to H, is a linear homeomorphism. Moreover, it follows from condition (a2) that $A^{-1}: H \to H$ is γ^{-1}-ball contractive and $S \equiv A^{-1/2}: H \to H$ is $\gamma^{-1/2}$-ball contractive (see Stuart [272]). This and (b2) imply that λA^{-1} is k-ball contractive with $k = | \lambda | \gamma^{-1} < 1$, the operator $L \equiv I - \lambda A^{-1}$ has a nontrivial null space $N(L)$, and $u \in N(L)$ if and only if $Au - \lambda u = 0$. Furthermore, $F = SNS: H \to H$ is such that $F = F_1 + F_2$ with $F_2 = SN_2 S$ compact and $F_1 = SN_1 S$ monotone and continuous by (c2).

Remark 5.11 It is easy to see that $u \in D(A)$ is a solution of (5.27) if and only if $v = A^{1/2} u$ and v is a solution of

$$Lv + Fv = S(f) \qquad [v \in H, S(f) \in H, F = SNS]. \tag{5.28}$$

We are now in the situation where we can apply to (5.27), or equivalently to (5.28), some of the results of Sections 5.1 and 5.2. To accomplish this we set $Y = X = H$, $\Gamma = \Gamma_1 = \{X_n, P_n\}$, and $T = L + F$. Corollary I.2.2 shows that L is A-proper w.r.t. Γ_1 and therefore Fredholm with $\text{ind}(L) = 0$. In view of the discussion above concerning $F = F_1 + F_2$: $H \to H$, it follows from Corollary 3.1 or [275, Lemma 2.8] that $T = L + F$ is A-proper w.r.t. Γ_1. Moreover, the last part of (c2) implies that $\| Fv \|/ \| v \| \to 0$ as $\| v \| \to \infty$. Indeed, since $\| Fv \| \leq \| S \| \| N(Sv) \|$ for each v in H, to prove the latter assertion it suffices to show that $\| N(Sv) \|/\| v \| \to 0$ as $\| v \| \to \infty$. But since $u = Sv \in H_0$ and $\| u \|_0 = \| A^{1/2} SV \| = \| v \|$, it follows from the last assumption in (c2) that

$$\frac{\| N(Sv) \|}{\| v \|} = \frac{\| Nu \|}{\| u \|_0} \to 0 \qquad \text{as} \quad \| v \| = \| u \|_0 \to \infty.$$

We are now in the situation to which Theorem 5.2 and Remark 5.1 (i.e., the A-proper mapping version of Theorem 5.2) apply.

Theorem 5.4 Suppose that conditions (a2), (b2), and (c2) hold. Then (5.27) has a solution $u \in D(A)$ for a given f in H if the following condition holds:

(H2) Either $\overline{\lim}(Nu_j, u_0) > (f, u_0)$ or $\underline{\lim}(Nu_j, u_0) < (f, u_0)$ whenever $\{u_j\} \subset H_0$ is such that $\| u_j \|_0 \to \infty$ and $u_j / \| u_j \|_0 \to u_0$ in H_0 with $u_0 \in N(A - \lambda I)$.

Proof. Since the solvability of (5.27) is equivalent to the solvability of (5.28) in H, in view of the preceding discussion and Theorem 5.2 with Remark 5.1, it suffices to show that either $\overline{\lim}(Fv_j, v_0) > (Sf, v_0)$ or $\underline{\lim}(Fv_j, v_0) < (Sf, v_0)$ whenever $\{v_j\} \subset H$ is such that $\| v_j \| \to \infty$ and $v_j / \| v_j \| \to v_0 \in N(L)$. To show this, let $\{v_j\} \subset H$ be such that $\| v_j \| \to \infty$ and $v_j / \| v_j \| \to v_0 \in N(L)$. Now since $A^{1/2}$ is a linear homeomorphism of H_0 onto H, there exist $u_0 \in H_0$ and $u_j \in H_0$ for each $j \in N$ such that $v_0 = A^{1/2} u_0$ and $v_j = A^{1/2} u_j$ with $\| u_j \|_0 = \| v_j \| \to \infty$ and

$$\left\| \frac{u_j}{\| u_j \|_0} - u_0 \right\|_0 = \left\| A^{1/2} \left(\frac{u_j}{\| u_j \|_0} - u_0 \right) \right\| = \left\| \frac{v_j}{\| v_j \|} - v_0 \right\| \to 0$$

as $j \to \infty$. Moreover, since $v_0 = A^{1/2} u_0 \in N(L)$ and $L = I - \lambda A^{-1}$, it follows that $0 = L(A^{1/2} u_0) = A^{1/2} u_0 - \lambda A^{-1/2} u_0$, so $Au_0 - \lambda u_0 = 0$. Thus $\{u_j\} \subset H_0$ is such that $\| u_j \| \to \infty$ and $u_j / \| u_j \|_0 \to u_0$ in H_0 with $u_0 \in N(A - \lambda I)$ and therefore, by (H2), either $\overline{\lim}(Nu_j, u_0) > (f, u_0)$ or $\underline{\lim}(Nu_j, u_0) < (f, u_0)$. Suppose that $\overline{\lim}(Nu_j u_0) > (f, u_0)$. In view of this and the fact that $u_j = Sv_j$, $u_0 = Sv_0$ and $S: H \to H$ is bounded and self-adjoint, we obtain the desired asymptotic positivity

$$(v_0, Sf) = (u_0, f) < \overline{\lim}(u_0, Nu_j) = \overline{\lim}(Sv_0, NSv_j) = \overline{\lim}(v_0, Fv_j)$$

when Theorem 5.2 (with Remark 5.1) is applied to (5.28). Similarly, one shows that $\underline{\lim}(Fv_j, v_0) < (f, Sv_0)$ follows from $\underline{\lim}(Nu_j, u_0) < (f, u_0)$.
                                                                                Q.E.D.

Problem 5.3 Let $Q \subset R^n$ be a bounded domain with sufficiently smooth boundary so that the Sobolev embedding theorem holds. Let $W_2^m(Q) \equiv W_2^m$ be the Sobolev space of all real functions u such that u and its generalized derivatives $D^\alpha u \in L_2(Q)$ for $| \alpha | \leq m$, where $\alpha = (\alpha_1, \ldots, \alpha_n)$ is the multi-index with $| \alpha | = \alpha_1 + \cdots + \alpha_n$. W_2^m is a separable Hilbert space with the inner product

$$(u, v)_m = \sum_{|\alpha| \leq m} \int_Q D^\alpha u D^\alpha v \, dx.$$

Let $\overset{\circ}{W}{}_2^m$ be the subspace of W_2^m that is the completion in the W_2^m-norm of $C_0^\infty(Q)$, the set of infinitely differentiable functions with compact support in Q.

As an application of Corollary 5.2 we consider the following boundary value problem:

$$\sum_{|\alpha|,|\beta|\leq m} (-1)^{|\alpha|} D^\alpha(a_{\alpha\beta}(x)D^\beta u) + h(x,u(x)) = f(x) \qquad (x \in Q)$$

$$D^\alpha u = 0 \qquad (x \in \partial Q, \quad |\alpha| \leq M - 1) \tag{5.29}$$

where $a_{\alpha\beta}(x)$ and $h(x,u)$ satisfy the following conditions:

(a3) $a_{\alpha\beta}(x) \in L^\infty(\overline{Q})$ for $|\alpha|, |\beta| \leq m$, $a_{\alpha\beta}(x) \in C(\overline{Q})$ for $|\alpha| = |\beta| = m$, and $\sum_{|\alpha|,|\beta| = m} a_{\alpha\beta}(x)\eta^\alpha\eta^\beta \geq d|\eta|^{2m} \, \forall \, \eta \in R^n$ and some $d > 0$.

(b3) h: $Q \times \mathbb{R} \to \mathbb{R}$ is continuous and $|h(x,s)| \leq a(x) \, \forall \, x \in Q$, $s \in \mathbb{R}$ and some $a \in L_2$.

For a given $f \in L_2(Q)$, by a *weak solution* of (5.29) we mean a function $u \in \overset{\circ}{W}{}_2^m$ such that

$$a(u,v) + b(u,v) = \langle f,v \rangle \qquad \forall \, v \in \overset{\circ}{W}{}_2^m, \tag{5.30}$$

where $a(u,v)$ and the $b(u,v)$ are generalized Dirichlet forms on $\overset{\circ}{W}{}_2^m$ given by

$$a(u,v) + \sum_{|\alpha|,|\beta|\leq m} \langle a_{\alpha\beta}D^\beta u, D^\alpha v \rangle, \; b(u,v) = \langle h(x,u),v \rangle, \tag{5.31}$$

where $\langle , \rangle = (,)_0$ and $\|\cdot\| = \|\cdot\|_0$ is the inner product and the norm in $L_2(Q)$, respectively. Condition (a3) implies that $a(u,v)$ is a bounded bilinear form on $\overset{\circ}{W}{}_2^m$ and hence there exists a unique mapping $L \in L(\overset{\circ}{W}{}_2^m, \overset{\circ}{W}{}_2^m)$ such that

$$(Lu,v)_m = a(u,v) \qquad \forall \, u,v \in \overset{\circ}{W}{}_2^m \tag{5.32}$$

and L satisfies the Gärding inequality (see [87]):

$$(Lu,u)_m \geq c_1\| u \|_m^2$$
$$- c_2\| u \|^2 \qquad \forall \, u \in \overset{\circ}{W}{}_2^m \; \text{ and some } \; c_1 > 0, c_2 \geq 0. \tag{5.33}$$

Similarly, condition (b3) implies that $b(u,v)$ is a continuous linear functional of v in $\overset{\circ}{W}{}_2^m$, so it determines a unique bounded continuous map N: $\overset{\circ}{W}{}_2^m \to \overset{\circ}{W}{}_2^m$ such that $(Nu,v)_m = b(u,v)$ for all $u,v \in \overset{\circ}{W}{}_2^m$. It was shown in [225] that since the embedding of $\overset{\circ}{W}{}_2^m$ into L_2 is compact, it follows from (5.33) that L: $\overset{\circ}{W}{}_2^m \to \overset{\circ}{W}{}_2^m$ is A-proper with respect to any given projectionally complete scheme $\Gamma_0 = \{X_n, P_n\}$ for $\overset{\circ}{W}{}_2^m$. The latter exists since $\overset{\circ}{W}{}_2^m$ is separable. It is obvious that in view of our conditions on $h(x,s)$ and the Sobolev embedding theorem, N is compact and $\| Nu \|_m / \| u \|_m \to$

0 as $\| u \|_m \to \infty$. Let $w_f \in \mathring{W}_2^m$ be such that $\langle f,v \rangle = (w_f,v)_m$ for all v in $\mathring{W}_2^m$. Then the equation

$$Lu + Nu = w_f \qquad (u \in \mathring{W}_2^m, \ w_f \in \mathring{W}_2^m) \qquad (5.34)$$

is equivalent to the conditions (5.30) for the weak solution of (5.29). It follows from Lemma 5.1 that if $N(L) = \{0\}$, then (5.29) has a weak solution for every $f \in L_2(Q)$. However, if $N(L) \neq 0$, then (5.34) need not have a solution for every $w_f \in \mathring{W}_2^m$. To apply our results of Section 5.1 to the solvability of (5.34) when $N(L) \neq \{0\}$, note first that $L^*: \mathring{W}_2^m \to \mathring{W}_2^m$ is also A-proper w.r.t. Γ_0 since L^* satisfies the same inequality as L. Consequently, by Theorem II.3.3(A2), the A-proper map L is Fredholm with $\mathrm{ind}(L) = 0$. Moreover, $\phi(u) = (Lu,u)_m$ is weakly lower semicontinuous. The last fact will be used elsewhere.

To state our existence results for (5.29) or (5.34), let us introduce the following symbols:

$$\underline{h}(\pm\infty) = \liminf_{s \to \pm\infty} h(x,s), \qquad \overline{h}(\pm\infty) = \limsup_{s \to \pm\infty} h(x,s) \qquad (x \in Q)$$

$$Q^+(w) = \{x \in Q \mid w(x) > 0\}, \qquad Q^-(w) = \{x \in Q \mid w(x) < 0\}$$
$$(w \in N(L))$$

Theorem 5.5   Suppose that (a3) and (b3) hold and suppose further that $N(L^*) = N(L)$. Then, given f in $L_2(Q)$, equation (5.29) has a weak solution provided that either

$$\int_Q fw \, dx < \int_{Q^+(w)} \underline{h}(\infty)w \, dx + \int_{Q^-(w)} \overline{h}(-\infty)w \, dx \qquad (5.35)$$
$$\forall \, w \in N(L), \quad \| w \|_m = 1$$

or

$$\int fw \, dx > \int_{Q^+(w)} \overline{h}(\infty)w \, dx + \int_{Q^-(w)} \underline{h}(-\infty)w \, dx \qquad (5.36)$$
$$\forall \, w \in N(L), \quad \| w \|_m = 1$$

Proof.   In view of the preceding discussion and Remark 5.1, Theorem 5.5 will follow from Theorem 5.2 (with $Y = X = \underline{H}$) if we can show that (5.35) implies B_1^+ of Remark 5.0 (i.e., $\underline{\lim} \int_Q h(x,u_n)w \, dx > \int_Q fw \, dx$ whenever $\{u_n\}$ is a sequence in $\mathring{W}_2^m$ such that $\| u_n \|_u \to \infty$ and $u_n/\| u_n \|_m \to w \in N(L)$) or that (5.36) implies B_1^- of Remark 1.0.

We first claim that (5.35) implies B_1^+ of Remark 5.0 with $Y = X = \mathring{W}_2^m$ and $M = I$. If not, there would exist a sequence $\{u_n\} \subset \mathring{W}_2^m$ with $t_n = \| u_n \|_m \to \infty$ and $w_n = t_n^{-1}u_n \to w \in N(L)$ with $\| w \|_m = 1$ such that

$$\underline{\lim} \int_Q h(x,u_n(x))w(x) \, dx \le \int_Q f(x)w(x) \, dx. \qquad (5.37)$$

Suppose that $\{u_n\}$ and w are as above and note that

$$\varliminf \int_Q h(x,u_n)w \, dx = \varliminf \left\{ \int_{Q^+(w)} h(x,u_n(x))w(x) \, dx \right.$$

$$+ \left. \int_{Q^-(w)} h(x,u_n(x))w(x)dx \right\} \tag{5.38}$$

$$\geq \varliminf \int_{Q^+(w)} h(x,u_n(x))w(x) \, dx$$

$$+ \varliminf \int_{Q^-(w)} h(x,u_n(x))w(x) \, dx.$$

Since $\{w_n\}$, or at least a subsequence, converges pointwise a.e. on Q to w, it follows that if $x \in Q^+(w)$, then $u_n(x) = t_n w_n(x) \to +\infty$, while if $x \in Q^-(w)$, then $u_n(x) = t_n w_n(x) \to -\infty$ as $n \to \infty$. In view of this and the boundedness of h, an application of the Lebesgue–Fatou lemma leads to

$$\varliminf \int_{Q^+(w)} h(x,u_n)w \, dx$$

$$+ \varliminf \int_{Q^-(w)} h(x,u_n)w \, dx \geq \int_{Q^+(w)} \underline{h}(\infty)w \, dx + \int_{Q^-(w)} \overline{h}(-\infty)w \, dx.$$

This together with (5.37) and (5.38) implies that

$$\int_Q fw \, dx \geq \int_{Q^+(w)} \underline{h}(\infty)w \, dx + \int_{Q^-(w)} \overline{h}(-\infty)w \, dx,$$

$$\| w \|_m = 1, \quad w \in N(L),$$

in contradiction to (5.35).

In a similar way, one shows that (5.30) implies B_1^- of Remark 5.0.

$$\text{Q.E.D.}$$

Theorem 5.5 is related to the results of Ambrosetti and Mancini obtained under different conditions on h by means of the Lyapunov–Schmidt method (see [18,77,186,212] and others where similar problems are treated).

Problem 5.4 In our final problem we use an approach similar to that of Problem 5.1 to study the solvability of the boundary value problem

$$\begin{cases} -\Delta u(x) - g(x,\nabla u(x), \Delta u(x)) + h(u(x)) = f(x) & (x \in Q) \\ \dfrac{\partial u}{\partial \eta}(x) = 0 & (x \in \partial Q), \end{cases} \tag{5.39}$$

which was studied in [77] under more restrictive conditions on g and the Sobolev spaces. We present it here to contrast the applicability of our Theorem 5.2 for A-proper maps with the results in [77] for condensing maps. To state the conditions on g and h under which (5.39) has a solution in $W_2^2(Q)$ for each $f \in L_2(Q)$, we let $X = \{u \in W_2^2(Q), \partial u/\partial \eta = 0$ for $x \in \partial Q\}$ and $Y = L_2(Q)$. Let $L: X \to Y$ be defined by $Lu = -\Delta u$ for $u \in X$. It is known (see [87]) that $N(L)$ consists of constant functions, $R(L)$ consists of those functions whose mean value is zero, $\text{ind}(L) = 0$, $Y = N(L) \oplus R(L)$, and $K = L + I: X \to Y$ is a linear homeomorphism, where $I: X \to Y$ is the inclusion map which is compact by the Sobolev embedding theorem. In what follows we use $\| \cdot \|_0$ to denote the equivalent norm in X given by $\| u \|_0 = \| Ku \|$.

Let $\{X_n\} \subset X$ be such that $\text{dist}(u, X_n) \to 0$ as $n \to \infty$ for each u in X and let $Y_n = K(X_n)$ for each n. Then, if $P_n: X \to X_n$ and $Q_n: Y \to Y_n$ denote the orthogonal projections, $\Gamma_0 = \{X_n, P_n; Y_n, Q_n\}$ is complete for (X, Y). A direct calculation shows that

$$(Lu, Ku) \geq \| u \|_0^2 - \| u \|_1^2 \qquad \text{for all} \quad u \in X. \tag{5.40}$$

Since W_2^1 is compactly embedded in W_2^2, it follows from (5.40) that $L: X \to Y$ is A-proper w.r.t. Γ_0. Suppose that the functions g and h satisfy the following conditions:

(a4) $g: Q \times \mathbb{R}^{n+1} \to \mathbb{R}$ is continuous and there exist $\alpha \in (0,1)$ and $\phi(x) \in L_2(Q)$ such that $| g(x,r,s) | \leq \phi(x)$ for $(x,r,s) \in Q \times \mathbb{R}^n \times \mathbb{R}$ and $(g(x,r,s_1) - g(x,r,s_2))(s_1 - s_2) \geq -\alpha| s_1 - s_2 |^2$ for $(x,r) \in Q \times \mathbb{R}^n$ and $s_1, s_2 \in \mathbb{R}$.

(b4) $h: \mathbb{R} \to \mathbb{R}$ is continuous and there are $a, b \subset \mathbb{R}^+$ and $\delta \in (0,1)$ such that $| h(s) | \leq a + b | s |^\delta$ for $s \in \mathbb{R}$ and $\lim_{s \to \infty} h(s) = +\infty$, while $\lim_{s \to -\infty} h(s) = -\infty$.

We define the operators $C, G: X \to Y$ by $Cu = h(u(x))$ and $Gu = g(x, \nabla u(x), \Delta u(x))$ for $u \in X$. First, (b4) implies that C is compact and $\| Cu \|/\| u \|_0 \to 0$ as $\| u \|_0 \to \infty$. Second, (a4) implies that G is continuous, $R(G)$ is bounded and for $u, v \in X$

$$(Gu - Gv, K(u - v)) \geq -\alpha\| \Delta(u - v) \|^2 - 2p\| u - v \| \tag{5.41}$$
$$+ (g(x, \nabla u, \Delta v) - g(x, \nabla v, \Delta v), \Delta u - \Delta v).$$

where $p = \| \phi \|$. From (5.40) and (5.41) follows

$$((L + G)u - (L + G)v, K(u - v)) \geq (1 - \alpha)\| u - v \|_0^2$$
$$- \| u - v \|_1^2 - 2p\| u - v \| + (g(x, \nabla u, \Delta u) \tag{5.42}$$
$$- g(x, \nabla v, \Delta v), \Delta u - \Delta v) \qquad \text{for} \quad u, v \in X.$$

In view of (5.42), the results in [225] imply that $L + G$ is A-proper w.r.t. Γ_0 and so is the map $T = L + N$ with $N = G + C$ because C is compact. We are now in the situation to which Theorem 5.2 applies. Indeed, if $\{u_n\}$ $\subset X$ is such that $t_n \equiv \|\,u_n\,\|_0 \to \infty$ and $w_n \equiv u_n/\|\,u_n\,\|_0 \to w \in N(L)$ as $n \to \infty$, then since w is a nonzero constant and $\{w_n\}$, or at least a subsequence, converges pointwise a.e. on Q to w, it follows that $h(u_n) = h(t_n w_n(x)) \to +\infty$ a.e. when $w > 0$ and $h(u_n) \to -\infty$ a.e. when $w < 0$. Hence, in either case, $\lim_{n\to\infty}(Nu_n,w) = +\infty$. Thus we have the following improvement of [77, Theorem 3.7].

Theorem 5.6 Suppose that g and f satisfy (a4) and (b4), respectively. Then the problem (5.39) has a solution in X for each $f \in L_2$.

In [77] the solvability of (5.39) for each $f \in L_p$ was proved by means of the degree theory of condensing maps under the more restrictive and essential conditions that $p > n$, $|\,g(x,r,s_1) - g(x,r,s_2)\,| \leq \alpha|\,s_1 - s_2\,|$ for some $\alpha \in (0,1)$ and all $(x,r) \in Q \times \mathbb{R}^n, s_1, s_2 \in \mathbb{R}$, and that $g(x,r,s)$ is continuous in $r \in \mathbb{R}^n$ uniformly with respect to $(x,s) \in Q \times \mathbb{R}$.

IV
Generalized Degree for A-Proper Mappings and Applications

If D is an open and bounded subset of X with $\overline{D}$ and ∂D its closure and boundary in X, $T: \overline{D} \to Y$ a given mapping that satisfies suitable conditions, and $f \notin T(\partial D)$, then the topological degree of T on D over f, $\deg(T,D,f)$, is in principle an algebraic count of the number of solutions $x \in D$ of the equation $Tx = f$. For this count to be useful, it must have several crucial properties: additivity on the domain D, invariance under suitable homotopies on T, existence of a solution x in D of $Tx = f$ if $\deg(T,D,f) \neq 0$, and so on.

In [153] Leray and Schauder extended the finite-dimensional Brouwer degree to mappings $T: D \subset X \to X$ acting in infinite-dimensional Banach spaces X of the form $T = I - C$ with $C: \overline{D} \to X$ compact. To carry through this extension, they used the uniform approximation of C by finite-dimensional mappings. In the recent development of nonlinear functional analysis and, in particular, in connection with the theory of nonlinear operators such as operators of strongly monotone and accretive type, operators of ultimately compact and condensing and semicondensing type, and operators of A-proper type, it has become an important objective to widen the scope of the degree theory beyond that of the Leray–Schauder degree. Degree theories of more general types were constructed for the general classes of mappings above in [36,37,40,140,262] for monotone, accretive, and (S_+) type, in [115,163,193,238,253,260] for ultimately compact and condensing or semicondensing type, and in [41,42,74,

158,234,302,306] for A-proper type. For a recent survey of some of these theories, see [15,98,234,314]. For the degree theory of quasilinear Fredholm maps, see [84]. For other degree results, see [46,251,305].

In [41,42] Browder and Petryshyn introduced and studied the notion of the generalized degree for an A-proper mapping $T: \overline{D} \subset X \to Y$ in the context of constructive functional analysis. We generalized the Leray–Schauder method by replacing uniform approximation by finite-dimensional maps by much more general approximation methods of Galerkin type. The interesting feature of the generalized degree is that, although in general it is a multivalued function, it still possesses most of the properties of the Brouwer degree. Since A-proper maps may act between different spaces in many cases, they can be used to provide existence results for boundary value problems avoiding the tedious procedure of looking for Green functions and transforming these problems into the equivalent integral form. In addition to the author, this degree theory was then studied further and used in various applications by a number of mathematicians whose contributions and references will be given in the course of our discussion.

The purpose of this chapter is to present the generalized degree theory of A-proper mappings and then discuss some of its applications to the solvability of abstract and differential equations. In Section 1 we outline the definitions and state those properties of the Leray–Schauder [55] degree for compact vector fields and for strict set-contractive vector fields developed independently by Nussbaum [192] and Sadovskii [253]. We state only those of their properties that are needed in this book. For the sake of completeness and the convenience of the reader, we also prove a general fixed-point theorem for 1-set and 1-ball contractions due to the author [230] and indicate various classes of mappings that this theorem covers. In Sections 2 and 3 we present the generalized degree theory for A-proper mappings as developed by Browder and Petryshyn in [41,42]; in Sections 4 and 5 we consider further properties and applications.

1. DEFINITIONS OF TOPOLOGICAL DEGREES FOR COMPACT AND STRICT SET-CONTRACTIVE VECTOR FIELDS AND SOME OF THEIR PROPERTIES

It was mentioned in Chapter III that the finite-dimensional Brouwer degree theory provides the basis for the development of the generalized degree for A-proper mappings $T: \overline{D} \subset X \to Y$. However, in our study of the latter theory we also utilize some properties of the Leray–Schauder degree for the compact vector fields $T = I - C: \overline{D} \subset X \to X$ developed in [153] (see also [55,136,155,190,250]) and the topological degree for the condensing vector fields $T = I - F: \overline{D} \subset X \to X$ developed independently in [192,193] and in [253]. Hence, for the convenience of readers, in Section 1.1 we introduce the notion of the Leray–Schauder degree and mention only those of its properties that will be needed. In Section 1.2 the same is done for the Sadovskii–Nussbaum degree.

1.1 Leray–Schauder Degree and Some of Its Properties

Let D be a bounded open subset of a real Banach space X. We first note that if T is a continuous mapping of $\overline{D}$ into X and y is an element in X such that $y \notin T(\partial D)$, then in general it is not possible to define a degree function $\deg(T,D,y)$ for which the basic properties (i)–(iii) of Theorem III.1.1 are satisfied. Indeed, as in [62], consider the following:

Example 1.1 Let $X = c_0$, $D = B(0,1)$, and $T: \overline{D} \to X$ be given by

$$Tx = \tfrac{1}{2}(1 + \| x \|)e_1 + \sum_{i \geq 1} x_i(1 - 2^{-i-1})e_{i+1}, \tag{1.1}$$

where $\{e_i\}$ is the usual basis in c_0. It is easy to see that $\| Tx \| \leq 1$ for $x \in \overline{D}$ and $\| Tx - Ty \| < \| x - y \|$ for $x \neq y$. Thus T is a continuous mapping of $\overline{D}$ into $\overline{D}$. Now, if for some $x \in \overline{D}$ we would have the equality $x = Tx$, then $x_1 = \tfrac{1}{2}(1 + \| x \|)$ and $x_n = x_1 \prod_{i=1}^n (1 - 2^{-i-1})$ and, consequently, $\| x \| = 1$ and $x_n \geq 2(1 - \sum_{i=1}^n 2^{-i-1}) \geq 1$ for each $n \in Z_+$, by mathematical induction (i.e., $x \notin c_0$). Hence T has no fixed point in $\overline{D}$.

Since the Brouwer fixed-point theorem (Theorem III.1.4) for $\overline{B}(0,1)$ was deduced in Section III.1.2 by using only properties (i)–(iii) of Theorem III.1.1, we see from Example 1.1 that the degree function *cannot* be defined for each continuous mapping $T: \overline{D} \subset X \to X$ when X is infinite dimensional. However, Leray and Schauder have shown in [153] that when T is of the form $T = I - C$, where $C: \overline{D} \to X$ is compact, it is possible to define $\deg(T,D,y)$ for $y \notin T(\partial D)$, which has the properties stated in Theorems III.1.1, III.1.2, III.1.3, and others. To carry through this extension, they used the fact that the compact map $C: \overline{D} \to X$ can be approximated uniformly by finite-dimensional mappings. Indeed, by virtue of the latter fact, if we let $\hat{T} = I - \hat{C}$, where $\hat{C}$ is a continuous mapping defined on $\overline{D}$ with finite-dimensional range so chosen that $\sup_{\overline{D}} \| \hat{C}x - Cx \| < \text{dist}(y,T(\partial D))$ and choose V to be a finite-dimensional linear subspace of X which contains y and $\hat{C}(\overline{D})$, then setting $D_V = D \cap V$, we can define the *Leray–Schauder degree of T on D over y* by the equality

$$\deg_{LS}(T,D,y) = \deg(\hat{T},D_V,y). \tag{1.2}$$

One easily shows that $\deg_{LS}(T,D,y)$ defined by (1.2) is independent of the particular $\hat{C}$ chosen to approximate C.

Using (1.2) and the properties of the Brouwer degree stated in Section III.1.1 it is not hard to prove the following:

Theorem 1.1   Let D be a bounded open subset of X, $C: \overline{D} \to X$ compact and $y \notin T(\partial D)$ with $T = I - C$. Then the Leray–Schauder degree $\deg_{LS}(T,D,y)$ given by (1.2) is well defined and possesses properties (i) to (xii) of the Brouwer degree enumerated in Theorems III.1.1 to III.1.3.

The Leray–Schauder degree has proved to be a powerful tool for the study of fixed points of C or of the properties of the map $T = I - C$, C compact. It has had significant applications to the solvability of differential and integral equations.

As an illustration we show how Theorem 1.1 can be used to prove the following well-known analog of Theorem III.1.5 proved in [153] and whose constructive version has been given in Section III.2.

Theorem 1.2   Let $D \subset X$ be bounded and open and $C: \overline{D} \to X$ compact. If there is $w \in D$ such that

(LS) $Cx - w \neq \eta(x - w)$ for all $x \in \partial D$ and all $\eta > 1$,

then C has at least one fixed point in $\overline{D}$.

Proof.   Assuming without loss of generality that $Cx \neq x$ for $x \in \partial D$, as in Theorem III.1.5, one shows that the compact homotopy $H: [0,1] \times \overline{D} \to X$ given by $H(t,x) = x - tCx - (1 - t)w$ is such that $0 \notin H([0,1],\partial D)$. Hence by Theorem 1.1(iii), $\deg(I - C, 0) = \deg(I - w, D, 0) = 1$. Hence by Theorem 1.1(i), there exists $x \in D$ such that $Cx = x$.   Q.E.D.

Since (LS) holds whenever D is convex and $C(\partial D) \in \overline{D}$, it follows that Theorem 1.2 includes the fixed-point theorems of Schauder [256], Rothe [250], Altman [6], Krasnoselskii [136], and others.

We complete this section with the statement of the following theorem of Leray and Schauder [153] (see also [136]), which we will utilize in Chapter V.

Theorem 1.3   Let $L \in L(X,X)$ be compact, λ^{-1} not an eigenvalue of L, and $D \subset X$ an open bounded set with $0 \in D$. Then

$$\deg_{LS}(I - \lambda L, D, 0) = (-1)^{\beta(\lambda)}, \tag{1.3}$$

where $\beta(\lambda)$ is the sum of the algebraic multiplicities of eigenvalues μ of L such that $\mu\lambda > 1$. If no such eigenvalues exist, we have $\beta(\lambda) = 0$.

We recall that if μ is an eigenvalue of L, the algebraic multiplicity $m(\mu)$ of μ is by definition the dimension of the linear subspace $N(\mu) = \cup_{n \geq 1} N(\mu I - L)^n$ of X.

1.2 Topological Degree for *k*-Ball and *k*-Set-Contractive Vector Fields with *k* < 1 and Some of Its Properties

Let $\Phi(Q)$ denote either the ball measure $\beta(Q)$ or the set measure $\gamma(Q)$ of noncompactness defined on bounded sets Q in X. We recall that a continuous bounded mapping $F: \overline{D} \subset X \to X$ is said to be k-Φ-contractive iff $\Phi(F(Q)) \leq k\Phi(Q)$ for each bounded set $Q \subset \overline{D}$ and some $k \geq 0$ [i.e., $T \in \sum_k(\overline{D}) = \{T: \overline{D} \to X; T = I - F, F \text{ a } k\text{-}\Phi\text{-contraction}\}$]. In this section we outline briefly the definition of the topological degree for vector fields $T \in \sum_k(\overline{D})$ with $k < 1$, state some of its properties, and then deduce from them the analog of the fixed-point theorem (Theorem 1.3) for 1-Φ-contractive mappings obtained by the author in [230]. The degree for maps T in $\sum_k(\overline{D})$ with $k < 1$ was defined by Nussbaum [192,193] (and also in Sadovskii [253]) by using the fixed-point index for compact mappings. To stay within the framework of the degree theory, in this outline we follow Lloyd's [155] treatment since it is somewhat simpler and is based on the Leray–Schauder degree.

Let D be a bounded open subset of X and $F: \overline{D} \to X$ such that $T = I - F \in \sum_1(\overline{D})$ with $k < 1$. We define the sequence $\{K_n\}$ by $K_1 = \overline{co}F(\overline{D})$ and, inductively, $K_n = \overline{co}F(K_{n-1} \cap \overline{D})$ for $n = 2, 3, \ldots$ and we let $K = \cup_{n=1}^{\infty} K_n$. We notice first that K is a closed convex set and $\{K_n\}$ is a decreasing sequence. Since $F(K_n \cap \overline{D}) \overline{co}F(K_n \cap \overline{D}) = K_{n+1} \subset K_n$ for each n, it follows that $F(K \cap \overline{D}) \subset K$. Now, since $T \in \sum_k(\overline{D})$ with $k < 1$, it follows that $\Phi(K_n) \leq k^n\Phi(\overline{D}) \to 0$ as $n \to \infty$. Hence, by a theorem of Kuratowski [144], K is nonempty and compact. By Dugundji's extension of Tietze's theorem (see [65]) there is a mapping F_1 from $\overline{D}$ into K that coincides with F on $\overline{D} \cap K$. Since the range of F_1 is compact, it is possible to speak of the Leray–Schauder degree of $I - F_1$ relative to D. If there is $x \in \overline{D}$ such that $F_1 x = x$, then clearly, $x \in K$ and $Fx = x$. Hence F and F_1 have the same fixed points, so that $T = I - F$ is unchanged on its zero set—the set that matters in this context. Also, $\deg_{LS}(I - F_1, D, 0)$ is defined provided that $0 \notin T(\partial D)$. Let $F_2: \overline{D} \to K$ be another continuous function that agrees with F on $\overline{D} \cap K$, and consider the homotopy $H(t,x) = x - tF_1 x - (1 - t)F_2 x$ for $x \in \overline{D}$ and $t \in [0,1]$. Since $F_1, F_2: \overline{D} \to K$ and K is convex, it follows that $H(t,x) = 0$ only if $x \in K$. But $F_1 = F_2 = F$ on K, so that $H(t,x) = 0$ only if $Fx = x$. Hence $0 \notin H(t,\partial D)$ for $t \in [0,1]$. Thus, by the homotopy invariance of Leray–Schauder degree, $\deg_{LS}(I - F_1, D, 0) = \deg_{LS}(I - F_2, D, 0)$.

Consequently, if $T \in \sum_k(\overline{D})$ with $k < 1$ and $0 \notin T(\partial D)$, we can define the degree of $T = I - F$ on $\overline{D}$ over 0 by

$$\deg_\Phi(T,D,0) = \deg_{LS}(T_1,D,0), \qquad (1.4)$$

where $T_1 = I - F_1$ and $F_1 \colon \overline{D} \to K$ is any extension of $F|_{\overline{D} \cap K}$ (as described above). For $y \notin T(\partial D)$, define $\deg_\Phi(T,D,y)$ to be $\deg_\Phi(T - y,D,0)$, where $T - y$ denotes the mapping $x \mapsto Tx - y$.

The degree $\deg_\Phi(T,D,0)$ defined by (1.4) possesses most of the properties of the Leray–Schauder degree. Using the definition above and the fact (not hard to prove) that when $\tilde{K}$ is any compact convex set containing K such that $F \colon \overline{D} \cap \tilde{K} \to \tilde{K}$ and $\tilde{F} \colon \overline{D} \to \tilde{K}$ agrees with F on $\overline{D} \cap \tilde{K}$, then $\deg_\Phi(T,D,0) = \deg_{LS}(I - \tilde{F},D,0)$, one can prove the following:

Theorem 1.4 Let D be a bounded open subset of X, $F \colon \overline{D} \to X$ k-Φ-contractive with $k < 1$, $T = I - F$, and $y \notin T(\partial D)$. Then:

(a) $\deg_\Phi(I,D,y) = \deg_\Phi(I - y, D, 0) = 1$ if $y \in D$.

(b) If D_1, D_2 are disjoint open subsets of D such that $y \notin T(\overline{D}\backslash D_1 \cup D_2)$, then $\deg_\Phi(T,D,y) = \deg_\Phi(T,D_1,y) + \deg_\Phi(T,D_2,y)$.

(c) If $H \colon [0,1] \to \sum_k(\overline{D})$ $(k < 1)$ is continuous and $y \notin H(t)(\partial D)$ for $t \in [0,1]$, then $\deg_\Phi(H(t),D,y)$ is independent of $t \in [0,1]$.

(d) If $T \in \sum_k(\overline{D})$ $(k < 1)$ and $\deg_\Phi(T,D,y) \neq 0$, then $T^{-1}(y) \neq \varnothing$.

A proof of Theorem 1.4, as well as some other properties enjoyed by $\deg_\Phi(T,D,y)$, can be found in [193] for $\Phi = \gamma$. It follows from its definition that if F is compact, then $\deg_\Phi(T,D,y) = \deg_{LS}(T,D,y)$.

1.3 Fixed-Point Theorems for Some 1-Φ-Contractive Mappings

It was already noted in [191] that it was Darbo's proof of his theorem in [59] that provided the idea for the procedure used in Section 1.2 in pacing from the Leray–Schauder degree to the degree for $T \in \sum_k(\overline{D})$ with $k < 1$. Indeed, it was shown in [59] that if D is also closed and convex, $F(D \subset D$, and $\{K_n\} \equiv \{D_n\}$ is constructed by $D_n = \overline{co}(F(D_{n-1}))$ for $n \geq 1$ with $D_0 \equiv D$ then $\{D_n\} \subset D$ for $n \geq 0$, $\gamma(D_n) \leq k^n \gamma(D) \to 0$ as $n \to \infty$, $K \subset D$, K is compact and nonempty, and $F(K) \subset K$. Hence F has a fixed point in $K \subset D$ by Schauder's theorem. In view of this, we have the following companion to Theorem 1.6 below.

Theorem 1.5 Let $D \subset X$ be closed, bounded, convex, and $F \colon D \to D$ be k-γ-contractive with $k < 1$. Then F has a fixed point in D. The same conclusion holds where $k = 1$ if F also satisfies condition:

(c) If $\{x_n\} \subset D$ and $x_n - F(x_n) \to 0$ as $n \to \infty$, then there is $x \in D$ so that $x = Fx$.

We now use Theorem 1.4 to prove the main result in [230] when D is open.

Theorem 1.6 Let $D \subset X$ be bounded and open and let $F: \overline{D} \to X$ be a 1-Φ-contractive map for which the following hold:

(LS) There is $w \in D$ such that $Fx - w \neq \eta(x - w)$ for all $x \in \partial D$ and $\eta > 1$.
   (C) If $\{x_n\} \subset \overline{D}$ is such that $x_n - Fx_n \to 0$, then there is $x \in \overline{D}$ with $Fx = x$.
   Then F has at least one fixed point in $\overline{D}$.

Proof. Note that without loss of generality we may assume that (LS) holds for $\eta \geq 1$. For each $\lambda_n \in (0,1)$ with $\lambda_n \to 1$, consider the mapping $F_n: \overline{D} \to X$ defined by $F_n(x) = \lambda_n F(x) + (1 - \lambda_n)w$. Then the properties of Φ imply that F_n is λ_n-Φ-contractive with $\lambda_n < 1$ for each $n \in Z_+$. Furthermore, F_n satisfies condition (LS) on ∂D. Indeed, if this were not the case, there would exist $x_0 \in \partial D$ and $\eta_0 > 1$ such that $F_n(x_0) - w = \eta_0(x_0 - w)$. But $F_n(x_0) - w = \lambda_n(Fx_0 - w) = \eta_0(x_0) - w$ [i.e., $Fx_0 - w = \eta_0\lambda_n^{-1}(x_0 - w)$ with $\eta_0\lambda_n^{-1} > \eta_0 > 1$ for each n, in contradiction to (LS)]. Now consider the homotopy $H_n(t,x) = x - w - t(F_nx - w)$ for $x \in \overline{D}$ and $t \in [0,1]$. It follows from (LS) that $H_n(t,x) \neq 0$ for $t \in [0,1]$, $x \in \partial D$, and $n \in Z_+$. Furthermore, the map $H_n(t) = t(T_n - w): [0,1] \to \sum_{\lambda_n}(\overline{D})(\lambda_n < 1)$ is continuous. Hence, by the homotopy part (b) of Theorem 1.4, $\deg_\Phi(H_n(1,\cdot),D,0) = \deg_\Phi(H_n(0,\cdot),D,0)$ [i.e., $\deg(I - F_n,D,0) = \deg_\Phi(I - w,D,0) = 1$ by (a)]. Hence for each $n \in Z_+$ there exists $x_n \in \overline{D}$ such that $x_n - F_n(x_n) = 0$. Consequently, since F and $\{x_n\} \subset \overline{D}$ are bounded and $\lambda_n \to 1$ as $n \to \infty$, it follows that $x_n - F(x_n) = (\lambda_n - 1)F(x_n) + (1 - \lambda_n)w \to 0$ as $n \to \infty$. This and the condition (C) imply the existence of an $x \in D$ such that $x - Fx = 0$.

Remarks on Theorem 1.5 and 1.6 1.1. Condition (LS) of Theorem 1.6 is certainly satisfied if D is convex and $F(\partial D) \subset \overline{D}$. It is also satisfied if

$$\| Fx - x \|^2 \geq \| Fx - w \|^2 - \| x - w \|^2 \qquad \text{for all} \quad x \in \partial D \qquad (1.5)$$

used in [6] when $w = 0 \in D$. In case X is a Hilbert space, then (1.5) reduces to

$$(Fx - x, w - x) \geq 0 \qquad \text{for all} \quad x \in \partial D \qquad (1.6)$$

used in [136] when $w = 0 \in D$. In this case the latter condition can be weakened to the assumption:

There exists some map $G: \overline{D} \to X$ such that $(Gx - x, w \quad (1.7)$
$- x) \geq 0$ and $\| Gx - Fx \| \leq \| Gx - x \|$ for all $x \in \partial D$.

used by the author in [229] when $w = 0$. The condition (1.7) reduces to
(1.6) when $G = F$.

1.2. Condition (C) is satisfied if F is Φ-condensing or a k-Φ-contraction
with $k < 1$, since in both cases the mapping $I - F$ is closed. When X is
reflexive, (C) also holds when $I - F$ is demiclosed. In various versions
Theorem 1.5 was proved in [59,94,192,249,253].

1.3. A special case of Theorem 1.5 is by now a well-known result which
asserts that if $F: D \to X$ is of the form $F = S + C$, where $S: D \to X$ is
k-contractive with $k < 1$ and C is compact, then F has a fixed point in D
if $F(D) \subset D$ since F is k-set contractive. Study of the fixed-point theorem
for the latter class of mappings was initiated by Krasnoselskii [136], who
first established the existence of the fixed point of F under the more
restrictive condition:

$$Sx + Cy \in D \quad \text{for all} \quad x,y \in D \quad \text{with} \quad D \quad \text{convex and closed.} \quad (1.8)$$

His study was continued by many authors (see [230,185,249,297] for many
references).

1.4. Theorem 1.5(b) is not valid if condition (C), or something similar,
is missing. Indeed, it was shown by Browder [36] that if $X = l_2$ and
$D = \overline{B}(0,1) \subset l_2$, then $S: D \to l_2$ and $C: D \to l_2$ given by

$$S(x) = (0,x_1,x_2, \ldots), \qquad C(x) = (1 - \| x \|^2,0,0, \ldots)$$

is nonexpansive and compact, respectively, and $F = S + C$ maps D into
D; that is, F is 1-set contractive and satisfies condition (LS) but F has no
fixed points in D, for $F(x) = x$ implies that $x_1 = 1 - \| x \|^2$ and $x_1 = x_2$
$= x_3, \ldots$, the latter can occur only if $x_i = 0$ for $i = 1, 2, \ldots$, and this
contradicts the former. Consequently, some additional condition has to
be imposed for $F = S + C$ to have a fixed point. Condition (C) appears
to be the weakest such condition.

1.5. Theorem 1.6 yields fixed-point theorems for many classes of mappings that do not appear immediately to fall under its scope. A few of
these will be mentioned. For full accounts and details, see [230].

If D is convex and closed, $S: D \to X$ nonexpansive (i.e., $\| Sx - Sy \|$
$\leq \| x - y \|$ for $x,y \in D$) and X uniformly convex, then $I - S$ is demiclosed
(see Browder [28]). Hence if $C: D \to X$ is completely continuous, then F
$= S + C$ is 1-set contractive and $T = I - F$ satisfies condition (C) and
thus F has a fixed point in D if $F(D) \subset D$. This result includes the fixed-
point theorem of Browder–Gölde–Kirk [28,102,131] when $C = 0$ and $T(D)$
$\subset D$ and of Browder [31] when $C = 0$. For various special cases of the
result above for $F = S + C$ obtained by various authors, see [230,
249,297].

If we assume only that X is reflexive and strengthen the condition on $S: \overline{D} \to X$ by requiring it to be a generalized contraction, that is,

$$\| Sx - Sy \| \le \alpha(x) \| x - y \| \qquad \text{for} \quad x,y \in \overline{D} \quad \text{with} \quad \alpha(x) < 1,$$

(1.9)

then S is 1-set contractive and $I - S$ satisfies condition (C) if $S(\overline{D}) \subset \overline{D}$, as was shown in Chapter III. Thus the fixed-point theorem in [11] for such maps follows from Theorem 1.6. If $C: \overline{D} \to X$ is completely continuous and $Sx + Cy \in \overline{D}$ for $x,y \in D$, then F is 1-set contractive and $I - F$ satisfies condition (C) so that the theorem applies. If we assume that $S: X \to X$ and (1.9) holds for $x \in X$ and $y \in \overline{D}$, then $F = S + C: \overline{D} \to X$ is 1-set contractive and $I - F$ satisfies condition (C) even if $C: \overline{D} \to X$ is compact. Hence in this case, F has a fixed point in $\overline{D}$ if (LS) holds on ∂D. We add that the main motivation for the study of generalized contractions stems from the fact (see Kirk [133]) that if $S: D \to X$ is continuously Fréchet differentiable on D, then S is a generalized contraction on D if and only if $\| S'_x \| < 1$ for each $x \in D$, where S'_x is the Fréchet derivative of S at $x \in D$.

Nussbaum introduced in [191] a class of maps that he called LANE (locally almost nonexpansive). Let X be a uniformly convex space and $D \subset X$ be bounded closed convex. A continuous map $F: D \to X$ is LANE if for $x \in D$ and $\epsilon > 0$, there is a neighborhood $U(x)$ of x in the weak topology such that

$$\| Fx - Fy \| \le \| x - g \| + \epsilon \qquad [x,y \in U(x)]. \tag{1.10}$$

Examples of LANE mappings are semicontractive mappings introduced by Browder [31] [F is *semicontractive* if there exists a mapping $V: D \times D \to X$ such that $Fx = V(x,x)$ for $x \in D$ and for each $x \in D$, $V(\cdot,x)$ is nonexpansive on D and $V(x,\cdot)$ is completely continuous from $\overline{D}$ to X, uniformly for $x \in D$]. It was shown in [191] that if X is uniformly convex, then a LANE mapping F is 1-set contractive and $(I - F)(D)$ is closed. Under these conditions, therefore, F has a fixed point in D if $F(D) \subset D$, by Theorem 1.5.

Finally, we add that various other classes of semicontractive type introduced by Browder [31,36] (see also Kirk [132]) are all covered by the fixed-point theorem (Theorem 1.6) (see [230] for complete details).

2. GENERALIZED DEGREE FOR A-PROPER MAPPINGS

Unless stated otherwise, we assume in this section that (X, Y) is a pair of real Banach spaces provided with an admissible scheme $\Gamma =$

$\{X_n, V_n; E_n, W_n\}$ with respect to which the generalized degree of an A-proper map $T: \overline{D} \subset X \to Y$ will be defined and some of its basic properties will be studied. The interesting feature of this degree is that it is a multivalued function which, however, possesses most of the properties of the Brouwer degree. It is shown in Section 2.2 that for special classes of A-proper maps the generalized degree is single-valued and includes the Leray–Schauder degree, Browder–Nussbaum degree, Nussbaum–Sadovskii degree for ball measure, and others as shown in Section 3.

2.1 Definition and Some Properties of the Generalized Degree

To define the generalized degree we need the following:

Lemma 2.1 Let $D \subset X$ be bounded and open, $T: \overline{D} \to Y$ A-proper w.r.t. Γ and $f \notin T(\partial D)$. Then there exists an $n_0 \in Z_+$ and $d > 0$ such that $\| T_n(x) - W_n f \| \geq d$ for $x \in \partial D_n$ and $n > n_0$.

Proof. Suppose that the assertion of Lemma 2.1 is false. Then there exist sequences $\{n_j\} \subset Z_+$ and $\{x_{n_j} \mid x_{n_j} \in \partial D_{n_j}\}$ such that $T_{n_j}(x_{n_j}) - W_{n_j} f \to 0$ as $j \to \infty$. Since T is A-proper, $\{x_{n_j}\}$ is bounded and $\partial D_n \subset \partial D$ for each $n \in Z_+$, there exist a subsequence $\{x_{n_{j(k)}}\}$ and $x \in \overline{D}$ such that $x_{n_{j(k)}} \to x$ in X as $k \to \infty$ and $Tx = f$ with $x \in \partial D$. This contradiction yields the validity of Lemma 2.1. Q.E.D.

Definition 2.1 Let D be a bounded open subset of X, $T: \overline{D} \to Y$ A-proper w.r.t. Γ, $f \notin T(\partial D)$, and $Z' = Z \cup \{\pm\infty\}$. We define $\mathrm{Deg}(T,D,f)$, the *generalized degree of T on D over F w.r.t. Γ,* as a subset of Z' given by:

(1) The integer $m \in \mathrm{Deg}(T,D,f)$ iff there is a sequence $\{n_j\} \subset Z_+$ such that $\deg(T_{n_j}, D_{n_j}, W_{n_j} f)$ is well defined and equals m for each $j \in Z_+$.

(2) $+\infty \ (-\infty) \in \mathrm{Deg}(T,D,f)$ iff there is $\{n_j\} \subset Z_+$ such that $\deg(T_{n_j}, D_{n_j}, W_{n_j} f)$ is well defined and $\lim_j \deg(T_{n_j}, D_{n_j}, W_{n_j} f) = +\infty \ (-\infty)$.

The degree $\deg(T_n, D_n, W_n f)$ used in Definition 2.1 is the Brouwer degree for mappings of oriented finite-dimensional spaces of the same dimension.

The degree mapping given by Definition 2.1 is in general multivalued, and even in the case of an A-proper linear homeomorphism, it need not be single valued. Indeed, if, for example, $Y = X$, then $\mathrm{Deg}(-I,D,0) = \{+1, -1\}$, where $I: X \to X$ is the identity mapping and D is an open bounded set containing 0, since $\deg(-I, D_n, 0) = +1$ (respectively, -1)

when dim X_n is even (respectively, odd). Nevertheless, using the properties of the Brouwer degree and of A-proper mappings, it will now be shown that the generalized degree $Deg(T,D,f)$ enjoys most of the useful properties of the Brouwer topological degree, as the following theorems show.

Theorem 2.1 Let $D \subset X$ be bounded and open, $T: \overline{D} \to Y$ A-proper w.r.t. Γ and $f \notin T(\partial D)$. Then:

(1) There exists $n_0 \in Z_+$ such that $deg(T_n,D_n,W_nf)$ is well defined for each $n \geq n_0$ and, in particular, $Deg(T,D,f)$ is a nonempty subset of Z'.

(2) If $Deg(T,D,f) \neq \{0\}$, there is $x \in D$ such that $Tx = f$. Moreover, there exists $r > 0$ such that $B(f,r) \subset T(D)$.

(3) If $H: [0,1] \times D \to Y$ is an A-proper homotopy and $f \notin H(t,\partial D)$ for $t \in [0,1]$, then $Deg(H(t,\cdot),D,f)$ is constant in $t \in [0,1]$.

(4) If $D_1 \subset D$ is open and $f \notin T(\overline{D} \backslash D_1)$, then $Deg(T,D,f) = Deg(T,D_1,f)$.

(5) If $D = D_1 \cup D_2$, $D' = (D_1 \cap D_2) \cup \partial D_1 \cup \partial D_2$, and $f \notin T(D')$, then $Deg(T,D,f) \subseteq Deg(T,D_1,f) + Deg(T,D_2,f)$, with equality holding if either $Deg(T,D_1,f)$ or $Deg(T,D_2,f)$ is single valued. [If $A_1,A_2 \subset Z'$, then $A_1 + A_2 = \{a \mid a = a_1 + a_2, a_1 \in A_1, a_2 \in A_2\}$ and we use the convention that $+\infty (-\infty) = Z'$.]

(6) Let $0 \in D$ and let D also be symmetric about 0. If $T: \overline{D} \to Y$ is odd on ∂D [i.e., $T(-x) = -Tx$ for $x \in \partial D$] and $0 \notin T(\partial D)$, then $Deg(T,D,0)$ is odd [i.e., $2m \notin Deg(T,D,0)$ for any $m \in Z_+$] and, in particular, $0 \notin Deg(T,D,0)$, so that $Tx = 0$ is feebly approximation solvable w.r.t. Γ.

Before we proceed with the proof of Theorem 2.1, we recall that if Q is any set in X, then $H: [0,1] \times Q \to Y$ is called an *A-proper homotopy* (w.r.t. Γ) if $H_n \equiv W_nH: [0,1] \times Q_n \to Y_n$ is continuous for each $n \in Z_+$ and if $\{t_{nj}\} \subset [0,1]$, $\{x_{nj} \mid x_{nj} \in Q_{nj}\}$ is bounded, and $W_{nj}H(t_{nj},x_{nj}) - W_{nj}g \to 0$ in Y for some g in Y, then there exists subsequences $\{t_{nj(k)}\}$ and $\{x_{nj(k)}\}$ such that $t_{nj(k)} \to t_0 \in [0,1]$, $x_{nj(k)} \to x_0 \in Q$, and $H(t_0,x_0) = g$.

Proof. (1) By Lemma 2.1, there exists $n_0 \in Z_+$ such that $T_n(x) \neq W_nf$ for all $x \in \partial D_n$ and $n \geq n_0$. Since $T_n: D_n \to Y_n$ is continuous, the Brouwer degree, $deg(T_n,D_n,W_nf)$, is well defined for each $n \geq n_0$. Thus either the sequence $\{deg(T_n,D_n,W_nf)\}$ is bounded, in which case at least one finite integer m appears in $Deg(T,D,f)$, or there exists an infinite subsequence converging to $\{+\infty\}$ or to $\{-\infty\}$. In either case, $Deg(T,D,f) \neq \emptyset$.

(2) If $Deg(T,D,f) \neq \{0\}$, there exists an infinite subsequence $\{n_j\} \subset Z_+$ with $n_j \to \infty$ such that $deg(T_{nj},D_{nj},W_{nj}f) \neq 0$ for each $j \in Z_+$. Hence by

Theorem III.1.1 there exists $x_{n_j} \in D_{n_j}$ such that $T_{n_j}(x_{n_j}) = W_{n_j}f$. Since T is A-proper, $\{x_{n_j} \mid x_{n_j} \in D_{n_j}\}$ is bounded and $T_{n_j}(x_{n_j}) - W_{n_j}f = 0 \rightarrow 0$, we may assume by passing to a subsequence that $x_{n_j} \rightarrow x \in \overline{D}$ and $Tx = f$. Since $x \notin \partial D$, it follows that $x \in D$. To prove the second part, let $B(f,r)$ be an open ball in Y with radius $r \in (0, d/\beta)$, where d is the constant in Lemma 1.1 and $\beta \geq \| W_n \|$ for all $n \in Z_+$. Then, if $g \in B(f,r)$,

$$\| T_n(x) - W_n g \| \geq \| T_n(x) - W_n f \| - \| W_n f - W_n g \|$$

$$\geq d - \beta \| f - g \| > d - \beta d/\beta = 0$$

for all $n \geq n_0$, so for each such n, $H_n(t,x) = T_n(x) - (1 - t)W_n f - tW_n g \neq 0$ for $x \in \partial D_n$. Consequently, by Theorem III.1.1 and the assumption in (2), $\deg(T_n - W_n g, D_n, 0) \neq 0$, from which our conclusion follows.

(3) It follows from the definition of an A-proper homotopy that $H(t, \cdot)$: $\overline{D} \rightarrow Y$ is A-proper for each $t \in [0,1]$, and hence since $f \notin H(t, \partial D)$ for $t \in [0,1]$, it follows from Lemma 2.1 that $\mathrm{Deg}(H(T, \cdot), D, f)$ is well defined for each $t \in [0,1]$. To show that $\mathrm{Deg}(H(t, \cdot), D, f)$ is independent of $t \in [0,1]$, it suffices to show that there exists an $n_1 \in Z_+$ such that $W_n f \notin H_n(t, \partial D_n)$ for all $t \in [0,1]$ and $n \geq n_1$, for this then implies that $\deg(H_n(t, \cdot), D_n, W_n t)$ is independent of $t \in [0,1]$ for each $n \geq n_1$. So suppose that the assertion is false. Then there would exist sequences $\{n_j\} \subset Z_+$, $\{x_{n_j} \mid x_{n_j} \in D_{n_j}\}$ and $\{t_{n_j}\} \subset [0,1]$ such that $H_{n_j}(t_{n_j}, x_{n_j}) = W_{n_j}f$ for each $j \in Z_+$. Since H is an A-proper homotopy, it follows that there exist subsequences $\{t_{n_{j(k)}}\}$ and $\{x_{n_{j(k)}}\}$ such that $t_{n_{j(k)}} \rightarrow t_0 \in [0,1]$, $x_{n_{j(k)}} \rightarrow x_0 \in \partial D$ as $k \rightarrow \infty$ and $H(t_0, x_0) = f$, in contradiction to our hypothesis.

(4) If, for infinitely many $n \in Z_+$, there exists $x_n \in D_n \backslash D_{1n}$ with $T_n(x_n) = W_n f$, then by the A-properness of T there are x_0 and a further subsequence $\{x_{n_j}\}$ such that $x_{n_j} \rightarrow x_0 \in \overline{D}$ and $Tx_0 = f$. But $x_{n_j} \in D \backslash D_1$, whence $f \in T(\overline{D} \backslash D_1)$, contrary to our hypothesis. Thus we deduce that $W_n f \notin T_n(D_n \backslash D_{1n})$ for large n. Since $W_n f \notin T_n(\partial D_n)$ for large enough n, we have that $W_n f \notin T_n(\overline{D}_n \backslash D_{1n})$, whence $\deg(T_n, D_n, W_n f) = \deg(T_n, D_{1n}, W_n f)$ for sufficiently large n by Theorem III.1.1. This and Definition 2.1 imply that $\mathrm{Deg}(T, D, f) = \mathrm{Deg}(T, D_1, f)$.

(5) *Case 1.* If $D_1 \cap D_2 = \emptyset$, $\overline{D} = \overline{D}_1 \cup \overline{D}_2$, and if $f \notin T(\partial D_1) \cup T(\partial D_2)$, then for n sufficiently large, $W_n f \notin T_n(\partial D_{1n}) \cup T_n(\partial D_{2n})$, and since $D_{1n} \cap D_{2n} = \emptyset$, Theorem III.1.1 implies that

$$\deg(T_n, D_n, W_n f) = \deg(T_n, D_{1n}, W_n f) + \deg(T_n, D_{2n}, W_n f).$$

In this case, if $\gamma = \lim_j (T_{n_j}, D_{n_j}, W_{n_j}f)$, then

$$\gamma = \lim_j \{\deg(T_{n_j}, D_{1n_j}, W_{n_j}f) + \deg(T_{n_j}, D_{2n_j}, W_{n_j}f)\}.$$

If $| \deg(T_{nj}, D_{1nj}, W_{nj}f) | \nrightarrow \infty$, we may pass to an infinite subsequence and assume that for all j,

$$\deg(T_{nj}, D_{1nj}, W_{nj}f) = m_1 \in \mathrm{Deg}(T, D_1, f).$$

For the same subsequence,

$$\lim_j \deg(T_{nj}, D_{2nj}, W_{nj}f) = \gamma - m_1 \equiv m_2 \in \mathrm{Deg}(T, D_2, f),$$

so that $\gamma = m_1 + m_2$ with $m_i \in \mathrm{Deg}(T, D_i, f)$ for $i = 1, 2$.

In the other case, we may find a similar subsequence such that $\deg(T_{nj}, D_{1nj}, W_{nj}f) \to \pm\infty$. Now, if γ is finite, it follows that $\deg(T_{nj}, D_{2nj}, W_{nj}f) \to \mp\infty$ and we have $\gamma = +\infty + (-\infty) \in \{\mathrm{Deg}(T, D_1, f) + \mathrm{Deg}(T, D_2, f)\}$. If $\gamma = \pm\infty$, then $\gamma = \pm\infty - m_2$ for any $m_2 \in \mathrm{Deg}(T, D_2, f)$ so that $\gamma \in \{\mathrm{Deg}(T, D_1, f) + \mathrm{Deg}(T, D_2, f)\}$. Finally, if $\gamma = \mp\infty$, then $\deg(T_{nj}, D_{2nj}, W_{nj}f) \to \pm\infty$ and the same conclusion holds. Hence

$$\mathrm{Deg}(T, D, f) \subseteq \mathrm{Deg}(T, D_1, f) + \mathrm{Deg}(T, D_2, f).$$

Case 2. If $D_1 \cap D_2 \neq \varnothing$ while $f \notin T(D_1 \cap D_2)$, we first note that $D = (D_1 \backslash \overline{D}_2) \cup (D_2 \backslash \overline{D}_1) \cup (D_1 \cap D_2)$, the three sets are pairwise disjoint, and $\partial(D_1 \backslash \overline{D}_2) \cup \partial(D_2 \backslash \overline{D}_1) \cup \partial(D_1 \cap D_2) \subset \partial D_1 \cup \partial D_2$. Hence, by Case 1, which is valid for $\overline{D}$ as a union of the closures of a finite number of disjoint open sets, we have

$$\mathrm{Deg}(T, D, f) \subseteq \mathrm{Deg}(T, D_1 \backslash \overline{D}_2, f) + \mathrm{Deg}(T, D_2 \backslash \overline{D}_1, f)$$

since $\mathrm{Deg}(T, D_1 \cap D_2, f) = \{0\}$ by (2) of Theorem 2.1. The last equality also implies that $\mathrm{Deg}(T, D_1, f) = \mathrm{Deg}(T, D_1 \backslash \overline{D}_2, f)$ and $\mathrm{Deg}(T, D_2, f) = \mathrm{Deg}(T, D_2 \backslash \overline{D}_1, f)$. Consequently,

$$\mathrm{Deg}(T, D, f) \subseteq \mathrm{Deg}(T, D_1, f) + \mathrm{Deg}(T, D_2, f).$$

Finally, we remark that if either $\mathrm{Deg}(T, D_1, f)$ or $\mathrm{Deg}(T, D_2, f)$ is a singleton, the equality holds. For instance, in Case 1, if $\mathrm{Deg}(T, D_1, f) = \{m\}$, we may assume that for $n \geq n_2$, $\deg(T_n, D_{1n}, W_n f) = m$. Hence the two sequences $\{\deg(T_n, D_n, W_n f)\}$ and $\{\mathrm{Deg}(T_n, D_{2n}, W_n f)\}$ converge together and their difference equals m for $n \geq n_2$.

(6) Suppose that T is odd on ∂D and that $0 \notin T(\partial D)$. Then, by Lemma 2.1, $0 \notin T_n(\partial D_n)$ for $n \geq n_0$ and T_n is odd on ∂D_n for each n. Therefore, by Theorem III.1.2, $\deg(T_n, D_n, 0)$ is an odd integer. Hence the set of limit points of the sequence $\{\deg(T_n, D_n, 0)\}$ does not include an even integer and therefore $2m \notin \mathrm{Deg}(T, D, 0)$ for any $m \in Z$. In particular, $0 \notin \mathrm{Deg}(T, D, 0)$. The last relation and Definition 2.1 imply the existence of $n_0 \in Z_+$ such that $\deg(T_n, D_n, 0) \neq 0$ for $n \geq n_0$, so for each such n, there

exists $x_n \in D_n$ such that $T_n(x_n) = 0$. Since T is A-proper, there exist a subsequence $\{x_{n_j}\}$ and $x \in \overline{D}$ such that $x_{n_j} \to x$ and $Tx = 0$ (i.e., $Tx = 0$ is feebly approximation solvable). Q.E.D.

Remark 2.1 It is not hard to show that if $H: [0,1] \times \overline{D} \to Y$ is continuous and such that $H_t(x) = H(t,x)$ is continuous in $t \in [0,1]$ uniformly w.r.t. $x \in \partial D$, and $H_t: \overline{D} \to Y$ is A-proper w.r.t. Γ for each $t \in [0,1]$, then H is an A-proper homotopy w.r.t. Γ.

Remark 2.2 Since the sum of two A-proper mappings need not be A-proper, the homotopy property (3) is not as effective as the Leray–Schauder degree for compact vector fields. However, the proof of Theorem 2.1 (3) suggests that instead of the homotopy $H: [0,1] \times \overline{D} \to Y$ one may consider the finite-dimensional homotopy $H_n: [0,1] \times \overline{D}_n \to Y_n$ for sufficiently large n and obtain the following useful result.

Proposition 2.1 Let $T_0, T_1: \overline{D} \to Y$ be A-proper w.r.t. Γ with $f \notin T_0(\partial D)$ and $f \notin T_1(\partial D)$. If there is $n_2 \in Z_+$ such that for each $n \geq n_2$ there exists a homotopy H_n from T_{0n} to T_{1n} such that $W_n f \notin H_n([0,1] \times \partial D_n)$ for $n \geq n_2$, then $\mathrm{Deg}(T_0, D, f) = \mathrm{Deg}(T_1, D, f)$.

Theorem 2.1 has as its corollaries a number of interesting fixed-point and mapping theorems, some of which will be deduced in subsequent sections. For the present we establish some further properties of the generalized degree that will prove to be useful in various applications.

Theorem 2.2 Let D, Γ, T, and f be as in Theorem 2.1. Then:

(7) If $C: \overline{D} \to Y$ is compact and $\| Cx \| < \| Tx - f \|$ for $x \in \partial D$, then $\mathrm{Deg}(T, D, f) = \mathrm{Deg}(T + C, D, f)$.

(8) There exists $\gamma > 0$ such that if $A: \overline{D} \to Y$ is A-proper, $f \notin A(\partial D)$ and $\| Ax - Tx \| \leq \gamma$ for $x \in \partial D$, then $\mathrm{Deg}(T, D, f) = \mathrm{Deg}(A, D, f)$. In particular, the conclusion holds if $A|_{\partial D} = T|_{\partial D}$.

(9) If T is continuous and f_0 and f_1 lie in the same component of $Y \backslash T(\partial D)$, then $\mathrm{Deg}(T, D, f_0) = \mathrm{Deg}(T, D, f_1)$.

(10) $\mathrm{Deg}(T, D, f) = 0$ for $f \notin T(\overline{D})$ and $\mathrm{Deg}(T - f, D, 0) = \mathrm{Deg}(T, D, f)$.

Proof. (7) It follows from Theorem I.1.2 that for each $t \in [0,1]$, the map $H(t, \cdot) = T + tC: \overline{D} \to Y$ is A-proper and continuous in $t \in [0,1]$, uniformly w.r.t. $x \in \overline{D}$, since $H(t,x) - H(0,x) = (t - s)Cx$. Hence by Remark 2.1,

H is an A-proper homotopy. Since

$$\| H(t,x) - f \| \geq \| Tx - f \| - t\| Cx \|$$
$$\geq \| Tx - f \| - \| Cx \| > 0 \qquad \text{for} \quad x \in \partial D,$$

it follows from Theorem 2.1 (3) that $\text{Deg}(T,D,f) = \text{Deg}(T + C, D, f)$.

(8) Since $f \notin T(\partial D)$ and T is A-proper, Lemma 2.1 implies the existence of a number $a > 0$ and $n_0 \in Z_+$ such that $\| T_n(x) - W_n f \| > a$ for $x \in \partial D_n$ and $n \geq n_0$. Let $b \equiv \sup_n \| W_n \|$ and set $\gamma = a/b$. Suppose that the A-proper map $A \colon \overline{D} \to Y$ is such that $\| Ax - Tx \| \leq \gamma$ for $x \in \partial D$. Then, since $\partial D_n \subset \partial D$ for all $n \in Z_+$, we see that for $n \geq n_0$ and $x \in \partial D_n$, $\| A_n(x) - T_n(x) \| \leq b\| Ax - Tx \| \leq b\gamma = a$. Now, for each $n \geq n_0$, consider the homotopy $H_n \colon [0,1] \times \overline{D}_n \to Y_n$ given by $H_n(t,x) = tA_n(x) + (1 - t)T_n(x) - W_n f$. It follows from the preceding discussion that for all $t \in [0,1]$, all $x \in \partial D_n$, and each $n \geq n_0$, we have

$$\| H_n(t,x) \| \geq \| T_n(x) - W_n f \| - t\| T_n(x) - A_n(x) \| > a - a = 0.$$

Hence, by Theorem III.1.1, $\text{deg}(T_n,D_n,W_n t) = \text{deg}(A_n,D_n,W_n f)$. Since this is true for all $n \geq n_0$, Definition 2.1 implies that $\text{Deg}(T,D,f) = \text{Deg}(A,D,f)$. The second part follows from the first since $T(x) = A(x)$ for $x \in \partial D$.

(9) Let U be a component of $Y\backslash T(\partial D)$ such that $f_0, f_1 \in U$. Then there exists a continuous curve $f(t) \subset U$, $0 \leq t \leq 1$, such that $f(0) = f_0$ and $f(1) = f_1$ and $f(t) \notin T(\partial D)$ for $t \in [0,1]$. It is easy to show that $H \colon [0,1] \times \overline{D} \to Y$ given by $H(t,x) = T(x) - f(t)$ is an A-proper homotopy w.r.t. Γ and, as was noted above, $H(t,x) \neq 0$ for $x \in \partial D$ and $t \in [0,1]$. Hence the conclusion follows from Theorem 2.1 and property (10) proved below.

(10) The first assertion follows from Theorem 2.1, while the second follows from Theorem III.1.1. Indeed, if an integer $m \in \text{Deg}(T,D,f)$, there is an infinite sequence $\{n_j\} \subset Z_+$ such that $m = \text{deg}(T_{n_j},D_{n_j},W_{n_j}f)$ for each $j \in Z_+$. But then, by Theorem III.1.1, $\text{deg}(T_{n_j},D_{n_j},W_{n_j}f) = \text{deg}(T_{n_j} - W_{n_j}f, D_{n_j}, 0)$ for all $j \in Z_+$ so that $m \in \text{Deg}(T - f, D, 0)$. If ∞ $(-\infty)$ lies in $\text{Deg}(T,D,f)$, then there exists $\{n_j\} \subset Z_+$ such that $\lim_j \text{deg}(T_{n_j},D_{n_j},W_{n_j}f) = +\infty$ $(-\infty)$. Hence

$$\lim_j \text{deg}(T_{n_j} - W_{n_j}f, D_{n_j}, 0)$$

$$= \lim_j \text{deg}(T_{n_j} - W_{n_j}f, D_{n_j}, W_{n_j}f) = +\infty \ (-\infty)$$

[i.e., $+\infty$ $(-\infty)$ lies in $\text{Deg}(T - f, D, 0)$. Thus $\text{Deg}(T,D,f) \subseteq \text{Deg}(T - f, D, 0)$. Similarly, one shows the opposite inclusion and thus proves (10).   Q.E.D.

Remark 2.3 The following observation concerning the definition of the generalized degree Deg(T,D,f) and its properties derived above will prove to be especially useful in our study of densely defined semilinear A-proper mappings (see Remark 4.1-1 in [225]). Going carefully over the proofs of Lemma 2.1 and Theorems 2.1 and 2.2, we see that in defining Deg(T,D,f) and in proving its properties (1) to (10), we *did not* actually use the assumption that D is an open set in X or that $\overline{D}$ and ∂D are the respective closure and the boundary of D in X. The only conditions that have been used are that for each $n \in Z_+$, $D_n \equiv D \cap X_n$ is an open set in X_n with its closure $\overline{D}_n$ and boundary ∂D_n in X_n and that $\overline{D}_n \subset \overline{D}$ and $\partial D_n \subset \partial D$ for all $n \in Z_+$, where $\overline{D}$ is not necessarily the closure or ∂D the boundary of D in X but they have the property that if $\{x_j\} \subset \partial D$ is such that $x_j \to x$ in X and x lies in $\overline{D}$, then $x \in \partial D$.

For example, let V be a linear subspace of X that is dense in X, let $\{X_n\} \subset V$ for each $n \in Z_+$, and let G be a bounded open subset of X with closure $\overline{G}$ and boundary ∂G in X. If we set $G_V \equiv G \cap V$, $\overline{G}_V \equiv \overline{G} \cap V$, $\partial G_V \equiv \partial G \cap V$, and $G_n \equiv G \cap X_n$ for each $n \in Z_+$, we see that G_n is a bounded open subset of X_n with its closure $\overline{G}_n$ and boundary ∂G_n in X_n, $G_n \subset G_V$, $\overline{G}_n \subset \overline{G}_V$, and $\partial G_n \subset \partial G_V$ for each $n \in Z_+$; moreover, if $\{x_j\} \subset \partial G_V$ is such that $x_j \to x$ in X and $x \in \overline{G}_V$, then $x \in \partial G_V$. But G_V is not an open set in X nor is $\overline{G}_V$ the closure or ∂G_V the boundary of G_V in X. Thus, if in Lemma 2.1 and Definition 2.1 we replace D by G_V, $\overline{D}$ by $\overline{G}_V$, and ∂D by ∂G_V, then Deg(t,G_V,f) is well defined by Definition 2.1 and possesses properties (1) to (10). For later use we summarize the discussion above in the following:

Theorem 2.1′ Let $V, G_V, \overline{G}_V$, and ∂G_V be as above, $T: \overline{G}_V \to Y$ A-proper w.r.t. Γ and $f \notin T(\partial G_V)$. Then the generalized degree Deg(T,G_V,f) given by (1) and (2) of Definition 2.1 is well defined and has the properties (1) to (10) stated in Theorems 2.1 and 2.2.

We note that the facts mentioned in Remark 2.3 were explicitly stated by Petry [200], who also obtained some existence results for quasi-A-proper mappings.

We complete this section by proving the following three results, which will prove useful in the study and application of the generalized degree theory.

Proposition 2.2 If $T \in L(X,Y)$ is A-proper w.r.t. Γ and injective, then for any bounded open set $D \subset X$, Deg(T,D,f) $\subset \{\pm 1\}$ for any $f \in T(D)$.

Proof. Since $T \in L(X,Y)$ is A-proper and injective, it follows from Theorem II.1.1 that there is $n_0 \in Z_+$ such that for each $n \geq n_0$ there is $x_n \in X_n$ such that $T_n(x_n) = W_n f$, $x_n \to x_0$ in X and $Tx_0 = f$. Let D be any bounded open set in X with $x_0 \in D$. Then $Tx \neq f$ for $x \in \partial D$ and there is $n_1 \in Z_+$ such that $x_n \in D_n \equiv D \cap X_n$ and $T_n(x) \neq W_n f$ for all $x \in \partial D_n$ and $n \geq n_1$. Since T_n is a linear bijective map of X_n onto E_n, $T_n^{-1}(W_n f) \in D_n$ but $T_n(x) \neq W_n f$ for all $x \in \partial D_n$ and each $n \geq n_2 = \max\{n_1, n_0\}$, it follows from Theorem III.1.3 that $\deg(T_n, D_n, W_n f) = \pm 1$ for each $n \geq n_2$. Hence, by Definition 2.1, $\text{Deg}(T,D,f) \subset \{\pm 1\}$ for any $f \in T(D)$.   Q.E.D.

Our next result, which depends on Proposition I.2.2 concerning the duality mappings $J: X \to X^*$, will be particularly useful in the generalized degree theory for A-proper maps from subsets of X to X^* in a similar way as the identity I from subsets of X into X is useful in the Leray–Schauder degree theory for compact vector fields (see [222]). For the use of J in other maping theories, see [45,53,126,147,179,180].

Proposition 2.3 If X is a separable reflexive space with X and X^* having Property (H) and $J: X \to X^*$ is the duality mapping with gauge function ϕ, then for each given $r > 0$ and any fixed w in X^* with $J_w x = Jx - w \neq 0$ for $x \in \partial B(0,r)$, $\text{Deg}(J_w, B(0,r), 0)$ is well defined and such that

$$\text{Deg}(J_w, B(0,r), 0) \neq \{0\} \quad \text{if } \|w\| < \phi(r) \quad \text{and} \quad \text{Deg}(J_w, B(0,r), 0)$$
$$= \{0\} \quad \text{if } \|w\| > \phi(r).$$

In particular, $J(X) = X^*$.

Proof. It follows from Proposition I. 2.2 that J is a single-valued continuous duality map, which is A-proper w.r.t. $\Gamma_I = \{X_n, V_n; X_n^*, V_n^*\}$ and odd. Now let $r > 0$ be any given real number, w an element in X^*, and $J_{tw}(x) = Jx - tw$ a continuous mapping of $\overline{B}(0,r) \times [0,1]$ into X^*. Then for each fixed $t \in [0,1]$ and w with $\|w\| < \phi(r)$, J_{tw} is an A-proper map that is continuous, uniformly for x in $\overline{B}(0,r)$, and $J_{tw}(x) \neq 0$ for all $x \in \partial B(0,r)$ and $t \in [0,1]$. Hence, by the homotopy property (3) of Theorem 2.1, it follows that $\text{Deg}(J_{tw}, B(0,r), 0)$ is independent of $t \in [0,1]$. Since $J_{0w} = J$ and $J_{1w} = J - w$, it follows that $\text{Deg}(J - w, B(0,r), 0) = \text{Deg}(J, B(0,r), 0) \neq \{0\}$ by (6) of Theorem 2.1 since J is odd. The last assertion of Proposition 2.3 follows from the fact that for $\|w\| > \phi(r)$, $Jx \neq w$ for all $x \in \partial B(0,r)$ and hence $\text{Deg}(J_w, B(0,r), 0)$ is well defined and must equal $\{0\}$, for otherwise, by (2) of Theorem 2.1, the equation $Jx = w$ would have a solution x in $B(0,r)$, which is impossible for $\|w\| > \phi(r)$.   Q.E.D.

Let us add that Proposition 2.3 remains valid if Γ_I is replaced by the admissible projectional scheme $\Gamma_D = \{X_n, P_n; R(P_n^*), P_n^*\}$.

As our last result in this section, we obtain the following restricted product formula for the generalized topological degree proved by the author in [228].

Theorem 2.3 Suppose that $A, C \in L(X, Y)$ are such that A is injective and A-proper w.r.t. Γ, C is compact, and $A - C$ injective. Set $K \equiv CA^{-1}$ and $L \equiv I - K: Y \to Y$. Let $r > 0$ be any fixed number and set $G \equiv A(B(0,r))$.

Then there exist $N_1 \in Z^+$, independent of r, such that for $n \geq N_1$,

$$\deg(A_n - C_n, B_n(0,r), 0) = \deg(A_n, B_n(0,r), 0) \deg_{LS}(L, G, 0). \quad (2.1)$$

Proof. Since A and $A - C$ are injective, $Ax \neq 0$ and $Ax - Cx \neq 0$ for $x \in \partial B(0,r)$ and any $r > 0$ and thus $Ly \equiv y - Ky \neq 0$ for $y \in \partial G$. Our assumption on A, $A - C$, and (1.3) of Theorem II.1.1 imply the existence of a constant $d > 0$ and $N_0 \in Z^+$ such that $\| A_n x \| \geq d\| x \|$ and $\| A_n x - C_n x \| \geq d\| x \|$ for all $x \in X_n$ and $n \geq N_0$. Moreover, since $K = CA^{-1}$: $Y \to Y$ is compact and $\Gamma_Y \equiv \{Y_n, Q_n\}$ is projectionally complete for Y, it follows that $L = I - K$ is injective and A-proper w.r.t. Γ_Y, so there exist $N_1 \in Z^+$ such that L_n is injective for $n \geq N_1$. Thus, by Theorem 2.4 in the next section,

$$\deg(L_n, G_n, 0) = \deg_{LS}(L, G, 0) \qquad \text{for each} \quad n \geq N_1. \quad (\#)$$

It follows from this discussion that if we take $N_2 = \max\{N_0, N_1\}$, the degrees appearing in (2.1) are well defined for $n \geq N_2$.

To establish the equality (2.1) for each $n \geq N_2$, we first consider the homotopy

$$H_n(t,x) = t(A_n(x) - C_n(x)) + (1 - t)L_n A_n \qquad (x \in B_n(0,r), t \in [0,1]).$$

Now we claim that $H_n(t,x) \neq 0$ for $x \in \partial B_n(0,r)$, $t \in [0,1]$, $n \geq N_2$. If not, there are sequences $\{x_{n_j} \mid x_{n_j} \in \partial B_{n_j}(0,r)\}$ and $\{t_{n_j}\} \in [0,1]$ such that $H_{n_j}(x_{n_j}, t_{n_j}) = 0$ for each j. Without loss of generality we may assume that $t_{n_j} \to t \in [0,1]$. Hence

$$t(A_{n_j}(x_{n_j}) - C_{n_j}(x_{n_j})) + (1 - t)L_{n_j}A_{n_j}(x_{n_j}) \to 0 \qquad \text{as} \quad j \to \infty$$

Replacing L by $I - K$, one easily shows that

$$A_{n_j}(x_{n_j}) - C_{n_j}(x_{n_j}) - (1 - t)Q_{n_j}KA_{n_j}(x_{n_j}) \to 0 \qquad \text{as} \quad j \to \infty.$$

Since $\{A_{n_j}(x_{n_j})\}$ is bounded in Y and K is compact, we may suppose that

$K(A_{n_j}(x_{n_j})) \to w$ in Y for some $w \in Y$ and, therefore, as $j \to \infty$, we see that

$$A_{n_j}(x_{n_j}) - tC_{n_j}(x_{n_j}) - Q_{n_j}(1 - t)w \to 0 \qquad \text{as} \quad j \to \infty.$$

Since $A - tC$ is A-proper, there exists a subsequence $\{x'_{n_j}\}$ and x' such that $x'_{n_j} \to x'$ in X and $Ax' - tBx' - (1 - t)w = 0$ with $x' \in \partial B(0,r)$. This shows that $w = KAx$. Thus $Ax - tCx - (1 - t)KAx = 0$ or $Ax - tCx - (1 - t)Cx = 0$ since $K = CA^{-1}$. Hence $Ax - Cx = 0$ with $\| x \| = r$, in contradiction to the injective property of $A - C$. Thus, for each $n \geq N_2$, $H_n(t,x) \neq 0$ for $x \in \partial B_n(0,r)$ and $t \in [0,1]$. Therefore, by the homotopy theorem for the Brouwer degree, for each $n \geq N_2$ we have

$$\deg(A_n - B_n, B_n(0,r), 0) = \deg(L_nA_n, B_n(0,r), 0).$$

Since, for fixed $n \geq N_2$, $L_n: Y_n \to Y_n$ and $A_n: X_n \to X_n$ are injective, it follows from the product formula for the Brouwer degree that for each such n,

$$\deg(L_nA_n, B_n(0,r), 0) = \deg(A_n, B_n(0,r), 0) \deg(L_n, G_n, 0).$$

The latter equality and (#) imply the validity of (2.1).   Q.E.D.

It follows from Theorem 2.3 that if A_n is orientation preserving, the product formula (2.1) can be written in the form

$$\mathrm{Deg}(A - C, B(0,r), 0) = \{\deg_{\mathrm{LS}}(I - CA^{-1}, G, 0)\} = \{(-1)^{\eta}\},$$
$$(2.1')$$

where η is the sum of the algebraic multiplicities, $\cup_{n=1} (N(\lambda - CA^{-1}))^n$, of eigenvalues $\lambda > 1$ of CA^{-1} [i.e., of characteristic values $\mu \in (0,1)$ of $Ax - \mu Cx = 0$].

2.2 The Uniqueness Theorem for the Generalized Degree

In this section we consider the second and technically more delicate part of the generalized degree theory, in which we give a general sufficient condition for $\mathrm{Deg}(T,D,f)$ to be single-valued. The subclass of A-proper mappings for which this is the case will prove to be important in various applications as well as in the constructive solvability of a given equation. It is also shown in this section and in Section 3 that for the type of spaces considered in this book, the single-valued generalized degree includes the Leray–Schauder degree, the Browder–Nussbaum degree, the Nussbaum–Sadovskii degree, and others. Moreover, even in case of the single-valued generalized degree we are able to treat various classes of mappings to which the degree theories of the other authors are not applicable.

In this section we assume that the admissible scheme $\Gamma = \{X_n, V_n; E_n, W_n\}$ for (X,Y), given by Definition I.1.1, has the following additional property (cf. condition (b_2) of Definition 2.1 [42]).

(p1) For any given finite-dimensional subspace F of Y there exists an $n_F \in Z_+$ such that $W_n|_F: F \to E_n$ is injective for each $n \geq n_F$.

The following proposition shows that (p1) holds if Γ is any one of the special schemes considered in Section I.1.

Proposition 2.3' If Γ is either an injective scheme $\Gamma_I = \{X_n, V_n; X_n^*, V_n^*\}$ for (X,X^*), a complete projection scheme $\Gamma_\alpha = \{X_n, P_n\}$ for (X,X), or an admissible projection scheme $\Gamma_P = \{X_n, V_n; Y_n, Q_n\}$ for (X,Y) with $Q_n^* w \rightharpoonup w$ in Y^* for each w in Y^*, then in each case F has property (p1).

Proof. Suppose first that Γ_I is an injective scheme for (X,X^*) and let F be any given finite-dimensional subspace of X^*. If (p1) were not true, there would exist a sequence $\{f_j\} \subset F$ with $\| f_j \| = 1$ and a sequence $\{n_j\} \subset Z_+$ with $n_j \to \infty$ such that and $V_{n_j}^*(f_j) = 0$ for each $j \in Z_+$. Since F is finite-dimensional, we may assume that $f_j \to f$ in F and $\| f \| = 1$. Let x be any element in X and $\{x_n \mid x_n \in X_n\}$ such that $x_n \to x$ in X. Then

$$(f,x) = \lim_j (f_j, x_{n_j}) = \lim_j (f_j, V_{n_j} x_{n_j}) = \lim_j (V_{n_j}^* f_j, x_{n_j}) = 0.$$

Hence $f = 0$, in contradiction to the fact that $\| f \| = 1$ [i.e., (p1) holds].

Suppose now that Γ_α is a projectional scheme for (X,X) and F is any given finite-dimensional subspace of X. If (p1) were not true, then, as before, we could find $\{f_j\} \subset F$ and $\{n_j\} \subset Z_+$ such that $\| f_j \| = 1$, $P_{n_j} f_j = 0$ for each j, and $f_j \to f$ in F with $\| f \| = 1$. Since $P_n(x) \to x$ for each x in X, it follows that for any given $w \in X^*$, we have $(P_n^* w, x) \to (w, x)$ for each x in X. Hence, for each w in X^*,

$$(w,f) = \lim_j (P_{n_j}^* w, f_j) = \lim_j (w, P_{n_j} f_j) = 0.$$

This implies that $f = 0$, in contradiction to the fact that $\| f \| = 1$. Thus (p1) holds in this case also.

Finally, suppose that Γ_P is an admissible projection scheme for (X,Y) with $Q_n^* w \rightharpoonup w$ in Y^* for each w in Y^* and let F be an arbitrary subspace of Y. If (p1) were not true, then, as before, there exist sequences $\{f_j\} \subset F$ and $\{Q_{n_j}\}$ such that $\| f_j \| = 1$, $f_j \to f$ in F, and $Q_{n_j} f_j = 0$ for each j. Since, for any $w \in Y^*$, $(Q_{n_j}^* w, f_j) = (w, Q_{n_j} f_j) = 0$ for each j, it follows that $0 = \lim_j (Q_{n_j}^* w, f_j) = (w, f)$. Hence $f = 0$, so (p1) holds in this case as well. Q.E.D.

Remark 2.4 If $\Gamma_P^* = \{X_n, V_n; R(P_n^*), P_n^*\}$ is an admissible projection scheme for (X, X^*), then $P_n x \to x$ in X for each x in X, and as in the case of Γ_I, one shows that Γ_P^* satisfies condition (p1). Note that, in general, the admissibility of Γ_P^* for (X, X^*) does not imply that $P_n^* w \to w$ in X^* for each w in X^*. However, if X is reflexive and Γ_P^* is nested (i.e., $X_n \subset X_{n+1}$ for each $n \in Z_+$), it is not hard to show that in this case the admissibility of Γ_P^* for (X, X^*) implies also that $P_n^* w \to w$ in X^* for each w in X^*. In this case the scheme Γ_P^* is often written in the form $\Gamma_P = \{X_n, P_n; R(P_n^*), P_n^*\}$.

The main result in this section is the following theorem, proved in [42] in a more general setting.

Theorem 2.4 Let D be a bounded open subset of X, $T: \overline{D} \to Y$ a continuous A-proper mapping w.r.t. $\Gamma = \{X_n, V_n; E_n, W_n\}$ and $f \notin T(\partial D)$. Suppose that $T = A + C$, where $C: \overline{D} \to Y$ is compact and A maps D homeomorphically onto an open subset $A(D)$ of Y, carrying $\overline{D}$ homeomorphically onto $\overline{A(D)}$. Let $T_n = A_n + C_n: \overline{D}_n \to E_n$ and suppose that A_n is an orientation-preserving homeomorphism of $\overline{D}_n$ onto $\overline{A_n(D_n)} \subset E_n$ such that

$$\| A_n(x) - A_n(y) \| \geq \alpha(\| x - y \|) \tag{2.2}$$
$$\text{for all} \quad x, y \text{ in } \overline{D}_n \text{ and all } n \in Z_+,$$

where $\alpha: R^+ \to R^+$ is continuous with $r_i \to 0$ whenever $\alpha(r_i) \to 0$.
Then there exists $n_0 \in Z_+$ such that for $n \geq n_0$,

$$\deg(T_n, D_n, W_n f) = \deg_{LS}(I + CA^{-1}, A(D), f).$$

In particular, $\mathrm{Deg}(T, D, f) = \{\deg_{LS}(I + CA^{-1}, A(D), f)\}$.

Proof. To prove Theorem 2.4 we first establish the following:

Lemma 2.2 There exists an $\epsilon > 0$ and an $n_1 \in Z_+$ such that if $C': \overline{D} \to Y$ is another compact mapping with the property that $\| Cx - C'x \| < \epsilon$ for all x in $\overline{D}$ and if $T' = H + C'$, then

(i) $\deg_{LS}(I + C'A^{-1}, A(D), f) = \deg(I + CA^{-1}; A(D), f)$.
(ii) $\deg(T_n, D_n, W_n f) = \deg(T_n', D_n, W_n f)$ for all $n \geq n_1$.

Proof. (i) This assertion follows from the invariance of the Leray–Schauder degree under small perturbations.

(ii) Suppose that (ii) were false. Then there would exist sequences $\{C^{(k)}\}$, $\{n_k\} \subset Z_+$, and $\{\epsilon_k\}$ such that $\epsilon_k \to 0$, $n_k \to \infty$ as $k \to \infty$, $\| C_x^{(k)} - Cx \| < \epsilon_k$ for all $x \in \overline{D}$, and

$$\deg(T_{n_k}^{(k)}, D_{n_k}, W_{n_k} f) \neq \deg(T_{n_k}, D_{n_k}, W_{n_k} f),$$

where $T^{(k)} = A + C^{(k)}$. For each k and t in $[0,1]$ put

$$C^{(k),t}(x) = (1 - t)C^{(k)}x + tCx, \quad T^{(k),t}(x)$$
$$= Ax + C^{(k),t}(x) \qquad (x \in \overline{D}).$$

Under our assumption of unequal degrees, there exist sequences $\{t_k\} \subset [0,1]$ and $\{x_{n_k} \mid x_{n_k} \in \partial D_{n_k}\}$ such that $T_{n_k}^{(k),t_k}(x_{n_k}) = W_{n_k}$ (f). In view of the equalities above, a simple manipulation shows that

$$W_{n_k}(f) = T_{n_k}^{(k),t_k}(x_{n_k}) = T_{n_k}(x_{n_k}) + (1 - t_k)(C_{n_k}^{(k)}(x_{n_k}) - C_{n_k}(x_{n_k})),$$

whence, using our conditions on $\{C^{(k)}\}$ and $\{W_n\}$, we derive the relation $\| T_{n_k}(x_{n_k}) - W_{n_k}(f) \| \le 2c_0\epsilon_k \to 0$ as $k \to \infty$. By the A-properness of T, there exists a subsequence $\{x_{n_{k(j)}}\}$ and $x \in \overline{D}$ such that $x_{n_{k(j)}} \to x$ and $Tx = f$. Since $\{x_{n_k}\} \subset \partial D$, it follows that $x \in \partial D$, in contradiction to the hypothesis that $f \notin T(\partial D)$. Q.E.D.

Proof of Theorem 2.4 continued. By Lemma 2.1, $\deg(T_n, D_n, W_n f)$ is well defined for all $n \ge n_0$. Since A_n is an orientation-preserving homeomorphism of D_n into E_n, we have the equality

$$\deg(T_n, D_n, W_n f) = \deg(I + C_n A_n^{-1}, A_n(D_n), W_n f).$$

By Lemma 2.2 we may assume that C maps $\overline{D}$ into a compact subset K of a finite-dimensional subspace F of Y. There is no loss in generality in assuming that $f \in F$, for otherwise we would consider the space spanned by $\{f, F\}$. Now, by the properties of the Leray–Schauder degree (see [55,136,153,250])

$$\deg(I + C_n A_n^{-1}, A_n(D_n), W_n f)$$
$$= \deg(I + C_n A_n^{-1}, A_n(D_n) \cap W_n F, W_n f)$$

since $C_n A_n^{-1}$ maps $A_n(D_n)$ into $W_n F$ while $W_n f \in W_n F$. By condition (p1) on $\{W_n\}$, there exists $n_2 \in Z_+$ such that W_n is an injective map of F into E_n for each $n \ge n_2$. Now $W_n|_F$ has an inverse and we can extend this inverse to a linear mapping R_n of E_n into Y such that $W_n R_n = I$ on E_n and $R_n W_n f = f$ for $f \in F$. Similarly, the assumption on C and the properties of the Leray–Schauder degree show that

$$\deg(I + CA^{-1}, A(D), f) = \deg(I + CA^{-1}, A(D) \cap F, f),$$

and for all $n \ge n_2$,

$$\deg(I + C_n A_n^{-1}, A_n(D_n) \cap W_n F, W_n f)$$
$$= \deg(I + R_n C_n A_n^{-1} W_n, R_n(A_n(D_n)) \cap F, f).$$

We note that $A_n(D_n) = W_n A(D \cap X_n)$, $R_n(A_n(D_n)) \cap F \subset W_n^{-1}(W_n A(D \cap X_n)) \cap F$ and $R_n C_n A_n^{-1} W_n = R_n W_n C A_n^{-1} W_n = C A_n^{-1} W_n$, so that

$$\deg(T_n, D_n, W_n f)$$
$$= \deg(I + C A_n^{-1} W_n, W_n^{-1}(W_n A(D \cap X_n)) \cap F, f).$$

To continue with the proof we need the following lemmas.

Lemma 2.3 Let $K_1 = \{v \mid v \in A(D), v = Au,$ where $Tu = f\}$ and let U be a neighborhood of K_1 in Y. Then there is $n_3 \in Z_+$ such that for $n \geq n_3$, any v in $W_n^{-1}(A_n(D_n))$ such that $v + C A_n^{-1} W_n v = f$, must lie in U.

Proof. Suppose not. Then there exists a sequence $\{n_j\} \subset Z_+$ with $n_j \to \infty$ and a corresponding sequence $\{v_{n_j}\}$ with $v_{n_j} \in (W_{n_j}^{-1}(A_{n_j}(D_{n_j})\backslash U)$ such that $v_{n_j} + C A_{n_j}^{-1} W_{n_j} v_{n_j} = f$. Put $x_{n_j} = A_{n_j}^{-1} W_{n_j} v_{n_j}$ and note that $x_{n_j} \in D_{n_j}$, $A_{n_j} x_{n_j} = W_{n_j} v_{n_j}$ and $v_{n_j} + C x_{n_j} = f$. Hence $W_{n_j} v_{n_j} + W_{n_j} C x_{n_j} = W_{n_j} f$ and therefore $T_{n_j} x_{n_j} = A_{n_j} x_{n_j} + C_{n_j} x_{n_j} = W_{n_j} f$. Since T is A-proper, we may assume that $x_{n_j} \to x$ in X, $Tx = f$, and $A x_{n_j} \to Ax$ in Y by continuity of A. Since $C x_{n_j} \to Cx$ and $Tx = f$, it follows that $v_{n_j} = f - C x_{n_j} \to f - Cx = Ax$ and $Ax \in K_1$. This is impossible because $v_{n_j} \in Y\backslash U$ and U is a neighborhood of K_1. Thus the assertion of Lemma 2.3 is true.   Q.E.D.

Lemma 2.4 (a) $K_1 = \{v \mid v \in A(D) \cap F, v + C A^{-1} v = f\}$.

(b) There exists $n_4 \in Z_+$ and a neighborhood U_1 of K_1 in $A(D) \cap F$ such that $U_1 \subset W_n^{-1}(A_n(D_n))$ for $n \geq n_4$.

Proof. (a) $K_1 = A(K)$, where $K = \{u \mid Tu = f\}$. Hence $K_1 = \{v \mid TA^{-1}v = v + C A^{-1} v = f\}$.

(b) Suppose that the assertion were false. Then there would exist $\{n_j\} \subset Z_+$ with $n_j \to \infty$ and $\{v_{n_j}\} \subset A(D) \cap F$ such that $\mathrm{dist}(v_{n_j}, K_1) \to 0$ and $W_{n_j} v_{n_j} \notin A_{n_j}(D_{n_j})$. Since $A(D) \cap F$ is relatively compact, we may assume that $v_{n_j} \to v$ with $v \in K_1$. For each j, $v_{n_j} = A(\cdot_j)$ for some $x_j \in D$, and $x_j = A^{-1}(v_{n_j}) \to A^{-1} v$ as $j \to \infty$. Thus the set $\{x_j\}$ is relatively compact in $\bar{D}$ and hence $\mathrm{dist}(x_j, D_{n_j}) = \epsilon_j \to 0$ as $j \to \infty$. In particular, we may find $w_{n_j} \in D_{n_j}$ such that $\| x_j - w_{n_j} \| < 2\epsilon_j$. Thus $w_{n_j} \to A^{-1} v$ and $A w_{n_j} \to v$. Since $v_{n_j} \to v$, $\| A w_{n_j} - v_{n_j} \| \to 0$. By the uniform boundedness of $\{W_n\}$, $\| W_{n_j} A w_{n_j} - W_{n_j} v_{n_j} \| \to 0$ and, therefore, $\mathrm{dist}(W_{n_j} v_{n_j}, A_{n_j}(D_{n_j})) \to 0$. Since $W_{n_j} v_{n_j} \notin A_{n_j}(D_{n_j})$, it follows that $\mathrm{dist}(W_{n_j} v_{n_j}, \partial A_{n_j}(D_{n_j})) \to 0$, while $\partial A_{n_j}(D_{n_j}) = A_{n_j}(\partial D_{n_j})$. Hence there exists $u_{n_j} \in \partial D_{n_j}$ such that $\| W_{n_j} v_{n_j} - A_{n_j} u_{n_j} \| \to 0$. In particular, $\| A_{n_j} u_{n_j} - W_{n_j} v \| \to 0$. Since A is A-proper w.r.t. Γ, we may assume that $u_{n_j} \to u$ in X with $Au = v$ for some v in K_1 and $u \in \partial D$. Since $Au \in K_1$, $Tu = f$, which contradicts the fact that $u \in \partial D$.   Q.E.D.

By Lemmas 2.3 and 2.4 and the preceding discussion, we know that for $n \geq n_5$ ($n_5 \in Z_+$ sufficiently large),

$$\deg(T_n, D_n, W_n f) = \deg(I + CA_n^{-1} W_n, W_n^{-1}(A_n(D_n)) \cap F \cap A(D), f)$$

and we may choose a neighborhood U_1 of K_1 in $F \cap A(D)$ such that $\deg(T_n, D_n, W_n f) = \deg(I + CA_n^{-1} W_n, U_1, f)$ and $\deg(I + CA^{-1}, A(D), f) = \deg(I + CA^{-1}, U_1, f)$. By the properties of the Leray–Schauder degree, the desired equality of degrees will then follow if we establish the following lemma.

Lemma 2.5 The quantity $\| CA_n^{-1} W_n u - CA^{-1} u \| \to 0$ as $n \to \infty$, uniformly for all u in U_1, a bounded set in $F \cap A(D)$.

Proof. Set $w_n = A_n^{-1} W_n u$, $w = A^{-1} u$ and note that $w_n \in D_n$, $A_n w_n = W_n u$ and $w \in D \cap A^{-1}(F \cap \overline{D})$. The element u lies in the compact set $K_2 = F \cap \overline{A(D)}$ and w lies in the compact set $K_3 = A^{-1}(K_2)$. Since K_3 is compact, $\epsilon_n = \mathrm{dist}(K_3, D_n) \to 0$ as $n \to \infty$. Thus for w in K_3, there exists x_n in D_n such that $\| w - x_n \| < \epsilon_n + n^{-1}$. Since A is continuous on $\overline{D}$ and hence uniformly continuous on K_3, $\| Aw - Av_n \| \leq \delta_n \to 0$, where δ_n is independent of the choice of w in K_3. Since $A_n w_n = W_n u$, $u = Aw$, we have

$$\| W_n Aw - W_n Ax_n \| \leq c_0 \delta_n$$

and $\quad \| A_n w_n - Ax_n \| \leq c_0 \delta_n \to 0 \quad$ as $\quad n \to \infty$.

By the hypothesis (2.1) of Theorem 2.4, $\alpha(\| w_n - x_n \|) \leq c_0 \delta_n \to 0$. Hence $\| w_n - x_n \| \to 0$ as $n \to \infty$ uniformly on K_2, so $\| w - w_n \| \leq \| w - x_n \| + \| x_n - w_n \| \to 0$. Finally, C is uniformly continuous at all points w of the compact set K_3. Hence there exists a sequence $\zeta_n \to 0$ such that for all w, as above,

$$\| Cw - Cw_n \| \leq \zeta_n \to 0, \quad \text{i.e.,} \quad \| CA^{-1} u - CA_n^{-1} W_n u \| \leq \zeta_n \to 0$$

for all $u \in U_1$.   Q.E.D.

Remark 2.5 If in Theorem 2.4 the condition that A_n is orientation preserving is omitted, the conclusion of Theorem 2.4 holds in the form

$$| \deg(T_n, D_n, W_n f) | \tag{2.3}$$
$$= | \deg_{LS}(I + CA^{-1}, A(D), f) | \quad \text{for} \quad n \geq n_0.$$

We omit the detailed proof of this fact since it follows literally from the proof of Theorem 2.4.

An immediate consequence is the following corollary of Theorem 2.4.

Corollary 2.1 Suppose that $Y = X$ and $\Gamma_\alpha = \{X_n, P_n\}$ is projectionally complete for (X,X). Then the conclusion of Theorem 2.4 holds if either (i) or (ii) is satisfied:

(i) $A = I$, with I an identity on X.
(ii) $A = I - S$ with S a c-contraction on X, $c < 1$ and $\alpha = 1$.

Proof. The claim is obvious when (i) holds. To establish its validity when (ii) is satisfied, it suffices to show that $A_n = P_n A|_{X_n}$ is orientation preserving since all other conditions are satisfied because $S: X \to X$ is c-contractive with $c < 1$ and $\| P_n \| = 1$.

Now it is well known that orientation is preserved under homotopies and that the identity I on X_n is orientation preserving. Since $A_n(t,x) = x - tP_n S$ for $x \in X_n$ and $t \in [0,1]$ is such that

$$(A_n(t,x) - A_n(t,y), J(x - y))$$

$$= \| x - y \|^2 - t(P_n Sx - P_n Sy, J(x - y)) \geq (1 - c) \| x - y \| > 0$$

for all $x,y \in X_n$ with $x \neq y$, it follows that $A_n(t,\cdot)$ is a homotopy between $A_n(1,\cdot) = A_n$ and $A(0,\cdot) = I$. Hence $A_n: X_n \to X_n$ preserves orientation since I does.   Q.E.D.

Remark 2.6 It follows from Corollary 2.1 that when X is a π_1-space, then $\mathrm{Deg}(A + C, D, f)$ coincides with the Leray–Schauder degree $\deg_{LS}(I + C, D, f)$ when $A = I$ and with the Browder–Nussbaum degree when $A = I - S$.

Other and more general cases of Theorem 2.4 are discussed in Section 3.

3. SINGLE-VALUED GENERALIZED DEGREE FOR COMPACT AND k-BALL-CONTRACTIVE PERTURBATIONS OF FIRMLY MONOTONE AND ACCRETIVE MAPS

In Section 3.1 we use Theorem 2.4 to study the single-valuedness of the generalized degree for maps $T + C: D \subseteq X$ into X^*, where X is a separable reflexive space and T is firmly monotone and C is either compact or k-ball contractive, while in Section 3.2 the same problem is studied for maps T of X into X, where X has a projectionally complete scheme and T is firmly accretive. We add that all the results in Section 3.2 and some in 3.1 concerning the generalized degree are new.

3.1 Compact and *k*-Ball-Contractive Perturbations of Firmly Monotone Mappings

Suppose that X is a separable reflexive Banach space and Γ is an injective scheme $\Gamma_I = \{X_n, V_n; X_n^*, V_n^*\}$ which is nested (i.e., $X_n \subset X_{n+1}$ for each $n \in Z_+$). To deduce here our first consequence of Theorem 2.4, we recall that Γ_I satisfies the additional condition (p1). Furthermore, it is not hard to show that $R: G_n \subset X_n \to X_n^*$ is an orientation-preserving homeomorphism if R is a homeomorphism of G_n into X_n^* such that $(Rx - Ry, x - y) > 0$ for all x and y in G_n with $x \neq y$.

Theorem 3.1 Let X be reflexive, D a bounded open subset of X, $C: \overline{D} \to X^*$ compact, and $A: \overline{D} \to X^*$ a bounded continuous mapping such that

$$(Ax - Ay, x - y) \geq c(\| x - y \|) \qquad \text{for all} \quad x,y \in \overline{D}, \tag{3.1}$$

where $c: R^+ \to R^+$ is an increasing continuous function such that $c(0) = 0$, $c(r) > 0$, if $r > 0$ and $r_j \to 0$ whenever $c(r_j) \to 0$.

Then $T = A + C: \overline{D} \to X^*$ is A-proper w.r.t. Γ_I and $\mathrm{Deg}(T,D,f) = \{\deg_{LS}(I + CA^{-1}, A(D), f)\}$ for $f \notin T(\partial D)$.

Proof. By virtue of Theorem 2.4, (3.1), and the remarks preceding Theorem 3.1, it suffices to show that $\| A_n x - A_n y \| \geq \alpha(\| x - y \|)$ for all x and y in $\overline{D}_n$, where $A_n = V_n^* A|_{D_n}$ and $\alpha(r) = c(r)/r$ for $r > 0$, that A is an A-proper (w.r.t. Γ_I) homeomorphism of $\overline{D}$ into X^*, and that $A(D)$ is open in X^*.

It follows trivially from (3.1) that A_n satisfies the foregoing inequality on $\overline{D}_n$ for each $n \in Z_+$ with $\alpha(r)$ such that $\alpha(r) > 0$ for $r > 0$ and $r_j \to 0$ whenever $\alpha(r_j) \to 0$ and that A is a homeomorphism of $\overline{D}$ into X^* since $\| Ax - Ay \| \geq \alpha(\| x - y \|)$ for all x,y in $\overline{D}$. We now prove that $A: \overline{D} \to X^*$ is A-proper w.r.t. Γ_I. Indeed, suppose that $\{x_{n_j} \mid x_{n_j} \in \overline{D}_{n_j}\}$ is such that $V_{n_j}^* A x_{n_j} - V_{n_j}^* g \to 0$ for some g in X^*. We suppose that $\{n_j\}$ increases with j and we let $k > j$. Then

$$c(\| x_{n_k} - x_{n_j} \|) \leq (Ax_{n_k} - Ax_{n_j}, x_{n_k} - x_{n_j})$$
$$= (V_{n_k}^* A x_{n_k}, x_{n_k} - x_{n_j}) + (V_{n_j}^* A x_{n_j}, x_{n_j}) - (Ax_{n_j}, x_{n_j})).$$

Since D is bounded, for each fixed j we have

$$| (V_{n_k}^* A x_{n_k}, x_{n_k} - x_{n_j}) - (V_{n_k}^* g, x_{n_k} - x_{n_j}) |$$
$$\leq \| V_{n_k}^* A x_{n_k} - V_{n_k}^* g \| \| x_{n_k} - x_{n_j} \| \to 0 \qquad \text{as} \quad k \to \infty.$$

Similarly, as $j \to \infty$,

$$| (V_{n_j}^* A x_{n_j} - V_{n_j}^* g, x_{n_j}) | \leq \| X_{n_j}^* A x_{n_j} - V_{n_j}^* g \| \| x_{n_j} \| \to 0.$$

Since X is reflexive and $\{x_{n_j}\}$ is bounded, we may assume that $x_{n_j} \rightharpoonup x$ in X as $j \to \infty$. Hence letting $k \to \infty$ and using the results above, we get

$$c(\| x_{n_j} - x \|) \leq \liminf_k c(\| x_{n_j} - x_{n_k} \|)$$

$$\leq (g, x - x_{n_j}) + (V_{n_j}^* A x_{n_j}, x_{n_j}) + (A x_{n_j}, x)$$

since $(V_{n_k}^* g, x_{n_k} - x_{n_j}) = (g, x_{n_k} - x_{n_j}) \to (g, x - x_{n_j})$. We note that $A x_{n_j} \rightharpoonup g$ since, by assumption, $\{A x_{n_l}\}$ is uniformly bounded for all l and for any element v in the dense set $\cup_j x_{n_j}$ ($v \in X_{n_l}$ for some l) we have, for $j > l$,

$$(g, v) - (A x_{n_j}, v) = (V_{n_j}^* g - V_{n_j}^* A x_{n_j}, v) \to 0 \qquad \text{as} \quad j \to \infty.$$

Hence the arguments above imply that, as $j \to \infty$,

$$(g, x - x_{n_j}) + (V_{n_j}^* A x_{n_j}, x_{n_j}) - (A x_{n_j}, x)$$

$$= (g, x) + (V_{n_j}^* A x_{n_j} - V_{n_j}^* g, x_{n_j}) - A x_{n_j}, x) \to (g, x) + 0 - (g, x) = 0,$$

and consequently, $c(\| x_{n_j} - x \|) \to 0$ as $j \to \infty$. Thus $x_{n_j} \to x \in \overline{D}$ and, by the continuity of A, $A x_{n_j} \to Ax$. Since $A x_{n_j} \rightharpoonup g$, $Ax = g$ (i.e., A is A-proper w.r.t. Γ_l).

Finally, we prove that $A(D)$ is open in X^*. Let $x_0 \in D$ and $f_0 = A x_0 \in A(D)$. Then for $x \in D$ near x_0, we have $(Ax - f_0, x - x_0) \geq c(\| x - x_0 \|)$. Let $r > 0$ be such that $B(x_0, r) \subset D$. Since $\text{dist}(x_0, X_n) \to 0$, there exists an $x_n \in X_n$ such that $x_n \to x_0$. Then for n sufficiently large (e.g., $n \geq n_0$), $x_n \in B(x_0, r) \cap X_n$, and therefore for all $x \in D_n$ near x_n, we have

$$(V_n^* Ax - V_n^* A x_n, x - x_n) \geq c(\| x - x_n \|) \qquad (n \geq n_0). \tag{3.2}$$

Now, choose s so small that $0 < s < c(r)/r$ and let $f \in B(f_0, s)$. Then for all $x \in D_n$ such that $\| x - x_n \| = r$ we have

$$(A_n x - V_n^* f, x - x_n) = (A_n x - V_n^* A x_n, x - x_n)$$
$$+ (V_n^* A x_n - V_n^* f_0, x - x_n) + (V_n^* f_0 - V_n^* f, x - x_n).$$

Since $x_n \to x_0$ in X, A is continuous, and since $s < c(r)/r$, there exists n_1 ($\geq n_0$) so that $\| A x_n - A x_0 \| < (c(r)/r - s)$ for $n \geq n_1$. Thus the preceding equality and (3.2) imply that for all $x \in D_n$ such that $\| x - x_n \| = r$ and all $n \geq n_1$, we have

$$(A_n x - V_n^* f, x - x_n) \geq c(r) - \{\| A x_n - A x_0 \| \tag{3.3}$$
$$+ s\} r > c(r) - \frac{c(r)}{r} r = 0.$$

To complete the proof of Theorem 3.1, we need the following finite-dimensional result.

Lemma 3.1 Let V be a finite-dimensional Banach space. If S is a continuous mapping of $B(v_0,r) \subset V$ into V^* and if for a given $g \in V^*$, $(Sv - g, v - v_0) > 0$ for $\partial B(v_0,r)$, then there exists an $u \in B(v_0,r)$ such that $Su = g$.

Proof. Since V is finite dimensional, there exists a bicontinuous linear map R of a Hilbert space W onto V. Let R^* be the dual map of V^* onto $W^* = W$. Consider $S' = R^*SR$ and set $G = R^{-1}B(v_0,r)$, $h_0 = R^{-1}v_0$, $h = R^{-1}v$, and $z = R^*g$. Then G is an open, bounded, and convex set in W such that for all $h \in \partial G \; [= R^{-1}(\partial B(v_0,r))]$, we have

$$(S'h - z, h - h_0) = (R^*SRh - R^*g, h - h_0)$$

$$= (Sv - g, v - v_0) > 0.$$

Thus we may assume that $V = V^*$ is a finite-dimensional Hilbert space and consider the homotopy $S_t(r) = tSv + (1 - t)(v - v_0 - g)$ for $v \in \bar{B}(v_0,r)$ and $t \in [0,1]$. Now, for all $v \in \partial B(v_0,r)$ and $t \in [0,1]$, $(S_tv - g, v - v_0) = t(Sv - g, v - v_0) + (1 - t)\| v - v_0 \|^2 > 0$, and therefore $\deg(S_t,B(v_0,r),g)$ is independent of $t \in [0,1]$. Since, for $t = 0$, $S_0v = v - v_0 + g$ and $\deg(S_0,B(v_0,r),g) = 1$, the conclusion of Lemma 3.1 follows.

Now, by virtue of Lemma 3.1, the inequality (3.3) implies that for each $n \geq n_1$ there exists an element $u_n \in B(x_n,r) \cap X_n$ such that $A_n u_n - V_n^* f = 0$. Since, as was proved above, A is A-proper w.r.t. Γ_1, we may choose a subsequence $\{u_{n_j}\}$ such that $u_{n_j} \to u$, $u \in B(x_0,r) \subset D$ and $Au = f$. Hence $B(f_0,s) \subset A(D)$ and $A(D)$ is open in X^*.   Q.E.D.

In conjunction with Remark 2.4, let us add that exactly the same arguments as those used to deduce Theorem 3.1 from Theorem 2.4 show the validity of the following.

Theorem 3.2 Suppose that X is a reflexive space with a nested projectionally complete scheme $\Gamma_P = \{X_n,P_n;R(P_n^*),P_n^*\}$ for (X,X^*). If $A,C: \bar{D} \subset X \to X^*$ satisfy the same conditions as in Theorem 3.1, then $T = A + C$ is also A-proper w.r.t. Γ_P and

$$\text{Deg}(T,D,f) = \{\deg_{LS}(I + CA^{-1}, A(D), f\}   \text{for each}   f \notin T(D).$$

We now use the results of Section 1.2 and Theorem 2.4 to obtain the following new result by using the argument of Webb [291].

Theorem 3.3 Let X and Γ_P be as in Theorem 3.2, with $\| P_n \| = 1$, $A: X \to X^*$ a bounded continuous mapping such that

$$(Ax - Ay, x - y) \geq \mu\| x - y \|^2 \tag{3.4}$$
$$\text{for all } x,y \in X \text{ and some } \mu > 0,$$

$F: X \to X^*$ a k-ball-contractive map with $k < \mu$, D a bounded open subset of X, and $f \notin (A + F)(\partial D)$. Then

$$\text{Deg}(A + F, D, f) = \{\text{deg}_\beta(I + FA^{-1}, A(D), f)\}. \tag{3.5}$$

Proof. First, we claim that there exists $n_0 \in Z_+$ such that for each $n \geq n_0$, $H(t,x) \equiv Ax + tP_nFx + (1 - t)Fx - f \neq 0$ for $x \in \partial D$ and $t \in [0,1]$. If not, there would exist sequences $\{n_j\} \subset Z_+$, $\{x_j\} \subset \partial D$ and $\{t_j\} \subset [0,1]$ with $n_j \to \infty$ and $t_j \to t_0 \in [0,1]$ such that

$$Ax_j + t_jP_{n_j}Fx_j + (1 - t_j)Fx_j - f = 0. \tag{3.6}$$

Since $t_j \to t_0$ and $\{Fx_j\}$ is bounded, we see that

$$g_j \equiv Ax_j + t_0P_{n_j}Fx_j + (1 - t_0)Fx_j - f \to 0 \quad \text{as } j \to \infty. \tag{3.7}$$

Now, since A is bijective and (3.4) implies the inequality

$$\| Ax - Ay \| \geq \mu\| x - y \| \quad \text{for all } x,y \in X, \tag{3.8}$$

we deduce easily from (3.8) that $\beta(\{Ax_j\}) \geq \mu\beta(\{x_j\})$. Hence it follows from this (3.7) and Corollary II.2.1 that

$$\mu\beta(\{x_j\}) \leq \beta(\{Ax_j\}) \leq \beta\{g_j + f - t_0P_{n_j}Fx_j + (1 - t_0)Fx_j\}$$
$$\leq 0 + t_0\beta(\{P_{n_j}Fx_{n_j}\}) + (1 - t_0)\beta(\{Fx_j\})$$
$$\leq \beta(\{Fx_j\}) \leq k\beta(\{x_j\}),$$

which leads to a contradiction unless $\beta(\{x_j\}) = 0$. We conclude that $\{x_j\}$ has a convergent subsequence, which we again denote by $\{x_j\}$, such that $x_j \to x \in \partial D$ as $j \to \infty$. It follows from this, the continuity of A and F, the completeness of Γ_P, and (3.7) that $Ax + Fx = f$ with $x \in \partial D$, in contradiction to $f \notin (A + F)(\partial D)$.

Now, since P_nF is compact, F is k-ball contractive with $k < \mu$, A surjective, and

$$\| P_n^*Ax - P_n^*Ay \| \geq \mu\| x - y \| \tag{3.9}$$
$$\text{for all } x,y \in X \text{ and } n \in Z_+,$$

it follows from Theorem I.2.2 that $H(t,\cdot)$ is A-proper w.r.t. Γ_P for each $t \in [0,1]$ and continuous in t, uniformly for $x \in \overline{D}$. Hence $H: [0,1] \times \overline{D}$

$\rightarrow X^*$ is an A-proper homotopy (w.r.t. Γ_P) and therefore, by Theorem 2.1,

$$\text{Deg}(A + F, D, f) = \text{Deg}(A + P_nF, D, f) \qquad \text{for each } n \geq n_0.$$

$$(3.10)$$

Now, since P_nF $(n \geq n_0)$ is compact and A satisfies all the conditions of Theorem 2.4, it follows from the latter that

$$\text{Deg}(A + P_nF, D, f) = \{\deg_{\text{LS}}(I + P_nFA^{-1}, A(D), f)\}$$

$$(3.11)$$
$$\text{for } n \geq n_0.$$

Since $A^{-1}: X^* \to X$ is such that $\| A^{-1}u - A^{-1}v \| \leq (1/\mu)\| u - v \|$ for all $u, v \in X^*$ and $F: X^* \to X$ is k-ball contractive, it follows that $\tilde{F} \equiv FA^{-1}: X^* \to X^*$ is $\tilde{k}$-ball contractive with $\tilde{k} = k/\mu < 1$. Hence if for each fixed $n \geq n_0$ we let

$$F(t,u) = tP_n\tilde{F}u + (1 - t)\tilde{F}u, \qquad u \in \overline{A(D)}, \quad t \in [0,1],$$

then since $P_n\tilde{F}$ is compact and $\tilde{F}$ $\tilde{k}$-ball contractive, $F(t,\cdot)$ is also $\tilde{k}$-ball contractive for each $t \in [0,1]$ and continuous in t, uniformly for $u \in \overline{A(D)}$. Further, since $\tilde{F} = FA^{-1}$, it follows from the discussion above that $u + F(t,u) \neq f$ for $u \in \partial A(D) = A(\partial D)$. Hence, by the homotopy theorem for the $\tilde{k}$-ball-contractive vector fields with $\tilde{k} < 1$, $\deg_{\text{B}}(I + F(0,\cdot), A(D), f) = \deg_{\text{B}}(I + F(1,\cdot), A(D), f)$, $A(D), f)$, that is,

$$\deg_{\text{B}}(I + \tilde{F}, A(D), f) = \deg_{\text{B}}(I + P_n\tilde{F}, A(D), f)$$

$$= \deg_{\text{LS}}(I + P_n\tilde{F}, A(D), f) \qquad \text{for } n \geq n_0.$$

$$(3.12)$$

Combining (3.12) with (3.11) and (3.10), we get (3.5).   Q.E.D.

3.2 Compact and k-Ball-Contractive Perturbations of Firmly Accretive Mappings

In this section we consider mappings from X to X, where X is a Banach space with a projectionally complete scheme $\Gamma_1 = \{X_n, P_n\}$ for (X, X). As in the preceding section, we note that Γ_1 satisfies condition (p1) and one can show (e.g., see Browder [36]) that if G_n is any open subset of X_n and $R: G_n \to X_n$ is continuous and accretive, then R is orientation preserving. Before we prove the analogs of Theorems 3.2 and 3.3 for mappings acting in X, we must establish some results for mappings of accretive type which are of interest in their own right.

The aim of this section is to discuss briefly firmly accretive mappings and, in particular, to deduce from Theorem 2.4 the following new result and its special cases.

Theorem 3.4   Let X and X^* be uniformly convex Banach spaces, $\Gamma_1 = \{X_n, P_n\}$ a nested projectionally complete scheme for (X,X), D a bounded open subset of X, $C: \overline{D} \to X$ compact, and $A: X \to X$ a continuous or at least demicontinuous strongly accretive mapping.
   Then $T = A + C: \overline{D} \to X$ is A-proper w.r.t. Γ_1 and

$$\mathrm{Deg}(T + C, D, f) = \{\deg_{LS}(I + CA^{-1}, A(D), f)\} \quad \text{for} \quad f \notin T(\partial D).$$

To verify the conditions imposed on A and A_n in Theorem 2.4 when $Y = X$ and $\Gamma = \Gamma_1$, we must utilize some mapping theorems for accretive operators which are important in their own right for reasons mentioned in Section III.4.
   The main result from the theory of accretive mappings that we need is Theorem III.4.3.

Proof of Theorem 3.4   First, since X^* is uniformly convex, it follows from Lemma I.2.5 that the normalized duality map $J: X \to X^*$ is single-valued, uniformly continuous on bounded sets in X, and $P_n^* J x = J x$ for all x in X_n and each $n \in Z^+$. It follows from the inequalities

$$(Ax - Ay, J(x - y)) \geq \mu \| x - y \|^2 \quad \text{for all} \quad x,y \in X \qquad (3.13)$$

$$\| Ax - Ay \| \geq \mu \| x - y \| \quad \text{for all} \quad x,y \in X, \qquad (3.14)$$

that A is A-proper by Theorem III.4.3 and also that A is a homeomorphism of X onto X such that for any bounded open subset $D \subset X$, A is a homeomorphism of $\overline{D}$ into X, and $A(D)$ is open in X, by results in [63]. Moreover, since $P_n^* J x = J x$ for all $x \in X_n$, it follows from (3.18) that

$$(P_n Ax - P_n Ay, J(x - y))$$
$$\geq \mu \| x - y \|^2 \quad \text{for all} \quad x,y \in X_n \quad \text{and} \quad n \in Z_+, \qquad (3.15)$$

and consequently,

$$\| P_n Ax - P_n Ax \|$$
$$\geq \mu \| x - y \| \quad \text{for all} \quad x,y \in X_n \quad \text{and} \quad n \in Z_+. \qquad (3.16)$$

The inequalities above and the accretiveness of $A_n = P_n A|_{X_n}$ imply that for each n, A_n is an orientation-preserving homeomorphism of $D_n = D \cap X_n$ onto the open subset $A_n(D_n)$ of X_n and A_n maps D_n homeomorphically onto $\overline{A_n(D_n)}$. Hence, by Theorem 2.4, there exists an $n_1 \in Z_+$ such that $\deg(T_n, D_n, P_n f) = \deg_{LS}(I + CA^{-1}, A(D), f)$ for each $f \notin T(\partial D)$

and each $n \geq n_1$ and, in particular, $\mathrm{Deg}(T,D,f)$ is the singleton integer $\{\mathrm{deg}_{LS}(I + CA^{-1}, A(D), f)\}$. Q.E.D.

Exercise 3.1 Show that if V is a finite-dimensional real Banach space, G an open subset of V, and F an accretive mapping of G into V, then F is orientation preserving.

In view of Theorems 3.5 and Theorems I.1.5 and I.2.2, using Theorem 3.5 and exactly the same argument as that in the proof of Theorem 3.4, we have the following extension.

Theorem 3.5 Let X, D, and $\Gamma_1 = \{X_n, P_n\}$ be as in Theorem 3.4, $A: X \to X$ continuous and strongly accretive, that is,

$$(Ax - Ay, J(x - y)) \qquad (3.17)$$
$$\geq \mu\| x - y \|^2 \quad \text{for all} \quad x,y \in X \quad \text{and some} \quad \mu > 0,$$

$F: X \to X$ k-ball contractive with $k < \mu$, and $f \notin (A + F)(\partial D)$. Then $T = A + F: \overline{D} \to X$ is A-proper w.r.t. Γ_1 and $\mathrm{Deg}(T + F, D, f) = \{\mathrm{deg}_B(I + FA^{-1}, A(D), f)\}$.

Remark 3.1 (i) Note that if $S: X \to X$ is c-contractive with $c < 1$ (i.e., $\| Sx - Sy \| \leq c\| x - y \| \forall x,y \in X$), then $A = I - S$ satisfies (3.13), so in this case Theorem 3.6 is an extension of Corollary 2.9(ii).

(ii) If $A = I$, then Theorem 3.5 reduces to the corresponding result of Webb [291], which in a somewhat different form was given earlier by Nussbaum [191].

(iii) It should be noted that some of these results could be extended somewhat if one used the approach and results of Deimling [63].

4. SOME APPLICATIONS OF THE GENERALIZED DEGREE

This section is devoted to the application of the generalized degree to existence and mapping theorems involving A-proper mappings $T: D \subset X \to Y$ defined on proper subsets of X. We are concerned primarily with three types of results: existence and antipodes theorems, structure of solution set theorems, and invariance of domain theorems.

Remark 4.1 Before we state some applications of the generalized degree given by Definition 2.1, we should point out that unlike other topological degrees, it embodies in its very structure a constructive aspect when applied to the solvability of a given equation $Tx = f$ for $x \in D$. This is obviously true when $\text{Deg}(T,D,f)$ is single valued, but it also happens in more general situations. For example, if $0 \notin \text{Deg}(T,D,f)$, there exists $n_0 \in Z_+$ such that $\deg\{T_n,D_n,W_nf\} \neq \{0\}$ for each $n \geq n_0$. Hence for each $n \geq n_0$ there is a $x_n \in D_n$ such that $T_n(x_n) - W_nf = 0$. This and the A-properness of T w.r.t. Γ imply the existence of a subsequence $\{x_{n_j}\}$ of $\{x_n\}$ and an $x_0 \in D$ such that $x_{n_j} \to x_0$ as $j \to \infty$ and $Tx_0 = f$, and in some cases (see Theorem 3.1A in [225]) we can even assert that $x_n \to x_0$ as $n \to \infty$. This is, of course, more than just the assertion that $Tx = f$ has a solution in D of $0 \notin \text{Deg}(T,D,f)$.

4.1   Zeros, Fixed Points, and Antipodes Theorems

In this section we use the generalized degree theory developed in Section 2 to obtain the existence of zeros for A-proper mappings, fixed point for P_1-compact mappings, and antipodes theorems for A-proper mappings of convex type.

Our first result is the following theorem obtained in [202].

Theorem 4.1   Let $D \subset X$ be bounded and open and suppose that $T: \overline{D} \to X$ is a mapping such that:

(A1) $T + \lambda I$ is A-proper w.r.t. Γ_α for each $\lambda \geq 0$ and $T(\partial D)$ bounded.
(A2) There is $w \in D$ such that $Tx + \lambda(x - w) \neq 0$ for $x \in \partial D$ and $\lambda > 0$.

Then there exists $x_0 \in \overline{D}$ such that x_0 satisfies the equation

$$Tx = 0. \tag{4.1}$$

If (A2) holds for $\lambda \geq 0$, then (4.1) is feebly approximation solvable w.r.t. Γ_α and strongly if x_0 is unique.

Proof.   We may assume that $Tx \neq 0$ for $x \in \partial D$, for otherwise there would be nothing further to prove. For each x in $\overline{D}$ and $t \in [0,1]$ we define the mapping $T_t(x) = tTx + (1 - t)(x - w)$. Since, for $x \in \partial D$ and t and s in $[0,1]$, $\| T_tx - T_sx \| \leq | t - s | \| Tx - x + w \|$ and $T(\partial D)$ is bounded, it follows that $T_t(x)$ is continuous in t, uniformly for $x \in \partial D$. Furthermore, for each $t \in [0,1]$, T_t is A-proper. Indeed, for any $t \in (0,1]$, the map

$$T_t = t\left[T + \left(\frac{1 - t}{t}\right) I - \left(\frac{1 - t}{t}\right) w \right]$$

is A-proper since $T + \lambda I - \lambda w$ [with $\lambda = (t - 1)/t > 0$] is A-proper, by (A1), and A-properness is invariant under multiplication by nonzero reals. For $t = 0$, $T_1 = I - w$ is obviously A-proper. To complete the verification of the hypotheses of Theorem 2.1(3), we have still to show that $T_t(x) \neq$

0 for $x \in \partial D$ and $t \in [0,1]$. Suppose that $T_t(x) = 0$ for some $x \in \partial D$. Clearly, the equality above cannot hold when $t = 0$ or $t = 1$. If $t \in (0,1)$, then $Tx + \lambda(x - w) = 0$ for $\lambda = (t - 1)/t > 0$, in contradiction to (A2). Hence Theorem 2.1 implies that $\mathrm{Deg}(T_t,D,0)$ is independent if $t \in [0,1]$. Consequently, since $T_0 = I - w$ and $T_1 = T$, $\mathrm{Deg}(T,D,0) = \mathrm{Deg}(I - w, D, 0) = \{1\}$ because $w \in D$.

Now, since $\mathrm{Deg}(T,D,0) = \{1\}$, it follows from Definition 2.1 that there is an $n_0 \in Z_+$ such that $\deg(T_n,D_n,0) \neq 0$ for each $n \geq n_0$. Hence there exists $x_n \in \overline{D}_n$ such that $T_n(x_n) = 0$ for $n \geq n_0$. Hence, by the A-properness of T, there exists a subsequence $\{x_n\}$, and $x_0 \in D$ such that $x_{n_j} \to x_0$ and $Tx_0 = 0$. Finally, if x_0 is unique, then the A-properness implies in this case that $x_n \to x_0$ in X [i.e., the Galerkin method is applicable to (4.1)]. Q.E.D.

As a consequence we deduce from Theorem 4.1 the following essentially constructive generalization of the fixed-point theorem (Theorem III.1.6) stated in [206] and proved in [236].

Theorem 4.2 Let $D \subset X$ be bounded and open and suppose that $F: \overline{D} \to X$ is P_1-compact, $F(\partial D)$ is bounded, and

(LS) There is $w \in D$ such that $Fx - w \neq \eta(x - w)$ for $x \in \partial$ and $\eta > 1$.

Then the conclusions of Theorem 4.1 hold for the equation

$$x - Fx = 0. \tag{4.2}$$

Proof. If in Theorem 4.1 we let $T = I - F$, then P_1-compactness of F means that $T + \lambda I$ is A-proper for each $\lambda \geq 0$, while the boundedness of $F(\partial D)$ implies the same for $T(\partial D)$ [i.e., (A1) holds]. On the other hand, the condition (LS) is equivalent to the condition $Fx - x \neq (\eta - 1)(x - w)$; that is, $Tx + \lambda(x - w) \neq 0$ for $x \in \partial D$ and $\lambda = \eta - 1 > 0$ [i.e., (A2) holds]. Note that $T_t(x) = tTx + (1 - t)(x - w) = x - w - t(Fx - w)$ for $x \in \overline{D}$ and $t \in [0,1]$. Hence Theorem 4.2 follows from Theorem 4.1. Q.E.D.

Remark 4.2 It was noted in Remark 1.1 that (LS) is satisfied if D is convex and $F(\partial D) \subset \overline{D}$ since otherwise there would exist $x \in \partial D$ and $\eta > 1$ such that $Fx - w = \eta(x - w)$ or $x = (1/\eta)Fx + (1 - 1/\eta)w$ with $1/\eta < 1$, contradicting the fact that $x \in \partial D$. Condition (LS) also holds if either (1.5), (1.6), or (1.7) holds. Thus Theorem 4.2 is valid when (LS) is replaced by any one of the boundary conditions above. Clearly, Theorem 4.2 holds if F is P_γ-compact for some $\gamma \in [0,1]$ or if F is ball condensing. Moreover, by virtue of Proposition III.2.1, the following holds.

Corollary 4.1 Suppose that $\Gamma_1 = \{X_n, P_n\}$ is nested. Let $D \subset X$ be bounded and open, $C: \overline{D} \to X$ compact, $S: \overline{D} \to X$ ball condensing, and $A: X \to X$ accretive, bounded, and continuous. If $F = S + C - A$ satisfies condition (LS) on ∂D, then $x - Fx = 0$ is feebly approximation solvable w.r.t. Γ_1 in, and in particular, F has a fixed point in D.

In Corollary 4.1, S can also be k-set contractive with $k < \frac{1}{2}$ or k-ball contractive with $k < 1$. For further extensions, see [297].

We know that, in general, a convex linear combination of two A-proper mappings is not A-proper. In view of this fact, the following definition will be convenient.

Definition 4.1 Let $D \subset X$ be open and symmetric about $0 \in D$. If $T: \overline{D} \to Y$ is a map such that $H: [0,1] \times \overline{D} \to Y$ given by

$$H_t(x) = \frac{1}{1 + t} T(x) - \frac{t}{1 + t} T(-x)$$

$$\text{for} \quad x \in \overline{D} \quad \text{and} \quad t \in [0,1] \tag{4.3}$$

is A-proper w.r.t. $\Gamma = \{X_n, V_n; E_n, W_n\}$ for each $t \in [0,1]$, then T is said to be of *convex A-proper type*.

Remark 4.3 It is obvious that if $F: \overline{D} \to X$ is k-ball contractive with $k < 1$, then $T = I - F$ is of convex A-proper type w.r.t. $\Gamma_1 = \{X_n, P_n\}$.

It is not difficult to show that if $A: X \to X$ is continuous and strongly accretive [i.e., $(Ax - Ay, J(x - y)) \geq \mu \| x - y \|$ for all $x,y \in X$ and some $\mu > 0$] and $F: \overline{D} \subset X \to X$ is k-ball contractive with $k < \mu$, then $T = A - F: \overline{D} \to X$ is of convex A-proper type w.r.t. Γ_1 provided that $X_n \subset X_{n+1}$ for each n. This follows from the fact that $(A_t(x) - A_t(y), J(x - y)) \geq \mu \| x - y \|^2$ for all $x,y \in X$ and $t \in [0,1]$ and $\beta(F_t(Q)) \leq k\beta(Q)$ for each $Q \subset \overline{D}$, where

$$A_t(x) = \frac{1}{1 + t} A(t) - \frac{t}{1 + t} A(-x)$$

$$\text{and} \quad F_t(x) = \frac{1}{1 + t} F(x) - \frac{t}{1 + t} F(-x).$$

It was shown in Theorem I.2.4 that if X is reflexive and $T: \overline{D} \subset X \to Y$ is demicontinuous and of type (KS_+), then under certain conditions on $K: X \to Y^*$ and $K_n: X_n \to D(W_n^*)$, the operator T is of convex A-proper type. In view of Theorem I.2.3, the same is true when $T: X \to Y$ is continuous and of modified type (KS_+). This is also the case, by Corollary I.2.5, when $T: X \to Y$ satisfies the inequality (2.9) of that corollary.

Consequently, all results proved below for mappings that are of convex A-proper type are valid for the various classes of mappings mentioned in Remark 4.3.

Using Theorem 2.1 and Remark 4.1, we now establish a couple of results that properly include some basic classical theorems for compact vector fields and recent theorems for k-ball-contractive vector fields. These results also indicate the importance of finding conditions for a given mapping T to be of convex A-proper type. See the author's paper [229] for an exhaustive study of mappings of convex A-proper type.

We start with the following essentially constructive *antipodes theorem* established in [229], which includes the classical result for compact vector fields and others.

Theorem 4.3 Let $D \subset X$ be bounded, open, and symmetric about $0 \in D$. If $T: \overline{D} \to Y$ is a mapping of convex A-proper type w.r.t. Γ such that $\| T(x) + T(-x) \| \le M$ for $x \in \partial D$ and

$$Tx \ne \lambda T(-x) \quad \text{for} \quad x \in \partial D \quad \text{and} \quad \lambda \in [0,1], \tag{4.4}$$

then equation (4.1), $Tx = 0$, is feebly approximation solvable w.r.t. Γ and strongly approximation solvable if T is one-to-one.

Proof. Consider the homotopy

$$H(t,x) = \frac{1}{1+t} T(x) - \frac{t}{1+t} T(-x) \quad \text{for} \quad x \in \overline{D} \quad \text{and} \quad t \in [0,1].$$

Since T is of convex A-proper type, $A(t,\cdot)$ is A-proper w.r.t. Γ for each $t \in [0,1]$. The definition of H and (4.4) imply that $0 \notin H_t(\partial D)$ for each $t \in [0,1]$. Hence $\mathrm{Deg}(H_t,D,0)$ is well defined for each $t \in [0,1]$. To show that $\mathrm{Deg}(H_t,D,0)$ is independent of $t \in [0,1]$, by Remark 2.1, it suffices to show that $H_t(x)$ is continuous in $t \in [0,1]$, uniformly for $x \in \partial D$. This is, however, the case since $\| T(x) + T(-x) \| \le M$ for all $x \in \partial D$ and

$$H_t(x) - H_s(x) = \frac{s-t}{(1+s)(1+t)} \{T(x) + T(-x)\}$$

$$\text{for} \quad x \in \partial D \quad \text{and} \quad t,s \in [0,1].$$

Thus, by Theorem 2.1, $\mathrm{Deg}(H_0,D,0) = \mathrm{Deg}(T,D,0) = \mathrm{Deg}(H_1,D,0)$, where $H_1(x) = \frac{1}{2}T(x) - \frac{1}{2}T(-x)$ is an odd mapping on $\overline{D}$. Hence $0 \notin \mathrm{Deg}(T,D,0)$ and therefore, by Remark 4.1, equation (4.1) (i.e., $Tx = 0$) is feebly approximation solvable w.r.t. Γ, and strongly if T is injective on $\overline{D}$. Q.E.D.

By virtue of Remark 4.3, Theorem 4.3 implies the validity of the following:

Corollary 4.2   Let $D \subset X$ be as in Theorem 4.3 and $T: \overline{D} \to Y$ a mapping such that (4.4) holds. Then the conclusions of Theorem 4.3 hold provided that any one of the following hypotheses are satisfied:

(h1) $T = I - F$, where $F: \overline{D} \to X$ is k-ball contractive with $k < 1$ (in particular, F is compact) and $\Gamma = \Gamma_1 = \{X_n, P_n\}$.

(h2) $F: \overline{D} \to X$ is k-ball contractive and $A: X \to X$ a continuous, bounded, and accretive mapping such that $k < \mu$ and Γ_1 is nested.

(h3) X is reflexive and $T: \overline{D} \to Y$ is a demicontinuous, bounded mapping of type (KS_+) for which the conditions of Theorem I.2.4 hold.

(h4) X is reflexive and $T: X \to Y$ is a continuous, bounded mapping that is either of modified type (KS_+), for which the conditions of Theorem I.2.3 hold, or T satisfies the inequality (2.9) of Corollary I.2.5.

Note that Corollary 4.2, under assumption (h1), contains the classical antipodes theorem when F is compact and a result of Nussbaum [191] when F is k-ball contractive with $k < 1$. Note also that the coresponding results of [36] when T is of type $(S+)$ are also special cases of Theorem 4.3. Another consequence of Theorem 4.3 is the following result of the author [229], whose first part extends some results of Leray and Schauder [153] (see Corollary 4.2 below) and whose second part extends the classical theorem of Borsuk (see [136,261]).

Theorem 4.4   Let $D \subset X$ be as in Theorem 4.3 and $T: \overline{D} \to Y$ a mapping of convex A-proper type with $\| T(x) + T(-x) \| \le M$ for $x \in \partial D$. Suppose that either (A3) or (A4) holds, where

(A3) There is an odd map $A: \overline{D} \to Y$ such that $\| Tx - Ax \| < \| Ax \|$ for $x \in \partial D$.

(A4) $Tx \ne 0$ and $Tx/\| Tx \| \ne T(-x)/\| T(-x) \|$ for $x \in \partial D$.

Then, in either case, Eq. (4.1) is feebly approximation solvable in D.

If $f \in Y$ is such that $\| Tx \| > \| f \|$ for $x \in \partial D$, the equation $Tx = f$ is feebly approximation solvable in D (in particular, there is $x_0 \in D$ such that $Tx_0 = f$) if either (A2) or (A4) holds.

Proof.   To prove Theorem 4.4 it suffices to show that (A3) and (A4) imply the inequality (4.4). Suppose first that (A3) holds. Then it follows from the oddness of A that for all $x \in \partial D$ and $\lambda \in [0,1]$, we have

$$\| T(x) - \lambda T(-x) \|$$
$$= \| T(x) - A(x) - \lambda\{T(-x) - A(-x)\} + (1 + \lambda)A(x) \|$$
$$\geq (1 + \lambda)\| Ax \| - \| Tx - Ax \| - \lambda\| T(-x) - A(-x) \|$$
$$> (1 + \lambda)\|Ax\| - \| Ax \| - \lambda\| Ax \| = 0$$

[i.e., (4.4) holds and hence Theorem 4.4 follows from Theorem 4.3 if (A3) holds]. Suppose now that (A4) holds. Then (4.4) clearly holds since the equality $Tx_0 = \lambda_0 T(-x_0)$ for some $\lambda_0 \in [0,1]$ and $x_0 \in \partial D$ would contradict assumption (A4). So again Theorem 4.4 follows from Theorem 4.3 when (A4) holds.

To prove the second part of Theorem 4.4, note that $H(t,x)$ given by (4.3) is an A-proper homotopy w.r.t. Γ, $H(t,x) \neq 0$ for $x \in \partial D$ and $t \in [0,1]$ and $H(1,x) = \frac{1}{2}T(x) - \frac{1}{2}T(-x)$ is odd. Hence $0 \notin \mathrm{Deg}(H(1,\cdot),D,0)$ $= \mathrm{Deg}(T,D,0)$. Now, for a given f in Y such that $\| Tx \| > \| f \|$ for $x \in \partial D$, consider the homotopy $F(t,x) = Tx - tf$ for $x \in \overline{D}$ and $t \in [0,1]$. It is easy to see that F_t is an A-proper homotopy and $0 \notin F(t,\partial D)$ for $t \in [0,1]$. Hence $\mathrm{Deg}(F(t,\cdot),D,0)$ is well defined and independent of $t \in [0,1]$. Consequently,

$$\mathrm{Deg}(F(1,\cdot),D,0) = \mathrm{Deg}(F(0,\cdot), D,0) = \mathrm{Deg}(T,D,0) \not\ni 0,$$

from which the assertion of the second part of Theorem 4.4 follows.

$$\text{Q.E.D.}$$

An immediate corollary of Theorem 4.4 [when (A3) holds] is the following result, which includes the Leray–Schauder theorem, where F is compact (see [136]).

Corollary 4.3 Suppose that $Y = X$, $\Gamma = \Gamma_1 = \{X_n, P_n\}$, and $F: \overline{D} = \overline{B}(0,r) \to X$ is a k-ball-contractive mapping with $k < 1$ such that there exists a linear mapping $L: X \to X$ with the property that for some fixed $\lambda \neq 0$ we have

(A3a) $\| Fx - \lambda Lx \| < \| x - \lambda Lx \|$ for $x \in \partial D$.
Then $Fx - x = 0$ is feebly approximation solvable on D.

Proof. If in Theorem 4.4 we put $Y = X$, $T = F - I$, and $A = I - \lambda L$, then A is odd and T is a bounded mapping of $\overline{D}$ into X such that for each

$t \in [0,1]$, the mapping

$$H(t,x) = \frac{1}{1+t} T(x) - \frac{t}{1+t} T(-x)$$

$$= \frac{1}{1+t} F(x) - \frac{t}{1+t} F(-x) - x \qquad (x \in \overline{D}, t \in [0,1])$$

is A-proper for each $t \in [0,1]$ and, of course, continuous in $t \in [0,1]$, uniformly for $x \in \partial D$. Moreover, the hypotheses (A3) of Theorem 4.4 holds since, by (A3a), we have

$$\| Tx - Ax \| = \| Fx - \lambda Lx \| < \| x - \lambda Lx \| = \| Ax \| \qquad \text{for} \quad x \in \partial D.$$

Hence Corollary 4.4 follows from the first part of Theorem 4.4.

$$\text{Q.E.D.}$$

Some applications of Theorems 4.1 to 4.4 will be given in subsequent sections (see also [229]).

4.2 On the Structure of the Set of Solutions of Nonlinear Equations

For a given $T: D \subset X \to Y$ and f in Y, consider the equation

$$T(x) = f \qquad (x \in D). \tag{4.5}$$

When (4.5) does have a solution, it is of interest to know the structure of the set of solutions. Clearly, the simplest situation occurs when (4.5) has only one solution. In this section we give conditions which will guarantee that the set of solutions of (4.5) is a *continuum* (i.e., it is nonempty, compact, and connected). The degree approach used here was initiated by Krasnoselskii and Sobelevskii [138] when $f = 0$ and $T = I - C$ with C compact. It was then extended by Deimling [61] to P_1-compact and by the author (see [76]) to k-set-contractive mappings. In [81,76] Fitzpatrick extended this approach to (4.5) (with $f = 0$) when T is an A-proper mapping or even a uniform limit of such mappings. In view of the fact that the class of A-proper mappings is quite large, the results of [76,81] unified and extended the corresponding results in [61,138] with some results in [36] and in Vidossich [285].

We start with the following general result proved in [76].

Theorem 4.5 Let $D \subset X$ be open and bounded and $T: D \to Y$ A-proper w.r.t. Γ with the inverse images of points compact. Suppose that there exists a sequence $T^k: \overline{D} \to Y$ of A-proper mappings such that

$$r_k = \sup\{\| T^k(x) - T(x) \| \mid x \in \overline{D}\} \to 0 \quad \text{as} \quad k \to \infty. \tag{4.6}$$

$$T^k \text{ is one-to-one when restricted to } (T^k)^{-1}(B(f,r_k)). \tag{4.7}$$

Then, if $\text{Deg}(T,D,f) \neq \{0\}$, $T^{-1}(f)$ is a continuum.

Proof. By Theorem 2.1(2) we see that $T^{-1}(f) \neq \varnothing$, and by our assumption on T, $T^{-1}(f)$ is compact and hence closed. Denote $T^{-1}(f)$ by S. Our proof that S is connected will be by contradiction. Suppose that S is not connected. Then there exist disjoint closed sets S_1 and S_2 with $T^{-1}(f) = S_1 \cup S_2$, and open sets D_1 and D_2 lying in D, with disjoint closures, such that $S_i \subset D_i$ for $i = 1,2$. Since $Tx \neq f$ for $x \in D' \equiv D - D_1 \cup D_2$, it follows that $\text{Deg}(T,D',f) = \{0\}$ and hence, by Theorem 2.1(5),

$$\text{Deg}(T,D,f) \subseteq \text{Deg}(T,D_1,f) + \text{Deg}(T,D_2,f).$$

We shall now show that $\text{Deg}(T,D_i,f) = \{0\}$ for $i = 1,2$, and hence obtain a contradiction. Let us first show that $\text{Deg}(T,D_1,f) = 0$. Since $f \notin T(\partial D_1)$, it follows from Theorem 2.2(8) that there exists a constant $\gamma > 0$ such that if $A: \overline{D}_1 \to Y$ is A-proper and $\| Ax - Tx \| < \gamma$ for $x \in \partial D_1$, then $\text{Deg}(A,D_1,f) = \text{Deg}(T,D_1,f)$. Now choose $k_0 \in Z_+$ such that $2\delta_{k_0} < \gamma$, choose $a \in S_2$ and define the map $A: \overline{D}_1 \to Y$ by $A(x) = T^{k_0}(x) - T^{k_0}(a) + f$ for $x \in \overline{D}_1$. Note that

$$\| Ax - Tx \| \leq \| T^{k_0}(x) - T(x) - (T^{k_0}(a) - Ta) \|$$
$$\leq \| T^{k_0}(x) - Tx \| + \| T^{k_0}(a) - T(a) \| \leq 2r_{k_0} < \gamma,$$

and hence, by the preceding remark, $\text{Deg}(T,D_1,f) = \text{Deg}(A,D_1,f)$. Now it follows from this that $\text{Deg}(T,D_1,f) = \{0\}$, for if this were not so, then $\text{Deg}(A,D_1,f) \neq \{0\}$. Thus there would exist $x_1 \in D_1$ such that $A(x_1) = f$ with $x_1 \neq a$ [i.e., $T^{k_0}(x_1) - T^{k_0}(a) + f = f$]. Since $A = T(a)$, this implies that $\| T^{k_0}(x_1) - f \| = \| T^{k_0}(a) - T(a) \| \leq r_{k_0}$. This contradicts the fact that T^{k_0} is one-to-one on $(T^{k_0})^{-1}(B(f,r_{k_0}))$ and therefore $\text{Deg}(T,D_1,f) = \{0\}$. In a similar fashion one shows that $\text{Deg}(T,D_2,f) = \{0\}$. Thus we see that $\text{Deg}(T,D,0) = \{0\}$, so we have obtained the contradiction under the assumption that $T^{-1}(f)$ is disconnected. Hence $T^{-1}(f)$ is connected; since $T^{-1}(f) \neq \varnothing$ and is compact, it is a continuum. Q.E.D.

Let us add that Deimling [61] proved earlier a result similar to Theorem 4 (when $f = 0$) under the assumption that T is, in addition to being A-proper, continuous and maps $\overline{D}$ into X. Since, as Theorem I.11 shows, continuity and A-properness imply properness, the result of [61] follows from Theorem 4.5.

To illustrate the generality of Theorem 4.5 we shall now deduce from it a number of important corollaries, some of which are new results even in classical theory. We follow the arguments of [76].

Our first new result, which together with Corollary 4.5 was a motivation for Theorem 4.5, was proven by Vidossich [285] for the case when $D = B(x_0, r)$.

Corollary 4.4 Suppose that $D \subset X$ is open and bounded, $C: \overline{D} \to X$ compact, and $T = I - C$ is such that $f \notin T(\partial D)$, where $f = Tx_0$ for some $x_0 \in D$. Suppose that there exists a sequence $\{T^k\}$, with $T^k = I - C^k$: $\overline{D} \to X$ for each $k \in Z_+$, such that C^k is compact and each T^k is a homeomorphism. Then if $\{T^k\}$ converges uniformly to T on $\overline{D}$, $T^{-1}(f)$ is a continuum.

Proof. Since $Tx \neq f$ for $x \in \partial D$, we may, by Theorem 2.2(8), choose $\gamma > 0$ such that if $A: \overline{D} \to X$ is A-proper w.r.t. Γ_2 and $\| Ax - Tx \| < \gamma$ for $x \in \partial D$, then $\mathrm{Deg}(T, D, f) = \mathrm{Deg}(A, D, f)$. Now choose $k_0 \in Z_+$ such that $\sup\{\| Tx - T^{k_0}(x)\| \mid x \in \overline{D}\} < \gamma/2$ and define the map $A: \overline{D} \to X$ by $A(x) = T^{k_0}(x) - T^{k_0}(x_0) + f$ for $x \in \overline{D}$. Note that $\| Ax - Tx \| \leq \| T^{k_0}(x) - T(x) \| + \| T^{k_0}(x_0) - Tx_0 \| < \gamma/2 + \gamma/2 = \gamma$. Hence $\mathrm{Deg}(T, D, f) = \mathrm{Deg}(A, D, f)$ since $Ax = x - C^{k_0}(x) - T^{k_0}(x_0) + f$ is a compact vector field from $\overline{D}$ to X such that A is a one-to-one mapping of $\overline{D}$ onto $A(\overline{D})$ and $A(x_0) = x_0 - C^{k_0}(x_0) - x_0 + C^{k_0}(x_0) + f = f$, it follows from the analog of Theorem III.1.e for the Leray–Schauder degree that $\deg_{LS}(A, D, f) = \pm 1$. This and Theorem 2.4 imply that $\mathrm{Deg}(A, D, f) = \{\deg_{LS}(A, D, f)\} \neq \{0\}$, and consequently, $\mathrm{Deg}(T, D, f) \neq \{0\}$. Since all the hypotheses of Theorem 4.5 are satisfied, we know that $T^{-1}(f)$ is a continuum.   Q.E.D.

An immediate consequence of Theorem 4.5 is the following result proved by the author in 1970 for ball-condensing and set-condensing mappings, which includes the classical result of [138] when the maps are compact and $f = 0$.

Corollary 4.5 Let $D \subset X$ be open and bounded and $F: \overline{D} \to X$ ball condensing and such that $\mathrm{Deg}(I - F, D, f)$ is well defined and $\neq \{0\}$ for some $f \in (I - F)(D)$. Suppose that there exists a sequence $\{F^k\}$ of ball-condensing mappings of $\overline{D}$ into X such that $\| F^k(x) - F(x) \| \leq r_k$ for $x \in \overline{D}$ and $k \in Z_+$ with $r_k \to 0$ as $k \to \infty$, and $T^k = I - F^k$ is one-to-one when restricted to $(T^k)^{-1}(B(f, r_k))$. Then $(I - F)^{-1}(f)$ is a continuum.

Our next consequence of Theorem 4.5 is the following result, proved in [61] when $f = 0$.

Proposition 4.1 Let $D \subset X$ be open and bounded and $F: \overline{D} \to X$ P_1-compact with $T = I - F$ accretive w.r.t. some duality map J and proper. Then if Deg (T,D,f) for some $f \in T(D)$ is well defined and $\neq \{0\}$, $(I - F)^{-1}(f)$ is a continuum.

Proof. Since F is P_1-compact we see that $\lambda I - F$ is A-proper for each $\lambda \geq 1$. Now define $T^k: \overline{D} \to X$ by $T^k(x) = (1 + 1/k)(x) - F(x)$. Thus, for each $k \in Z_+$, T^k is A-proper and $(T^k(x) - T^k(y),w) \geq (1/k) \phi(\| x - y \|)\| x - y \|) > 0$ for $x,y \in \overline{D}$, $w \in J(x - y)$, where $x \neq y$. Hence T^k is one-to-one for each k. Since T^k converges uniformly to $T = I - F$ on $\overline{D}$, we may invoke Theorem 4.5 to conclude that $(I - F)^{-1}(f)$ is a continuum.   Q.E.D.

If $f = 0$, then as in [76], one deduces the following corollary of Proposition 4.1.

Corollary 4.6 Suppose that $F: \overline{D} \to X$ is P_1-compact, $F(\partial D)$ is bounded, and $T = I - F$ is proper and accretive. Then, if $(I - F)^{-1}(0) \neq \varnothing$ and $Fx \neq x$ for $x \in \partial D$, $(I - F)^{-1}(0)$ is a continuum.

Proof. To apply Proposition 4.1, it suffices to show that Deg$(I - F, D, 0) \neq \{0\}$. Let $x_0 \in (I - F)^{-1}(0)$ and let $H: [0,1] \times \overline{D} \to X$ be defined by $H(t,x) = t(x - Fx) + (1 - t)(x - x_0)$. Since $H(t,x_0) = 0$ for all $t \in [0,1]$, $x_0 - Fx_0 = 0$, $x - Fx \neq 0$ for $x \in \partial D$, and $I - F$ is accretive, it follows easily that $H((t,x),w) \geq (1 - t)\phi(\| x - x_0 \|)\| x - x_0 \| > 0$ for $w \in J(x - x_0)$ and $t \in [0,1)$. Hence $H(t,x) \neq 0$ for $t \in [0,1]$ and $x \in \partial D$, so, by Theorem 2.1, Deg$(I - F, D, 0) = $ Deg$(I - x_0, D, 0) = \{1\}$.
                                                                    Q.E.D.

Another consequence of Theorem 4.5 is the following result proved in [61] for D convex and in [76] for general D.

Proposition 4.2 Let $D \subset X$ be open and bounded and $F: \overline{D} \to X$ continuous and P_1-compact, and let $F(\partial D)$ be bounded. Assume that

(LS) There is $x_0 \in D$ such that $Tx - x_0 \neq \eta(x - x_0)$ for $x \in \partial D$ and $\eta \geq 1$.

Suppose that there exists a sequence $\{T^k\}$ of continuous A-proper mappings of $\overline{D}$ to X such that $r_k = \sup\{\| T^k(x) - T(x) \| \mid x \in \partial D\} \to 0$ as $k \to \infty$ and T^k is injective on $(T^k)^{-1}(B(0,r_k))$.
    Then $(I - F)^{-1}(0)$ is a continuum.

Proof. All we need is to show that Deg$(I - F, D, 0) \neq \{0\}$ and then use Theorem 4.5.

Define $H: [0,1] \times \overline{D} \to X$ by $H(t,x) = (x - x_0) - t(Fx - x_0)$ for $t \in [0,1]$ and $x \in \overline{D}$. Then $H(t,\cdot)$ is A-proper for each $t \in [0,1]$, and H is continuous in $t \in [0,1]$, uniformly for $x \in \partial D$. Further, condition (LS) implies that $H(t,x) \neq 0$ for $x \in \partial D$ and $t \in [0,1]$. Hence, by Theorem 2.1, $\mathrm{Deg}(I - F, D, 0) = \mathrm{Deg}(I - x_0, D, 0) = \{1\}$.   Q.E.D.

Remark 4.4 Propositions 4.1 and 4.2 and Corollary 4.6 are applicable to mappings $F: \overline{D} \to X$ of the form $F = S + C - A$, where $S: \overline{D} \to X$ is ball condensing, $C: \overline{D} \to X$ is compact, and $A: X \to X$ is continuous, accretive, and bounded since F is P_1-compact and $I - F$ is proper.

We complete this section with the following observation. If $F: D \to X$ is nonexpansive (i.e., $\| Fx - Fy \| \leq \| x - y \|$ for $x,y \in D$) and J is any duality map defined on X, then $I - F$ is accretive. Thus the results we have derived above are applicable to the study of fixed points of nonexpansive mappings. However, this is of interest only when X is not strictly convex, for when X is strictly convex and D is convex, closed, and bounded, Schaefer [255] has shown that the fixed points of nonexpansive mappings form not only a connected, but also a convex set. On the other hand, De Marr has shown that if X is not strictly convex, this is not always the case. In view of these remarks, Proposition 4.2 implies the following (see [76]).

Proposition 4.3 Let $D \subset X$ be bounded, open, and $0 \in D$. If $F: \overline{D} \to X$ is nonexpansive, P_1-compact, and satisfies (LS) with $x_0 = 0$, then $(I - F)^{-1}(0)$ is a continuum.

Finally, Corollary 4.6 and the remarks above imply the following result (see [285] when F is compact).

Corollary 4.7 Suppose that $F: D \to X$ is ball condensing and nonexpansive. If $(I - F)^{-1}(0) \neq \varnothing$ and $x \neq Fx$ for $x \in \partial D$, then $(I - F)^{-1}(0)$ is a continuum.

4.3 Invariance of Domain Theorems

In this section we first establish the invariance of domain theorem for a locally A-proper mapping $T: D \subset X \to Y$ obtained by the author in [222] with the additional assumption that T is continuous. We then deduce from it some classical as well as some recent invariance of domain theorems for compact and noncompact vector fields. We complete this section by proving the invariance of domain theorem due to Fitzpatrick [74] for uniform limits of A-proper mappings.

Suppose that (P) denotes a certain property and let D be an open set in X. We say that $T: D \to Y$ has a local property (P) if to each $x_0 \in D$ there exists a ball $\overline{B}(x_0,r) \subset D$ such that $T: \overline{B}(x_0,r) \to Y$ has property (P).

Our first basic result is the following (see [222]).

Theorem 4.6 (Invariance of domain) Let $D \subset X$ be open and T a map of D into Y. If to each $x_0 \in D$ there exists a neighborhood $B \equiv B(x_0,r)$ with $\overline{B} \subset D$ such that $T: \overline{B} \to Y$ is A-proper w.r.t. Γ with $\mathrm{Deg}(T,B,Tx_0)$ well defined and $\neq \{0\}$, then $T(D)$ is open in Y.

Proof. To prove Theorem 4.6 we must show that to each $g \in T(D)$ there exists an open ball $B(g,s)$ about g such that $B(g,s) \subset T(D)$. Let g be any element in $T(D)$ with $g = Tx_0$ for some $x_0 \in D$. Let $r > 0$ be such that $\mathrm{Deg}(T,B(x_0,r),g)$ is well defined and not equal to $\{0\}$. Since the degree above is well defined, it follows that $Tx \neq g$ for $x \in \partial B$, where $B = B(x_0,r)$. Hence, by Lemma 2.1, there exist a constant $q > 0$ and $n_0 \in Z_+$ such that $\| T_n(x) - W_n g \| \geq q$ for $x \in B_n$ and $n \geq n_0$. Let $b = \sup_n \| W_n \|$, $s = q/2b$, and let $h \in B(g,s)$, and for each $n \geq n_0$, define the continuous mapping $H_n: [0,1] \times \overline{B}_n \to Y_n$ by $H_n(t,x) = T_n(x) - tW_n(g) - (1 - t)W_n(h)$. Note that $H_n(t,x) \neq 0$ for $t \in [0,1]$, $x \in \partial B_n$, and $n \geq n_0$ since

$$\| H_n(t,x) \| \geq \| T_n(x) - W_n g \| - (1 - t) \| W_n(g - h) \|$$

$$\geq q - b \| g - h \| \geq q - b \cdot s = q - b \frac{q}{2b} = \frac{1}{2} q.$$

Hence, by Theorem III.1.1, $\deg(H_n(t,\cdot),B_n,0)$ is constant in $t \in [0,1]$ for each $n \geq n_0$ and, in particular, $\deg(H_n(0,\cdot),B_n,0) = \deg(H_n(1,\cdot),B_n,0)$; that is, $\deg(T_n - W_n h, B_n, 0) = \deg(T_n - W_n g, B_n, 0)$ or $\deg(T_n,B_n,W_n h) = \deg(T_n,B_n,W_n g)$. Since this is true for each $n \geq n_0$, it follows from Definition 2.1 that $\mathrm{Deg}(T,B(x_0,r),h) = \mathrm{Deg}(T,B(x_0,r),g) \neq \{0\}$. Hence, by Theorem 2.1, there exists $x \in B(x_0,r)$ such that $Tx = h$ [i.e., $T(D)$ is open]. Q.E.D.

For Theorem 4.6 to be useful in applications, we need some simple conditions which would imply that $\mathrm{Deg}(T,B(x_0,r),Tx_0) \neq \{0\}$. It is easy to see that the next theorem serves that purpose.

Theorem 4.7 Let $D \subset X$ be open and $T: D \to Y$ such that

(H) To each $x_0 \in D$ there exist a ball $\overline{B}(x_0,r) \subset D$ and an A-proper (w.r.t. Γ) homotopy $H: [0,1] \times \overline{B}(x_0,r) \to Y$ such that $H_0(x) \equiv H(0,x) = Tx$ for $x \in \overline{B}(x_0,r)$ and $H_t(x) \neq Tx_0$ for $x \in \partial B(x_0,r)$ and $t \in [0,1]$.

If $\mathrm{Deg}(H_1,B(x_0,r),Tx_0) \neq \{0\}$, then $T(D)$ is open.

Proof. Since $H: [0,1] \times \bar{B}(x_0,r) \to Y$ is an A-proper homotopy, it follows that $H_n: [0,1] \times \bar{B}_n(x_0,r) \to Y_n$ is continuous and since $H(t,x) \neq Tx_0$ for all $x \in \partial B$ and $t \in [0,1]$, there exists $n_0 \in Z_+$ such that $H_n(t,x) \neq W_n T(x_0)$ for all $x \in \partial B_n(x_0,r)$, $t \in [0,1]$, and $n \geq n_0$. Hence, by Theorem III.1.1, $\deg(H_n(1,\cdot),B_n(x_0,r),W_nTx_0) = \deg(H_n(0,\cdot),B_n(x_0,r),W_nTx_0)$ for all $n \geq n_0$. This, Definition 2.1 and our hypothesis imply that

$$\mathrm{Deg}(H_n,B(x_0,r),Tx_0) = \mathrm{Deg}(T,B(x_0,r),Tx_0) = \mathrm{Deg}(H_1,B,Tx_0) \neq \{0\}.$$

Hence, by Theorem 4.6, $T(D)$ is open.   Q.E.D.

It follows from Theorem 4.7 that the simplest way to verify conditions of Theorem 4.6 for a particular A-proper mapping is to show that (H) holds. In view of this, we now deduce, as the first corollary of Theorem 4.7, the following result for local ϵ-mappings established by Granas [105] as a generalization of the invariance of domain theorems of Schauder [257] and Leray [151]. Recall that $T: D \to Y$ is called a *local ϵ-map* if to each $x \in D$ there exists $\bar{B}(x,\epsilon_x) \subset D$ such that

$$Tx_1 = Tx_2 \Rightarrow \| x_1 - x_2 \| < \epsilon_x \qquad \text{for any} \quad x_1,x_2 \in \bar{B}(x,\epsilon_x) \qquad (4.8)$$

Corollary 4.8 If $D \subset X$ is open and $T = I + C: D \to X$ is a local ϵ-map with $C: D \to X$ locally compact, then $T(D)$ is open in in X.

Proof. By our condition on T, to each $x_0 \in D$ there exists a ball $B(x_0,\epsilon_0)$ with $\bar{B}(x_0,\epsilon_0) \subset D$ such that $T(\bar{B}(x_0,\epsilon_0))$ is precompact in X and T is an ϵ_0-mapping on $\bar{B}(x_0,\epsilon_0)$ [i.e., (4.8) holds on $\bar{B}(x_0,\epsilon_0)$]. Since the assumed properties are invariant under translation, there is no loss in generality in assuming that $x_0 = 0$ and $T(0) = 0$. Now define the continuous mapping

$$H_t(x) \equiv H(t,x) = T\left(\frac{x}{1+t}\right) - T\left(\frac{-tx}{1+t}\right)$$

$$= x + C_t(x), \qquad x \in \bar{B} = B(0,\epsilon_0), \quad t \in [0,1].$$

Note first that in view of (4.8), $T_t(x) \neq 0$ for $x \in \partial B$ and $t \in [0,1]$. Indeed, if not, then there would exist $x \in \partial B$ and $t \in [0,1]$ such that $H_t(x) = 0$, that is,

$$x + C_t(x) = x + C\left(\frac{x}{1+t}\right) - C\left(\frac{-tx}{1+t}\right) = 0.$$

Hence

$$\frac{-tx}{1+t} + C\left(\frac{-tx}{1+t}\right)$$

$$= \frac{x}{1+t} + C\left(\frac{x}{1+t}\right) \quad \text{and} \quad \left\| \frac{x_0}{1+t} - \frac{-tx_0}{1+t} \right\| = \| x \| = \epsilon_0,$$

in contradiction to (4.8). Further, $H_t(\cdot)$ is an A-proper (w.r.t. $\Gamma_\alpha =$ $\{X_n, P_n\}$) homotopy since $C_t(x)$ is a compact mapping of $[0,1] \times \bar{B} \to X$, i.e., if $\{x_{n_j} \mid x_{n_j} \in B_{n_j}\}$ and $\{t_{n_j}\} \subset [0,1]$ are any sequences such that $x_{n_j} + C(t_{n_j}, x_{n_j}) \to g$ for some g in X, then there exist subsequences $\{t_{n_{j(k)}}\}$ and $\{x_{n_{j(k)}}\}$ such that $t_{m_{j(k)}} \to t_0 \in [0,1]$, $x_{n_{j(k)}} \to x_0 \in \bar{B}$, and $x_0 +$ $C(t_0, x_0) = g$. Hence, by Theorem 2.1, $\mathrm{Deg}(H_t, B, 0)$ is well defined and is independent of $t \in [0,1]$. Since H_1 is odd on $\bar{B} = \bar{B}(0, \epsilon_0)$, we see that $\mathrm{Deg}(H_0, B, 0) = \mathrm{Deg}(T, B, 0) = \mathrm{Deg}(H_1, B, 0) \neq \{0\}$. Thus Corollary 4.8 follows from Theorem 4.7 and therefore $T(D)$ is open. Q.E.D.

Using similar arguments, we get the following generalization of Corollary 4.8, which contains the invariance of domain theorem obtained in [193] (see also [291]) in case of k-ball-contractive mappings.

Corollary 4.9 Let X be a π_1-space. If $D \subset X$ is open and $T = I - F: D \to X$ is a local ϵ-map with $F: D \to X$ locally k-ball contractive with $k < 1$, then $T(D)$ is open.

Proof. In view of the preceding discussion and Theorem 4.7, to deduce Corollary 4.9 from Theorem 4.7, it suffices to know that

$$H_t(x) = x + F\left(\frac{x}{1+t}\right) - F\left(\frac{-tx}{1+t}\right)$$

is an A-proper homotopy w.r.t. Γ_1 from $[0,1] \times \bar{B}(0,r)$ into X. So let $\{x_n \mid x_n \in B_n\}$ and $\{t_n\} \subset [0,1]$ be any sequences such that

$$g_n \equiv x_n + P_n F\left(\frac{x_n}{1+t_n}\right)$$

$$- P_n F\left(\frac{-t_n x_n}{1+t_n}\right) \to g \qquad \text{for some } g \text{ in } X.$$

To show that there exist convergent subsequences $\{x_{n_j}\}$ and $\{t_{n_j}\}$, we argue as follows. We may assume that $t_n \to t \in [0,1]$, and therefore to any $\epsilon > p$, there exist $n(\epsilon) \in Z_+$ such that

$$\left| \frac{1}{1+t_n} - \frac{1}{1+t} \right| < \epsilon \quad \text{and} \quad \left| \frac{t_n}{1+t_n} - \frac{t}{1+t} \right| < \epsilon$$

$$\text{for all } n \geq n(\epsilon).$$

Consider the sets

$$A = \left\{ \frac{x_n}{1+t_n} \;\middle|\; n \geq n(\epsilon) \right\}, \qquad \tilde{A} = \left\{ \frac{x_n}{1+t} \;\middle|\; n \geq n(\epsilon) \right\},$$

$$B = \left\{ \frac{t_n x_n}{1+t_n} \;\middle|\; n \geq n(\epsilon) \right\}, \qquad \tilde{B} = \left\{ \frac{tx_n}{1+t} \;\middle|\; n \geq n(\epsilon) \right\}.$$

It follows that $A \subset N_\epsilon(\tilde{A})$ and $B \subset N_\epsilon(\tilde{B})$ and hence by Lemma I.2.3,

$$\beta(A) \leq \beta(N_\epsilon(\tilde{A}) + \epsilon$$

$$= \frac{1}{1 + t} \beta(A_0) + \epsilon \quad \text{and} \quad \beta(B) \leq \frac{t}{1 + t} \beta(A_0) + \epsilon,$$

where $A_0 = \{x_n \mid n \geq n(\epsilon)\}$. Clearly, to show that $\{x_n\}$ has a convergent subsequence, it suffices to show that $\beta(A_0) = 0$. Suppose that $\beta(A_0) > 0$. Then since $g_n \to g$ and for all $n \geq n(\epsilon)$, $A_0 = \{x_n\} \subseteq \{g_n\} + \{P_n F(B)\} + \{P_n F(A)\}$, it follows from Corollary I.3.1 that

$$\beta(A_0) \leq 0 + \beta(\{P_n F(B)\}) + \beta(\{P_n F(A)\}) \leq \beta(F(B)) + \beta(F(A))$$

$$\leq k\beta(B) + k\beta(A) \leq \frac{k}{1 + t} \beta(A_0) + k\epsilon + \frac{kt}{1 + t} \beta(A_0) + k\epsilon$$

$$= k\beta(A_0) + 2k\epsilon.$$

Since $\epsilon > 0$ is arbitrarily small and $k < 1$, we have the contradiction when $\beta(A_0) > 0$. Thus we must conclude that $\beta(A_0) = 0$, so there exists a subsequence $\{x_{n_j}\}$ of $\{x_n\}$ and $x \in B(0,r)$ such that $x_{n_j} \to x$, $t_{n_j} \to t$, and $H_t(x) = g$ by continuity of $H_t(x)$ [i.e., $H_t(\cdot)$: $[0,1] \times \bar{B}(0,r) \to X$ is an A-proper homotopy].   Q.E.D.

Since a locally one-to-one map is clearly an ϵ-map, a consequence of Corollary 4.9 is the following result, which includes Schauder's invariance of domain theorem in [257] when X is reflexive, S is contractive and weakly continuous, C is completely continuous, and $T = I + S + C$ is one-to-one.

Corollary 4.10   Let X be a π_1-space. If $D \subset X$ is open and $T = I + S + C$ is locally one-to-one with $S: D \to X$ and $C: D \to X$ locally contractive and locally compact, respectively, then $T(D)$ is open.

Proof.   Let $x_0 \in D$. We wish to find $r > 0$ such that $B(Tx_0, r) \subset T(D)$. To that end it suffices to show that (H) holds for $T = I + S + C$. Without loss of generality we may assume that $x_0 = 0$ and $T(x_0) = 0$. Let $q > 0$ and $k_0 \in (0,1)$ be such that $S: \bar{B}(0,q) \to X$ is k_0-contractive, $C: \bar{B}(0,q) \to X$ compact, and $T: \bar{B}(0,q) \to X$ injective. It was shown by Nussbaum [193] that there exists $q_1 < q$ such that $I + S: \bar{B}(0,q_1) \to X$ is A-proper w.r.t. Γ_1 and so is $T = I + S + C$ since C is compact on $\bar{B}(0,q_1)$. Moreover, since any covering of an arbitrary set $A \subset B(0,q/2)$ must have centers in $B(0,q)$, it follows that $I + S: B(0,q/2) \to X$ and $T = I + S + C: B(0,q/2) \to X$ are k_0-ball contractive. In view of this and the fact that T is also one-to-one on $\bar{B}(0,q/2)$, Corollary 4.10 follows from Corollary 4.9 when $F = S + C$ and $k = k_0$.   Q.E.D.

Let us add that Corollary 4.10 can also be deduced from the invariance of domain theorem of Nussbaum [193] for k-set contractions with $k < 1$.

We indicate now some other classes of mappings to which Theorems 4.6 and 4.7 apply.

Theorem 4.8 Let $D \subset X$ be open, $T_\lambda = T + \lambda I: D \to X$ locally A-proper w.r.t. $\Gamma_2 = \{X_n, P_n\}$ for each $\lambda \geq 0$, and suppose that one of the following two conditions holds:

(a) $T_\lambda = T + \lambda I$ is locally injective for each $\lambda \geq 0$.
(b) T is locally injective and locally accretive on D.
Then, in either case, $T_\lambda(D)$ is open in X for each $\lambda \geq 0$.

Proof. It suffices to prove Theorem 4.8 when (a) holds since, as is not hard to see, (b) implies (a).

Now we shall show that (a) implies (H) of Theorem 4.7. Indeed, for a given $x_0 \in D$, let $B(x_0,r)$ be a sufficiently small ball with $\overline{B}(x_0,r) \subset D$ so that $T_\lambda: \overline{B}(x_0,r) \to X$ is A-proper w.r.t. Γ_α for each $\lambda \geq 0$, $T(\overline{B}(x_0,r))$ is bounded and T_λ is one-to-one on $\overline{B}(x_0,r)$ for each $\lambda \geq 0$. Now define a continuous map of $[0,1] \times \overline{B}(x_0,r)$ to X by

$$T_{\lambda t}(x) = (1 - t)T_\lambda(x) + tT_\lambda(x_0) + t(x - x_0) \qquad (\lambda \geq 0) \qquad (4.9)$$

and observe that $T_{\lambda 0}(x) = T_\lambda(x)$ on $\overline{B}(x_0,r)$, $T_{\lambda t}(\cdot)$ is A-proper for each $t \in [0,1]$ and continuous in $t \in [0,1]$, uniformly for $x \in \overline{B}(x_0,r)$. Indeed, if $t = 0$, $T_{\lambda 0}(x) = T_\lambda(x)$ is A-proper on $\overline{B}(x_0,r)$ for $\lambda \geq 0$, by hypothesis; if $t = 1$, $T_{\lambda 1}(x) = x + T_\lambda(x_0) - x_0$ is A-proper since I is A-proper and so is $I + T_\lambda(x_0) - x_0$ for each $\lambda \geq 0$; if $t \in (0,1)$, then $T_{\lambda t}(\cdot)$ is A-proper since

$$(1 - t)^{-1}T_{\lambda t}(x) = T(x) + \lambda x + \frac{t}{1 - t}(T_\lambda(x_0) - x_0)$$

is A-proper with $\lambda' = \lambda + t/1 - t \geq 0$. The fact that for each fixed $\lambda \geq 0$, $T_{\lambda t}(x)$ is continuous in $t \in [0,1]$, uniformly for $x \in \overline{B}(x_0,r)$, follows from the definition of $T_{\lambda t}(x)$ and the boundedness of $T(\overline{B}(x_0,r))$. Now, if we define $H_{\lambda t}(\cdot): \overline{B}(x_0,r) \times [0,1]$ into X by $H_{\lambda t}(x) = T_{\lambda t}(x) - T_\lambda(x_0)$, then $H_{\lambda t}(\cdot)$ satisfies (H) of Theorem 4.7. To see this, all we need to do is to verify that $H_{\lambda t}(x) \neq 0$ for $t \in [0,1]$ and $x \in \partial B(x_0,r)$, $\lambda \geq 0$ fixed. Indeed, suppose that $H_{\lambda(t)}(x) = 0$ for some $x \in \partial B(x_0, r)$ and $t \in [0,1]$. Then $t = 1$, $0 = H_{\lambda 0}(x) = x - x_0$, in contradiction to the fact that $x \in \partial B(x_0,r)$; if $t \in [0,1)$, then $Tx + \lambda'x = Tx_0 + \lambda'x_0$ for $\lambda' = \lambda + t/1 - t \geq 0$, contradicting the one-to-one property of $T_{\lambda'}$ on $\overline{B}(x_0,r)$. Hence $\text{Deg}(H_{\lambda t}, B(x_0,r),0)$ is independent of $t \in [0,1]$ and, in particular, $\text{Deg}(H_{\lambda 1}, B(x_0,r),0) = \text{Deg}(I - x_0, B(x_0,r), 0) = \{1\} \neq \{0\}$ for each fixed $\lambda \geq 0$. Thus Theorem 4.8 follows from Theorem 4.7. Q.E.D.

Note that $F: D \to X$ is locally P_1-compact iff $T + \lambda I$ is locally A-proper for each $\lambda \geq 0$, where $T = I - F$.

By virtue of Theorem I.3.1 (when $Y = X$ and $K = J$) we deduce from Theorem 4.8 the following simple result, which extends the Hilbert space invariance of domain theorems of Minty and Browder for locally strongly monotone mappings and which, we believe, was the first invariance of domain theorem for continuous monotone mappings, proved in [27,178] under rather restrictive conditions. We mention it here only for historical reasons, since the more general results were given later in [36] and in [74], culminating with Theorem III.3.3.

Corollary 4.11 Let X be a reflexive space with a single-valued weakly continuous duality mapping J of X to X^*. If $D \subset X$ is open and $T: D \to X$ is a continuous locally accretive and locally one-to-one mapping that satisfies the local modified condition (S), then $T_\lambda(D)$ is open for each $\lambda \geq 0$.

An analog to Corollary 4.11 follows from Theorem 4.7 for maps from $T: D \subset X \to X^*$ when X is reflexive and X and X^* have property (H). We do not state it here since a more general result due to Fitzpatrick [74] will be deduced from the following extension of Theorem 4.6, proved in [74] by using the degree theory for uniform limits of A-proper mappings, which we do not use here.

Theorem 4.9 Let $D \subset X$ be open, $T: D \to Y$ such that $T(\overline{B}(x_0,r))$ is closed in Y for each $\overline{B}(x_0,r) \subset D$, and suppose that there exists a bounded map $G: D \to Y$ such that $T + \mu G$ is A-proper w.r.t. Γ for each $\mu > 0$. Then, if to each $x_0 \in D$ there exist constants $r_{x_0} > 0$, $c_{x_0} > 0$, and $\gamma_{x_0} > 0$ such that $\| Tx - Tx_0 \| \geq c_{x_0}$ for $x \in \partial B(x_0, r_{x_0})$ and $\text{Deg}(T + \mu G, B(x_0, r_{x_0}), Tx_0) \neq \{0\}$ for all $\alpha \in (0, \gamma_{x_0})$, then $T(D)$ is open in Y.

Proof. Let $g \in T(D)$ and let $x_0 \in D$ such that $g = Tx_0$. Hence by our hypotheses, $\| Tx - g \| \geq c_{x_0}$ for $x \in \partial B_0$, where $B_0 \equiv B(x_0, r_{x_0})$, $\text{Deg}(T + \alpha G, B_0, g) \neq \{0\}$ for $\alpha \in (0, \gamma_{x_0})$. Let $\gamma > 0$ be such that $\gamma \leq \gamma_{x_0}$ and $\gamma \| Gx \| < G_{x_0}/4$ for $x \in \partial B_0$, let $y \in Y$ be such that $\| y - g \| < c_{x_0}/2$, and for fixed $\mu \in (0, \gamma)$, consider the homotopy

$$H_t^\mu(x) = Tx + \mu Gx - g + t(g - y) \qquad \text{for} \quad x \in \overline{B}_0, \, t \in [0,1].$$

Then, obviously H_t^μ is an A-proper homotopy and $H_t^\mu(x) \neq 0$ for $x \in \partial B_0$, $t \in [0,1]$, and $\mu \in (0, \gamma)$, since

$$\| H_t^\mu(x) \| \geq \| Tx - g \| - u \| Gx \| - \| g - y \| \geq c_{x_0}/4 - c_{x_0}/2 = c_{x_0}/4.$$

Hence, by Theorem 2.1 and the hypothesis on the degree,

$$\text{Deg}(T + \mu G - y, B_0, 0) = \text{Deg}(T + \mu G - g, B_0, 0)$$
$$= \text{Deg}(T + \mu G, B_0, g) \neq \{0\}$$

for each $\mu \in (0,\gamma)$. Let $n_0 \in Z_+$ be such that $(1/n) \in (0,\gamma)$ for $n \geq n_0$. Then for each such n, $\text{Deg}(T + (1/n)G, B_0, y) \neq \{0\}$, so there exists $x_n \in B_0$ such that $Tx_n + (1/n)G(x_n) = y$. Since G is bounded, we see that $Tx_n = y - (1/n)G(x_n) \to y$ as $n \to \infty$, so, since $T(\overline{B}_0)$ is closed in Y, there exists $x \in \overline{B}_0 = \overline{B}(x_0, r_{x_0})$ such that $y = Tx$ [i.e., $T(D)$ is open in Y].

Q.E.D.

Some special cases of Theorem 4.9 will be discussed later.

5. CALCULATION OF GENERALIZED DEGREE

So far we only know that when D is a bounded open subset of X with $0 \in D$ and an A-proper mapping $T: \overline{D} \subset X \to Y$ is homotopic to some odd mapping $A: X \to Y$, its generalized degree is different from zero (in fact, it does not contain 0). This is the case, for example, when $Y = X$ and $A = I$ or when $Y = X^*$ and $A = J$, a single-valued duality map. Furthermore, we saw in Theorem 1.3 (the Leray–Schauder formula) that when $L \in L(X,X)$ is compact and $\lambda = 1$ is not an eigenvalue of L, then $\text{Deg}(I - L, D, 0) = \{(-1)^\eta\}$ by Corollary 2.7, where η is the sum of algebraic multiplicities of eigenvalues $\lambda > 1$ of L. The same formula holds when L is k-ball contractive with $k < 1$ in view of Theorem 3.9 and the result of Stuart and Toland [274] and O'Neil and Thomas [197].

For applications of the powerful tool "generalized degree theory," it is important to know the comprehensive classes of maps for which the degree can be computed explicitly. This problem is studied in the next two sections.

5.1 Fréchet Differentiable A-Proper Mappings

In the first part of this section we outline some basic results obtained in [76,228,236] for F-differentiable P-compact and A-proper mappings. The main result of this section is a theorem due to Fitzpatrick [76], which asserts that if the A-proper mapping $T: \overline{D} \subset X \to Y$ has the Fréchet derivative T'_{x_0} at $x_0 \in D$, which is A-proper and injective, there exists $r > 0$ such that

$$\text{Deg}(T, B(x_0, r), Tx_0) = \text{Deg}(T'_{x_0}, B(0, r), 0).$$

This is the extension of the classical result for compact vector fields due to Leray and Schauder [153], as well as of the result of Wong [307] for P-compact mappings. It should be added that Wong uses a different definition of the degree of A-proper mappings than that given by Definition 2.1.

Recall first that if $D \subset X$ is open and $T: D \rightarrow Y$, then T is said to be *F-differentiable at* $x \in D$ if there exists $T'_x \in L(X,Y)$, called the *F-derivative of* T *at* x, such that for all $h \in X$ will $\| h \|$ sufficiently small, one has

$$T(x + h) - Tx = T'_x h + w(x,h) \tag{5.1}$$

$$\text{with} \quad \frac{\| w(x,h) \|}{\| h \|} \rightarrow 0 \quad \text{as} \quad \| h \| \rightarrow 0.$$

We first prove the following lemma due to Nussbaum [195], which will play an important role in what follows. See also [58], where a more general measure of noncompactness is used.

Lemma 5.1 Let $D \subset X$ be an open set. If $F: D \rightarrow Y$ is k-Φ-contractive for some $k \geq 0$ and F-differentiable at $x \in D$, then F'_x is also k-Φ-contractive, where $\Phi(Q)$ is either $\gamma(Q)$ or $\beta(Q)$.

Proof. Let $B(x,r)$ be such that $B(x,r) \subset D$. Let Q be a subset of $B(0,r)$, $0 < t \leq 1$, and $d = \sup\{\| h \| \mid h \in Q\}$. Then $x + tQ \subset B(x,r)$ and

$$\Phi(F'_x(tQ)) \leq \Phi(F(x + tQ)) + \Phi(-Fx) + \Phi(-w(x,tQ))$$

$$\leq k\Phi(x + tQ) + \Phi(B(0,\phi(td)) \leq k\Phi(tQ) + \phi(td)\Phi(B(0,1)),$$

where

$$w(x,h) = F(x + h) - Fx - F'_x h,$$

$$\phi(0) = \sup\{\| w(x,h) \| \mid \| h \| \leq s\}.$$

Hence

$$\Phi(F'_x(Q)) \leq k\Phi(Q) + \inf_{0<t\leq1} \frac{\phi(td)}{t} \Phi(B(0,1)) = k\Phi(Q).$$

Let now Q be any bounded set in X. Then there exists $t > 0$ such that $tQ \subset B(0,r)$. This and the last inequality imply that $\Phi(F'_x(Q)) = t^{-1}\Phi(F'_x(tQ)) \leq t^{-1}k\Phi(tQ) = k\Phi(Q)$.   Q.E.D.

Lemma 5.1 contains the classical result which asserts that if $C: D \rightarrow Y$ is compact, then C'_x is also compact. It was shown in [280] that the converse of this result is false in general. However, Vainberg has shown in [280] that if $C: X \to Y$ is F-differentiable at each $x \in X$, C_x' is compact, and $C': X \to L(X,X)$ defined by $x \to C_x'$ is compact, then C is compact. It was noted in [76] that an example of S. Yamamuro shows that $T: X \to X$ can have an A-proper (in fact, P-compact) F-derivative T_x' at each $x \in X$ without T itself being A-proper.

Example 5.1 Let $X = l_2$ and $C(x) = \sum_{i+1}^{\infty} (\phi_i, x)^2 \phi_i$ for $x \in l_2$, where $\{\phi_n\}$ is the natural basis for l_2. Then $I - C$ is not A-proper w.r.t. $\Gamma_1 = \{X_n, P_n\}$ determined by $\{\phi_n\}$ since $T\phi_n = 0$ and $P_n T\phi_n = 0 \to 0$ as $n \to \infty$, but $\{\phi_n\}$ has no convergent subsequence. Since $C_x'(y) = \sum_{i=1}^{\infty} (x, \phi_i) (y, \phi_i)\phi_i$ for each $y \in l_2$, C_x' is compact for each $x \in l_2$ and thus $T_x' = I - C_x'$ is A-proper.

An analog of Vainberg's result was obtained by Petryshyn and Tucker [236] for P-compact mappings $F: X \to X$ under strengthened continuity conditions on F_x'. In fact, it was shown in [236] that if X is reflexive, $F: X \to X$ is F-differentiable at each $x \in X$, and if $F': X \to L(X,X)$ is completely continuous [i.e., $F_{x_j}' \to F_x'$ in $L(X,X)$ if $x_j \to x$ in X], then F_x' is P-compact at each $x \in X$ iff F is P-compact.

Using the technique of [236], it was shown by Fitzpatrick [76] that an analogous result holds for an F-differentiable A-proper mapping $T: X \to Y$ with X reflexive under the following slightly weaker continuity assumption:

$$T_{y_n}'(x - z_n) - T_x'(x - z_n) \to 0$$

$$\text{whenever} \quad y_n \to x \quad \text{and} \quad z_n \to x \quad \text{in} \quad X.$$

Since both of these assumptions are very difficult to verify in various applications, we shall not reproduce here the somewhat lengthy proofs of these results. Interested readers should consult the original papers.

To state the basic results in this section, we first recall that a point $x_0 \in D$ is said to be an *isolated solution* of the equation $T(x) = f$ $(x \in D)$, or an *isolated f-point* of T provided that $Tx_0 = f$ and there exists $r > 0$ such that $\bar{B}(x_0, r) \subset D$ and $Tx \neq f$ for all $x \in \bar{B}(x_0, r)$, $x \neq x_0$.

It was shown in [236] that if $F: X \to X$ has a fixed point $x_0 \in X$ and F is F-differentiable at x_0 with F_{x_0}' P-compact and $I - F_{x_0}'$ injective, then x_0 is an isolated fixed point of F. We shall not prove this result since a more general one is proved below.

The following two theorems and their corollaries are due to Fitzpatrick [76] in case (X,Y) has a projectionally complete scheme Γ_p. We prove these results here for (X,Y) having an admissible scheme $\Gamma = \{X_n, V_n; E_n, W_n\}$ following essentially the argument of [76].

Theorem 5.1 Let Γ be an admissible scheme for (X,Y), $D \subset X$ open and bounded, $T: \overline{D} \to Y$ A-proper w.r.t. Γ with an A-proper F-derivative $A \equiv T'_{x_0}$ at $x_0 \in D$. If A is also injective, then x_0 is an isolated solution of the equation $Tx = Tx_0$ $(x \in D)$, and there exists $r_1 > 0$ such that for any $r \in (0,r_1]$,

$$\text{Deg}(T,B(x_0,r),Tx_0) = \text{Deg}(T'_{x_0},B(x_0,r),T'_{x_0}(x_0)). \tag{5.2}$$

Proof. We shall first show that under the conditions on $A = T'_{x_0}$ the point $x_0 \in D$ is an isolated solution of $Tx = f$, $x \in D$ and $f = Tx_0$. Now, since $A \in L(X,Y)$ is A-proper w.r.t. Γ and injective, Theorem II.1.2 implies the existence of a constant $c_0 > 0$ such that

$$\| Ay \| \geq c_0 \| y \| \qquad \text{for all} \quad y \in X. \tag{5.3}$$

Let $\overline{B}(x_0,r_0) \subset D$ such that

$$\| Tx - f - A(x - x_0) \| < \tfrac{1}{3}c_0\| x - x_0 \| \qquad \text{for} \quad x \in \overline{B}(x_0,r_0). \tag{5.4}$$

It is clear from (5.3) and (5.4) that x_0 is an isolated solution of $Tx = f$.

To establish the validity of (5.2), note first that since A is A-proper and injective, Theorem II.1.3 implies the existence of a constant $c > 0$ and $n_1 \in Z_+$ such that

$$\| W_n Ay \| \geq c \| y \| \qquad \text{for all} \quad y \in X_n \quad \text{and} \quad n \geq n_1. \tag{5.5}$$

Let $a = \sup_n\| W_n \|$ and choose $r_1 > 0$ with $r_1 \leq r_0$ such that $\overline{B}(x_0,r_1) \subset D$ and

$$\| w(x_0,h) \| \leq \frac{c}{4a} \| h \| \qquad \text{if} \quad \| h \| \leq r \quad \text{for any fixed} \quad r \in (0,r_1]. \tag{5.6}$$

Since $\text{dist}(x,X_n) \to 0$ for each $x \in X$, there exists $y_n \in X_n$ such that $y_n \to x_0$ as $n \to \infty$. Choose $n_2 \in Z_+$ such that

$$\| y_n - x_0 \| < \min\left\{\frac{re}{8a\| A \|}; \frac{r}{2}\right\} \qquad \text{for} \quad n \geq n_2. \tag{5.7}$$

Now, for a fixed $n \geq \max\{n_1,n_2\}$, consider the homotopy

$$H_n(t,x) = (1 - t)W_n T(x) + tW_n A(x - x_0)$$
$$- (1 - t)W_n T(x_0), \qquad x \in \overline{B}(x_0,r), \quad t \in [0,1].$$

It is easy to see that $H_n(t,x)$ can be written in the form

$$H_n(t,x) = W_n A(x - y_n)$$
$$+ (1 - t)W_n[Tx - Tx_0 - A(x - x_0)] + W_n[Ay_n - Ax_0].$$

It follows from (5.1), (5.5), (5.6), and (5.7) that for all $x \in \partial B_n(x_0,r)$ and $t \in [0,1]$, where $B_n(x_0,r) \equiv B(x_0,r) \cap X_n$, we have

$$\| H_n(t,x) \| \geq \| W_n A(x - y_n) \| - a\| Tx - Tx_0 - A(x - x_0) \|$$
$$- a\| A \| \| y_n - x_0 \|$$
$$\geq c\| x - y_n \| - \frac{car}{4a} - \frac{a\| A \| rc}{8a\| A \|}$$
$$\geq c\| x - x_0 \| - c\| x_0 - y_n \| - \frac{cr}{4} - \frac{cr}{8} \geq \frac{cr}{8}.$$

Hence, for each $n \geq \max\{n_1,n_2\}$, $\deg(H_n(t,\cdot),B_n(x_0,r),0)$ is independent of $t \in [0,1]$ and therefore

$$\deg(H_n(0,\cdot),B_n(x_0,r),0) = \deg(H_n(1,\cdot),B_n(x_0,r),0),$$

that is,

$$\deg(T_n - W_n T(x_0), B_n(x_0,r), 0) = \deg(A_n - A_n(x_0), B_n(x_0,r), 0).$$

and thus, by property VI of Theorem III.4.1, for $n \geq \max\{n_1, n_2\}$, $\deg(T_n, B_n(x_0, r), WT(x_0)) = \deg(A_n, B_n(x_0, r), A_n(x_0))$. This implies (5.2).

Q.E.D.

Definition 5.1 Let $T: \overline{D} \subset X \to Y$ be A-proper w.r.t. Γ. The *index* of an isolated f-point $x_0 \in D$ of T is defined by $\operatorname{ind}(T,x_0,f) = \operatorname{Deg}(T,D,f)$, where $D = B(x_0,r)$ with $r > 0$ so small that T has no other f-points in D.

Theorem 5.2 Suppose that $T \in L(X,Y)$ is A-proper w.r.t. Γ and injective. Then if $x_0 \in X$ and $r > 0$, we have

$$\operatorname{Deg}(T,B(x_0,r),Tx_0) = \operatorname{Deg}(T,B(0,r),0). \tag{5.8}$$

Proof. Choose $x_0 \in X$ and $r > 0$. Since $Tx \neq Tx_0$ for $x \in \partial B(x_0,r)$ we may choose $c > 0$ and $n_0 \in Z_+$ such that

$$\| W_n Tx - W_n Tx_0 \| \geq c \quad \text{for} \quad x \in \partial B_n(x_0,r) \quad \text{and} \quad n \geq n_0.$$

Choose $n_1 \in Z_+$ and $x_n \in B_n(x_0,r)$ such that

$$\| W_n Tx_0 - W_n Tx_n \| < c \quad \text{for} \quad n \geq n_1.$$

Choose $n_2 \in Z_+$ such that $\| x_n - x_0 \| < r$ for $n \geq n_2$. Let $n_3 = \max\{n_i \mid i = 0,1,2\}$, and for the remainder of the proof let $n \in Z_+$, with $n \geq n_3$, be fixed.

By our choice of n_0 and n_1 and by Theorem III.1.1 we see that

$$\deg(T_n - W_n Tx_0, B_n(x_0,r), 0) = \deg(T_n - T_n(x_n), B_n(x_0,r), 0). \tag{5.9}$$

We also know that

$$\deg(T_n - T_n(x_n), B_n(x_0,r), 0) = \deg(T_n, B_n(x_0,r), T_n(x_n)).$$

Now, let $R_n \colon \overline{B}_n(0,r) \to X_n$ be defined by $R_n(x) = x + x_n$. Then R_n is a continuous injective mapping and $R_n(0) = x_n$. Hence we see that since R_n preserves orientation,

$$\deg(R_n, B_n(0,r), x_n) = 1.$$

By our choice of n_2, $x_n \in B_n(x_0,r)$ for each $n \geq n_2$. Consequently, by Theorem III.1.2, we have

$$\deg(T_n \cdot R_n, B_n(0,r), T_n x_n) = \deg(R_n, B_n(0,r), x_n) \cdot \deg(T_n, B_n(x_0,r), T_n x_n).$$

Since T_n is linear, $(T_n \cdot R_n)(x) = T_n x + T_n x_n$, and hence

$$\deg(T_n + T_n(x_n), B_n(0,r), T_n(x_n)) = \deg(T_n, B_n(x_0,r), T_n(x_n)).$$

But

$$\deg(T_n + T_n(x_n), B_n(0,r), T_n(x_n)) = \deg(T_n, B_n(0,r), 0),$$

and hence

$$\deg(T_n, B_n(0,r), 0) = \deg(T_n, B_n(x_0,r), T_n(x_n)).$$

Using this last equality, together with (5.9), we get

$$\deg(T_n, B_n(0,r), 0) = \deg(T_n - W_n Tx_0, B_n(0,r), 0),$$

and since this is true for all $n \geq n_3$, we have

$$\mathrm{Deg}(T, B(0,r), 0) = \mathrm{Deg}(T - Tx_0, B(x_0,r), 0) = \mathrm{Deg}(T, B(x_0,r), Tx_0).$$

Q.E.D.

We may now combine Theorems 5.1 and 5.2 to obtain the following generalization of the Leray–Schauder theorem.

Corollary 5.1 Let $D \subset X$ be open and bounded, $T \colon \overline{D} \to Y$ A-proper w.r.t. Γ with A-proper F-derivative T'_{x_0} at $x_0 \in D$ and T'_{x_0} injective. Then there exists $r_1 > 0$ such that for each $r \in (0, r_1]$

$$\text{Deg}(T, B(x_0, r), Tx_0) = \text{Deg}(T'_{x_0}, B(0, r), 0). \tag{5.10}$$

Now it is clear that if $T: D \subset X \rightarrow Y$ is a continuous proper mapping and if $f \in Y$ is such that the solutions of $Tx = f$ $(x \in D)$ are isolated, the set of solutions is finite. Since, by Theorem I.1.1, continuous A-proper mappings, when restricted to bounded closed sets, are proper, we may combine the last observation with the first part of Theorem 5.1 to obtain the following:

Corollary 5.2 Let $D \subset X$ open and bounded, $T: \overline{D} \rightarrow Y$ a continuous A-proper map, and $f \in Y$ such that if $x_0 \in \overline{D}$ is such that $T(x_0) = f$, then T'_{x_0} exists and is A-proper and injective. Then there are at most a finite number of solutions in D of $Tx = f$.

It was noted by the author in [228] that in some applications (e.g., bifurcation theory) one is interested in the assertion of Corollary 5.1 in the case when T has either an F-derivative at 0 with $T(0) = 0$ or T has an asymptotic derivative T'_∞ at ∞, that is,

$$Tx = T'_\infty(x) + Q_\infty(x) \quad \text{where} \quad \frac{\| Q_\infty(x) \|}{\| x \|} \rightarrow 0 \quad \text{as} \quad \| x \| \rightarrow \infty. \tag{5.11}$$

In these cases one gets (5.10) in a very simple way, as the following shows.

Proposition 5.1 Let $T: X \rightarrow Y$ be A-proper w.r.t. Γ.

(a) If $T(0) = 0$ and T has an injective A-proper F-derivative T_0 at 0, then there is $r_0 > 0$ such that for each $r \in (0, r_0]$,

$$\text{Deg}(T, B(0, r), 0) = \text{Deg}(T'_0, B(0, r), 0). \tag{5.12}$$

(b) If T has an injective A-proper asymptotic derivative T'_∞ at ∞, then there exists $r_\infty > 0$ such that for each $r \geq r_\infty$,

$$\text{Deg}(T, B(0, r), 0) = \text{Deg}(T'_\infty, B(0, r), 0). \tag{5.13}$$

Proof. We shall prove (a) and (b) simultaneously by using β to denote either 0 or ∞ and letting $a = \sup_n \| W_n \|$. Now, since $T'_\beta \in L(X, Y)$ is injective and A-proper w.r.t. Γ, Theorem II.1.1 implies the existence of $c_\beta > 0$ and $n_\beta \in Z_+$ such that

$$\| W_n T'_\beta(x) \| \geq c_\beta \| x \| \quad \text{for all} \quad x \in X_n \quad \text{and} \quad n \geq n_\beta. \tag{5.14}$$

Choose $r_\beta > 0$ such that for all $x \in X_n$ with $\| x \| \geq r_\infty$, respectively, $\| x \| \leq r_0$, we have

$$\| T(x) - T'_\beta(x) \| < c_\beta(2a)^{-1} \| x \|. \tag{5.15}$$

Then we claim that for every $r \geq r_0$, respectively, $0 < r \leq r_0$, and for each $n \geq n_\beta$ and $t \in [0,1]$ we have

$$H_n(t,x) = W_n T'_\beta(x) + t[W_n T(x)$$
$$- W_n T'_\beta(x)] \neq 0 \qquad \text{for} \quad x \in \partial B_n(0,r).$$

Indeed, it follows from (5.14) and (5.15) that for each fixed $n \geq n_\beta$, $t \in [0,1]$ and $x \in \partial B_n(0,r)$ one has the inequality $\| H_n(t,x \| \geq \| W_n T'_\beta(x)\| - \| W_n T(x) - W_n T'_\beta(x) \| \geq c_\beta r - ac_\beta(2a)^{-1}r = \frac{1}{2}r > 0$. Hence, by Theorem III.1.1 it follows that $\deg(H_n(1,\cdot),B_n(0,r),0) = \deg(H_n(0,\cdot),B_n(0,r),0)$ for each $n \geq n_\beta$ [i.e., $\deg(W_n T,B_n(0,r),0) = \deg(W_n T'_\beta,B_n(0,r),0)$]. Since this is true for each $n \geq n_\beta$, we have the equality $\text{Deg}(T,B(0,r),0) = \text{Deg}(T'_\beta,B(0,r),0)$ for $\beta = 0$ or $\beta = \infty$.   Q.E.D.

The value of Corollary 5.1 and Proposition 5.1 is that they reduce the calculation of the degree of a mapping to that of the degree of a linear mapping.

5.2 Generalization of the Leray–Schauder Formula to Linear P_1-Compact Mappings and Its Applications

Let $\Gamma_\alpha = \{X_n,P_n\}$ be a projectionally complete scheme for (X,X), $L \in L(X,X)$ is P_1-compact (i.e., $\lambda I - L$ is A-proper w.r.t. Γ_α for each $\lambda \geq 1$) and assume also that (a) $I - L$ is injective, and (b) the spectrum $\sigma(L)$ of L in $(1,\infty)$ consists of a finite number of points, each of finite multiplicity. We shall also need a technical hypothesis which, because of its nature, we introduce in the course of the argument.

Let X_1 be the vector space spanned by the set $\{x \mid (L - \lambda I)^k x = 0$, some k and $\lambda > 1\}$. By hypothesis, X_1 is finite dimensional; therefore, X_1 has a topological complement in X, say X_2, so $X = X_1 \oplus X_2$. Clearly, X_1 is invariant under L. We make the additional hypothesis: (c) X_2 can be chosen so that it is invariant under L.

We are now in the position to prove the following extension of Theorem 1.3, which contains the recent result of Toland [275] as well as the earlier result of Stuart, and of Stuart and Toland [274] when L is a k-ball contraction with $k < 1$. Our argument is similar to that in [274].

Theorem 5.3 Let $L \in L(X,X)$ be P_1-compact and satisfy conditions (a), (b), (c) and D a bounded neighborhood of 0. Then

$$\text{Deg}(I - L, D, 0) = \{(-1)^{\eta}\}, \tag{5.16}$$

where η is the sum of the multiplicities of eigenvalues $\lambda > 1$ of L.

Proof. Let P be a projection of X onto X_1 and $\check{L} = LP$. Then $\check{L}$ is compact, and since L is a bounded P_1-compact mapping, $H: [0,1] \times \overline{D} \to X$ defined by $H(t,x) = x - \check{L}x - (1 - t)Lx$ is an A-proper homotopy. Moreover, $H(t,x) \neq 0$ for $x \in \partial D$ and $t \in [0,1]$. Indeed, if this were not the case, there would exist $x \in \partial D$ and $t \in [0,1]$ such that $H(t,x) = x - t\check{L}x - (1 - t)Lx = 0$. Since $x = x_1 + x_2$ with $x_1 \in X_1$ and $x_2 \in X_2$ and $\check{L}x = LP(x_1 + x_2) = Lx_1$, it follows that $x_1 + x_2 = tLx_1 + (1 - t)Lx_1 + (1 - t)Lx_2$. This and (c) imply that $x_1 = Lx_1$ and $x_2 = (1 - t)Lx_2$. Since $I - L$ is injective, it follows that $x_1 = 0$ and $x_2 \neq 0$. Hence $Lx_2 = \lambda x_2$ with $\lambda = 1/(1 - t) > 1$, in contradiction to the fact that L, restricted to X_2, has no eigenvalues greater than 1. Thus, by Theorem 2.1, $\text{Deg}(I - L, D, 0) = \text{Deg}(I - \check{L}, D, 0)$.

But $\check{L}$ is compact and $\text{Deg}(I - \check{L}, D, 0) = \{\text{deg}_{LS}(I - \check{L}, D, 0)\}$ by Corollary 2.1. Hence, using the Leray–Schauder formula, we see that $\text{Deg}(I - L, D, 0) = \{(-1)^{\eta}\}$, where η is the sum of the multiplicities of eigenvalues $\lambda > 1$ of $\check{L}$.

To complete the proof we need only show that $V_n = \check{V}_n$ for each $n \in Z_+$, where

$$V_n = \{(x,\lambda) \mid x \in X, x \neq 0, \lambda > 1 \quad \text{and} \quad (\lambda I - L)^n x = 0\}$$

$$\check{V}_n = \{(x,\lambda) \mid x \in X, x \neq 0, \lambda > 1 \quad \text{and} \quad (\lambda I - \check{L})^n x = 0\}.$$

Suppose that $(x,\lambda) \in V_n$ and put $x = x_1 + x_2$, where $x_1 \in X_1$ and $x_2 \in X_2$. Then $(\lambda I - L)^n x_1 = -(\lambda I - L)^n x_2$. Then $x_2 = 0$, so $Lx = Lx_1 = \check{L}x$. This shows that $V_n \subset \check{V}_n$. Suppose that $(x,\lambda) \in \check{V}_n$. Then $(\lambda I - \check{L})x_1 = -(\lambda I - \check{L})^n x_2 = -\lambda^n x_2$, where $x = x_1 + x_2$ with $x_1 \in X_1$ and $x_2 \in X_2$. Hence $x_2 = 0$ and $\check{L}x = LPx = Lx$. This shows that $\check{V}_n \subset V_n$, and thus the proof is complete.   Q.E.D.

In view of Definition 5.1, we see that combining Theorem 5.3 with Corollary 5.1, we have the following basic result, which will prove to be useful in various applications.

Proposition 5.2 Let $D \subset X$ be open and bounded, $F: \overline{D} \to X$ P_1-compact with a P_1-compact F-derivative $L = F'_{x_0}$ at $x_0 \in D$. Suppose that conditions (a), (b), and (c) of Theorem 5.3 hold. Then x_0 is an isolated fixed point of F (i.e., a 0-point of $I - F$) and

$$\operatorname{ind}(I - F, x_0, 0) = (-1)^\eta \tag{5.17}$$

where η is the sum of the multiplicities of eigenvalues $\lambda > 1$ of L.

In virtue of Lemma 5.1, Proposition 5.2 includes the earlier results of Stuart, Stuart–Toland [274], and Danes [58] when F is k-ball contractive with $k < 1$.

To discuss the classes of operators for which conditions (a) and (c) hold and to deduce further results from Theorem 5.3 and Proposition 5.2, we now outline some relevant recent results on the radius of the essential spectrum of a linear operator L.

Now it is useful to associate with a bounded linear operator $L: X \rightarrow X$ the nonnegative numbers $\gamma(L) = \inf\{k \mid L \text{ is } k\text{-set contraction}\}$ and $\beta(L) = \inf\{k \mid L \text{ is } k\text{-ball contraction}\}$. It is obvious that $\gamma(L)$, $\beta(L) \leq \| L \|$, and if L is a nontrivial compact map, then $0 = \beta(L) < \| L \|$ and $0 = \gamma(L) < \| L \|$. Moreover, when X is a Hilbert space H, then $\gamma(L) = \beta(L)$, and as was shown in [115], at least for a bounded self-adjoint operator in a complex or real Hilbert space it turns out that $\gamma(L)$ [and $\beta(L)$] is determined by the spectrum $\sigma(L)$ of L. Let $F_k(\overline{D})$ be the class of k-Φ contractions $F: \overline{D} \rightarrow X$ with $k < 1$.

Let $L: X \rightarrow X$ be a closed, densely defined linear operator with X real or complex. Browder [20] defined the *essential spectrum* of L, $\sigma_e(L)$, to be the set $\lambda \in \sigma(L)$ such that at least one of the following holds:

(e1) $R(\lambda I - L)$ is not closed.
(e2) λ is a limit point of $\sigma(L)$.
(e3) $\cup_{j \geq 1} N(\lambda I - L)^j$ is infinite dimensional.

The set $\sigma_d = \sigma(L) \backslash \sigma_e(L)$ is called the *discrete* spectrum of L. Clearly, the set $\sigma_e(L)$ is closed and the *radius of the essential spectrum*, denoted by $r_e(L)$, is defined by $r_e(L) = \max\{|\lambda| \mid \lambda \in \sigma_e(L)\}$.

Remark 5.1 It was shown by Nussbaum in [195] that if $L \in L(X,X) \cap F_k(X)$ is such that $r_e(L) \leq 1$, there exists a finite-dimensional subspace X_1 and a closed subspace X_2 such that $X = X_1 \oplus X_2$, where X_1 and X_2 are both invariant under L and $(I - tL)|_{X_2}$ is a homomorphism of X_2 onto itself for each $t \in [0,1]$. Thus for $L \in L(X,X)$ for which $r_e(L) < 1$, conditions (b) and (c) hold. By virtue of this fact, Theorem 5.3 and Proposition 5.2 reduce to the corresponding results of Toland [275] if instead of (b) and (c) we assume that $r_e(L) < 1$ and X is a Hilbert space.

Remark 5.1 indicates the need of finding the way to estimate $r_e(L)$ as accurately as possible. As was shown by Hetzer [115], the same problem of calculating $r_e(L)$ arises when we try to solve a semilinear equation with a linear part having a nontrivial null space. It was shown by Nussbaum [194] that if $L \in L(X,X) \cap F_k$, then $r_e(L) \le \gamma(L)$. Stuart [273] improved this result in case X is Hilbert by proving the following:

Proposition 5.3 Let $\bar{H}$ be a complex Hilbert space and $L: \bar{H} \to \bar{H}$ self-adjoint. Then:

(i) If L is bounded, $r_e(L) = \gamma(L) = \beta(L)$.
(ii) If $\sigma_e(L) \subset \{\lambda \in R | \, |\lambda| \ge q\}$ and $0 \notin \sigma(L)$, then L^{-1} exists, is bounded, and $\gamma(L^{-1}) = \beta(L^{-1}) \le q^{-1}$.

Proof. Let $E(\lambda)$ be the resolution of the identity corresponding to L.

(i) It follows immediately from the results of [194] that $r_e(L) \le \gamma(L)$. Let $k = r_e(L)$. Then, for $\epsilon > 0$, we have

$$L = \int_{|\lambda| < k + \epsilon} \lambda \, dE(\lambda) + \int_{|\lambda| \ge k + \epsilon} \lambda \, dE(\lambda) = S + C.$$

Clearly, $\| S \| \le k + \epsilon$, and C is compact since it is of finite rank. Hence L is a $(k + \epsilon)$-set and $(k + \epsilon)$-ball contraction, so $\gamma(L) = \beta(L) \le k$. This establishes (i).

(ii) Since $0 \notin \sigma(L)$, L^{-1} exists, is bounded, and is given by

$$L^{-1} = \int_{|\lambda| > q - \epsilon} \lambda^{-1} \, dE(\lambda) + \int_{|\lambda| \le q + \epsilon} \lambda^{-1} \, dE(\lambda) = S + C$$

for any $\epsilon > 0$. Clearly, $\| S \| \le (q - \epsilon)^{-1}$, and C is compact since it is of finite rank. Hence L^{-1} is a $(q - \epsilon)^{-1}$-set and $(q - \epsilon)^{-1}$-ball contraction, so $\gamma(L^{-1}) = \beta(L^{-1}) \le q^{-1}$.   Q.E.D.

Remark 5.2 Let $L: H \to H$ be self-adjoint with H real. Let $\bar{H}_c$ be the complexification of H and $\bar{L}_c$ the complexification of L. It is easy to check that $\bar{L}_c: \bar{H}_c \to \bar{H}_c$ is self-adjoint and $\sigma(\bar{L}_c) = \sigma(L)$. For future reference we state the following consequence of Proposition 5.3.

Corollary 5.3 Let H be real Hilbert space and $L: H \to H$ self-adjoint with $0 \notin \sigma(\bar{L}_c) [= \sigma(L)]$ and $\sigma_e(\bar{L}_c) \subset [q, \infty)$ for some $q > 0$. Then $\gamma(L^{-1}) = \beta(L^{-1}) \le q^{-1}$.

For later use we state the following result from [273]:

Proposition 5.4 Let $L: H \to H$ be a positive self-adjoint operator from H (real or complex) into H. Then $\gamma(L^{1/2}) = \{\gamma(L)\}^{1/2}$.

A second formula for $r_e(L)$, where $L: D(L) \subset H \to H$ is a densely defined self-adjoint operator, is due to Hetzer [115] and plays an important role in the solvability of semilinear equations. We start with the following result due also to Hetzer [115], which is of interest in its own right and will be used later.

Let X, Y be Banach spaces and $L: D(L) \subset X \to Y$ a closed linear operator. Then L is said to be a Φ_+-operator iff dim $N(L) < \infty$ and $R(L)$ is closed. Of course, an Φ_+-operator is Fredholm if additionally dim($Y/R(L)$) $< \infty$. Further, we set:

$$l(L) = \sup\{r \mid r \in R^+, r\gamma(Q) \le \gamma(L(Q)) \tag{5.18}$$
$$\text{for each bounded} \quad Q \subset D(L)\}.$$

$$\tilde{l}(L) = \sup\{r \mid r \in R^+, r\beta(Q) \le \beta(L(Q)) \tag{5.19}$$
$$\text{for each bounded} \quad Q \subset D(L)\}.$$

The following result from [115] will prove to be useful.

Theorem 5.4 Suppose that $L: D(L) \subset X \to Y$ is as above. Then $l(L) > 0$ [or $\tilde{l}(L) > 0$] iff L is a Φ_+-operator.

Proof. Suppose first that $L: D(L) \subset X \to Y$ is a Φ_+-operator. Then there is a closed linear subspace $X_1 \subset X$ such that $N(L) \oplus X_1 = X$. Setting $L_1 = L|_{D(L_1)}$, where $D(L_1) = D(L) \cap X_1$, we see that L_1 is injective, closed, $R(L_1) = R(L)$ and L_1^{-1} is bounded by the closed graph theorem. Hence there exists $r > 0$ such that

$$\| L_1 x - L_1 y \| \ge r\| x - y \| \quad \text{for} \quad x, y \in D(L_1). \tag{5.20}$$

If $\tilde{Q} \subset D(L_1)$ is bounded and $\{D_1, \ldots, D_n\}$ is a covering of $L_1(\tilde{Q})$ with diam(D_j) $\le d$ for $1 \le j \le n$, then (5.20) implies that r diam($L_1^{-1}(D_j) \le d$ for $1 \le j \le n$. Hence for each bounded $\tilde{Q} \subset D(L_1)$, we get

$$r\gamma(\tilde{Q}) \le \gamma(L_1(\tilde{Q})). \tag{5.21}$$

Now let P be a projection of X onto $N(L)$ and Q a bounded set in $D(L)$. Using Lemma I.2.1 and (5.21), we get

$$r\gamma(Q) \le r\gamma((I - P)(Q) + P(Q))$$
$$\le r\gamma((I - P)(Q)) \le \gamma(L_1 \cdot (I - P)(Q))$$
$$\le \gamma(L(Q)) + \gamma(L \cdot P(Q)) \le \gamma(L(Q)) \quad [\text{i.e.,} \ l(L) \ge r > 0].$$

Suppose now that $l(L) > 0$. Let B_1 be the unit ball of X. Then, in view of (5.18), for $Q = B_1 \cap N(L)$ we have $r\gamma(Q) \leq \gamma(\{0\}) = 0$. Lemma I.2.1 then implies that Q is compact and therefore dim $N(L) < \infty$. Thus there is a closed subspace X_1 of X such that $X = N(L) \oplus X_1$. Clearly, $L_1 = L|_{D(L) \cap X_1}$ is injective. If L_1^{-1} is not continuous, there exists a sequence $\{x_n\} \subset X_1$ such that $\| x_n \| = 1$ and $L_1(x_n) \to 0$. Hence $\gamma(\{Lx_n\}) = 0$; that is, $\{x_j\}$ has a convergent subsequence $\{x_{n_j}\}$ such that $x_{n_j} \to x$ for some $x \in X_1$ with $\| x \| = 1$. Since L is closed and $L_1(x_n) = L(x_n) \to 0$, it follows that $x \in D(L)$ and $Lx = 0$. This is a contradiction to $x \in X_1$. So L_1^{-1} is continuous and therefore $R(L_1)$ is closed and $R(L_1) = R(L)$ (i.e., L is a Φ_+-operator). The same argument shows that L is a Φ_+-operator iff $\bar{l}(L) > 0$, where $\bar{l}(L)$ is given by (5.19). Q.E.D.

In dealing with applications, a calculation of $l(L)$ or $\hat{l}(L)$ for a given Fredholm operator L is necessary. Direct estimates can be given in case of ODEs (see [115,234]), but they fail, in case we deal with PDEs. As was shown by Hetzer [115], a calculation of $l(L)$ or $\bar{l}(L)$ is possible if the problem involves a self-adjoint Fredholm operator in a Hilbert space.

Now if $Y = X$ and X is the Hilbert space H and $L: D(L) \subset H \to H$ is a self-adjoint operator, the essential spectrum $\sigma_e(L)$ of L is defined by

$$\sigma_e(L) = \{\lambda \mid \lambda \in \sigma(L), \lambda \text{ is not an isolated}$$

$$\text{eigenvalue of finite multiplicity}\}. \quad (5.22)$$

Note that there are many definitions of the essential spectrum, but that they all coincide, when L is self-adjoint. We now prove the following result from [115].

Theorem 5.5 Let H be a real or complex Hilbert space and $L: D(L) \subset H \to H$ closed and self-adjoint. Then $l(L) = \bar{l}(L) = r_e(L) = \inf\{|\lambda| \mid \lambda \in \sigma_e(L)\}$.

Proof. We first show that $l(L) \leq q$, where $q = r_e(L)$. It is known that for $\lambda \in \sigma_e(L)$ there exists a sequence $\{x_n\} \subset D(L)$ with $\| x_n \| = 1$ for each $n \in Z_+$ and $\lim_n(\lambda x_n - Lx_n) = 0$, and $\{x_n\}$ has no convergent subsequence. Hence $\gamma(\{x_n\}) > 0$ and $\gamma(\{Lx_n\}) = \gamma(\{\lambda x_n\}) = |\lambda| \gamma(\{x_n\})$ showing that $l(L) \leq |\lambda|$ for each $\lambda \in \sigma_e(L)$ [i.e., $l(L) \leq q$].

Next we show that $l(L) \geq q$. If $0 \in \sigma_e(L)$, the assertion is obvious. Otherwise, $q > 0$ and L is a Fredholm operator with $\text{ind}(L) = 0$ because L is assumed to be self-adjoint. We first consider the case when H is complex. Since L is self-adjoint, L is reduced by $H_1 \equiv N(L)$ and by $H_1^\perp$. Set $L_1 = L|_{D(L_1)}$, where $D(L_1) = H_1^\perp \cap D(L)$, and $L_1: D(L_1) \to H_1^\perp$ injective and self-adjoint. It follows from $\sigma(L) = \sigma(L|_{H_1}) \cup \sigma(L_1)$

and $\sigma(L|_{H_1}) = \{0\}$ that $\sigma_e(L) = \sigma_e(L_1)$. Since $q > 0$, Proposition 5.3(ii) asserts that L_1^{-1} is k-set contractive with $k \leq q^{-1}$. We show that $l(L_1) \geq q$. Let $Q \subset D(L_1)$ be bounded. We can assume that $\gamma(L_1(Q)) < \infty$, and obtain $\gamma(L_1^{-1}(L(Q))) \leq k\gamma(L(Q))$ [i.e., $k^{-1}\gamma(Q) \leq \gamma(L(Q))$], which implies that $k^{-1} \leq l(L_1)$ or $l(L_1) \geq q$. If P is the orthogonal projection of H onto $N(L)$, then for each bounded $Q \subset D(L)$ we have

$$\gamma(Q) = \gamma[(P + I - P)(Q)] \leq \gamma[P(Q) + (I - P)(Q)]$$
$$\leq \gamma(P(Q)) + \gamma((I - P)(Q)) = \gamma((I - P)(Q)) \leq \gamma(Q),$$

using the fact that $I - P$ is nonexpansive and P is compact. Hence $\gamma(Q) = \gamma((I - P)(Q))$. Since H_1 and $H_1^\perp$ reduce L, $L \cdot (1 - P) = (1 - P) \cdot L$ and we conclude analogously that

$$\gamma(L(Q)) = \gamma[L \cdot (P + I - P)(Q)] \leq \gamma(L(I - P)(Q))$$
$$= \gamma((I - P)L(Q)) = \gamma(L(Q)).$$

This implies that $\gamma(L(Q)) = \gamma(L_1(I - P)(Q))$. Both assertions together ensure that $l(L) = l(L_1)$, which establishes the claim when H is complex.

When H is real, we consider the complexification $\tilde{H}_c$ and the operator $\tilde{L}_c$ induced by L. We know that $\tilde{L}_c$ is self-adjoint and $\sigma_e(L) = \sigma_e(\tilde{L}_c)$. Therefore, $l(\tilde{L}_c) \geq q$. On the other hand, we obtain for $\epsilon > 0$ and bounded $Q \subset D(L)$ the relation

$$\gamma(L(Q)) = \gamma(\tilde{L}_c(Q \times \{0\})) \geq (l(\tilde{L}_c) - \epsilon)\gamma(Q \times \{0\}) = (q - \epsilon)\gamma(Q),$$

which shows that $l(L) \geq q$ [i.e., $l(L) = r_e(L)$].

In the same way one proves that $\tilde{l}(L) = r_e(L)$.   Q.E.D.

6. BIFURCATION AND ASYMPTOTIC BIFURCATION FOR EQUATIONS INVOLVING A-PROPER MAPPINGS WITH APPLICATIONS

6.1 Outline

Let R be a real line, X a real separable Banach space, and $C: X \rightarrow X$ a compact map with $C(0) = 0$. The classical bifurcation problem, whose rigorous study was initiated by Krasnoselsky [136], concerns itself with the existence of nontrivial solutions of arbitrarily small norms of the equation

$$x = \lambda C(x) \quad \text{with} \quad C = A + B, \tag{6.1}$$

where A is a compact linear operator and $B: X \rightarrow X$ is compact with $\| Bx \|/\| x \| \rightarrow 0$ as $\| x \| \rightarrow 0$. In this context, Krasnoselskii [136] has shown that if $\lambda_0 \in R$ is a characteristic value of A of odd multiplicity, then λ_0 is a bifurcation point for equation (6.1). On the other hand, the asymptotic bifurcation problem concerns itself with the existence of nontrivial solutions of (6.1) of arbitrarily large norms. It was also shown in [136] that if $\| Bx \|/\| x \| \to 0$ as $\| x \| \to \infty$, then each characteristic value $\lambda_\infty \in R$ of A of odd multiplicity is an asymptotic bifurcation point for (6.1).

Since bifurcation phenomena occur in many parts of physics (see [130]), the existence of bifurcation solutions became a subject of extensive study under various conditions on C. (See the survey paper of Rabinowitz [245] for the literature up to 1971.)

Recently, the basic bifurcation results of Krasnoselskii have been extended in two directions. First, Rabinowitz [245,246] has investigated the nature of bifurcation and asymptotic bifurcation from characteristic values of A of odd multiplicity and has shown that strong results hold on the topological nature of the bifurcation phenomena involved. The global results of [245,246] were extended by Stuart [272] to the case when $C = A + B$ is k-set-contractive, B is compact, and A is either a Fréchet derivative of C at 0 or an asymptotic derivative of C. The global bifurcation results for k-set contraction $C = A + B$ with A not necessarily a Fréchet derivative of C at 0 has also been obtained by Nussbaum [196]. Bifurcation and asymptotic bifurcation results for k-set-contractive gradient operators have recently been obtained by Toland [275]. Further results have been obtained in [54,57,130,198,276] and others.

Second, assuming that X is a separable reflexive Banach space, Skrypnik [262] studied the local bifurcation problem for the equation

$$Tx = \lambda C \qquad (x \in X) \tag{6.2}$$

where $C: X \to X^*$ is compact with Fréchet derivative at 0, and $T: X \to X^*$ is Fréchet differentiable at 0 and satisfies condition (α) (the latter condition is essentially equivalent to condition (S_+) of Browder [36]). If X is a Hilbert space and $T: X \to X$ is strongly monotone and asymptotically linear, the asymptotic bifurcation for equation (6.2) has been studied by Aldarwish and Langenbach [3] (see also [149]). Most of the authors mentioned above applied their abstract results to elliptic PDEs and ODEs.

In [78,80] (and in [4] with others), Fitzpatrick obtained a general global bifurcation theorem for a family of A-proper maps depending on a single parameter under the condition that the parity of the curve of linearizations about the trivial branch is negative. His results contained those in [228,300,303]. Although our bifurcation results in [228] are less general and only local, we present them here because of the simplicity of their proofs, conditions, and practical usefulness.

First, after outline, in Section 4.2 we extend Theorems IV.2.1 and

IV.3.1 of Krasnoselskii [136] to an equation of the form

$$Tx = \lambda Cx \qquad (x \in X, \lambda \in R), \tag{6.3}$$

where T is an A-proper mapping of the Banach space X into another Banach space Y, $C: X \to Y$ is compact and such that $T = T_0' + Q_0$ and $C = C_0' + P_0$, where T_0' and C_0' are Fréchet derivatives of T and C, respectively, with T_0' A-proper, or $T = T_\infty' + Q_\infty$ and $C = C_\infty' + P_\infty$, where T_∞' and C_∞' are asymptotic derivatives of T and C, respectively, with T_∞' A-proper. Note that, in either case, equation (6.3) can be written in the form

$$(T_\alpha' + Q_\alpha)x = \lambda(C_\alpha'x + P_\alpha x), \qquad T_\alpha', C_\alpha' \in L(X,Y), \tag{6.4}$$

where α is either 0 or ∞. Since, in general, Q_α is neither compact nor k-set contractive, the bifurcation results of the above-mentioned authors are not applicable to equation (6.4). On the other hand, since the class of A-proper mappings with A-proper Fréchet or asymptotic derivatives is quite large, our bifurcation and asymptotic bifurcation results (Theorems 6.1 and 6.2) will contain as special cases some of the local bifurcation results of the above-mentioned authors. See Remarks for special cases.

Second, in Section 6.3 we first apply our Theorem 6.2 to the variational asymptotic bifurcation problem involving quasilinear elliptic operators in divergence form of order $2m$ and then deduce the result of [3] under much weaker conditions than those used in [3]. In the second part of Section 6.3 we use Theorem 6.1 to study the bifurcation problem involving second-order elliptic quasilinear operators as well as the operators of order $2m$ acting in Sobolev spaces. The ordinary differential equations are also treated. Finally, let us add that problems I and II in this section can be handled neither by the classical results in [136] nor by the more recent results in [196,246,276]. Moreover, problems III, IV, and V show that even in those cases where the classical bifurcation results are applicable (via integral equations) the bifurcation results based on the A-proper mapping theory offer a more direct alternative approach to certain differential bifurcation problems.

6.2 Bifurcation and Asymptotic Bifurcation Theorems

In this section we use the finite-dimensional topological Brouwer degree, the Leray–Schauder fixed-point index theorem (see [136]) and the generalized degree theory for A-proper mappings developed by Browder–Petryshyn [42] to establish the existence of bifurcating solutions for the equation

$$Tx = \lambda Cx, \tag{6.5}$$

where $C: X \to Y$ is compact, $T: X \to Y$ is A-proper w.r.t. $\Gamma = \{X_n, V_n; E_n, W_n\}$ and $\lambda \in R$. A point $0 \neq \lambda \in R$ is called an *eigenvalue* of equation (6.5) if there exists $0 \neq x \in X$ such that (6.5) holds. Here we are interested in establishing the existence of nontrivial solutions of (6.5) of arbitrarily small and/or large norms. Thus our bifurcation results are essentially of a local character. To the best of our knowledge the bifurcation problems for (6.5) when T is A-proper has not been studied by any other author.

Bifurcation points   To define and study the bifurcation problem associated with (6.5) [i.e., the existence of nontrivial solutions of (6.5) with small norms], we impose the following hypotheses on T and C:

(i) $T(0) = 0$ and T has the Fréchet derivative T_0' at $x = 0$ which is A-proper w.r.t. Γ.

(ii) $C(0) = 0$ and C has the Fréchet derivative C_0' at $x = 0$.

Definition 6.1   A point $\lambda_0 \in R$ is called a *bifurcation point* (BP) for (6.5) if to each sufficiently small $\epsilon > 0$ there corresponds a number $r_0(\epsilon) > 0$ such that for any $r \in (0, r_0(\epsilon))$, equation (6.5) has a solution (λ, x_λ) with $|\lambda - \lambda_0| < \epsilon$ and $0 < \|x_\lambda\| = r$.

Since the bifurcation problem involves the existence of solutions of equation (6.5) of small norms, the entire theory developed below remains valid where the mappings T and C are defined only on a closure of some open bounded neighborhood of the origin. This actually will be the case in some applications in Section V.3.

It will be shown in Proposition 6.1(a) in Section 6.3 that a BP for equation (6.5) can only be an eigenvalue of the linearized problem

$$T_0'(x) - \lambda C_0'(x) = 0. \tag{6.6_0}$$

But as in the theory of compact operators (see [136]), bifurcation need not occur at each eigenvalue of (6.6_0). When we additionally assume that T_0 is injective, then by Theorem II.1.1, T_0' is a homeomorphism of X onto Y and since, by Lemma II.4.1 in [136], C_0' is compact, it follows that $C_0'T_0'^{-1}: Y \to Y$ is compact and thus λ is an eigenvalue of (6.6_0) if and only if λ is a characteristic value of the compact map $C_0'T_0'^{-1}$. In this case by the *multiplicity* of the eigenvalue λ of equation (6.6_0) we mean the dimension of $\cup_{j \geq 1} N((I - \lambda C_0'T_0'^{-1})^j)$, where $N(A)$ denotes the null space of the linear operator A.

Before we state our first main result in this section, we note that if $A \in L(X, Y)$ is injective and A-proper w.r.t. Γ, then by Theorem II.1.2, there

exists a constant $c > 0$ and an integer $n_0 > 1$ such that $\| A_n(x) \| \geq c\| x \|$
for all $x \in X_n$ and each $n \geq n_0$ and, consequently, for each fixed $n \geq n_0$,
A_n is a linear homeomorphism of X_n onto E_n. In view of this, we say that
an injective A-proper map $A \in L(X,Y)$ has a *constant sign* $S_\Gamma(A)$ if either
$S_\Gamma(A) = \deg(A_n, B_n(0,r),0) = +1$ for all $n \geq n_0$ or $S_\Gamma(A) =$
$\deg(A_n, B_n(0,r),0) = -1$ for all $n \geq n_0$ with $r > 0$ arbitrary, where $B_n(0,r)$
$\equiv B(0,r) \cap X_n$ and $\deg(A_n, B_n(0,r),0)$ is the Brouwer degree. The following
result, which extends Theorem IV.2.1 of Krasnoselskii [136], is our first
bifurcation theorem for maps that are differentiable at 0.

Theorem 6.1 Suppose that $T: X \to Y$ is A-proper w.r.t. the admissible
scheme $\Gamma = \{X_n, V_n; E_n, W_n\}$ and $C: X \to Y$ is compact. Suppose further
that conditions (i) and (ii) hold with T_0' injective.

(a) If T_0' also has a constant sign $S_\Gamma(T_0')$, then each eigenvalue of the
linearized problem (6.6_0) of odd multiplicity is a BP for equation (6.5).

(b) If T and T_0' are A-proper w.r.t. the projectionally complete scheme
$\Gamma_P = \{X_n, V_n; Y_n, Q_n\}$, then the conclusion of (a) holds without the additional assumption that T_0' has a constant sign w.r.t. Γ_P.

Asymptotic bifurcation points To study the asymptotic bifurcation problem associated with equation (6.5) [i.e., the existence of nontrivial solutions of (6.5) with very large norms] we assume that T and C are *asymptotically linear* with the *asymptotic derivatives* T_∞' and C_∞', respectively;
that is, we assume the following:

(j) There is $T_\infty' \in L(X,Y)$ such that $\| T(x) - T_\infty'(x) \|/\| x \| \to 0$ as $\| x \|$
$\to \infty$ and T_∞' is A-proper w.r.t. Γ.

(jj) There is $C_\infty' \in L(X,Y)$ such that $\| C(x) - C_\infty'(x) \|/\| x \| \to 0$ as $\| x \|$
$\to \infty$.

Definition 6.2 A point $\lambda_\infty \in R$ is called an *asymptotic bifurcation point*
(ABP) for equation (6.5) if to each sufficiently small $\epsilon > 0$ there corresponds a number $r_\infty(\epsilon) > 0$ such that for any $r > r_\infty(\epsilon)$, equation (6.5) has
a solution (λ, x_λ) with $| \lambda_\infty - \lambda | < \epsilon$ and $\| x_\lambda \| = r$.

It will be shown in Proposition 6.1(b) in Section 1.3 that an ABP for
(6.5) can only be an eigenvalue of the linearized problem

$$T_\infty'(x) - \lambda C_\infty'(x) = 0. \qquad (6.6_\infty)$$

It was shown in [136] (for the case when $Y = X$ and $T_\infty' = I$) that asymptotic
bifurcation need not occur at each eigenvalue of (6.6_∞). However, our

second basic result shows that the following analog of Theorem 6.1, which extends Theorem IV.3.1 of [136], is valid for asymptotically linear maps (i.e., for maps that are differentiable at ∞).

Theorem 6.2 Suppose that $T: X \to Y$ is A-proper w.r.t. Γ, $C: X \to Y$ is compact, and conditions (j) and (jj) hold with T'_∞ injective.

(a) If T'_∞ also has a constant sign $S_\Gamma(T'_\infty)$, then each eigenvalue of (6.6_∞) of odd multiplicity is an ABP for (6.5).

(b) If T and T'_∞ are A-proper w.r.t. $\Gamma_P = \{X_n,V_n;Y_n,Q_n\}$, then the conclusion of (a) holds without the additional assumption that T'_∞ has a constant sign w.r.t. Γ_P.

Proof of the theorems. The proof of Theorems 6.1 and 6.2 is based on the following propositions, some of which are of independent interest. Since in some parts the arguments are identical in both cases, we shall state and prove these propositions simultaneously for maps T and C which are differentiable either at 0 or at ∞.

Proposition 6.1 Let $\Gamma = \{X_n,V_n;E_n,W_n\}$ be an admissible scheme for (X,Y), $T: X \to Y$ A-proper w.r.t. Γ and $C: X \to Y$ compact.

(a) Suppose that conditions (i) and (ii) hold. If λ_0 is not an eigenvalue of the linearized problem (6.6_0), then λ_0 is not a BP for (6.5).

(b) Suppose that conditions (j) and (jj) hold. If λ_∞ is not an eigenvalue of the linearized problem (6.6_∞), then λ_∞ is not an ABP for (6.5).

Proof. To shorten the proofs and to clarify the notation in what follows, it is always assumed that either $\alpha = 0$ or $\alpha = \infty$. That is, $\alpha = 0$ corresponds to Theorem 6.1 while $\alpha = \infty$ corresponds to Theorem 6.2.

First note that conditions (i) and (j) can be put in the form

$$T(x) = T'_\alpha(x) + Q_\alpha(x) \quad \text{with} \quad \| Q_\alpha(x) \| \tag{6.7_α}$$
$$= o(\| x \|) \quad \text{as} \quad \| x \| \to \alpha,$$

where $T'_\alpha \in L(X,Y)$ is A-proper w.r.t. Γ, while (ii) and (jj) can be put in the form

$$C(x) = C'_\alpha(x) + P_\alpha(x) \quad \text{with} \quad \| P_\alpha(x) \| \tag{6.8_α}$$
$$= o(\| x \|) \quad \text{as} \quad \| x \| \to \alpha,$$

where $C'_\alpha \in L(X,Y)$ and α is either 0 or ∞.

Since, as was shown in [136], C'_α is compact and, by our assumption, T'_α is A-proper w.r.t. Γ, it follows from Theorem I.1.2 that $A_\alpha \equiv T'_\alpha -$

$\lambda_\alpha C'_\alpha \in L(X,Y)$ is A-proper w.r.t. Γ. Moreover, since λ_α is not an eigenvalue of (6.6_α), the map A_α is injective and therefore, by Theorem II.1.1, A_α is a homeomorphism of X onto Y. Hence there exists a constant $a_\alpha > 0$ such that

$$\| A_\alpha(x) \| \ge a_\alpha \| x \| \qquad \text{for all} \quad x \text{ in } X \ (\alpha = 0 \quad \text{or} \quad \alpha = \infty) \qquad (6.9_\alpha)$$

Let $\lambda \in R$ be such that

$$| \lambda - \lambda_\alpha | < a_\alpha (4 \| C'_\alpha \|)^{-1} \equiv b_\alpha \qquad (\alpha = 0 \quad \text{or} \quad \alpha = \infty) \qquad (6.10_\alpha)$$

and choose $r'_\alpha > 0$ such that for all x in X with $\| x \| \ge r'_\infty$ (respectively, $\| x \| \le r'_0$), we have for $\alpha = 0$ or $\alpha = \infty$ the inequalities

$$\| Q_\alpha(x) \| = \| T(x) - T'_\alpha(x) \| \le a_\alpha(\tfrac{1}{4}) \| x \| \qquad (6.11_\alpha)$$

and

$$\| P_\alpha(x) \| = \| C(x) - C'_\alpha(x) \| \le a_\alpha(| \lambda_\alpha | + b_\alpha)^{-1} \| x \|. \qquad (6.12_\alpha)$$

Then, in view of (6.9_α) and (6.11_α)–(6.12_α) for every $\| x \| \ge r'_\alpha$ (respectively, $0 < \| x \| \le r'_0$), and every $\lambda \in R$ satisfying (6.10_α), we have

$$\begin{aligned}
\| T(x) - \lambda C(x) \| &= \| (T'_\alpha(x) - \lambda_\alpha C'_\alpha(x) \\
&\quad + (\lambda_\alpha - \lambda)C'_\alpha(x) + Q_\alpha(x) - \lambda P_\alpha(x) \| \\
&\ge \| A_\alpha(x) \| - | \lambda_\alpha - \lambda | \| C'_\alpha \| \| x \| \\
&\quad - \| Q_\alpha(x) \| - (| \lambda_\alpha | + b_\alpha) \| P_\alpha(x) \| \\
&\ge (a_\alpha - \tfrac{1}{4}a_\alpha - \tfrac{1}{4}a_\alpha - \tfrac{1}{4}a_\alpha) = \tfrac{1}{4}a_\alpha \| x \|.
\end{aligned}$$

The last inequality establishes the validity of the assertion (a) for the case when $\alpha = 0$ and of the assertion (b) for the case when $\alpha = \infty$.

$$\text{Q.E.D.}$$

As was mentioned above, it follows from Proposition 6.1(a) that a BP for equation (6.5) can occur only at an eigenvalue of the linearized problem (6.6_0), while Proposition 6.1(b) implies that an ABP for (6.5) can occur only at an eigenvalue of (6.6_∞).

Remark 6.1 To finish the proof of Theorems 6.1 and 6.2, we recall for completeness that when $T: X \to Y$ is A-proper with injective A-proper derivatives T'_0 and T'_∞ at 0 (with $T(0) = 0$) and at ∞ respectively, then as was shown above

(a) There exists $r_0'' > 0$ such that for each $r \in (0, r_0'')$

$$\text{Deg } (T, B(0,r), 0) = \text{Deg } (T_0', B(0,r), 0)$$

(6.13_α)

(b) There exists $r_\infty'' > 0$ such that for each $r \geq r_\infty''$

$$\text{Deg } (T, B(0,r), 0) = \text{Deg } (T_\infty', B(0,r), 0)$$

(6.14_α)

Proof of Theorems 6.1 and 6.2. Since the arguments are similar in both cases, we combine the proofs of Theorems 6.1 and 6.2 by setting $\alpha = 0$ in case of Theorem 6.1 and $\alpha = \infty$ in case of Theorem 6.2.

(a) Since $T_\alpha' \in L(X,Y)$ is injective A-proper map w.r.t. the admissible scheme Γ, T_α is a homeomorphism of X onto Y by Theorem II.1.1 and consequently, $\lambda \in R$ is an eigenvalue of equation (6.6_α) if and only if λ is a characteristic value of the compact operator $C_\alpha' T_\alpha'^{-1} \colon Y \to Y$. Thus the eigenvalues of (6.6_α), or equivalently, the characteristic values of $C_\alpha' T_\alpha'^{-1}$, form a discrete set. Let λ_α be a characteristic value of $C_\alpha' T_\alpha'^{-1}$ of odd multiplicity. Then there exists $\epsilon_\alpha > 0$ such that there are no eigenvalues of (6.6_α) in $[\lambda_\alpha - \epsilon_\alpha, \lambda_\alpha + \epsilon_\alpha]$ distinct from λ_α. Let ϵ be an arbitrary but fixed number in $(0, \epsilon_\alpha)$ and set $D_\alpha(\pm\epsilon) \equiv T_\alpha' - (\lambda_0 \pm \epsilon)C_\alpha'$. Then $T_\alpha', D_\alpha \in L(X,Y)$ are injective and A-proper w.r.t. Γ.

Now consider the mapping $H(x,t)\colon X \times [0,1] \to Y$ given by

$$H(x,t) = T(x) - (\lambda_\alpha - 2t\epsilon + \epsilon)C(x). \qquad (6.15_\alpha)$$

Then $H(\cdot,0) = T - (\lambda_0 + \epsilon)C$, $H(\cdot,1) = T - (\lambda_0 - \epsilon)C$, $H(x,t)$ is A-proper w.r.t. Γ for each fixed $t \in [0,1]$ and uniformly continuous in x in any bounded set of X with H_α the corresponding derivative of H. Moreover, $H_\alpha(\cdot,0) = D_\alpha(+\epsilon)$ and $H_\alpha(\cdot,1) = D_\alpha(-\epsilon)$ with $D_\alpha(+\epsilon)$ and $D_\alpha(-\epsilon)$ injective. Hence it follows from Proposition 5.1 that there exists a number $r_\alpha(\epsilon) > 0$ such that for any $r \in (0, r_0(\epsilon))$ [respectively, $r \in (r_\infty(\epsilon), \infty)$], $H(x,0) \neq 0$ and $H(x,1) \neq 0$ for $x \in \partial B(0,r)$ and

$$\text{Deg}(T - (\lambda_\alpha \pm \epsilon)C, B(0,r), 0) = \text{Deg}(D_\alpha(\pm\epsilon), B(0,r), 0). \qquad (6.16_\alpha)$$

Now, since T_α and $D_\alpha(\pm\epsilon) = T_\alpha' - (\lambda_\alpha \pm \epsilon)C_\alpha'$ are injective linear A-proper mappings, and by the additional assumption in (a) the map T_α' has a constant sign $S_\Gamma(T_\alpha')$, it follows from Theorem 2 of Browder–Petryshyn [41] that

$$\text{Deg}(D_\alpha(+\epsilon), B(0,r), 0)$$

$$= S_\Gamma(T_\alpha') \deg_{LS}(I - (\lambda_0 + \epsilon)C_\alpha' T_\alpha'^{-1}, T_\alpha'(B(0,r)), 0)$$

(6.17_α)

and

$$\text{Deg}(D_\alpha(-\epsilon),B(0,r),0) \tag{6.18$_\alpha$}$$
$$= S_\Gamma(T_\alpha') \, \text{deg}_{\text{LS}}(I - (\lambda_0 - \epsilon)C_\alpha'T_\alpha'^{-1}, \, T_\alpha'(B(0,r), 0).$$

On the other hand, by the Leray–Schauder formula for the degree of compact linear vector fields (see, e.g., Theorem II.4.6 in [136]), it follows from (6.17_α) and (6.18_α) that $\text{Deg}(D_\alpha(+\epsilon),B(0,r),0) = S_\Gamma(T_\alpha)(-1)^\beta$ and $\text{Deg}(D_\alpha(-\epsilon),B(0,r),0) = S_\Gamma(T_\alpha)(-1)^\gamma$, where β (respectively, γ) is the sum of multiplicities of the eigenvalues of $C_\alpha'T_\alpha'^{-1}$ whose sign is the same as that of $\lambda_\alpha + \epsilon$ (respectively, $\lambda_\alpha - \epsilon$) and whose absolute value is greater than $|\lambda_\alpha + \epsilon|^{-1}$ (respectively, $|\lambda_\alpha - \epsilon|^{-1}$). Since λ_α is of odd multiplicity, it follows from the above that

$$\text{Deg}(D_\alpha(+\epsilon),B(0,r),0) \neq \text{Deg}(D_\alpha(-\epsilon),B(0,r),0).$$

Thus (6.16_α) and the homotopy Theorem 2.1(3) imply the existence of an $t \in (0,1)$ and an $x \in \partial B(0,r)$ such that $H(x,t) = 0$ [i.e., $Tx - \lambda Cx = 0$ with $\lambda = \lambda_\alpha - 2t\epsilon + \epsilon \in (\lambda_\alpha - \epsilon_\alpha, \lambda_\alpha + \epsilon_\alpha)$ and $\|x\| = r$]. This establishes the assertion (a) of Theorem 6.1, which is the case when $\alpha = 0$, and of Theorem 6.2, which is the case when $\alpha = \infty$.

(b) Suppose now that T and T_α' are A-proper w.r.t. the projectionally complete scheme $\Gamma_P = \{X_n,V_n;Y_n,Q_n\}$ with T_α' not necessarily having a constant sign. As was noted above, it follows that Γ_P is not only admissible but is also such that $Q_n(y) \to y$ for each y in Y and, of course, $M = \sup_n\|Q_n\| < \infty$.

Let λ_α be a characteristic number of $C_\alpha'T_\alpha'^{-1}$ of odd multiplicity and let ϵ be a fixed number in $(0,\epsilon_\alpha)$ such that $D_\alpha(\pm\epsilon) \equiv T_\alpha' - (\lambda_\alpha \pm \epsilon)C_\alpha'$ is injective and for each x in X we set $D(\pm\epsilon)(x) = T(x) - (\lambda_0 \pm \epsilon)C(x)$. Then, if (i) and (ii) hold, $D(\pm\epsilon)(0) = 0$ and $D_0(\pm\epsilon) = T_0' - (\lambda_0 \pm \epsilon)C_0'$ is the Fréchet derivative of $D(\pm\epsilon)$ at 0, while if (j)–(jj) hold, $D_\infty(\pm\epsilon) = T_\infty' - (\lambda_0 \pm \epsilon)C_\infty'$ is the asymptotic derivative of $D(\pm\epsilon)$. Now, since $D_\alpha(\pm\epsilon) \in L(X,Y)$ is injective and A-proper w.r.t. Γ_P, there exists a number $c_\alpha(\epsilon) > 0$ and an integer $N_\alpha(\epsilon) \geq 1$ such that

$$\|Q_nD_\alpha(\pm\epsilon)x\| \geq c_\alpha(\epsilon)\|x\| \qquad \text{for all} \quad x \in X_n \quad \text{and} \quad n \geq N_\alpha(\epsilon).$$
$$\tag{6.19$_\alpha$}$$

Choose $\rho_\alpha(\epsilon) > 0$ such that for all x in X with $\|x\| \geq \rho_\infty(\epsilon)$ [respectively, $0 < \|x\| \leq \rho_0(\epsilon)$], we have

$$\|D(\pm\epsilon)(x) - D_\alpha(\pm\epsilon)(x)\| < (2M)^{-1}c_\alpha(\epsilon)\|x\|. \tag{6.20$_\alpha$}$$

Then we claim that for every $r \geq \rho_\infty(\epsilon)$ [respectively, $0 < r < \rho_0(\epsilon)$], and for each fixed $n \geq N_\alpha(\epsilon)$ and every $t \in [0,1]$ we have

$$F_n(x,t) \equiv Q_n D_\alpha(\pm\epsilon)(x) + t[Q_n D(\pm\epsilon)(x) - Q_n D_\alpha(\pm\epsilon)(x)] \neq 0$$

for all $x \in \partial B_n(0,r)$. Indeed, it follows from (6.19_α) and (6.20_α) that for $x \in \partial B(0,r) \cap X_n$ and $t \in [0,1]$ we have

$$\| F_n(x,t) \| \geq \| Q_n D_\alpha(\pm\epsilon)(x) \| - \| Q_n D(\pm\epsilon)(x) - Q_n D_\alpha(\pm\epsilon)(x) \|$$
$$\geq (c_\alpha - M(2M)^{-1}c_\alpha) \| x \| = \tfrac{1}{2}rc_\alpha.$$

Consequently, by the homotopy theorem for the Brouwer degree, we see that for each fixed $n \geq N_\alpha(\epsilon)$ we have

$$\deg(Q_n D(\pm\epsilon),B_n(0,r),0) = \deg(Q_n D_\alpha(\pm\epsilon),B_n(0,r),0). \qquad (6.21_\alpha)$$

Now the operator $D_\alpha(\pm\epsilon) \equiv T_\alpha' - (\lambda_\alpha \pm \epsilon)C_\alpha' \in L(X,Y)$ is such that T_α' is injective and A-proper, $(\lambda_\alpha \pm \epsilon)C_\alpha' \in L(X,Y)$ is compact, and $D_\alpha(\pm\epsilon)$ is A-proper and injective. Hence by Theorem 2.3 there exists an integer $N_\alpha (\geq N_\alpha(\epsilon))$ independent of r such that for each fixed $n \geq N_\alpha$ we have

$$\deg(Q_n D_\alpha(\pm\epsilon),B_n(0,r),0) \qquad (6.22_\alpha)$$
$$= \deg(Q_n T_\alpha',B_n(0,r),0) \deg_{LS}(I - (\lambda_\alpha \pm \epsilon)C_\alpha' T_\alpha'^{-1}, T_\alpha(B(0,r)).$$

Since $(\lambda_\alpha \pm \epsilon)$ is not a characteristic value of $C_\alpha' T_\alpha'^{-1}$ and λ_α is a characteristic value of $C_\alpha' T_\alpha'^{-1}$ of odd multiplicity, it follows from the Leray–Schauder formula used in the proof of (a) and the equalities (6.21_α) and $6.22_\alpha)$ valid for each fixed $n \geq N_\alpha$ that

$$\deg(Q_n D(\pm\epsilon),B_n(0,r),0) \neq \deg(Q_n D(-\epsilon),B_n(0,r),0). \qquad (6.23_\alpha)$$

Thus, by (6.23_α) for each $n \geq N_\alpha$ there exist $x_n \in \partial B_n(0,r)$ and $t_n \in (0,1)$ such that

$$t_n[Q_n T(x_n) - (\lambda_\alpha + \epsilon)Q_n C(x_n)]$$
$$+ (1 - t_n)[Q_n T(x_n) - (\lambda_0 - \epsilon)Q_n C(x_n)] = 0$$

or equivalently,

$$Q_n T(x_n) - (\lambda_\alpha - 2t_n\epsilon + \epsilon)Q_n C(x_n) = 0. \qquad (6.24_\alpha)$$

We may suppose without loss of generality that $t_n \to t \in [0,1]$ and observe that in this case

$$Q_n T(x_n) - (\lambda_\alpha - 2t\epsilon + \epsilon)Q_n C(x_n)$$
$$= (t - t_n)Q_n C(x_n) \to 0 \qquad \text{as} \quad n \to \infty.$$

Hence, by the A-properness of $H(\cdot,t) \equiv T - (\lambda_\alpha - 2t\epsilon + \epsilon)C$, there exist a subsequence $\{x_{n_j}\}$ and an $x \in X$ such that $x_{n_j} \to x$ as $i \to \infty$ and $T(x) - (\lambda_\alpha - 2t\epsilon + \epsilon)C(x) = 0$ with $\|x\| = r$ and $t \in (0,1)$ because it follows from (6.19_α), (6.20_α), and the property of $\{Q_n\}$ that

$$\| T(x) - (\lambda_\alpha \pm \epsilon)C(x) \| \geq \tfrac{1}{2}c_\alpha(\epsilon) \| x \|$$

$$\text{for all} \quad x \in X \quad \text{with} \quad \|x\| > \rho_\infty(\epsilon)$$

(when $\alpha = \infty$) [respectively, for $x \in X$ with $\|x\| < \rho_0(\epsilon)$ (when $\alpha = 0$)].

Thus for any fixed $\epsilon \in (0,\epsilon_\alpha)$ there exists $\rho_\alpha(\epsilon) > 0$ such that for any $0 < r < \rho_0(\epsilon)$ [respectively, $r > \rho_\infty(\epsilon)$], equation (1) has a solution (λ,x_λ) with $\lambda = \lambda_\alpha - 2t\epsilon + \epsilon \in (\lambda_\alpha - \epsilon, \lambda_\alpha + \epsilon)$ and $\|x_\lambda\| = r$. This establishes Assertion (b) of Theorem 6.1, which is the case when $\alpha = 0$, and of Theorem 6.2 when $\alpha = \infty$.   Q.E.D.

Remark 6.2   Going over the proof of Theorem 6.1 (i.e., the case when $\alpha = 0$) we note that Theorem 6.1, which treats the existence of nontrivial solutions of very small norms, remains valid when T and C are defined only on the closure of some bounded open neighborhood D of 0. Indeed, if this is the case, there exists $r_0 > 0$ such that $\bar{B}(0,r_0) \subset D$ and consequently, for a given $\epsilon \in (0,\epsilon_0)$, all we have to do is to choose $r_0(\epsilon) > 0$ such that $r_0(\epsilon) \leq r_0$ in case (a) and $\rho_0(\epsilon) > 0$ such that $\rho_0(\epsilon) \leq r_0$ in case (b). The corresponding homotopies are restricted to $\bar{D}x[0,1]$ in case (a) and to $D_n x[0,1]$ in case (b). This remark is important since as we shall see later, in some applications the maps are defined only on proper subsets of X [e.g., balls $B(0,r)$ containing 0].

Remark 6.3: (Special cases)   (1) It is obvious that the bifurcation Theorems IV.2.1 and IV.3.1 for the problem (6.1) of Krasnoselskii [136] follow from Theorems 6.1 and 6.2 when $Y = X$, $\Gamma = \{X_n, P_n\}$, and $T = I$, the identity on X.

(2) Since the derivatives T_0' and T_∞' of a k-ball-contractive mapping $F: X \to X$ are also k-ball-contractive, a second special case of Theorems 6.1 and 6.2 is the following.

Proposition 6.2   Suppose that the k-ball-contractive mapping $F: X \to X$ with $k < 1$ has derivatives F_0' and F_∞' with $F(0) = 0$. Then:

(a) If $T_0' = I - F_0'$ is injective, each characteristic value of $C_0'(T_0')^{-1}$ of odd multiplicity is a BP for the equation

$$x - F(x) = \lambda C(x). \tag{24}$$

(b) If $T'_\infty = I - F'_\infty$ is injective each characteristic value of $C'_\infty(T'_\infty)^{-1}$ of odd multiplicity is an ABP for equation (24).

(c) As the third special case we deduce the local bifurcation results involving A-proper maps $T: X \to Y$ which are F-differentiable and which are "semicoercive," as given by (6.26) below. In particular, we obtain bifurcation results when T is strongly monotone, or when T is strongly accretive.

Before we state our next result, we need the following:

Lemma 6.1 Let $K: X \to Y^*$ and $K_n: X_n \to E_n^*$ be such that $K(tx) = t^P K(x)$ for all $x \in X$, $t > 0$ and some $p \geq 1$ with $Kx \neq 0$ whenever $x \neq 0$, and for all g in Y and x in X_n,

$$(W_n g, K_n(x)) = (g, Kx) \quad \text{and} \quad \| K_n(x) \| = \| K(x) \|. \tag{6.25}$$

Suppose that $T: X \to Y$ is a map such that for all $x \in X$ and some $c_\alpha > 0$,

$$(Tx, Kx) \geq c_\alpha \| x \| \| Kx \| \tag{6.26}$$
$$- \phi_\alpha(\| x \|) \| Kx \| \quad \text{for some} \quad \phi_\alpha: R^+ \to R^+$$

with $\phi_\alpha(t)/t \to 0$ as $t \to \alpha$ ($\alpha = 0$ or $\alpha = \infty$). Suppose also that either

(a) $T(0) = 0$ and T has the Fréchet derivative T'_0 at 0, or
(b) T has the asymptotic derivative T'_∞.

Then, in either case, T'_α ($\alpha = 0$ or $\alpha = \infty$) is an injective A-proper map w.r.t. Γ provided that any one of the following hypotheses holds:

(H1) X is reflexive and K is continuous with $R(K)$ dense in Y^* or $\Gamma = \Gamma_p$ with $R(Q_n^*) \subset R(Q_{n+1}^*)$ for each $n \in Z^+$.

(H2) Either $\Gamma = \Gamma_p$ and $Q_n^* g \to g$ for each g in Y^* or $R(T_\alpha) = Y$ with $\alpha = 0$ or $\alpha = \infty$.

Proof. Since $T(x) = T'_\alpha(x) + G_\alpha(x)$ for all x in X with $\| G_\alpha(x) \|/\| x \| \to 0$ as $\| x \| \to \alpha$ with either $\alpha = 0$ or $\alpha = \infty$, it follows from (6.26) that for a given fixed $x \in X$ and any $t > 0$, $(T'_\alpha(tx), K(tx)) + (G_\alpha(tx), K(tx)) \geq c_\alpha \| tx \| \| K(tx) \| - \phi_\alpha(\| tx \|) \| K(tx) \|$. Since $(G_\alpha(tx), K(tx)) \leq t^p \| (G_\alpha(tx)) \| \| K(x) \|$, it follows that

$$(T_\alpha^* x, Kx) + (\| G_\alpha(\| tx \|) \|/\| tx \|) \| Kx \| \| x \| \geq c_0 \| x \| \| Kx \|$$
$$- (\phi_\alpha(\| tx \|)/\| tx \|) \| Kx \| \| x \|.$$

Since x is fixed, $\| G_\alpha(\| tx \|)\|/\| tx \| \to 0$ and $\phi_\alpha(\| tx \|)/\| tx \| \to 0$ as $t \to \alpha$ ($\alpha = 0$ or $\alpha = \infty$), taking the limit in the inequality above as $t \to \alpha$, we obtain the inequality

$$(T'_\alpha x, Kx) \geq c_\alpha \| x \| \| Kx \| \qquad \forall x \in X \, (\alpha = 0 \quad \text{or} \quad \alpha = \infty). \qquad (6.27)$$

Now it follows from (6.27) that $A \equiv T'_\alpha$ ($\alpha = 0$ or $\alpha = \infty$) is obviously injective. To show that A is A-proper, we note that by (6.25) and (6.27), the linear map $A_n \colon X_n \to Y_n$ satisfies the inequality

$$\| A_n(x) \| \geq c_\alpha \| x \| \qquad \text{for all} \quad x \in X_n \quad \text{and each} \quad n. \qquad (6.28)$$

Hence it follows from Theorems II.1.3 and II.1.4 that A is A-proper because of (6.28) and one of conditions (H1) and (H2) holds.   Q.E.D.

In virtue of Lemma 6.1, as another special case of Theorems 6.1 and 6.2, one obtains the following new bifurcation results.

Theorem 6.3 Suppose that $T, C \colon X \to Y$ are such that C is compact, $K \colon X \to Y^*$ and $K_n \colon X_n \to E_n^*$ satisfy (6.25), and any one of the hypotheses in assumptions (H1) and (H2) holds. Suppose that

(E) T is A-proper and satisfies the inequality (6.26) of Lemma 6.1.

(a) If $T(0) = C(0) = 0$ and T and C have Fréchet derivatives T'_0 and C'_0 at 0, then each eigenvalue of

$$T'_0 x = \lambda C'_0 x \qquad (6.29)$$

of odd multiplicity is a BP for

$$Tx = \lambda C. \qquad (6.30)$$

(b) If T and C have asymptotic derivatives T'_∞ and C'_∞, respectively, then each eigenvalue of

$$T'_\infty x = \lambda C'_\infty x \qquad (6.31)$$

of odd multiplicity is an ABP for (6.30).

Remark 6.4 It is known that if $T \colon X \to Y$ is strongly K-monotone, that is, there exists a constant $c > 0$ such that

$$(Tx - Ty, K(x - y)) \geq c \| x - y \| \| K(x - y) \|, \qquad \forall x, y \in X, \qquad (6.32)$$

then under certain additional conditions on T and K and/or X and Y, the map T is A-proper. Let us mention here some situations for which this is the case.

(E1) If X is reflexive, $Y = X^*$, $K = I$, and T is demicontinuous, then T is A-proper with respect to Γ_I. In fact, a bounded demicontinuous map $T: X \to X^*$ of type (S) is A-proper with respect to Γ_I, so is a map of type (α) studied in [262,269].

(E2) If $Y = X$, $K = J$, and $T: X \to X$ is continuous, and $\Gamma_X = \{X_n, P_n\}$ is projectionally complete for (X,X) with $\| P_n \| = 1$, then T is A-proper w.r.t. Γ_X because T is a-stable and surjective by a theorem of Deimling [63]. If X and X^* are uniformly convex and $T: X \to X$ is only demicontinuous, then T is still A-proper by a theorem of Webb [295].

In view of Remark 6.4, immediate consequences of Theorem 6.3 are the following two corollaries, which will prove to be useful in our study of bifurcation points for differential equations.

Corollary 6.1 Suppose that all the conditions of Theorem 6.3 hold except for condition (E), which is replaced by the following:

(H) $T: X \to Y$ is strongly K-monotone and either (E1) or (E2) of Remark 6.4 holds.

Then if T and C satisfy condition (a) [or (b)] of Theorem 6.3 each eigenvalue λ of (6.29) [or of (6.31)] of odd multiplicity is a BP (or an ABP) of (6.30).

Corollary 6.2 Suppose that X is a Hilbert space, $C: X \to X$ compact, and $T: X \to X$ is a bounded demicontinuous map of type (S), such that

$$(Tx,x) \geq C_0 \| x \|^2 - C_2 \quad \text{for some constants} \quad C_0 > 0 \quad \text{and} \quad C_2.$$

(6.33)

(a) If $T(0) = C(0) = 0$, $C_2 = 0$ and T and C have Fréchet derivatives T_0' and C_0' at 0, then each eigenvalue of (6.29) of odd multiplicity is a BP for (6.30).
(b) If T and C have asymptotic derivatives T_∞' and C_∞', then each eigenvalue of (6.31) of odd multiplicity is an ABP for (6.30).

Remark 6.5   Since every bounded continuous, and strongly monotone map T from a separable Hilbert space H into H, which is asymptotically linear, is A-proper and has an A-proper asymptotic derivative, Theorems 8 and 10 of Aldarwish and Langenbach [3] are special cases of Corollary 6.1 as well as of Corollary 6.2.

V

Solvability of PDEs and ODEs and Bifurcation Problems

The purpose of Chapter V is to show how the theory developed in preceding chapters can be used to obtain approximation solvability and/or existence theorems for nonlinear PDEs and ODEs, which need not be of the form to which other abstract theories apply. Thus in Section 1 we treat the solvability of elliptic boundary value problems of order $2m$, while in Section 2 we deal with the approximation solvability of ordinary differential equations.

1. SOLVABILITY OF PARTIAL DIFFERENTIAL EQUATIONS

In this part of Chapter V we are concerned primarily with the solvability of nonlinear elliptic partial differential equations of order $2m$ defined on some bounded domain in $\mathbb{R}^n$.

1.1 Solvability of Quasilinear Elliptic Boundary Value Problems

The purpose of this section is to use the theory of A-proper mappings and their uniform limits to obtain in a simple way general variational approximation solvability and/or existence theorems for not necessarily coercive elliptic boundary value problems of the form

$$\begin{cases} A(u) = \sum_{|\alpha| \le m} (-1)^{|\alpha|} D^\alpha A_\alpha(x,u, \ldots, D^m u) = F \text{ in } Q \\ B_j(u) = 0 \quad \text{on} \quad \partial Q, \ 0 \le j \le m - 1, \end{cases} \tag{1.1}$$

where Q is a bounded domain in R^n, F is a given function that will be specified later, and B_j is a nonlinear differential operator of order $m - j - 1$.

When $A_\alpha(x,u \ldots, D^m u) \in L_q(Q)$ for every u in $W_p^m(Q)$, with $p > 1$ and $1/p + 1/q = 1$, we associate with $A(u)$ its generalized form

$$a(u,v) = \sum_{|a| \le m} \langle A_\alpha(x,u, \ldots, D^m u), D^\alpha v \rangle \qquad (u,v \in W_p^m(Q)) \tag{1.2}$$

and closed subspace V of $W_p^m(Q)$ with $V \supseteq \overset{\circ}{W}_m^p(Q)$, where $\langle f,g \rangle = \int_Q fg \, dx$ for f in $L_p(Q)$ and g in $L_q(Q)$. Problem (1.1) is said to be of *variational type* provided that its boundary conditions are implicitly verified by the constraints: $u \in V$ and $(Au,v) = a(u,v)$ for all v in V, where (g,v) denotes the value of g in V^* at v in V. By the repeated integration of (Au,v) we see that the condition $(Au,v) = a(u,v)$ implies that certain intergrals over the boundary ∂Q vanish. The space V is determined by the boundary conditions (1.1).

With the variational problem (1.1) one associates problem (A,V): Given $F \in V^*$, find $u \in V$ such that

$$a(u,v) = (F,v) \qquad \text{for all} \quad v \text{ in } V. \tag{1.3}$$

A solution of the problem (A,V) is said to be a *variational solution* of (1.1).

The approach of Section II.4.4 is used in Section 5.1 to establish the constructive solvability of the nonlinear problem (A,V) for each F in V^* when the leading part of $A(u)$ [i.e., the part $A(u)$ corresponding to $|\alpha| = |\beta| = m$] satisfies a nonlinear version of the strong ellipticity condition and some additional conditions (which appear to be weakest) are also assumed. In Theorem 1.2 a new existence result for the problem (A,V) for each F in V^* is proved under the condition which ensures that the operator $T: V \to V^*$ determined by the form $a(u,v)$ in (1.2) is a uniform limit of A-proper mappings and under certain additional assumptions which are general enough so as to extend and unify the existence results of Leray–Lions [152], Browder [25,26,30,32,33], Pohožayev [239], as well as the results in [16,39,67,75,211,262,269] and others. The solvability results for the problem (A,V) are deduced from much more general abstract results for T, mapping a Banach space X into Y, which is either A-proper (Theorem 4.1) or a uniform limit of A-proper maps (Theorem 4.2). These abstract theorems are also applicable to differential equations involving operators that need not be in divergence form.

For historical record we add that existence theorems for elliptic boundary value problems of the type above were first obtained by Višik [288] using compactness arguments and a priori estimates on the $(m + 1)$ derivatives. The theory of coercive monotone operators was first applied to (1.2) by Browder [30]. The existence theorem for problem (A,V) was extended by Leray–Lions [152] when $A(u)$ gives rise to a special case of a coercive pseudomonotone operator. Odd operators $A(u)$ satisfying strong monotonicity conditions were first studied by Pohožayev [239] and later by Browder [25,26] in case of monotone and semimonotone maps. The generalized degree for pseudomonotone maps, which is based on the degree theory for A-proper maps of Browder–Petryshyn [42], were first applied in [32]. Subsequently, the solvability of the problem (1.2) when $A(u)$ gives rise to maps $T: X \to X^*$ of monotone type have been studied by many authors (see [16,36,39,51,67,75,99] and [113,118,154,174,211, 269,314] for further references).

The direct application of the A-proper mapping theory to the solvability of linear and quasilinear ODEs and PDEs was initiated by Petryshyn in [225,231] and studied further in [83,175] and [211,212,220,221,226,232, 235]. For other references, see [234].

To apply the A-proper mapping theory to the solvability of the general variational boundary value problems for nonlinear elliptic partial differential operators, we first introduce some notions and definitions. Let Q be a bounded domain in R^n with boundary ∂Q so smooth that the Sobolev embedding theorem holds on Q. Let $C_0^\infty(Q)$ be the family of real- or complex-valued infinitely differentiable functions with compact support in Q. For a multi-index $\alpha = (\alpha_1, \ldots, \alpha_n)$ we denote by D^α the derivative $D_1^{\alpha_1} \cdots D_n^{\alpha_n}$ of order $|\alpha| = \alpha_1 + \cdots + \alpha_n$. For any real number p with $1 < p < \infty$, let $L_p(Q)$ be the Banach space with the norm $\| \cdot \|_p$ and for any integer $m \geq 0$ let $W_p^m \equiv W_p^m(Q) = \{u \mid u \in L_p(Q), D^\alpha u \in L_p(Q)$ for $|\alpha| \leq m\}$ be the Sobolev space with norm $\|u\|_{m,p} = \{\sum_{|\alpha| \leq m} \| D^\alpha u \|_p^p\}^{1/p}$, where $D^\alpha u$ denote the distribution derivatives of u. W_p^m is a uniformly convex and separable Banach space which for every p includes $C^\infty(Q)$. We let $\mathring{W}_p^m$ denote the closure of $C_0^\infty(Q)$ in W_p^m. Let $\langle u,v \rangle = \int_Q uv \, dx$ [and $\langle u,v \rangle = \int_Q u\bar{v} \, dx$ if $L_p(Q)$ is complex] denote the natural pairing between $u \in L_p$ and $v \in L_q$ with $q = p(p - 1)^{-1}$. When $p = 2$, $\langle \cdot,\cdot \rangle$ denotes the inner product in $L_2(Q)$. In case $p = 2$, W_2^m is a Hilbert space with the inner product $(u,v)_m = \sum_{|\alpha| \leq m} \langle D^\alpha u, D^\alpha v \rangle$ and norm $\|u\|_{m,2} = (u,u)_m^{1/2}$. For the study of such spaces, see [1,91,116].

Clearly, as in the linear case, different variational boundary conditions are determined by the different choices of V. In the case when $V =$

$\mathring{W}_p^m(Q)$, the condition $(Au,v) = a(u,v)$ is automatically satisfied. It is not hard to show that the constraint $u \in V$ implies that $B_j = \partial^j/\partial n^j$, where $\partial/\partial n$ denotes the normal derivative. These are the *generalized Dirichlet boundary conditions*. When $V = W_p^m(Q)$, the constraint in (1.3) yields

$$\sum_{j=0}^{m-1} \int_{\partial Q} C_j(u) \frac{\partial^j v}{\partial n^j} \, d\sigma = 0 \qquad \text{for all} \quad v \in V, \tag{1.4}$$

and the boundary conditions in (1.1) are given by $B_j = C_j$. These are the *generalized Neumann boundary conditions*. Finally, for the intermediate choice $\mathring{W}_p^m(Q) \subset V \subset W_p^m(Q)$, we must deal with boundary conditions of mixed type. Notice that all the boundary conditions except for the Dirichlet ones are nonlinear (see [2,87,240] for further study).

With the variational problem (1.1) one associates the *problem* (A,V): Given $F \in V^*$, find $u \in V$ such that

$$a(u,v) = (F,v) \qquad \text{for all} \quad v \text{ in } V. \tag{1.5}$$

A solution of the problem (A,V) is said to be a *variational solution* of the problem (1.1). If for each fixed u in V the function $a(u,\cdot): V \to R$ is linear and bounded, it defines a nonlinear operator $T: V \to V^*$ such that

$$a(u,v) = (Tu,v) \qquad \text{for all} \quad v \in V. \tag{1.6}$$

Hence a solution of (1.5) is also a solution of

$$Tu = F \qquad (u \in V, F \in V^*) \tag{1.7}$$

and conversely (see [30,152]).

To make precise statements of our results, we introduce the following notation: For a given integer $m \geq 1$, we let $\xi = \{\xi_\alpha: |\alpha| \leq m\}$, and set $\zeta = \{\zeta_\alpha: |\alpha| = m\}$, $\eta = \{\eta_\beta: |\beta| \leq m - 1\}$, where each ξ_α, ζ_α, and η_β is an element of R. The set of all ξ of the form above is a Euclidean space R^{sm}, and correspondingly, $\zeta \in R^{\acute{s}m}$, $\eta \in R^{sm-1}$. We also set $D^m u = \{D^\alpha u: |\alpha| = m\}$, $\delta u = \{D^\alpha u: |\alpha| \leq m - 1\}$, $\xi = (\eta,\zeta)$, and $A_\alpha(x,\xi) = A_a(x,\eta,\zeta)$ for $|\alpha| \leq m$.

For each α, let $A_\alpha: Q \times R^{sm} \to R$ be such that:
(a1) $A_\alpha(x,\xi)$ is measurable in x for fixed ξ and continuous in ξ for fixed x. For a given p with $1 < p < \infty$, there exist a constant $c > 0$ and $k(x) \in L_q(Q)$ such that

$$|A_\alpha(x,n,x)| \leq c[|\eta|^{p-1} + |\xi|^{p-1} + k(x)], \qquad \frac{1}{p} + \frac{1}{q} = 1. \tag{1.8}$$

Note that by virtue of (a1) the form $a(u,v)$ is well defined on $V \times V$ since $|a(u,v)| \leq \phi(\|u\|_{m,p}) \|v\|_{m,p}$ with $\phi(t) = c_0 t^{p-1} + \|k\|_q$, so there exists a bounded and continuous mapping $T: V \to V^*$ such that (1.6) holds. Consequently, the solvability of (1.5) is equivalent to that of (1.7).

Now let $\{X_n\}$ be a sequence of finite-dimensional subspaces of V such that $\mathrm{dist}(u,X_n) \to 0$ as $n \to \infty$ for each $u \in V$ and let V_n be a linear injection of X_n into V. Then $\Gamma_I = \{X_n, V_n; X_n^*, V_n^*\}$ is an admissible injective scheme for maps from V to V^*.

Definition 1.1 For $F \in V^*$, (1.5) is said to be *strongly* (respectively, *feebly*) *a-solable* iff there is $n_F \in Z_+$ such that the finite-dimensional problem

$$a(u_n,v) = (F,v) \qquad \forall\, v \in X_n, \quad n \geq n_F, \tag{1.9}$$

has a solution $u_n \in X_n$ such that $u_n \to u$ in V (respectively, $u_{n_j} \to u$ in V) and u satisfies (1.5).

Since $u_n \in X_n$ solves (1.9) iff u_n solves

$$V_n^* T(u_n) = V_n^* F \qquad (u_n \in X_n, V_n^* F \in X_n^*), \tag{1.10}$$

we see that Definition 1.1 is equivalent to the strong (respectively, feeble) a-solvability of (1.7) w.r.t. Γ_I.

Definition 1.2 The form $a(u,v)$ is said to satisfy *condition* $(+)$ iff $\{u_j\}$ is any sequence in V such that $|\,a(u_j,v) - (g,v)\,| \leq \epsilon_j \| v \|_v$ for some $g \in V^*$ and all $v \in V$ with $\epsilon_j \to 0$ as $j \to \infty$, the $\{u_j\}$ is bounded.

It is easy to see that $a(u,v)$ satisfies *condition* $(+)$ iff $\{u_j\}$ is bounded whenever $Tu_j \to g$ for some g in V^* [i.e., iff T satisfies condition $(+)$].

It is known that condition $(+)$ is important in establishing surjectivity theorems for various classes of nonlinear mappings. For exmple, improving on some surjectivity results of Browder [30] and Minty [177], it was shown by Rockafellar [252] (see also [75]) that for a demicontinuous monotone mapping $T: V \to V^*$ the *condition* $(+)$ *is not only sufficient for* $T(V) = V^*$ *but also necessary.* However, to establish the surjectivity results for other classes of maps of monotone or A-proper type, some further condition has to be imposed such as $(Tu,u) \geq 0$ or $\| Tu \| + (Tu,u)/\| u \| \geq 0$ for $\| u \| \geq r$ (see [26,39,67,75,175,231]). It seems that our condition (a3) [or (H3)] below is the *weakest* one among such additional conditions.

We now state our first theorem in this section which extends (A1) of Theorem II.4.1 to nonlinear problem (A,V).

Theorem 1.1 Let V be a closed subspace of $W_p^m(Q)$ with $\mathring{W}_p^m(Q) \subset V$ and suppose that the coefficients A_α of $A(u)$ in (1.1) satisfy (a1) and the conditions:

(a2) There exists a constant $\mu_0 > 0$ such that

$$\sum_{|\alpha|=m} [A_\alpha(x,\eta,\zeta) - A_\alpha(x,\eta,\zeta')](\zeta_\alpha - \zeta'_\alpha) \geq \mu_0 \sum_{|\alpha|=m} |\zeta_\alpha - \zeta'_\alpha|^p$$

for $x \in Q$ (a.e.), $\eta \in R^{sm-1}$, and $\zeta,\zeta' \in R^{s'm}$.

(a3) $Tu \neq \gamma Ju$ for all $u \in V - B(0,r)$, all $\gamma < 0$, and some $r > 0$, where $J: V \to V^*$ is a duality map given by $(Ju,u) = \|u\|^2$ and $\|Ju\| = \|u\|$.

Then if $a(u,v)$ satisfies condition $(+)$, equation (1.5) is feebly a-solvable for each $F \in V^*$ [and in particular, $T(V) = V^*$], and strongly if (1.5) is uniquely solvable. If T is odd on $V - B(0,r)$ [i.e., $T(-u) = -T(u)\ \forall\ u \in V - B(0,r)$], then the conclusion holds without condition (a3).

Remark 1.1 It is easy to show that condition $(+)$ holds if one of the following holds:

$(1+)\ \dfrac{a(u,u)}{\|u\|} \to \infty$ as $\|u\|_V \to \infty$ [i.e., $a(u,u)$ is coercive]

$(2+)\ \|Tu\| + \dfrac{a(u,u)}{\|u\|} \to \infty$ as $\|u\|_V \to \infty$

$(3+)$ There exists a continuous function $\psi: R^+ \to R^+$ with $\|u\|_V \leq \psi(\|F\|)$ for each solution u of (1.5).

Remark 1.2 Note that (a3) is satisfied if, for example, one of the following holds:

$(a3_1)\ a(u,u) \geq 0\ \forall\ u \in V - B(0,r)$ and some $r > 0$

$(a3_2)\ \|Tu\| + \dfrac{a(u,u)}{\|u\|_V} \geq 0\ \forall\ u \in V - B(0,r)$ and some $r > 0$.

Remark 1.3 In view of the remarks above, we see that Theorem 1.1 remains valid if instead of (a3) and $(+)$ we assume that

$$\|Tu\| + \frac{a(u,u)}{\|u\|_V} \to \infty \qquad \text{as} \quad \|u\|_V \to \infty. \tag{1.11}$$

We shall deduce Theorem 1.1 from the abstract Theorem III.4.1 concerning the solvability of

$$Tx = f \qquad (x \in X, f \in Y) \tag{1.12}$$

involving operators T from a Banach space X to a Banach space Y with Y not necessarily equal to X^*. Thus, in particular, the abstract results obtained here will be applicable to differential equations involving operators that are *not* in divergence form.

Let us first recall that if (X,Y) is a pair of Banach spaces, $\{X_n\}$ and $\{E_n\}$ sequences of oriented finite-dimensional spaces with $X_n \subset X$, V_n an inclusion map of X_n into X, and W_n a linear map of Y onto E_n, then $\Gamma = \{X_n, V_n; E_n, W_n\}$ is said to be an *admissible scheme* for (X,Y) provided that dim $X_n = $ dim E_n for each n, dist$(x, X_n) \to 0$ as $n \to \infty$ for each $x \in X$, and $\{W_n\}$ is uniformly banded.

A simple example of an admissible scheme is the injective scheme $\Gamma_I = \{X_n, V_n; X_n^*, V_n^*\}$ for (X, X^*).

In order to deduce Theorem 1.1 from Theorem III.4.1, set $X = V$, $Y = V^*$, $K = I$, $G = J$, $E_n = X_n^*$, $W_n = V_n^*$ and note that $\Gamma_I = \{X_n, V_n; X_n^*, V_n^*\}$ is admissible for (V, V^*).

Now, it follows from the definition of J that J is odd, bounded, and $(Ju, u) = \| Ju \| \| u \| > 0$ for $u \neq 0$. Furthermore, J is continuous since V is uniformly convex (see [128]), and as was shown above, J is A-proper w.r.t. Γ_I. Hence (H1) of Theorem III.4.1 holds, (H3) holds by assumption (a3), and T satisfies condition $(+)$ because $a(u, v)$ does. Thus to complete the proof, we must verify (H2) and prove that T is A-proper w.r.t. Γ_I. This is done in the following:

Lemma 1.1 If the coefficient functions $A_\alpha(x, \eta, \zeta)$ of $a(u, v)$ satisfy (a1) and (a2), then $T: V \to V^*$, given by (1.6), is A-proper w.r.t. Γ_I. The map $T + \mu J: V \to V^*$ is also A-proper w.r.t. Γ_I for each $\mu > 0$.

Proof. Since $T_n \equiv V_n^* T|_{X_n} : X_n \to X_n^*$ is continuous for each n, to show that T is A-proper, it suffices to prove that T satisfies condition (H). Now, T can be written in the form $T = T_1 + T_2$, where

$$(T_1 u, v) = a_1(u, v) = \sum_{|\alpha| = m} \langle A_\alpha(x, \delta u, D^m u), D^\alpha v \rangle \tag{1.13}$$

$$(T_2 u, v) = a_2(u, v) = \sum_{|\alpha| < m} \langle A_\alpha(x, \delta u, D^m u), D^\alpha v \rangle \tag{1.14}$$

with $T_2: V \to V^*$ compact in virtue of (a1) and the complete continuity of the embedding of W_p^m into W_p^j for $0 \leq j \leq m - 1$ (for an easy proof of this, see [221]). Since A-properness is invariant under compact perturbation, to prove Lemma 1.1, it suffices to show that T_1 is A-proper w.r.t. Γ_I. So let $\{u_{n_j} \mid u_{n_j} \in X_{n_j}\}$ be any bounded sequence such that $V_{n_j}^* T_1(u_{n_j}) - V_{n_j}^* g \to 0$ for some g in V^*. For simplicity of notation set

$u_{n_j} \equiv u_j$ for all $j \in Z_+$ and note that because $\{u_j\}$ is bounded and V is reflexive we may assume that $u_j \rightharpoonup u_0$ for some u_0 in V. Since $D^\alpha u_j \rightarrow D^\alpha u_0$ in L_p for each α with $|\alpha| < m$ by the Sobolev embedding theorem, to prove that $u_j \rightarrow u_0$ in V it suffices to show that $D^\alpha u_j \rightarrow D^\alpha u_0$ in L_p for each α with $|\alpha| = m$. Now, since $\text{dist}(u_0, X_n) \rightarrow 0$ as $n \rightarrow \infty$, there exist $w_j \in X_j$ such that $w_j \rightarrow u_0$ in V as $j \rightarrow \infty$. This and the boundedness of $\{T_1 u_j\}$ imply that $(T_1 u_j, w_j - u_0) \rightarrow 0$ as $j \rightarrow \infty$ and hence, since $(T_1 u_j, u_j - w_j) = (V_j^* T_1 u_j - V_j^* g, u_j - w_j) + (g, u_j - w_j) \rightarrow 0$ as $j \rightarrow \infty$, it follows that $(T_1 u_j, u_j - u_0) \rightarrow 0$ as $j \rightarrow \infty$. Consequently, $(T_1 u_j - T_1 u_0, u_j - u_0) \rightarrow 0$ as $j \rightarrow \infty$. In view of this and the inequality

$$(T_1 u_j - T_1 u_0, u_j - u_0)$$
$$= \sum_{|\alpha|=m} \langle A_\alpha(x, \delta u_j, D^m u_j) - A(x, \delta u_j, D^m u_0), D^\alpha u_j - D^\alpha u_0 \rangle$$
$$+ \sum_{|\alpha|=m} \langle A_\alpha(x, \delta u_j, D^m u_0) - A_\alpha(x, \delta u_0, D^m u_0), D^\alpha u_j - D^\alpha u_0 \rangle$$
$$\geq \mu_0 \sum_{|\alpha|=m} \| D^\alpha u_j - D^\alpha u_0 \|_p^p + \sum_{|\alpha|=m} \langle A_\alpha(x, \delta u_j, D^m u_0)$$
$$- A_\alpha(x, \delta u_0, D^m u_0), D^\alpha u_j - D^\alpha u_0 \rangle, \tag{1.15}$$

which is implied by (a2), it follows that $D^\alpha u_j \rightarrow D^\alpha u_0$ in L_p for each α with $|\alpha| = m$ since $A_\alpha(x, \delta u_j, D^m u_0) \rightarrow A_\alpha(x, \delta u_0, D^m u_0)$ in L_q. Thus T_1 is A-proper w.r.t. Γ_l and so is $T = T_1 + T_2$. The A-properness of $T + \mu J$ for each $\mu > 0$ is proved similarly. Thus the proof of Theorem 1.1 is complete.   Q.E.D.

Remark 1.4   Note that when $p = 2$ and when we let

$$A_\alpha(x, u, \ldots, D^m u) = \sum_{|\beta| \leq m} A_{\alpha\beta}(x) D^\beta u \qquad \text{with} \quad A_{\alpha\beta}(x) \in L_\infty(Q)$$

for $|\alpha| \leq m$ and $|\beta| \leq m$, the form $a(u,v)$ in (1.2) coincides with the bilinear form $B[u,v]$ in (4.7) (Ch. II), and (a2) reduces to

$$\sum_{|\alpha|=|\beta|=m} A_{\alpha\beta}(x) \zeta_\alpha \zeta_\beta \geq \mu_0 \sum_{|\alpha|=m} |\zeta_\alpha|^2 \tag{1.16}$$

for $x \in Q$ (a.e.), and all $\zeta = \{\zeta_\alpha : |\alpha| = m\} \in R^{s'_m}$, which is the same as (c2) in the linear case when all functions are real valued. Since the linear operator L ($= T$) is odd and L satisfies condition ($+$) iff L has a bounded inverse, which because L is A-proper is the case iff L is one-to-one, we see that assertion (A1) of Theorem II.4.1 follows from Theorem 1.1 when T is odd.

We now show how the A-proper mapping approach can still be used to establish a new solvability result of (1.5) for each $F \in V^*$ when instead of the (a2) we assume a much weaker condition,

$$\sum_{|\alpha| = m} [A_\alpha(x,\eta,\zeta) - A_\alpha(x,\eta,\zeta')](\zeta_\alpha - \zeta'_\alpha) \geq 0 \qquad (1.17)$$

for $x \in Q$ (a.e.), $\eta \in R^{sm-1}$, and $\zeta,\zeta' \in R^{sm}$, provided that $a(u,v)$ satisfies condition $(++)$ given by

Definition 1.3 The form $a(u,v)$ is said to satisfy *condition* $(++)$ iff when $\{u_j\} \subset V$ is any bounded sequence such that $a(u_j,v) \to (g,v)$ as $j \to \infty$ uniformly with respect to $v \in S_1 \equiv \partial B(0,1)$ for some $g \in V^*$, there exists a $u_0 \in V$ such that $a(u_0,v) = (g,v)$ for all $v \in S_1$.

Theorem 1.2 Let $\mathring{W}_p^m \subset V \subset W_p^m$, let the functions $A_\alpha(x,\eta,\zeta)$ satisfy the hypotheses (a1) and (1.17), and suppose that either T is odd on $V - B(0,r)$ or that condition (a3) of Theorem 1.1 holds. Suppose further that $a(u,v)$ satisfies condition $(++)$. Then if $a(u,v)$ satisfies condition $(+)$, (1.5) is solvable for each $F \in V^*$ [i.e., $T(V) = V^*$].

Proof. It suffices to show that hypotheses of Theorem 1.2 imply those of Theorem III.4.2 when $X = V$, $Y = V^*$, $K = I$, $G = J$, $\Gamma = \Gamma_1 = \{X_n, V_n; X_n^*, V_n^*\}$, and $T: V \to V^*$ is given by $(Tu,v) = a(u,v)$. Now, as in the proof of Theorem III.4.2, one checks that (H1), (H3), and (H4) hold. Further, (H2) (i.e., that $T + \mu J$ is A-proper for each $\mu > 0$) follows from the same argument as that used to prove Lemma 1.1 since (1.17) and the properties of J imply for $T_\mu = T + \mu J$ the inequality

$$(T_\mu(u) - T_\mu(v), u - v) \geq \mu(\| u \|_V - \| v \|_V)^2$$

$$+ \sum_{|\alpha| = m} \langle A_\alpha(x,\delta u, D^m u) - A_\alpha(x,\delta u, D^m v), D^\alpha u - D^\alpha v \rangle \qquad (1.18)$$

$$+ \sum_{|\alpha| < m} \langle A_\alpha(x,\delta u, D^m u) - A_\alpha(x,\delta v, D^m v), D^a u - D^a v \rangle$$

for $\mu > 0$ and for $u,v \in V$. Finally, since for any bounded $\{u_j\} \subset V$ we have that $a(u_j,v) \to (g,v)$ uniformly for $v \in S_1$ iff $Tu_j \to g$ for some $g \in V^*$, it follows that T satisfies $(++)$ because $a(u,v)$ does. Thus all the conditions of Theorem III.4.2 are verified and thus $T(V) = V^*$.

Q.E.D.

In applying Theorem 1.2 to problem (1.5), it is important to find some analytic conditions on $A_\alpha(x,\eta,\zeta)$ which would ensure that $a(u,v)$ satisfies condition $(++)$. As our first corollary to Theorem 1.2 we show that if

(1.17) is required to hold for all $|\alpha| \leq m$, then Theorem 1.2 remains valid without condition $(++)$.

Corollary 1.1 Let $A_\alpha(x,\eta,\zeta)$ satisfy (a1) and

$$\sum_{|\alpha| \leq m} [A_\alpha(x,\eta,\zeta) - A_\alpha(x,\eta,\zeta')](\zeta_\alpha - \zeta'_\alpha) \geq 0 \qquad (1.19)$$

for $x \in Q$ (a.e.), $\eta \in R^{sm-1}$ and $\zeta,\zeta' \in R^{sm}$. Suppose that

(H3.1) Either T is odd on $V - B(0,r)$ or $Tu \neq \gamma Ju$ for $u \in V - B(0,r)$ and $\gamma < 0$.

Then, if $a(u,v)$ satisfies condition $(+)$, $T(V) = V^*$.

Proof. To deduce Corollary 1.1 from Theorem 1.2 we need to show that $a(u,v)$ satisfies condition $(++)$ if (1.19) holds.

Let us write $a(u,v)$ in another notation which separates its dependence on lower-order derivatives of u from its dependence on the mth derivatives $D^m u$. Let us define $a(u,w;v)$ on V by

$$a(u,w;v) = \sum_{|\alpha| \leq m} \langle A_\alpha(x,\delta u, D^m w), D^\alpha v \rangle. \qquad (1.20)$$

It follows from (a1) that for fixed $u,w \in V$, $a(u,w;v)$ is a bounded linear functional in $v \in V$. Hence there exists a unique $S(u,w)$ in V^* such that

$$a(u,w;v) = (S(u,w),v) \qquad \forall\, v \in V. \qquad (1.21)$$

The operator T is then given by $T(u) = S(u,u)$ for all $u \in V$. It follows from (1.19) and (1.20)–(1.21) that

$$(S(u,u) - S(u,v),\, u - v)$$
$$= \sum_{|\alpha| \leq m} \langle A_\alpha(x,\delta u, D^m u) - A_\alpha(x,\delta u, D^m v), D^\alpha u - D^\alpha v \rangle \geq 0 \qquad (1.22)$$

for all $u,v \in V$. Furthermore, it is not hard to prove (e.g., see [25,152]) that by virtue of (1.20)–(1.21) and the Sobolev embedding theorem, we have:

(a) For fixed $u \in V$, $S(\cdot,u)$ is completely continuous from V to V^* [i.e., $S(u_j,w) \to S(u,w)$ in V^* if $u_j \rightharpoonup u$ in V].

(b) For fixed $u \in V$, $S(u,\cdot)$ is continuous from V to V^* [i.e., $S(u,w_j) \to S(u,w)$ in V^* if $w_j \to w$ in V].

The discussion above shows that $T(u) = S(u,u)$ is semimonotone in the sense of Browder [25,26]. Now, let $\{u_j\} \subset V$ be any bounded sequence such that for some $g \in V^*$, $a(u_j,v) \to (g,v)$ uniformly for $v \in S_1 = \partial B(0,1)$. Then, since $\|Tu_j - g\| = \sup\{|a(u_j,v) - (g,v)|: \|v\|_V = 1\}$, it follows that $Tu_j \to g$ in V^* and because $\{u_j\}$ is bounded and V reflexive, we may assume that $u_j \rightharpoonup u_0$ in V. Since, by (1.22),

$$(S(u_j,u_j) - S(u_j,u), u_j - u) \geq 0 \qquad \forall \quad u \in V \tag{1.23}$$

and $j \in Z_+$, we see that, in view of (a), the passage to the limit in (1.23) as $j \to \infty$ gives

$$(g - S(u_0,u),u_0 - u) \geq 0 \qquad \forall \quad u \in V. \tag{1.24}$$

It follows from (1.24) and Minty's lemma that $g = S(u_0,u_0) = Tu_0$; that is, T satisfies condition $(+ +)$, so Corollary 1.1 follows from Theorem 1.2. Q.E.D.

Remark 1.5 When $a(u,u)$ is coercive, Corollary 2.1 was proved by Browder [25] by applying the abstract theorem for semimonotone operators.

In a second corollary of Theorem 1.2, we show that if in (1.17) we require the inequality to be strictly greater than 0, then Theorem 1.2 remains valid without condition $(+ +)$.

Corollary 1.2 Suppose that in addition to (a1), the functions $A_\alpha(x,\eta,\zeta)$ satisfy the condition

$$\sum_{|\alpha| = m} [A_\alpha(x,\eta,\zeta) - A_\alpha(x,\eta,\zeta')](\zeta_\alpha - \zeta'_\alpha) > 0 \tag{1.25}$$

for all $\zeta,\zeta' \in R^{s_m}$ with $\zeta \neq \zeta'$, all $\eta \in R^{s_{m-1}}$, and $x \in Q$ (a.e.). Suppose also that (H3.1) of Corollary 1.1 holds. Then, if $a(u,v)$ satisfies condition $(+)$, $T(V) = V^*$.

Proof. Since (1.25) implies (1.17), it follows that (1.18) holds for each $\mu > 0$ and all u,v in V. This implies that $T + \mu J$ is A-proper w.r.t. Γ_t for each $\mu > 0$. Thus, all we need is to show that $a(u,v)$ satisfies condition $(+ +)$ if (1.25) holds.

Now, as in [152], let us define the operator $G: V \times V^*$ by

$$(G(u,w),v) = \sum_{|\alpha| = m} \langle A_\alpha(x,\delta u,D^m w),D^\alpha v \rangle$$
$$+ \sum_{|\alpha| < m} \langle A_\alpha(x,\delta u,D^m u),D^\alpha v \rangle. \tag{1.26}$$

Again, it follows from (a1) that G is bounded, continuous, and $G(u,u) = T(u)$ for all $u \in V$. By (1.25),

$$(G(u,u) - G(u,w), u - w)$$
$$= \sum_{|\alpha| = m} \langle A_\alpha(x,\delta u,D^m u) - A_\alpha(x,\delta u,D^m w), D^\alpha u - D^\alpha w \rangle \geq 0 \tag{1.27}$$

for $u, w \in V$. Using (1.25), we now show that if $u_j \rightharpoonup u_0$ in V and $(G(u_j, u_j) - G(u_j, u_0), u_j - u_0) \to 0$, then

$$
\begin{aligned}
(G(u_j, w), u_j - w) &= \sum_{|\alpha| = m} \langle A_\alpha(x, \delta u_j, D^m w), D^\alpha u_j - D^\alpha w \rangle \\
&\quad + \sum_{|\alpha| < m} \langle A_\alpha(x, \delta u_j, D^m u_j), D^\alpha u_j - D^\alpha w \rangle \\
&\to (G(u_0, w), u_0 - w) \tag{1.28}
\end{aligned}
$$

for each $w \in V$.

So let $\{u_j\} \subset V$ be such that $u_j \rightharpoonup u_0$ in V and

$$
\begin{aligned}
(Tu_j - G(u_j, u_0), u_j - u_0) &= \int_Q \sum_{|\alpha| = m} [A_\alpha(x, \delta u_j, D^m u_j) \\
&\quad - A_\alpha(x, \delta u_j, D^m u_0)](D^\alpha u_j - D^\alpha u_0)\, dx \to 0 \tag{1.29}
\end{aligned}
$$

as $j \to \infty$. Call this integrand $F_j(x)$. By (1.25), $F_j(x) \geq 0$ for $x \in Q$ (a.e.). Hence $F_j(x) \to 0$ strongly in $L_1(Q)$, so that there exists a subsequence $\{F_k(x)\}$ such that $F_k(x) \to 0$ for $x \in Q$ (a.e.). Now, since $u_j \rightharpoonup u_0$ in V, it follows from the Sobolev theorem that $D^\alpha u_j \to D^\alpha u_0$ in L_p for $|\alpha| < m$. We claim that $D^m u_k(x) \to D^m u_0(x)$ for $x \in Q$ (a.e.). To prove this, take a point x such that $F_k(x) \to 0$ and $\delta u_k(x) \to \delta u_0(x)$ and the hypothesis (1.26) holds at that point. If $\{D^m u_k(x)\}$ has an accumulation point ζ, we have in the limit

$$
\sum_{|\alpha| = m} [A_\alpha(x, \delta u_0(x), \zeta) - A_\alpha(x, \delta u_0(x), D^m u_0(x))](\zeta - D^\alpha u_0(x)) = 0.
$$

Hence $\zeta = D^m u_0(x)$ by (1.25). If, on the contrary, the numbers $\zeta_k \equiv \zeta(u_k) \equiv D^m u_k(x)$ are unbounded, we may assume, by passing to a subsequence if necessary, that $|\zeta_k - \zeta| \geq 1$ for all large k and

$$
\zeta_k^* \equiv (\zeta_k - \zeta)/|\zeta_k - \zeta| \to \zeta^* \neq 0 \qquad \text{where} \quad \zeta \equiv D^m u_0(x).
$$

Setting $\eta_k \equiv \delta u_k(x)$ and using the monotonicity condition (1.25) with respect to ζ, we get

$$
\begin{aligned}
&\sum_{|\alpha| = m} [A_\alpha(x, \eta_k, \zeta_k) - A_\alpha(x, \eta_k, \zeta)](\zeta_{k\alpha} - \zeta_\alpha) \\
&= \sum_{|\alpha| = m} [A_\alpha(x, \eta_k, \zeta_k) - A_\alpha(x, \eta_k, \zeta + \zeta_k^*)](\zeta_{k\alpha} - \zeta_\alpha) \\
&\quad + \sum_{|\alpha| = m} [A_\alpha(x, \eta_k, \zeta + \zeta_k^*) - A_\alpha(x, \eta_k, \zeta)](\zeta_{k\alpha} - \zeta_\alpha) \\
&\geq \sum_{|\alpha| = m} [A_\alpha(x, \eta_k, \zeta + \zeta_k^*) - A_\alpha(x, \eta_k, \zeta)](\zeta_{k\alpha} - \zeta_\alpha) \geq 0 \\
&\tag{1.30}
\end{aligned}
$$

since in view of the relation $|\zeta_k - \zeta| \geq 1$ for large k and the fact that $\zeta_k - (\zeta + \zeta_k^*) = (\zeta_k - \zeta)(1 - 1/|\zeta_k - \zeta|)$, the condition (1.25) implies that the first term is nonnegative; the same is true about the second term since $(\zeta + \zeta_k^*) - \zeta = \zeta_k^*$ and $\zeta_{k\alpha} - \zeta_\alpha = \zeta_{k\alpha}^* |\zeta_k - \zeta|$ for each $|\alpha| = m$ and all large k. This also implies that for all sufficiently large k we have

$$\sum_{|\alpha| = m} [A_\alpha(x, \eta_k, \zeta + \zeta_k^*) - A_\alpha(x, \eta_k, \zeta)](\zeta_{k\alpha} - \zeta_\alpha)$$

$$\geq \sum_{|\alpha| = m} [A_\alpha(x, \eta_k, \zeta + \zeta_k^*) - A_\alpha(x, \eta_k, \zeta)](\zeta_{k\alpha}^*) \geq 0. \tag{1.31}$$

It follows from (1.29), (1.30), (1.31), and the fact that $\zeta_k^* \to \zeta^*$ and $\eta_k \to \eta$ that

$$0 = \lim_k \sum_{|\alpha| = m} [A_\alpha(x, \eta_k, \zeta + \zeta_k^*) - A_\alpha(x, \eta_k, \zeta)](\zeta_{k\alpha}^*)$$

$$= \sum_{|\alpha| = m} [A_\alpha(x, \eta, \zeta + \zeta^*) - A_\alpha(x, \eta, \zeta)](\zeta_\alpha^*) = 0.$$

Hence $\zeta^* = 0$ by (1.25), which contradicts $\zeta^* \neq 0$, and therefore we conclude that $D^m u_k(x) \to D^m u_0 x$. It now follows that $A_\alpha(x, \delta u_k, D^m u_k) \to A_\alpha(x, \delta u_0, D^m u_0)$ for $x \in Q$ (a.e.). Since $\{A_\alpha(x, \delta u_k, D^m u_k)\}$ is also bounded in L_q, it follows that $A_\alpha(x, \delta u_k, D^m u_k) \to A_\alpha(x, \delta u_0, D^m u_0)$ in L_q. This and the fact that $D^\alpha u_k \to D^\alpha u_0$ in L_p for $|\alpha| < m$ imply that

$$\int_Q \sum_{|\alpha| < m} A(x, u_k, D^m u_k)(D^\alpha u_k - D^\alpha w) \, dx$$

$$\to \int_Q \sum_{|\alpha| < m} A_\alpha(x, \delta u_0, D^m u_0)(D^\alpha u_0 - D^\alpha w) \, dx.$$

On the other hand,

$$\int_Q \sum_{|\alpha| = m} A_\alpha(x, \delta u_j, D^m w)(D u_j - D w) \, dx$$

$$\to \int_Q \sum_{|\alpha| = m} A_\alpha(x, \delta u_0, D^m w)(D^\alpha u_0 - D^\alpha w) \, dx.$$

Hence

$$(G(u_j, w), u_j - w) \to (G(u_0, w), u_0 - w) \qquad \text{for each} \quad w \in V.$$

We now show that $a(u, v)$ satisfies condition $(+ +)$. Let $\{u_j\} \subset V$ be any bounded sequence such that $a(u_j, v) \to (g, v)$ for some $g \in V^*$, uniformly for $v \in S_1$ (i.e., $T u_j \to g$ in V^*). This and the nonrestrictive assumption that $u_j \to u_0$ for some $u_0 \in V$ imply that $(G(u_j, u_j) - G(u_j, u_0), u_j - u_0) \to 0$ and hence, by the preceding argument, $(G(u_j, w), u_j - w)$

$\rightarrow (G(u_0,w), u_0 - w)$ for each $w \in V$. Since, by (1.27), $(G(u_j,u_j) - G(u_j,w), u_j - w) \geq 0$ for each w in V, the passage to the limit in the last equality yields $(g - G(u_0,w), u_0 - w) \geq 0$ for each $w \in V$. It follows from this and Minty's lemma that $g = G(u_0,u_0) = Tu_0$ [i.e., condition $(++)$ holds].   Q.E.D.

Remark 1.6 Corollary 1.2 extends a surjectivity theorem of Leray and Lions [152] proved by them for the case when $a(u,v)$ is coercive and the following condition (2_2) also holds:

(2_2) $\sum_{|\alpha|=m} A_\alpha(x,\eta,\zeta)\zeta_\alpha / |\zeta| + |\zeta|^{b-1} \rightarrow \infty$ as $|\zeta| \rightarrow \infty$

uniformly for bounded η and $x \in Q$ (a.e.).

Let us add that comparing the method of our proof of Corollary 1.2 with that of Leray–Lions [152] and with Browder's proof (his Lemma 3 (or even Lemma 3') in [33, Appendix to Section 1]), we see that essentially the same arguments can be used to prove the validity of

Lemma 1.2 Assume that in addition to (a1) the following hypotheses hold:

$$\sum_{|\alpha|=m} [A_\alpha(x,\eta,\zeta) - A_\alpha(x,\eta,\zeta')](\zeta_\alpha - \zeta'_\alpha) > 0 \qquad (1.32)$$

for all $\eta \in R^{s_{m-1}}$, all $\zeta,\zeta' \in R^{s_m}$ with $\zeta \neq \zeta'$, and all $x \in Q$. There are constants $c_0 > 0$ and c_1 such that

$$\sum_{|\alpha|\leq m} A_\alpha(x,\xi)\xi^\alpha \geq c_0 |\xi|^p - c_1 \qquad \text{for } \xi \in R^{s_m}, x \in Q. \qquad (1.33)$$

Then the mapping $T: V \rightarrow V^*$ determined by (1.6) is continuous, bounded, coercive, satisfies condition (S) of Browder [35] (i.e., $u_n \rightharpoonup u$ in V and $(Tu_n - Tu, u_n - u) \rightarrow 0 \Rightarrow u_n \rightarrow u$ in Y).

Using Lemma 1.2 and Theorem 2.2 in [210], we have the following result.

Theorem 1.3 Suppose that $T: V \rightarrow V^*$ satisfies all the conditions of Lemma 1.2. Then the variational problem (1.5) [or (1.7)] corresponding to the quasilinear elliptic boundary value problem (1.1), is feebly approximation solvable with respect to Γ_l for each F in V^* and, in particular, $T(V) = V^*$. Moreover, if for a given F in V^*, equation (1.5) has at most one solution, it is strongly approximation solvable (i.e., the Galerkin method converges).

Proof. Since, by Lemma 1.2, T is continuous, bounded, and satisfies condition (S), it follows from Theorem 2.2 in [210] that T is A-proper w.r.t. Γ_I. We add that, subsequently, it was also proved by Browder in [36] that if $T: X \to X^*$ is continuous, bounded, and satisfies condition (S), then T is A-proper. For the sake of completion, we provide the proof of this fact here.

Let $\{x_{n_k} \mid x_{n_k} \in X_{n_k}\}$ be any bounded sequence such that $f_{n_k} \equiv V_{n_k}^*(Tx_{n_k}) - V_{n_k}^* g \to 0$ in V^* as $k \to \infty$. For the sake of simplified notation, let $x_k \equiv x_{n_k}$, so $f_k \equiv V_k^* Tx_k - V_k^* g \to 0$. Since V is reflexive and $\{x_k\}$ is bounded, we may assume that $x_k \rightharpoonup x$ weakly in V. Select $y_k \in X_k$ such that $y_k \to x$ in V as $k \to \infty$. Then, for each k, we have

$$(Tx_k - Tx, x_k - x) = (Tx_k, x_k - y_k) + (Tx_k, y_k - x)$$

$$+ (Tx, x_k - x)$$

$$= (V_k^* Tx_k, x_k - y_k) + (Tx_k, y_k - x)$$

$$- (Tx, x_k - x)$$

$$= (f_k + V_k^* g, x_k - y_k) + (Tx_k, y_k - x)$$

$$- (Tx, x_k - x)(f_k, x_k - y_k) + (g, x_k - y_k)$$

$$= (Tx_k, y_k - x) - (Tx, x_n - x).$$

Since $x_k \rightharpoonup x$, $y_k \to x$, $f_k \to 0$ as $k \to \infty$, and $\{Tx_k\}$ is bounded, it follows that each of the four sequences defined above converges to 0. Since $x_k \rightharpoonup x$ in V and T is of type (S), it follows that $x_k \to x$ in X as $k \to \infty$. This and continuity of T imply that $Tx_k \to Tx$ in V^* as $k \to \infty$. To show that $Tx = g$, we note that if $v \in \cup_n X_n$ is any element, then $v \in X_m$ for some $m \in Z^+$ and therefore

$$(Tx_m - g, v) = (V_m^* Tx_m - V_m^* g, v) = (f_m, v).$$

Taking the limit as $m \to \infty$ on both sides of the last equality and noting that $Tx_m \to Tx$, we get the equality $(Tx - g, v) = 0$ for any $v \in \cup_n X_n$. Since the latter union is dense in V, we see that $Tx = g$ (i.e., T is A-proper with respect to Γ_I).

Since, in view of condition (1.33), T is also coercive [i.e., $(Tx,x)/\| x \| \to \infty$ as $\| x \| \to \infty$], it follows from the earlier results that the variational problem (1.3) is feebly approximation solvable w.r.t. Γ_I for each F in V^* [and, in particular, $T(V) = V^*$]; that is, there exists an integer $n_0 \geq 1$ such that the Galerkin equation (1.9) or (1.10) has a solution $x_n \in X_n$ for each $n \geq n_0$ such that for some subsequence $\{n_k\}$ we have $u_{n_k} \to u_0$ as $k \to \infty$ and u_0 is a solution of (1.3) [i.e., $a(u_0,v) = (F,v)$ for all $v \in V$]. If equation (1.3) has at most one solution for a given F, then the entire sequence $u_n \to u_0$ in V as $n \to \infty$. Q.E.D.

We note that the existence part of Theorem 1.3 was first proved by Leray–Lions [152] under slightly more restrictive conditions, and subsequently by Browder [33] under present conditions. A more general result involving a lower-order perturbation of the operator $A(u)$ in (1.1) was recently obtained by Fitzpatrick–Massabo–Pejsachowicz [85].

In [258], Schumann and Zeidler used the finite difference method and Theorem 1.1 of Petryshyn [210] to study the strong approximation solvability of the boundary value problem (1.1) when $A(w)$ gives rise to a strongly monotone operator. For other studies of problem (1.1) under various conditions, see Zeidler [314] and Lions [154].

1.2 Application of the Extension Theory to PDEs

In this section we use the extension Theorem III.4.8 and, in particular, its Corollary 4.10 to establish the existence of strong solutions for partial differential equations, and in Section 2 we apply this theory together with some results obtained by Conjura and Petryshyn in [51] to the solvability of ordinary differential equations.

As our first application of Corollary 4.10 in this section we consider the Dirichlet boundary value problem for an elliptic nonlinear partial differential equation of second order. Let us add that some of the problems in elasticoplasticity [93,100,149] are described by differential equations of the type considered below.

Let Q be a bounded region in the n-space R^n with a smooth boundary Γ. Let L_2 be the Hilbert space of real-valued square-integrable functions $u(x)$, $x = (x_1, x_2, \ldots, x_n)$, defined on $\overline{Q} = Q + \Gamma$ with the inner product and norm

$$(u,v) = \int_Q uv \, dx, \qquad \| u \| = \left(\int_Q u^2 \, dx \right)^{1/2}. \tag{1.34}$$

Let $C_0^2(\overline{Q})$ denote the set of all $u(x) \in L_2$ that are twice continuously differentiable on $\overline{Q}$ and satisfy the boundary conditions

$$u|_\Gamma = 0. \tag{1.35}$$

Let P be the nonlinear partial differential operator of second order defined for all $u \in D_P = C_0^2(\overline{Q})$ by the expression

$$Pu = -\sum_{i=1}^{n} \frac{\partial \alpha_i(x_j, p_j)}{\partial xi} + b(x_j, u), \qquad p_j = \frac{\partial u}{\partial x_j}, \tag{1.36}$$

such that the following conditions are satisfied:

(i) P is elliptic, that is,

$$\sum_{i,k=1}^{n} \frac{\partial \alpha_i}{\partial p_k} \xi_i \xi_k \geq m \left(\sum_{i=1}^{n} \xi_i^2 \right), \qquad m > 0,$$

(ii) There exist three constants $l, C > 0$, $D > 0$ such that $| \partial \alpha_i / \partial p_k | \leq C$, $| \partial b / \partial u | \leq D$, and $\partial b / \partial u$ is bounded below by l so that $\eta \equiv m + l/d > 0$ if $l < 0$ and $\eta \equiv m$ if $l \geq 0$, where $d > 0$ is a constant determined by the Friedrichs inequality

$$\int_Q \sum_{i=1}^{n} \left(\frac{\partial h}{\partial x_i} \right)^2 dx \geq d \int_Q h^2 \, dx, \qquad h \in C_0^1(\overline{Q}). \tag{1.37}$$

Our task is to solve the boundary-value problem

$$-\sum_{i=1}^{n} \frac{\partial \alpha_i(x_j, p_j)}{\partial x_i} + b(x_j, u) = f(x_j), \qquad u |_{\Gamma} = 0, \tag{1.38}$$

where $f(x)$ is given function in L_2, or equivalently, the equation

$$Pu = f, \qquad f \in L_2. \tag{1.39}$$

If we chose the operators K and T to be such that $K = I$ and T is defined for all $u \in D_T = D_P = C_0^2(\overline{Q})$ by

$$Tu = -\Delta u = -\sum_{i=1}^{n} \partial^2 \frac{u}{\partial x_i^2}, \tag{1.40}$$

then, as is know [87], T is symmetric and positive definite on D_T, that is,

$$(Tu, u) = (-\Delta u, u)$$
$$= \int_Q \sum_{i=1}^{n} \left(\frac{\partial u}{\partial x_i} \right)^2 dx \geq \alpha \| u \|^2, \qquad \alpha > 0. \tag{1.41}$$

Furthermore, the space H_0 is obtained as a completion of $C_0^2(\overline{Q}) = D_T$ in the metric

$$[u, v] = (Tu, v) = (-\Delta u, v), \qquad | u | = [u, u]^{1/2} \tag{1.42}$$

whose norm $| \cdot |$ defines the same topology as the norm $\| \cdot \|_{W_2^1}$, and $\mathring{W}_1^2(Q) \subset W_2^1(Q) \subset L_2$ [87] and T has a self-adjoint positive definite extension, which we shall also denote by T or by $-\Delta$, mapping its domain $\mathring{W}_2^2 \equiv W_2^2 \cap \mathring{W}_2^1$ onto L_2. Thus the problem

$$Tu = -\Delta u = g \tag{1.43}$$

has a unique solution $u \in \mathring{W}_2^2$ for every $g \in L_2(Q)$.

Let us now verify that under conditions (i) and (ii) the operator P defined by (1.36) satisfies the conditions of Corollary 4.10. Indeed, for every $h \in D_P$

$$(Pu,h) = \sum_{i=1}^{n} \int_Q \alpha_i(x_j,p_j) \frac{\partial h}{\partial x_i} dx + \int_Q b(x_j,u)h \, dx, \qquad u \in D_P.$$

Consequently, for any u and v in D_P with $g_j \equiv \partial v/\partial x_j$

$$(Pu - Pv, u - v) = \sum_{i=1}^{n} \int_Q [\alpha_i(x_j,p_j) - \alpha_i(x_j,g_j)] \frac{\partial}{\partial x_i} (u - v) \, dx$$

$$+ \int_Q [b(x_j,u) - b(x_j,v)](u - v) \, dx. \qquad (1.44)$$

In view of our conditions (i) and (ii), we derive from (1.44) the relations

$$(Pu - Pv, u - v) \geq \eta \sum_{i=1}^{n} \left(\frac{\partial}{\partial x_i} (u - v) \right)^2 = \eta(T(u - v), u - v),$$

$$(1.45)$$

where $\eta = m + 1/d > 0$ if $l < 0$ and $\eta = m$ if $l \geq 0$, and

$$| (Pu - Pv, h) | \leq \beta(T(u - v), u - v)^{1/2}(Th,h)^{1/2}, \qquad h \in \mathring{W}_2^2,$$

$$(1.46)$$

where $\beta = \beta(C,D,d) > 0$. Thus Corollary 4.10 and the preceding discussion imply the validity of the following:

Theorem 1.4 Suppose that the operator $P: C_0^2(\overline{Q}) \rightarrow L_2(Q)$ given by (1.36) satisfies conditions (i) and (ii). Then P has a unique solvable extension $P_0: D_{P_0} \subset L_2(Q) \rightarrow L_2(Q)$ such that the boundary value problem (1.38) has a unique strong solution $u \in D_{P_0}$ for each given f in $L_2(Q)$.

Remark 1.7 Similar results can be obtained for the differential equation of type (1.36) if the a_i's are also functions of u, and b are also functions of the b_j's [i.e., $a_i = a_i(x_i,u,b_j)$, $b = b(x_j,u,b_j)$].

1.3 PDEs and Pseudo-A-Proper Mappings

In this section we apply the results of Section I.4 to the variational solvability of a boundary value problem in $W_p^m(Q)$ for quasilinear equation

$$\sum_{|\alpha|,|\beta| m} D^\beta[a_{\alpha\beta}(x,u,Du, \ldots ,D^{m-1}u)D^\alpha u] = f \qquad (f \in L^q(Q)),$$

$$(1.47)$$

which has been treated by Shinbrot in [259] by a different method for the case when $p = 2$ (i.e., in the case of Hilbert spaces). We call the reader's attention to the difference between equations (1.47) and (1.1). We study (1.47) under the following assumptions:

(b1) The coefficients $a_{\alpha\beta}$ are continuous functions of all of their variables and there exists a continuous function g such that

$$| a_{\alpha\beta}(x,u,du, \ldots ,D^{m-1}u) | \leq g(\| u \|_{m,p}) \tag{1.48}$$

$$\text{for } |\alpha| \leq m, |\beta| \leq m, x \in Q.$$

Let V be a closed subspace of $W_p^m \subseteq V$ and assume that
(b2) The linear embedding of V into W_p^{m-1} is compact.
Now, condition (b1) implies that the (nonlinear) form

$$a(u,v) = \sum_{|\alpha|,|\beta| \leq m} \left\langle a_{\alpha\beta}(x,u,Du, \ldots ,D^{m-1}u)D^\alpha u,D^\beta v \right\rangle \tag{1.49}$$

is defined for all u,v in V and it follows from the results on Neinytskii operators (see [136,281]) that the map $T: V \to V^*$ given by

$$a(u,v) = (Tu,v) \qquad (u,v \in V), \tag{1.50}$$

is bounded and continuous, where (F,v) denotes the value of the functional F in V^* at the element v in V. As before, we say that $u_0 \in V$ is a weak or variational solution of (1.47) corresponding to V if u_0 satisfies the equation

$$a(u_0,v) = \langle f,v \rangle \qquad \text{for all } v \text{ in } V. \tag{1.51}$$

Since to a given f in $L^q(Q)$ there corresponds a unique $w_f \in V^*$ by $\langle f,v \rangle = (w_f,v)$ for all v in V, we see that (1.51) is equivalent to the operator equation

$$Tu = w_f \qquad (u \in V, w_f \in V^*). \tag{1.52}$$

Equation (1.51) together with the restriction that u_0 lies in V has the force not only of requiring that u_0 satisfy (1.51) (at least in the weak sense) but also of imposing boundary conditions upon u_0. The choices $V = \overset{\circ}{W}_p^m$ and $V = W_p^m$ lead to the homogeneous Dirichlet and Neumann problem for (1.47), respectively. Using Theorem 4.1 in Section I.4 we now prove the following existence theorem for (1.47).

Theorem 1.5 Suppose that the functions $a_{\alpha\beta}$ satisfy condition (b1). If there exists a constant $d > 0$ such that $a(u,u) \geq \langle f,u \rangle$ for all u in V with $\| u \|_{m,p} = d$, then (1.47) has a weak solution in $\bar{B}(0,d) \subset V$.

If there is a function of R^+ into R such that $c(r) \to \infty$ as $r \to \infty$ and

$$a(u,u) \geq c(\| u \|_{m,p}) \qquad \text{for all} \quad u \text{ in } V,$$

then equation (1.47) has a weak solution for each f in $L^q(Q)$.

Proof. We shall deduce Theorem 1.5 from the abstract Theorem 4.1 in Section I.4 when we set there $X = V$, $Y = V^*$, $K = I$, $\Gamma'_p = \{X_n, P; X_n^*, P_n^*\}$ and $M_n = I_n$ (identity on X_n), and show that the mapping $T: V \to V^*$ determined by (1.49)–(1.50) is psuedo-A proper with respect to Γ'_p. For this, in view of Proposition 1.1(b) in Section I.4, it suffices to show that T is weakly continuous. So let $u_n \rightharpoonup u$ weakly in V. Since the linear injection of V into W_p^{m-1} is compact, it follows that $u_n \to u$ strongly in W_p^{m-1}. Thus u and all its derivatives of order less than m approach the corresponding derivatives u almost everywhere in Q. Since each coefficient is continuous, we find that

$$a_{\alpha\beta}(x, un, Du_n, \ldots, D^{m-1}u_n) \to a_{\alpha\beta}(x, u, Du, D^{m-1}u) \quad \text{in} \quad Q \text{ (a.e.)}.$$

$$(1.53)$$

Moreover, (1.48) and the uniform boundedness theorem imply that the approach in (1.53) is bounded. It is then an easy application of Lebesgue's theorem on bounded convergence to show that

$$a_{\alpha\beta}(x, u_n, Du_n, \ldots, D^{m-1}u_n)D^\alpha u_n$$
$$\to a_{\alpha\beta}(x, u, Du, \ldots, D^{m-1}u)D^\alpha u \qquad \text{in} \quad L^q(Q)$$

for every α with $|\alpha| \leq m$. This is exactly what we wanted to prove.

Since the discussion above shows that the mapping $T: V \to V^*$ is weakly continuous, the conclusions of Theorem 1.5 follow from the special case of Theorem 4.1 in Section I.4. Q.E.D.

1.4 Construction of Weak Solutions of Semilinear PDEs of Order 2*m*

In this section we construct weak solutions in a given closed subspace V of $W_2^m(Q)$ ($V \supseteq \mathring{W}_2^m$) of the semilinear elliptic boundary value problem

$$\sum_{|\alpha|, |\beta| \leq m} (-1)^{|\alpha|} D^\alpha(a_{\alpha\beta}(x)D^\beta u)$$
$$+ \sum_{|\alpha| \leq m} (-1)^\alpha D^\alpha b_\alpha(x, \eta(u), D^m u) = f(x), \quad (1.54)$$

where $\eta(u) \equiv \{u, Du, \ldots, D^{m-1}u)$ and the functions $a_{\alpha\beta}(x)$'s and b_α's are assumed to satisfy the following conditions:

(d1) $a_{\alpha\beta}(x) \in L^\times(Q)$ for all $|\alpha|, |\beta| \le m$ and there exists a constant $\mu_0 > 0$ such that

$$\sum_{|\alpha|=|\beta|=m} a_{\alpha\beta}(x)\zeta_\alpha\zeta_\beta \ge \mu_0(\sum_{|\alpha|=m} |\zeta_\alpha|^2) \quad \text{for} \quad x \in Q, \quad \zeta \in R^{s_m'}$$

(d2) For all $|\alpha| \le m$, $b_\alpha: Q \times R^{s_m} \to R^1$ satisfy the Carathéodory conditions and there exist a constant $K_0 > 0$ and $h \in L^2$ such that

$$|b_\alpha(x,\eta,\zeta)| \le K_0(h(x) + |\eta| + |\zeta|)$$
$$\text{for } (\eta,\zeta) \in R^{s_{m-1}} \times R^{s_m}, \quad |\alpha| \le m.$$

To apply Theorems 4.1 and 4.2 in Section III.4 to the solvability of the semilinear elliptic equation (1.54) in V, we need to impose different analytic conditions on the functions b_α when $|\alpha| = m$ in addition to (d2).

When (d1) holds, one such requirement is condition (d3) in Lemma 1.3 below. As we shall see, with these conditions satisfied, Theorem 4.1 is applicable to the weak solvability of equation (1.54). Others will be given later.

Definition 1.4 Let V be a closed subspace of W_2^m such that $V \supseteq \overset{\circ}{W}_2^m$. For a given $f \in L^2$, *the boundary value problem corresponding to V for* (1.54) is the problem of establishing the existence of an element $u \in V$, called a *weak solution* of (1.54), such that

$$a(u,v) + b(u,v) = (f,v) \quad \text{for all} \quad v \in V, \tag{1.55}$$

where $(\cdot,\cdot)$ denotes the inner product in L^2 and where $a(u,v)$ and $b(u,v)$ are the generalized forms on V associated with differential expressions in (1.54) by

$$a(u,v) = \sum_{|\alpha|,|\beta| \le m} (a_{\alpha\beta}D^\beta u, D^\alpha v),$$
$$b(u,v) = \sum_{|\alpha| \le m} (b_\alpha(x,\eta,u), D^m u), D^\alpha v). \tag{1.56}$$

Note that conditions (d1) on the $a_{\alpha\beta}$'s imply that $a(u,v)$ is a bounded bilinear functional on V. Hence there exists a unique $A \in L(V,V)$ such that

$$a(u,v) = \langle Au,v \rangle \quad \forall \quad u,v \in V, \tag{1.57a}$$

where $\langle \cdot,\cdot \rangle$ denotes the inner product on V. Similarly, the conditions (d2) on the b_α's imply that for each fixed u in V, $b(u,v)$ is a bounded linear functional of v in V. Hence there exists a unique $B(u) \in V$ such that

$$b(u,v) = \langle Bu, v \rangle \qquad \forall \quad u, v \in V. \tag{1.57b}$$

The map $B: V \to V$ is clearly bounded. It is also continuous since if $\{u_j\}$ $\subset V$ is such that $u_j \to u_0$ in V, then

$$\| B(u_j) - B(u_0) \|_{2m} = \sup\{| (b(u_j,v) - b(u_0,v) | : \| v \|_{2m} = 1\}$$
$$\leq \sum_{|\alpha| \leq m} \| B_\alpha(u_j) - B_\alpha - (u_0) \|_q \to 0$$

as $j \to \infty$ because, by Property 3.1 in Section I.3, $\| B_\alpha(u_j) - B_\alpha(u_0) \|_q$ $\to 0$ as $u_j \to u_0$ in V. If $w_f \in V$ is such that $(f,v) = \langle w_f, v \rangle$ for all v in V, the equation

$$Au + Bu = w_f \qquad (u \in V, \, w_f \in V) \tag{1.58}$$

is equivalent to the conditions (1.55) for weak solutions of (1.54).

To establish the foregoing claims, let $\Gamma_1 = \{X_n, P_n\}$ be any given projectionally complete scheme for (V,V). Such schemes exist since V is a separable Hilbert space.

We saw that if $a_{\alpha\beta}$'s satisfy (d1), then the bounded linear map $A: V \to V$, given by (1.57a), can be written as $A = A_1 + A_1'$, where

$$\langle A_1 u, v \rangle = \sum_{|\alpha| = |\beta| = m} (a_{\alpha\beta}(x) D^\beta u, D^\alpha v) \qquad \forall \quad u, v \in V \tag{1.59}$$

and $A_1' \in L(V,V)$ is compact in view of the complete continuity of the embeddings of W_2^m into W_2^k for $k = 0, 1, \ldots, m - 1$.

Now, the projection method for (1.58) consists in solving

$$P_n A(u_n) + P_n B(u_n) = P_n w_f \qquad (u_n \in X_n, \, P_n w_f \in X_n) \tag{1.60}$$

or, equivalently, the algebraic system

$$\sum_{|\alpha|,|\beta| \leq m} (a_{\alpha\beta} D^\beta u_n, D^\alpha v) + \sum_{|\alpha| \leq m} (b_\alpha(x, \eta(u_n), D^m u_n, D^\alpha v)$$
$$= (f,v) \qquad (v \in X_n) \tag{1.61}$$

for $u_n \in X_n$ such that $\{u_n\}$ (or a subsequence) converges in V to a solution of (1.58) or a weak solution of (1.54).

To apply Theorem III.4.1 to the solvability of (1.58), we must impose some further analytic conditions on the functions b_α's to ensure that $T = A + B: V \to V$ is A-proper w.r.t. Γ_1. In the next lemma we impose one such simple analytic condition. Other sufficient conditions will be discussed subsequently.

Lemma 1.3 Suppose that in addition to (d1) and (d2) we assume that

(d3) There is $\mu \in [0,\mu_0)$ so that for $x \in Q$ (a.e.) and $\eta \in R^{sm-1}$:

$$\sum_{|\alpha|=m} (b_\alpha(x,\eta,\zeta) - b_\alpha(x,\eta,\zeta^*))(\zeta_\alpha - \zeta_\alpha^*) \geq$$

$$- \mu \sum_{|\alpha|=m} |\zeta_\alpha - \zeta_\alpha^*|^2 \qquad \text{for} \quad \xi_\alpha, \xi_\alpha \in R^{sm}.$$

Then $T = A + B: V \to V$ is A-proper w.r.t. Γ_1.

Proof. Since T is bounded and continuous, to prove the A-properness of T, it suffices to show that T satisfies condition (S). Let $\{u_j\}$ be any sequence in V such that $u_j \rightharpoonup u$ in V and

$$\lim_j \langle Tu_j - Tu, u_j - u \rangle = 0. \qquad (1.62)$$

As in the preceding case, to show that $u_j \to u$ whenever $\{u_j\} \subset V$ is such that $u_j \rightharpoonup u$ in V and (d3) holds, it suffices to show that $D^\alpha u_j \to D^\alpha u$ in L^2 for $|\alpha| = m$. Now, since $D^\alpha u_j \to D^\alpha u$ in L^2 for $|\alpha| < m$ and $b_\alpha(x,\xi(u_n))$'s are bounded in L^2, it follows that

$$\lim_j [\sum_{|\alpha|<m} (b_\alpha(x,\xi(u_j)) - b_\alpha(x,\xi(u)), D^\alpha u_j - D^\alpha u)] = 0.$$

In view of this and the fact that $A_1'(u_j) \to A_1'(u)$ in V, it follows from (1.62) that

$$\lim_j [\sum_{|\alpha|=|\beta|=m} (a_{\alpha\beta} D^\beta(u_j - u), D^\alpha u_j - D^\alpha u) \qquad (1.63)$$

$$+ \sum_{|\alpha|=m} (b_\alpha(x,\eta(u_j),D^m u_j) - b_\alpha(x,\eta(u),D^m u), D^\alpha u_j - D^\alpha u)] = 0.$$

Since $\eta(u_j) \to \eta(u)$ in L^2, $b_\alpha(x,\eta(u_j),D^m u) \to b_\alpha(x,\eta(u),D^m u)$ in L^2 by Property 3.1 in Section I.3 and $D^\alpha u_j \rightharpoonup D^\alpha u$ in L^2 for $|\alpha| = m$, it follows that

$$\sum_{|\alpha|=m} (b_\alpha(x,\eta(u_j),D^m u) - b_\alpha(x,\eta(u),D^m u), D^\alpha u_j - D^\alpha u) \to 0.$$

This and (1.63) imply that

$$\lim_j [\sum_{|\alpha|=|\beta|=m} (a_{\alpha\beta} D^\beta(u_j - u), D^\alpha u_j - D^\alpha u)$$

$$+ \sum_{|\alpha|=m} (b_\alpha(x,\eta(u_j),D^m u_j) - b_\alpha(x,\eta(u_j),D^m u), D^\alpha u_j - D^\alpha u)] = 0.$$

Since in view of the inequalities in (d1) and (d3), the expression in brackets is $\geq (\mu_0 - \mu) \sum_{|\alpha|=m} |D^\alpha u_j - D^\alpha u|^2$ and $\mu_0 - \mu > 0$, it follows on passage to the limit as $j \to \infty$ that $D^{\alpha}u_j \to D^{\alpha}u$ in L^2 for $|\alpha| = m$ (i.e., $u_j \to u$ in V, so $T = A + B$ is A-proper w.r.t. Γ_1).   Q.E.D.

In view of Lemma 1.3, the following consequence of Theorem III.4.1 holds.

Theorem 1.6 Suppose that the assumptions (d1), (d2), and (d3) hold and suppose that V is a given closed subspace of W_2^2 with $\mathring{W}_2^m \subseteq V$. If we also assume that $T = A + B$: $V \to V$ satisfies condition $(+)$ and there exists $r > 0$ such that $(Tu,u) \geq 0$ for $|\alpha| \geq r$, equation (1.58) is feebly approximation solvable w.r.t. Γ_1 for each f in L^2 [and, in particular, (1.59) has a weak solution u in V for each f in L^2]. For a given $f \in L^2$, equation (1.58) is strongly approximation solvable if it has at most one solution in V [i.e., the solutions $u_n \in X_n$ of (1.61) converge strongly in V to a weak solution of (1.54)].

2. SOLVABILITY OF ORDINARY DIFFERENTIAL EQUATIONS

In the second part of Chapter V we are interested in applying our theoretical results obtained in preceding chapters to the solvability and/or the approximation solvability of the boundary value and periodic boundary value problems for ordinary differential equations. We restrict our attention to equations of order 2 and 3. Needless to say, the same arguments work for ODEs of higher order. For the results for equations of the latter type, see [234,235] and the references listed there, where, as here, we are concerned primarily with equations whose nonlinearities depend on the derivatives of the highest order. As is well known, the classical methods, depending on the Schauder fixed-point theorem or Leray–Schauder topological degree theory, are no longer applicable to such problems. Sometimes, however, one can use the condensing mapping theory.

2.1 Boundary Value Problems

In this section we first make use of Theorem III.4.1.

Solvability of the ODE of the Dynamics of Wires

Using the continuation principle for stably solvable maps of Furi–Martelli–Vignoli [95,96], it was shown by Sanches [254] that a differential equation which appears in the study of the dynamics of wires has a solution for each $f(t) \in L^2(0,b)$. Here we indicate how Theorem III.4.1, when T is odd, can be used to deduce the result of [254] in a simpler and sometimes constructive way. A perturbation problem to which the abstract results in [95,96] cannot be applied will also be treated (see [221]).

Let $b > 0$ and consider the nonlinear ODE

$$\begin{cases} u'' + r(t)u - p(t)u^3 + q(t)u' + |u'|u' = f(t) \\ u(0) = u(b), \quad u'(0) = u'(b), \end{cases} \tag{2.1}$$

where $f \in L_2(0,b)$ and the functions r, q, and p satisfy the following conditions:

(b1) $r,q \in L_\infty(0,b)$

(b2) $p \in W^1_\infty(0,b)$, $p(0) = p(b)$ and there exists $\alpha > 0$ such that $p(t \geq \alpha$ for $t \in [0,b]$.

To apply Theorem III.4.1 to the solvability of (2.1) we let $Y = L_2(0,b)$, $X = \{u \in W^2_2(0,b) \mid u(0) = u(b), u'(0) = u'(b)\}$ and define the map $T: X \to Y$ by

$$Tu = u'' + r(t)u - p(t)u^3 + q(t)u' + |u'|u'. \tag{2.2}$$

Let $\{Y_n\} \subset L_2(0,b)$ be finite-dimensional subspaces such that $\text{dist}(g,Y_n) \to 0$ for each $g \in L_2(0,b)$. It is easy to see that $L \in L(X,Y)$ defined by $Lu = u''$ is Fredholm of index 0 and that $K \in L(X,Y)$ defined by $Ku = u'' - u$ is a linear homeomorphism of X onto Y. Hence if for each n, we choose X_n in X to be such that $Y_n = K(X_n)$ and let Q_n be the orthogonal projection of Y onto Y_n, then $\Gamma_K = \{X_n,V_n;Y_n,Q_n\}$ is an admissible scheme for (X,Y), the operator L is A-proper w.r.t. Γ_K, and the operator $T: X \to Y$ defined by (2.2) is also A-proper w.r.t. Γ_K since the map $N: X \to Y$ given by $Nu = r(t)u - p(t)u^3 + q(t)u' + |u'|u'$ is compact, by the Sobolev embedding theorem. Since T is also odd, to apply Theorem III.4.1 to (2.1), all we need is to show that T satisfies condition $(+)$ [i.e., if $\{u_j\} \subset X$ is any sequence such that $Tu_j \to g$ for some g in Y, then $\| u_j \|_{2,2} \leq M$ for all j and some constant $M > 0$]. Thus, as in [254], we are led to derive certain a priori estimates; that is, it suffices to show that there exist a continuous function $\psi: R^+ \to R^+$ such that if $u \in X$ is a solution of

$$u'' + r(t)u - p(t)u^3 + g(t)u' + |u'|u' = f \tag{2.3}$$

for any given $f \in L_2(0,b)$, then $\| u \|_{2,2} \leq \psi(\| f \|)$. To prove this we will need the following inequalities from [111] for any $a > 0$ and $b > 0$:

$$ab \leq \frac{\epsilon}{2}a^2 + \frac{1}{2\epsilon}b^2, \quad \epsilon > 0. \tag{2.4}$$

$$ab \leq \epsilon a^{1/\alpha} + \left(\frac{\alpha}{\epsilon}\right)^{\alpha/(1-\alpha)}(1-\alpha)b^{1/(1-\alpha)}, \quad \epsilon > 0, \quad \alpha \in (0,1). \tag{2.5}$$

To get an estimate for $\| u'' \|_2$ we first note that $\int_0^b | u' | u'u'' \, dt = 0$ since $| u' | u'' = (\tfrac{1}{2}| u' | u')'$ and

$$\int_0^b | u' | u'u'' \, dt = \tfrac{1}{2}| u' |^3 \Big|_0^b - \tfrac{1}{2}\int_0^b | u' | u'u'' \, dt.$$

Thus multiplying (2.3) by u'', integrating over $[0,b]$, and using the Cauchy–Schwarz inequality, one gets

$$\| u'' \|_2^2 \le R\| u \|_2 \| u'' \|_2 + P\| u \|_6^3 \| u'' \|_2$$
$$+ Q\| u' \|_2 \| u'' \|_2 + \| f \|_2 \| u'' \|_2,$$

where R, P, and Q denote the $L_\infty(0,b)$-norms of r, p, and q and $\| \cdot \|_p$ is the $L_p(0,b)$-norm. Hence, in view of (2.4),

$$\| u'' \|_2^2 \le c\{\| u \|_6^6 + \| u' \|_2^2 + \| f \|_2 + 1), \tag{2.6}$$

where here and afterward c denotes various constants that are independent of u. Next multiply (2.3) by u' and integrate to get

$$\int_0^b ruu' \, dt + \frac{1}{4}\int_0^b p'u^4 \, dt + \int_0^b q(u')^2 \, dt + \int_0^1 | u' |^3 = \int_0^b fu' \, dt,$$

where we have used the fact that

$$\int_0^b u''u' \, dt = 0 \quad \text{and} \quad \int_0^b pu^3u' \, dt = -\frac{1}{4}\int_0^b p'u^4 \, dt.$$

Again, as above, we get from the preceding equality that

$$\| u' \|_3^3 \le c\{\| u' \|_2^2 + \| u \|_4^4 + \| f \|_2^2 + 1\}. \tag{2.7}$$

Finally, multiply (2.3) by u and integrate to obtain

$$\| u' \|_2^2 + \alpha\| u \|_4^4 \le R\| u \|_2^2 + Q\| u \|_2 \| u' \|_2 \tag{2.8}$$
$$+ \| f \|_2 \| u \|_2 + \int_0^b \| u' \|^2 | u | \, dt,$$

where we used the inequality $p(t) \ge \alpha$ and the fact that $\int_0^b u''u \, dt = - \| u' \|_2^2$. Now, using (2.5) with $\alpha = \tfrac{3}{4}$ we get

$$| u' |^2 | u | \le \epsilon| u' |^{8/3} + \left(\frac{1}{\epsilon}\frac{3}{4}\right)^3 \left(\frac{1}{4}\right) | u |^4 \qquad \text{for any} \quad \epsilon > 0$$

and therefore

$$\int_0^b | u' |^2 | u | \, dt \le \epsilon\| u' \|_{8/3}^{8/3} + \frac{1}{4}\left(\frac{3}{4\epsilon}\right)^3 \| u \|_4^4.$$

In view of this, it follows from (2.8) that

$$\| u' \|_2^2 + \| u \|_4^4 \le c\{\| u' \|_{8/3}^{8/3} + \| f \|_2^2 + 1\}. \tag{2.9}$$

Since

$$\| u' \|_{8/3}^{8/3} = \int_0^b | u' |^{8/3} \, dt \le c \left(\int_0^b | u' |^3 \right)^{8/9} = c(\| u \|_3^3)^{8/9},$$

it follows from (2.9) and (2.7) that

$$\| u' \|_2^2 + \| u \|_4^4 \le c\{[\| u' \|_2^2 + \| u \|_4^4 + \| f \|_2^2 + 1]^{8/9} + \| f \|_2^2 + 1\}. \tag{2.10}$$

This obviously implies that $\| u' \|_2^2 + \| u \|_4^4 \le M_0$, M_0 being a constant independent of u, and by virtue of the Sobolev theorem, we also have $\| u \|_6$ bounded. This and (2.6) imply that $\| u'' \|_2$ is also bounded. Hence $\| u \|_{2,2} \le \psi(\| f \|)$, where $\psi(t) = c\{t_2^2 + 1\}$.

The discussion above implies the validity of

Theorem 2.1 For any functions p, q, and r satisfying (b1) and (b2), the boundary value problem (2.1) is feebly a-solvable w.r.t. Γ_K for each f in $L_2(0,b)$ and, in particular, $T(X) = L_2$. If, for some f in $L_2(0,b)$, the problem (2.1) has a unique solution, then (2.1) is strongly a-solvable.

Using the result above, we again use Theorem III.4.1 to solve a perturbation problem to which the abstract results in [95,96] are not applicable.

Theorem 2.2 Suppose that the functions p, q, and r satisfy conditions (b1) and (b2) and suppose that

(b3) g: $[0,b] \times R^3 \to R$ is continuous and bounded and such that there exists $d \in [0,1)$ such that

$$[g(t,s,r,z_1) - g(t,s,r,z_2)](z_1 - z_2) \ge -d| z_1 - z_2 |^2$$
$$\text{if} \quad t \in [0,b], \, s,r,z_1,z_2 \in R$$

and also suppose that

$$g(t,s,r,z) = -g(t,-s,-r,-z) \quad \text{if} \quad t \in [0,1], \quad s,r,z \in R.$$

Then the conclusions of Theorem 2.1 hold for

$$\begin{cases} -u'' + g(t,u,u',u'') + N(u) = f(t) \\ u(0) = u(b), \quad u'(0) = u'(b) \end{cases}$$

for each $f \in L_2(0,b)$ with $N(u)$ the same as in Theorem 2.1.

Proof. Let $G: X \to Y$ be the mapping defined by $G(u) = g(t,u,u',u'')$. Condition (b3) implies that G is continuous, has a bounded range, is odd, and in view of Lemma 3.1 in [212], $L + G: X \to Y$ is A-proper w.r.t. Γ_K. Furthermore, since $N: X \to Y$ is compact, it follows that $L + G + N$ is A-proper. Since $G(X)$ is bounded, it follows that $L + G + N$ satisfies condition $(+)$. This and the oddness of $L + G + N$ allow us to invoke the assertion (A2) of Theorem III.4.1.   Q.E.D.

2.2 Solvability of ODEs via the Friedrichs Extension Theory

It was shown in [51] that the inequalities (4.46)–(4.47) in Theorem III.4.9 and the fact that $D(P_0) = D(T_0)$ play an essential role in providing estimates for the error $| u_n - u_0 |_0$ and the residual $\| Pu_n - h \|_H$. However, the following simple example shows that in general one cannot expect P or S, when P is of the form $P = T + S$, to satisfy the foregoing global Lipschitz conditions. This has been studied in [51], although the following simple example illustrates the point.

Consider the boundary value problem

(A) $\begin{cases} -u'''(t) + (u'(t))^2 u'(t) = f(t) & [t \in (0,1), \, f \in L_2(0,1)] \\ u(0) = u'(0) = u'(1) = 0. \end{cases}$

Let $Tu = -u'''$ on $D(T) = \{u \in C^3(0,1) \mid u(0) = u'(0) = u'(1) = 0\}$, $Nu = (u')^2 u'$, $Pu = Tu + Su$ on $D(P) = D(T)$, and $Ku = u'$ on $D(K) = \{u \in C^1(0,1) \mid u(0) = 0\}$. It is not hard to show (see [51]) that T is K-p.d. and K-symmetric and thus has an extension T_0 with $D(T_0) \subset H_0$, where H_0 is the completion of $D(T)$ with respect to $[u,v] = (Tu,Kv) = \int_0^1 u''v'' \, dt$, $| u |_0 = \sqrt{[u,u]}$, and $H_0 = \{u \in L_2(0,1) \mid u,u'$ are absolutely continuous (a.c.), $u'' \in L_2(0,1)$ and $u(0) = u'(0) = u'(1) = 0\}$. Moreover, as will be seen below, $D(T_0) = \{u \in H_0 \mid u''$ is a.c. and $u''' \in L_2(0,1)\}$. Using an inequality from [314], it is easy to see that

$(Pu - Pv, K(u - v)) \geq | u - v |_0^2 \qquad \forall \quad u,v \in D(P)$

[i.e., P satisfies condition (4.46)]. However, S does not satisfy any one of the foregoing Lipschitz global conditions on $D(P)$. But a simple calculation shows that S is Lipschitz on bounded sets in $D(P)$. In fact, if $u,v \in D(P)$ and $| u |_0, | v |_0 \leq M$, we have $\| Su - Sv \|_{L_2} \leq \frac{3}{8} M^2 | u - v |_0$.

The boundary value problem (A) (more complicated examples will be given below) shows that the following extensions of Corollary 4.11 in Section III.4 will prove to be useful in various applications, especially in the derivations of the estimates for errors and the residuals.

Proposition 2.1 Let T be K-p.d. and $P: D(P) = D(T) \subset H \to H$ satisfy (4.46) and the following condition:

(a) There exists a function $c: R^+ = \{t \geq 0\} \to R^+$ such that $| Pu - Pv, Kh) | \leq c(M) | u - v |_0 | h |_0$ for $u,v,h \in D(P)$ with $| u |_0, | v |_0 \leq M$, then the conclusion of Theorem III.4.9 holds.

Proof. Since $K_0: H_0 \to H$ is continuous and $D(T)$ is dense in H_0, it is easy to see that (4.47) of Theorem III.4.9 is implied by condition (a) and thus Proposition 2.1 is a consequence of Theorem III.4.9. Q.E.D.

Proposition 2.2 If T is K-p.d. and $P = T + S$ is such that (4.46) of Chapter III holds and

(b) There exists a function $b: R^+ \to R^+$ such that $\| Su - Sv \| \leq b(M)$ $| u - v |_0$ whenever $u,v \in D(P)$ and $| u |_0, | v |_0 \leq M$, then $D(P_0) = D(T_0)$ and $P_0 = T_0 + S_0$ with S_0 an extension of S to H_0.

Since (b) implies (a), Proposition 2.2 follows from Theorem III.4.9 and the same argument as the one used by the author in his proof of Corollary 3 in [217].

An immediate consequence of Proposition 2.2 is that for each $f \in L_2(0,1)$ the boundary value problem (A) has a strong solution $u \in D(P_0) = D(T_0)$.

The following result can be applied to problems in the theory of elasticoplasticity studied by Gajewski, Langenbach, Petryshyn, Skrypnik, and others (see [93,99,100,149,234,269] for references) under different and, in some cases, stronger conditions.

Proposition 2.3 Let P be a densely defined nonlinear operator such that the Gateaux derivative P'_u exists for each $u \in D(P)$ with P'_0 K-p.d. and for some $\alpha > 0$:

(c) $(P'_u(h),Kh) \geq \alpha(P'_0 h,Kh) \equiv \alpha | h |_0^2$ for all $u,h \in D(P)$.

Assume also that for $u,v,h \in D(P)$, the function $\phi(t) = (P'_{v+t(u-v)}(u - v), Kh)$ is continuous in t for $u,v,h \in D(P)$ and

(d) To each $M > 0$, there exists a $c(M) > 0$ such that if $u,w,h \in D(P)$ and $| w |_0 \leq M$, then $| (P'_w(u),Kh) | \leq c(M) | u |_0 | h |_0$.

Then the conclusions of Theorem III.4.9 hold.

Proof. It follows from the assumptions made on $\phi(t)$ and (c) that for all $u,v \in D(P)$ we have the inequality

$$(Pu - Pv, K(u - v))$$
$$= \int_0^1 (P'_{v+t(u-v)}(u - v), K(u - v)) \, dt \geq (P'_0(u - v), K(u - v)).$$

Also, if $u,v,h \in D(P)$ are such that $|u|_0, |v|_0 \leq M$, then $|(Pu - Pv, Kh)| = |\int_0^1 (P'_{v+t(u-v)}(u - v), Kh)\, dt|$. Since $|v + t(u - v)|_0 \leq M$, it follows from (d) that $|(Pu - Pv, Kh)| \leq c(M)|u - v|_0 |h|_0$. Thus Proposition 2.3 follows from Proposition 2.1.   Q.E.D.

Using Theorems III.4.8 and III.4.9, we now state a result proved in [51] which is not as sharp as Theorem III.4.8 but which is more useful in applications than Theorem III.4.9 when applied to the solvability of the operator equation

$$P(u) = f \qquad [u \in D(P), f \in H]. \tag{2.11}$$

We recall that for a given f in H, the unique solution $u \in D(P_0)$ of the equation $P_0(u) = f$ is called a *strong solution* of (2.11). Suppose now that for each u in $D(P)$, the equation

$$p(u,v) = (Pu, K_0 v) \qquad [u \in D(P), v \in H_0] \tag{2.12}$$

defines a bounded linear functional of v in H_0, which can be extended to all of H_0 [i.e., $p(u,v)$ is well defined for all u,v in H_0]. Then, for a given f in H, u is called a *weak solution* of (2.11) iff $p(u,v) = (f,K_0 v)$ for all v in H_0. It follows that u in H_0 is a weak solution of (2.11) iff u is a solution of

$$\overline{W}u = f_0 \qquad (f = T_0^{-1}f \in H_0). \tag{2.13}$$

Furthermore, if $\Gamma = \{H_n, \Pi_n\}$ is a projectionally complete scheme for (H_0, H_0), then the Galerkin approximant u_n in H_n is determined by the equation

$$p(u_n,v) = (f,Kv) \qquad (v \in H_n, n = 1,2,3, \ldots) \tag{2.14}$$

or, equivalently, by the operator equation

$$\Pi_n \overline{W}(u) = \Pi_n T_0^{-1}f \qquad (u \in H_n, T_0^{-1}f \in H_0). \tag{2.15}$$

We now state and prove the result from [51] which will be used in what follows and which provides the error estimate for the Galerkin method when it is applied to the approximate solvability of equation (2.11) in the form

$$Q_n P(u_n) = Q_n f \qquad [u_n \in X_n \subset D(P), Q_n f \in Y_n], \tag{2.16}$$

where Q_n is an orthogonal projection of H onto Y_n, $\{Y_n\} \subset H$.

Theorem 2.3 Suppose that the hypotheses of Theorem III.4.9 hold. Then:

(i) If $\{H_n, \Pi_n\}$ is projectionally complete for H_0 with $[H_n] \subset H_0$ monotonically increasing, then for each h in H there exists a unique u_0

$\in D(P_0)$, $u_n \in H_n$, such that $P_0(u_0) = h, u_n$ satisfies equation (2.14) or, equivalently, equation (2.15), and $\| u_n - u_0 |_0 \to 0$ as $n \to \infty$. Moreover, if $\overline{W}: H_0 \to H_0$ is either continuous or bounded and demicontinuous, we may drop the condition that $\{H_n\}$ is monotonically increasing.

(ii) Suppose we assume additionally that to each $M > 0$ (with $| u_0 |_0 \le M$) there exists a $c(M) > 0$ such that

$$| [\overline{W}u_0 - \overline{W}v, u - v] | \le c(M) | u_0 - v |_0 | u - v |_0 \tag{2.17}$$
$$\text{for} \quad u,v \in H_n, \quad | u |_0, | v |_0 \le M,$$

then, independent of any conditions on H_n,

$$| u_n - u_0 |_0 \le \left[1 + \frac{c(M)}{\eta} \right] \inf_{v \in H_n} | v - u_0 |_0. \tag{2.18}$$

(iii) If we assume that $\{H_n\} \subset D(T)$, the Galerkin scheme (2.15) is equivalent to the equation

$$Q_n P u_n = Q_n f \tag{2.19}$$
$$\text{or, equivalently, to} \quad (Pu_n, Kv) = (f, Kv) \quad \forall \quad v \in H_n,$$

where $Q_n: H \to K(H_n)$ is an orthogonal projection, and the condition (2.17) can be stated in the form

$$| (f - Pv, K(u - v)) | \le c(M) | u_0 - v |_0 | u - v |_0 \tag{2.20}$$
$$\text{for} \quad u,v \in H_n \quad \text{with} \quad | u |_0, | u |_0, | v |_0 \le M.$$

Proof. (i) Since (1.1) and (1.2) imply that $\overline{W}: H_0 \to H_0$ is strongly stable and demicontinuous, the assertion (i) follows from Theorems III.4.9 and III.4.3.

(ii) To prove (ii), note that it follows from (III.4.6 and the structure of P_0) that if $u_n \in H_n$ solves (2.15), then

$$| u_n |_0^2 \le \frac{1}{\eta} | [\Pi_n \overline{W} u_n - \Pi_n \overline{W}(0), u_n] | = \frac{1}{\eta} \left| [T_0^{-1}(f - P(0)), u_n] \right|$$

$$\le \frac{1}{\eta} | T_0^{-1}(f - P(0)) |_0 | u_n |_0 \le \frac{\sqrt{\alpha 2}}{\eta} \| f - P(0) \| | u_n |_0.$$

Thus the approximate solutions $\{u_n\}$ are bounded by $\overline{M} = (\sqrt{\alpha 2}/\eta) \| f - P(0) \|$. By the same arguments, $| u_0 |_0 \le \overline{M}$. Now, since $u_n \in H_n$ and $| u_n |_0 \le \overline{M}$, it follows that

$$\inf_{v \in H_n} | v - u_0 |_0 = \inf_{v \in H_n} | v - u_0 |_0, \quad | v |_0 \le 2\overline{M} \tag{2.21}$$

Let $M = 2\overline{M}$; then for $v \in H_m$ with $|v|_0 \le M$, by equations (2.17) and (III.4.6)

$$|u_n - v|_0^2 \le \frac{1}{\eta}|[\overline{W}u_n - \overline{W}v, u_n - v]|$$

$$= \frac{1}{\eta}|[\overline{W}u_0 - \overline{W}v, u_n - v]| \le \frac{c(M)}{\eta}|u_0 - v|_0|u_n - v|_0.$$

Thus

$$|u_n - v|_0 \le \frac{c(M)}{\eta}|u_0 - v|_0. \tag{2.22}$$

By (2.22), (2.21), the triangle inequality, and the fact that v was chosen arbitrarily, (2.18) follows.

(iii) Under the assumption that $\{H_n\} \subset D(T)$, the equivalence of (2.16) and (2.19) follows since $\Pi_n\overline{W}u_n = \Pi_n T_0^{-1}Pu_n$ for $u_n \in D(P)$ and u_n solves (2.16) if and only if $[T_0^{-1}Pu_n, v] = [T_0^{-1}h, v]$ for each $v \in H_n$. But the last expression implies that

$$(Pu_n, Kv) = (h, Kv) \qquad \text{for} \quad v \in H_n, \tag{2.23}$$

and (2.23) is equivalent to (2.19). The remainder of the proof uses the same arguments as above.   Q.E.D.

Note that in actual applications of Theorem 2.3(i), to obtain the approximant u_n to the solution u_0 of equation (2.11) one actually uses the Galerkin equation (2.16), since, in general, $\overline{W}$ and T_0^{-1} are not known in advance.

Among the class of operators that satisfy condition (2.17) are those that satisfy (a), (b), or (d). This is, in fact, the case for a number of concrete problems treated in Section 2.3. For a considerable improvement on Theorem 2.3, see [218].

2.3 Application to the Solvability of ODEs

We begin this section with the following simple boundary value problem for a third-order ordinary differential equation, which will illustrate the way in which the results of the preceding section are used. Let $h \in L_2(0,1)$,

$$Pu(t) = -(p(t)u''(t))' + f(t,u,u') = h(t) \qquad [t \in (0,1)] \tag{2.24}$$

with respect to either of the boundary conditions

$$u(0) = u'(0) = u'(1) = 0 \tag{2.25}$$

or

$$u(0) = u'(0) + u''(1) = 0. \tag{2.26}$$

We will assume that $p(t)$ is absolutely continuous on $[0,1]$ and $p'(t)$ is bounded with

(Ia) $p(t) \geq p_0 > 0$ for all $t \in [0,1]$ and some constant $p_0 > 0$, while $f: [0,1] \times R^2 \to R^1$ is continuous and

(Ib) $\dfrac{f(t,\xi_1,\eta_1) - f(t,\xi_2,\eta_2)}{\eta_1 - \eta_2} \geq \alpha > -\dfrac{\pi^2}{4} p_0.$

Before we state our main theorem concerning the approximation solvability of the boundary value problems (2.24)–(2.25) and (2.24)–(2.26) and the corresponding estimates for the error and the residual, we first prove the following lemma.

Lemma 2.1 Suppose that $Ku = u'$ on $D(K) = \{u \in C^1[0,1] \mid u(0) = 0\}$ and $Tu(t) = -(p(t)u''(t))'$ on $D(T) = \{u \in C^3[0,1] \mid u$ satisfies (2.25) [or (2.26)]\}, where p satisfies the conditions indicated above. Then

(i) T is K-p.d. and K-symmetric and T has a unique K-p.d. and K-symmetric extension T_0 with $D(T_0) \subset H_0 = \{u \in L_2(0,1) \mid u$ and u' are absolutely continuous (a.c.) on $[0,1]$, $u'' \in L_2(0,1)$ and u satisfies (2.25) $[u(0) = u'(0) = 0]\}$, where H_0 is the completion of $D(T)$ in the norm $\mid u \mid_0$ derived from the inner product $[u,v] = \int_0^1 p(t)u''v'' \, dt$, and

(ii) $D(T_0) = \{u \in H_0 \mid u''$ is a.c. on $[0,1]$, and $u''' \in L_2(0,1)$ [and $u''(1) = 0]\}$.

Proof. Let us first consider the operator T defined by (2.24) and boundary conditions (2.25). As defined, $D(T) \subset D(K)$, $D(T)$ is dense in $L_2(0,1)$, and $K(D(T))$ is dense in $L_2(0,1)$ [51]. For $u,v \in D(T)$, integrating by parts and using the boundary conditions, we see that T is K-symmetric since $(Tu,Kv) = \int_0^1 pu''v'' \, dt = (Ku,Tv)$. Moreover, for $u \in D(T)$, $(Tu,Ku) \geq p_0 \int_0^1 (u'')^2 \, dt$. But if y is any a.c. function on $[0,1]$ such that $y(0) = 0$, then [111] Hardy's inequality is satisfied, that is,

(h) $\displaystyle\int_0^1 (y')^2 \, dt \geq \frac{\pi^2}{4} \int_0^1 (y)^2 \, dt.$

Therefore, since u and u' are a.c. and $u(0) = u'(0) = 0$, it follows that T is K-p.d. with $\alpha_2 = 4/p_0\pi^2$ and $\alpha_1 = (\pi^4/16)p_0$. The existence of T_0 now follows from Theorem II.2.1. To complete the proof of part (i) we will only mention that the nature of H_0 follows in a straightforward way using the K-positive definiteness of T and standard results from the theory of functions.

To prove (ii), let $h \in L_2(0,1)$. Then there exists a unique $u \in D(T_0)$ such that for each $v \in H_0$, $[u,v]_0 = (h,Kv)$; that is,

$$0 = \int_0^1 (pu''v'' - hv') \, dt = \int_0^1 pu''v'' \, dt - \int_0^1 \left[\frac{d}{dt} \int_0^t h(s)ds \right] v' \, dt,$$

and

$$\int_0^1 \left[pu'' + \int_0^t h(s) \, ds \right] v'' \, dt = 0.$$

Now let $\langle 1 \rangle^\perp = \{u \in L_2(0,1) \mid \int_0^1 u(t) \, dt = 0\}$. Suppose that $z \in \langle 1 \rangle^\perp$; then let $w(x) = \int_0^x z(s) \, ds$ and $v(t) = \int_0^t w(x) \, dx$. Since $v(0) = v'(0) = v'(1) = 0$ and $v'' = z$, if we define $B: H_0 \rightarrow L_2(0,1)$ by $Bv = v''$, it follows that $\langle 1 \rangle^\perp \subseteq B(H_0)$. Conversely, if $v \in H_0$, then $\int_0^1 Bv \, dt = v'(1) - v'(0) = 0$; that is, $B(H_0) \subseteq \langle 1 \rangle^\perp$. Therefore, $B(H_0) = \langle 1 \rangle^\perp$. Now from the equality above we have that $(pu'' + \int_0^t h(s) \, ds, Bv) = 0$ for each $v \in H_0$. It therefore follows that for almost every $t \in [0,1]$,

$$p(t)u''(t) + \int_0^t h(s) \, ds = C \text{ (const).}$$

Thus u'' is a.c. on $[0,1]$, $u''' \in L_2(0,1)$, and $Tu(t) = h(t)$ for a.e. $t \in [0,1]$. We have so far shown that $D(T_0) \subseteq M = \{u \in L_2(0,1) \mid u,u'$, and u'' are a.c. on $[0,1]$, $u''' \in L_2(0,1)$, and u satisfies (2.25)$\}$. However, if $v \in M$, then v is certainly in H_0. Let $h = -(pu'')' \in L_2(0,1)$, and let $u \in D(T_0)$ be such that $T_0u = h$; then for each $z \in H_0$, $[v,z]_0 = \int_0^1 pv''z'' \, dt = \int_0^1 hz' \, dt = [u,z]_0$. Thus $v = u \in D(T_0)$, $M \subseteq D(T_0)$, and the results of (ii) follow relative to boundary conditions (2.25).

Let us complete the proof with an indication of the adjustments necessary when we consider T relative to boundary conditions (2.26); that is, $u(0) = u'(0) = u''(1) = 0$. Without proof, H_0 is as described in (i), but note that the natural boundary condition $u''(1) = 0$ is lost in H_0. As we will see, it is regained in $D(T_0)$. As before, if $h \in L_2(0,1)$ and $u \in D(T_0)$ such that $T_0u = h$, it follows that

$$\int_0^1 \left[pu'' + \int_0^t h(s) \, ds \right] v'' \, dt - \left[\int_0^t h(s) \, ds \right] v' \Big|_0^1 = 0.$$

Since $v'(0) = 0$ and $v'(1) = \int_0^1 v'' \, dt$, we obtain

$$\int_0^1 \left[pu'' + \int_0^t h(s) \, ds - \int_0^1 h(s) \, ds \right] v'' \, dt = 0.$$

But for $z \in L_2(0,1)$, if $w(x) = \int_0^x z(s) \, ds$ and $v(t) = \int_0^t w(x) \, dx$, then $v \in H_0$ [relative to boundary conditions (2.26)] and $Bv = v'' = z$, that is, $B(H_0) = L_2(0,1)$. Therefore, it follows from above that for a.e. $t \in [0,1]$,

$$p(t)u''(t) + \int_0^t h(s) \, ds - \int_0^1 h(s) \, ds = 0.$$

Since $p(1)u''(1) = 0$ and $p(1) \geq p_0 > 0$, then $u''(1) = 0$. Using the same arguments as before, the result follows. Q.E.D.

Note that $D(T_0)$ is independent of our choice of $p(t)$, which satisfies (Ia). Thus if $p(t) \equiv 1$, we have $Tu = -u'''$, which will be useful in applications. As we will see, the eigenfunctions of $Tu - \lambda Ku = 0$ are useful in solving both linear and nonlinear equations and will be used to obtain error estimates for the residuals. Relative to (2.25) the eigenvalues are given by $\lambda_n = [n\pi]^2$, where n is a positive integer and the eigenfunctions are $\phi_n = c_n[1 - \cos(n\pi t)]$. If $c_n = \sqrt{2}/n\pi$, then $\{K\phi_n\}$ is a complete orthonormal sequence in $L_2(0,1)$. Also, we should mention that the boundary conditions (2.26) are used concerning the equation $-u''' - \lambda qu' = 0$ when considering the question of determining the critical load in the stability problem of a compressed bridge belt and in the problem of buckling of a bar under distributed axial load.

Using Theorems III.4.9 and 2.3 and Lemma 2.1, we obtain for the boundary value problems above the following result:

Theorem 2.4 Let P be given by (2.24) and (2.25) [or (2.26)]. Assume that p and f satisfy the conditions indicated above. Then:

(i) For each $h \in L_2(0,1)$ there exists a unique strong solution $u_0 \in D(P_0) \subset H_0 = \{u \in L_2(0,1) \mid u, \text{ and } u' \text{ are absolutely continuous on } [0,1], u'' \in L_2(0,1) \text{ and } u \text{ satisfies } (2.25) \ [u(0) = u'(0) = 0]\}$ of (2.24)–(2.25) [(2.26)].

(ii) If $\{H_n\} \subset H_0$ is a sequence of finite-dimensional subspaces such that for each $h_0 \in H_0$, dist$(h_0, H_n) \to 0$ as $n \to \infty$, then there exist a unique $u_n \in H_n$ such that

$$\int_0^1 [pu_n''v'' + f(t, u_n, u_n')v'] \, dt = \int_0^1 hv' \, dt \tag{2.27}$$

for each $v \in H_n$ and $|u_n - u_0|_0 \to 0$ as $n \to \infty$.

(iii) If we assume that $| f(t,\xi_1,\eta_1) - f(t,\xi_2,\eta_2) | \leq C(M) | \xi_1 - \xi_2 | + \overline{K}(M) | \eta_1 - \eta_2 |$, where $C(M), \overline{K}(M) > 0$ whenever $| \xi_1|,| \xi_2 | \leq M$ and $| \eta_1 |,| \eta_2 | \leq M$, then

$$| u_n - u_0 |_0 \leq \left[1 + \frac{L(M)}{\eta} \right] \inf_{v \in H_n} | v - u_0 |_0$$

where $\quad L(M) = \left(1 + \frac{\sqrt{K(M)}}{p_0^{3/2}} \cdot \frac{2}{\pi} \right)$

and $\quad K(M) = 2[\max\{C^2(M), \overline{K}^2(M)\}] \cdot \left[\left(\frac{4}{\pi^2} \right)^2 + \frac{4}{\pi^2} \right]$

$$(2.28)$$

and η depends on p_0 and α. Also,

$D(P_0) = D(T_0)$

$\quad = \{u \in H_0 \mid u''$ is absolutely continuous on $[0,1]$

$\quad\quad$ and $u''' \in L_2[0,1]$ [and $u''(1) = 0]\}.$

$$(2.29)$$

(iv) If, additionally, we assume that $H_n = \text{span} \{\phi_1,\phi_2, \ldots ,\phi_n\}$, where $T\phi_i = \lambda_i K\phi_i$, then $\| Pu_n - h \| \to 0$ as $n \to \infty$ and

$$\| Pu_n - h \| \leq b(M) \left[1 + \frac{L(M)}{\eta} \right] \delta_n,$$

$$\text{where} \quad \delta_n \to 0 \quad \text{as} \quad n \to \infty \quad \text{and} \quad b(M) = L(M).$$

$$(2.30)$$

Moreover, $\| u_n''' - u_0''' \|_{L_2} \to 0$ as $n \to \infty$.

Proof. Let $Ku = u'$ and $D(K) = \{u \in C^1[0,1] \mid u(0) = 0\}$; then with $Tu = -(p(t)u'')'$ and $D(T) = \{u \in C^3[0,1] \mid u$ satisfies (2.25) $[(2.26)]\}$ it follows from (i) of Lemma 2.1 that T is K-p.d. and K-symmetric and H_0 is as described in (i). Now for $u,v \in D(P) = D(T)$,

$$(Pu - Pv, K(u - v)) = \int_0^1 p(t)(u'' - v'')^2 \, dt + \int_0^1 [f(t,u,u')$$
$$- f(t,v,v')](u' - v') \, dt$$
$$\geq \int_0^1 p(t)(u'' - v'')^2 \, dt + \alpha \int_0^1 (u' - v')^2 \, dt$$

by (Ib).

If we let $\eta > 0$ be such that $\alpha \geq -(\pi^2/4)p_0 + (\pi^2/4)p_0\eta$, then

$$(Pu - Pv, K(u - v)) \geq \eta \int_0^1 p(t)(u'' - v'')^2 \, dt$$

$$+ (1 - \eta) \int_0^1 p(t)(u'' - v'')^2 \, dt$$

$$+ \alpha \int_0^1 (u' - v')^2 \, dt \geq \eta| u - v |_0^2$$

$$+ \left[(1 - \eta)p_0 \frac{\pi^2}{4} + \alpha \right] \int_0^1 (u' - v')^2 \, dt$$

by Hardy's inequality (h) and the choice of η. Therefore, for $u,v \in D(P)$,

$$(Pu - Pv, K(u - v)) \geq \eta| u - v |_0^2. \tag{2.31}$$

Now let $| u_n - u |_0 \to 0$ as $n \to \infty$; then $| u_n^{(k)}(t) - u^{(k)}(t) | \leq c_0| u_n - u |_0^2$, $c_0 > 0$ for $k = 0,1$ and arbitrary $t \in [0,1]$. Therefore, $u_n^{(k)} \to u^{(k)}$ uniformly in t. Thus there exists a $C > 0$ such that $\| u_n^{(k)} \|_{\sup}, \| u^{(k)} \|_{\sup} \leq C, k = 0, 1$. By the uniform continuity of f on $[0,1] \times [-C,C] \times [-C,C]$, it follows that $Nu_n = f(\cdot,u_n,u_n') \to Nu = f(\cdot,u,u')$ uniformly in t and $\| Nu_n \|_{\sup} \leq M$, where $M > 0$. Thus $\| Nu_n - Nu \|_{L_2} \to 0$ as $n \to \infty$. In view of this and (2.31), Theorem III.4.9 implies the validity of assertion (i) of Theorem 2.4 while (ii) follows from Theorem 2.3. If the conditions of (iii) are satisfied, then for $u,v \in H_0$ such that $| u |_0,| v |_0 \leq M$, it follows that for some $C_0 > 0$, $\| u^{(k)} \|_{\sup}, \| v^{(k)} \|_{\sup} \leq C_0M(k = 0,1)$ and, by (iii) and Hardy's inequality, it follows that $\| Nu - Nv \|^2 \leq 2 \max\{C^2(M),\overline{K}^2(M)\} \int_0^1 [(u - v)^2 + (u' - v')^2] \, dt \leq 2 \max\{C^2(M),\overline{K}^2(M)\} [(4/\pi^2)^2 + (4/\pi^2)] \int_0^1 (u'' - v'')^2 \, dt$. Thus (2.28) and (2.29) of (iii) follow by direct calculations for Theorem 2.3, Proposition 2.2, and Lemma 2.1. Now, since $D(P_0) = D(T_0)$ and $u_0,u_n \in D(T_0)$, the additional conditions in (iv) imply, by [51, Theorem 1.6], that $\| Pu_n - h \|_{L_2} = \| Pu_n - P_0u_0 \|_{L_2} \to 0$ and that (2.30) holds. Finally, this and the equality $p(u_n''' - u_0''') = Pu_n - P_0u_0 + p'(u_n'' - u_0'') - (Nu_n - Nu_0)$ imply that $\| u_0''' - u_n''' \|_{L_2} \to 0$ as $n \to \infty$. Q.E.D.

The last assertion implies that $\{u_n\}$, $\{u_n'\}$, and $\{u_n''\}$ converge uniformly to u_0, u_0', and u_0'', respectively, while u_n''' converges to u''' in the L_2-norm. To illustrate what alternative assumptions might be imposed on p, we include, without proof, the following result.

Theorem 2.5 Let the hypotheses of Theorem 2.4 be satisfied except (Ia), which is replaced by the following condition. $p(t)$ is concave on $[0,1]$ with piecewise smooth derivative [i.e., $p''(t) \leq 0$ where it exists], $p'(0) \leq 0$, and $p(t) > 0$ on [0,1], while (Ib) is satisfied with $\alpha > (-\pi^2/4) \int_0^1 p(t) \, dt$.

Then the conclusions of Theorem 2.4 are satisfied with η depending on $\int_0^1 p(t) \, dt$ and α.

The proof of Theorem 2.5 is based on the inequality

$$\int_0^1 p(t)[u''(t)]^2 \, dt \geq \frac{\pi^2}{4} \left[\int_0^1 p(t) \, dt \right] \left[\int_0^1 [u'(t)]^2 \, dt \right]$$

obtained in [278] and the same arguments as those used in the proof of Theorem 2.4. We therefore omit it.

Let us now apply the results of Section 2.2 to the more complicated *singular* ordinary differential operators. Consider the problem

$$Pu(t) = -(p(t)u'(t))' + q(t)u(t) + f(t,u) = h(t) \qquad [t \in (0,\infty)],$$
$$u(0) = 0, \qquad (2.32)$$

where $h(t) \in L_2(0,\infty)$. We suppose that p, q, and f satisfy the conditions:

$p(t)$ is absolutely continuous on finite intervals, $p'(t)$ is bounded, and
$p_1 \geq p(t) \geq p_0 > 0$ for $t \in [0,\infty)$
for some constants $p_0 > 0$ and $p_1 > 0$. (2.33)

$q(t)$ is measurable on $[0,\infty)$ and $q_1 \geq q(t) \geq q_0 > 0$ for $t \in [0,\infty)$.
(2.34)

f satisfies the Carathéodary condition on (2.35)
$[0,\infty) \times R$, $f(t,0) \in L_2(0,\infty)$ and

$$\frac{f(t,\xi_1) - f(t,\xi_2)}{\xi_1 - \xi_2} \geq -\lambda_0 \qquad \text{where} \quad q_0 > \lambda_0.$$

$|f(t,\xi_1) - f(t,\xi_2)| \leq K(M) |\xi_1 - \xi_2|$, $t \in [0,\infty)$, $K(M) > 0$ (2.36)
whenever $|\xi_1|, |\xi_2| \leq M$.

With $Tu = -(pu')' + qu$, $Ku = u$, and $D(P) = D(T) = \{u \in C^2(0,\infty) \,|\, u(0) = 0$, and there exists a $C_u > 0$ such that $u(t) = 0$ for $t \geq C_u\} = \{u \in C_0^2(0,\infty) \,|\, u(0) = 0\}$, we can now derive the following result.

Theorem 2.6 Let P be defined by (2.32) and assume that conditions (2.33)–(2.36) are satisfied. Then

(i) For each $h \in L_2(0,\infty)$ there exists a unique strong solution $u_0 \in D(P_0) \subset H_0 = \{u \in W_2^1(0,\infty) \,|\, u(0) = 0\}$ of (2.32) where $[u,v] = \int_0^\infty (pu'v' + quv) \, dt$.

(ii) $D(P_0) = H_0 \cap W_2^2(0,\infty) = D(T_0)$.

(iii) If $\{H_n\} \subset H_0$ is a sequence of finite-dimensional spaces such that for each $h_0 \in H_0$, dist$(h_0,H_n) \to 0$ as $n \to \infty$, there exists a unique $u_n \in H_n$ such that

$$\int_0^\infty \{pu_n'v' + qu_nv + f(t,u_n)v\} \, dt = \int_0^\infty hv \, dt \tag{2.37}$$

$$\text{for each} \quad v \in H_n,$$

and $|u_n - u_0|_0 \to \infty$ as $n \to \infty$ with

$$|u_n - u_0|_0 \leq \left[1 + \frac{C(M)}{\eta}\right] \inf_{v \in H_n} |v - u_0|_0, \tag{2.38}$$

where η depends on q_0 and λ_0, and $C(M)$ is defined in terms of $K(M)$.

Proof. With T and K as defined above, it follows that T is K-p.d. and K-symmetric and H_0 is as described in (i). Let us show that $N: H_0 \to L_2(0,\infty)$, given by $Nu = f(\cdot,u)$, is well defined. If $u \in H_0$, since $u(0) = 0$, then for any $t \in [0,\infty)$,

$$u^2(t) = \int_0^t (u^2(s))' \, ds$$

$$= 2\int_0^t u'(s)u(s) \, ds \leq 2\| u' \|_{L_2} \| u \|_{L_2} \leq \int_0^\infty [(u')^2$$

$$+ (u)^2] \, dt \leq \max\left\{\frac{1}{p_0}, \frac{1}{q_0}\right\} \int_0^1 [p(u')^2 + qu^2] \, dt$$

$$= C^2 |u|_0^2.$$

Thus $\| u \|_{\sup} \leq C|u|_0$. By (2.36) there exists a $K(M) > 0$ such that

$$|f(t,u(t))| \leq |f(t,0)| + |f(t,u(t))$$

$$- f(t,0)| \leq |f(t,0)| + K(M)|u(t)|.$$

Since $f(t,0) \in L_2(0,\infty)$, it follows that $Nu \in L_2(0,\infty)$. Now if $u,v \in H_0$ are such that $|u|_0, |v|_0 \leq M$, then by the same arguments as used above, (2.36) implies that there exists a $C(M) > 0$ such that $\| Nu - Nv \|_{L_2} \leq C(M)|u - v|_0$. Finally, for $u,v \in D(T)$,

$$(Pu - Pv, u - v) \geq \int_0^\infty [p(t)(u' - v')^2$$

$$+ q(t)(u - v)^2 + (-\lambda_0)(u - v)^2] \, dt$$

by (2.35). Now let $\eta \in (0,1)$ be such that $(1 - \eta)q_0 > \lambda_0$. Therefore, $(1 - \eta)q(t) \geq (1 - \eta)q_0 > \lambda_0$ and $q(t) - \lambda_0 > \eta q(t)$ for $t \geq 0$. Thus

$$(Pu - Pv, u - v) \geq (1 - \eta) \int_0^\infty p(t)(u' - v')^2 \, dt$$

$$+ \eta \int_0^\infty \{p(t)(u' - v')^2 + q(t)(u - v)^2\} \, dt \geq \eta |u - v|_0^2.$$

The results of (i) and (iii) then follow directly from Theorems III.4.9 and 2.3, while (ii) follows from [66] and Proposition 2.2 since the self-adjoint extension of T given in [66] is in fact equal to T_0.   Q.E.D.

3. APPROXIMATION-SOLVABILITY OF PERIODIC BOUNDARY VALUE PROBLEMS

The fundamental purpose of this section is to show how certain results of A-proper mapping theory developed by the author in [227] can be used to obtain (in some cases constructive) existence of periodic solutions for boundary value problems for ODEs arising in some parts of physics, mechanics, elasticity, and other fields. A similar approach can also be used in the solvability of certain PDEs (see [83,212,226]) Many, but not all of the results presented in this section were obtained recently in a series of papers by the author and Yu (see [214,235]) and the author (see [224,234]).

Our discussion proceeds as follows. In Section 3.1 we state some relevant definitions and an abstract result of the author [214, Theorem 1.1] concerning the solvability of semilinear equations at resonance involving A-proper mappings. This result is used in Section 3.2 to establish the existence (sometimes constructive) of solutions to periodic boundary value problems:

$$p(t)x')' + f(t,x,x',x'') = y(t); \qquad x(0) = x(T), \quad x'(0) = x'(T).$$
$$(3.1)$$

The interesting feature of our solvability of equation (3.1) is that if we impose the one-sided conditions on $f(t,x,0,0)$ when $|x|$ is large, the nonlinearity $f(t,p,r,q)$ is allowed to grow linearly in all its variables p, r, and q. The results presented here for (3.1) properly extend the earlier results of Šlapak, Fitzpatrick [77], Petryshyn [212], and others. The sublinear growth of $f(t,p,r,q)$ imposed in [77,212] was essential to the argument used there.

3.1 A Useful Abstract Result

To obtain the solvability of equation (3.1) for a given function $y(t)$ when $(p(t)x')' = 0$ for some $x(t) \neq 0$ for $t \in [0,T]$, we have first to prove an abstract existence theorem obtained in [214]. For that, we first recall (see [66]) that a linear operator $L: X \to Y$ is called *Fredholm* if its range $R(L)$ is closed in Y, its null space $N(L)$ has dim $N(L) \equiv \alpha(L) < \infty$ and dim ($Y/R(L)$) $= \beta(L) < \infty$. In this case the index of L is defined by $i(L) = \alpha(L) - \beta(L)$. In what follows we denote by $\Phi_0(X,Y)$ the class of all Fredholm operators of index zero. Now if $L \in \Phi_0(X,Y)$, there exist closed subspaces X_1 of X and Y_2 of Y with dim $Y_2 = \alpha(L)$ such that $X = N(L) \oplus X_1$ and $Y = Y_2 \oplus R(L)$. Let P be a projection of X onto $N(L)$ and Q of Y onto Y_2 and $[\cdot,\cdot]$ a continuous bilinear form on $Y \times X$ such that $y \in R(L)$ iff $[y,x] = 0$ for all $x \in N(L)$. It was shown by Mawhin that if $\{\varphi_1, \ldots, \varphi_m\}$ is a basis in $N(L)$, the linear map $J: Y_2 \to N(L)$, given by $Jx = \sum_{j=1}^m [y,\varphi_j]\varphi_j$ is an isomorphism, and if $y = \sum_{i=1}^m c_i\varphi_i$, then $[J^{-1}\varphi_i,\varphi_j] = \delta_{ij}$ and $[J^{-1}y,\varphi_j] = c_j$ for $1 \leq i,j \leq m$. Now if we set $A = J^{-1}P: X \to Y_2$ and assume that $L: X \to Y$ is A-proper, then $L - A$ is A-proper and $\mathrm{Deg}(L - A, B(0,r), 0) \neq 0$ for any $r > 0$ bcause $L - A$ is injective.

In this section we state and prove the abstract approximation-solvability result obtained in [214, Theorem 1.1] which we shall use and which deals with the solvability of the semilinear equations

$$Lx - Nx = y \qquad (y \in Y, x \in X), \tag{3.2}$$

where X, Y are real Banach spaces, $L: X \to Y$ is a Fredholm map of ind(L) $= 0$ with $N(L) \neq \{0\}$ and $N: X \to Y$ is a nonlinear map such that for each $\lambda \in [0,1]$, $L - \lambda N: X \to Y$ is A-proper with respect to some admissible projection scheme $\Gamma = \{X_n, P_n; Y_n, Q_n\}$.

Using the theory of the generalized degree for A-proper mappings developed in Chapter IV, the following result which we use here has been proved in [214]. For the convenience of the reader, we outline its proof.

Theorem 3.1   Let $L: X \to Y$ be a linear Fredholm map of ind(L) $= 0$, which is A-proper w.r.t. Γ with $N(L) \neq \{0\}$ and suppose there exists a continuous bilinear form $[\cdot,\cdot]$ on $Y \times X$ mapping (y,x) into $[y,x]$ such that

$$y \in R(L) \quad \text{if and only if} \quad [y,x] = 0 \quad \text{for each} \quad x \text{ in } N(L). \tag{3.3}$$

Let $y \in Y$, let $G \subset X$ be a bounded open set in X with $0 \in G$, and let $N: \overline{G} \to Y$ be a nonlinear map such that

(a) $L - \lambda N: \overline{G} \to Y$ is A-proper w.r.t. Γ for each $\lambda \in (0,1]$.
(b) $Lx \neq \lambda Nx + \lambda y$ for $x \in \partial G$ and $\lambda \in (0,1]$.

(c) $QNx + Qy \neq 0$ for $x \in N(L) \cap \partial G$, where Q is a linear projection of Y onto Y_2.

(d) Either (i): $[QNx + Qy, x] \geq 0$ or (ii): $[QNx + Qy, x] \leq 0$ for $x \in N(L) \cap \partial G$.

Then equation (3.2) is feebly approximation solvable for each such y. If for some such y, equation (3.2) has at most one solution, then equation (3.2) is strongly approximation solvable.

Proof. For the convenience of the reader we sketch the proof of Theorem 3.1. Thus one first shows, using (b) and (c), that

$$\text{Deg}(L - N + y, G, 0) = \text{Deg}(L - QN + Qy, G_D, 0). \quad (3.4)$$

Then, using (b), (c), and (d1) or (d2), one shows that $H: [0,1] \times \overline{G} \to Y$, given by $H(\lambda,x) = Lx - (1 - \lambda)Ax - \lambda QNx + \lambda Qy$, is an A-proper homotopy with $H(\lambda,x) \neq 0$ for $\lambda \in [0,1]$ and $x \in \partial G$, where A is a linear compact map from X onto Y_2 defined by using $[\cdot,\cdot]$ (see [227]) such that $L - A: X \to Y$ is a bijection. Hence, by Theorem IV.2.1 $\text{Deg}(L - A, G, 0) = \text{Deg}(L - QN + Qy, G, 0)$ with $0 \notin \text{Deg}(L - A, G, 0)$. This and (3.4) imply that $0 \notin \text{Deg}(L - N + y, G, 0)$. Hence, by (6) of Theorem IV.2.1, there exists $N_y \in Z^+$ such that the Galerkin equation

$$Q_nLx - Q_nNx = Q_ny \quad (x \in G_n, Q_ny \in Y_n) \quad (3.5)$$

has a solution $x_n \in G_n$ for each $n \geq N_y$. Since $Q_ny \to y$ in Y and $L - N: \overline{G} \to Y$ is A-proper, there exist a subsequence $\{x_{n_j}\}$ and $x \in G$ such that $x_{n_j} \to x$ and $Lx - Nx = y$ [i.e., (3.2) is feebly a-solvable]. If x is the only solution of (3.2), then, using the A-properness of $L - N$ again, one shows that the entire sequence $x_n \to x$ in X, so (3.2) is strongly a-solvable [i.e., the Galerkin method applies to (3.2) in this case]. Q.E.D.

3.2 Solvability of Periodic Boundary Value Problem (3.1)

In Section 3.2, Theorem 3.1 is used to obtain the existence (in some cases constructive) of periodic solutions to boundary value problem (3.1). In this approach the main problem is to show that under suitable growth conditions on $f(t,x,r,q)$, there exists an open bounded set $G = B(0,r) \subset X$ such that (b), (c), and (d) of Theorem 3.1 hold. This involves the a priori estimates of solutions and their derivatives of the problem (3.1).

To state our first theorem for (3.1), we let $Y = C([0,T])$, $H = L^2(0,T)$ and $X = \{x \in C([0,T]), x^{(j)}(0) = x^{(j)}(T), j = 0, 1\}$—equipped with the norms $|\cdot|_0$, $\|\cdot\|$, $|\cdot|_2$, respectively.

Let $p \in C^1([0,T])$ be such that $p(0) = p(T)$, $p_0 = \min\{p(t) : 0 \leq t \leq T\} > 0$, and $p_1 = \max\{|p'(t)| : 0 \leq t \leq T\}$. If $L: D(L) = X \to Y$ is now defined by $L(x)(t) = (p(t)x')'$ for $t \in [0,T]$ and $x \in X$, then $L \in \Phi_0(X,Y)$, $N(L) = \{x \in X : x(t) \text{ constants}\}$, $R(L) = \{u \in Y : \int_0^T u \, dt = 0\}$, $X = N(L) \oplus X_1$, and $Y = N(L) \oplus R(L)$. Let $c > 0$ be any fixed constant such that $K_c = L - cI : X \to Y$ is a homeomorphism. Let $\{Y_n, Q_n\}$ be as above and let $\{X_n\} \subset X$ be such that $Y_n = K_c(X_n)$ for each n. Then $\Gamma_c = \{X_n, Y_n, Q_n\}$ is admissible and $L : X \to Y$ is A-proper w.r.t. Γ_c. Now we assume that f satisfies the following condition:

(C1) $f : [0,T] \times R^3 \to R$ is continuous and there are constants $A, B, C, D \in R^+$ such that $D < p_0$, $(p_0 B + \pi p_0 C + \pi p_1 D/p_0(p_0 - D)) < 2\pi^2$, and $|f(t,p,r,q)| \leq A + B|p| + C|r| + D|q|$ for $t \in [0,T]$ and $p,r,q \in R$.

Now it follows from (C1) that $N : X \to Y$, given by $Nx = f(t,x,x',x'')$ for $t \in [0,T]$ and $x \in X$, is well defined, continuous, and maps bounded sets in X into bounded sets in Y. We now state and prove the following theorem for (3.1) which generalizes slightly Theorem 3.1 in [235] and proves a good estimate for the solutions x of boundary value problem (3.1) which is needed in some applications and which was not given in [235].

Theorem 3.2 Let L be as above, and in addition to (C1) let f satisfy:

(C2) $L - \lambda N : X \to Y$ is A-proper w.r.t. Γ_c for each $\lambda \in (0,1]$.

(C3) To a given $y \in Y$ there is $M > 0$ such that $\int_0^T (f(t,x,x',x'') - y) \, dt \neq 0$ for $x \in X$ with $|x(t)| \geq M$ for $t \in [0,T]$.

(C4) There is $M_1 \geq M$ and $a,b \in R$ such that either (i) $a \geq b$, $x \in N(L)$ and $x \geq M_1 \Rightarrow f(t,x,0,0) \geq a$ if $t \in [0,T]$, $x \leq -M_1 \Rightarrow f(t,x,0,0) \leq b$ if $t \in [0,T]$, and $b \leq y_T \leq a$; or (ii) $a \leq b$, $x \in N(L)$ and $x \geq M_1 \Rightarrow f(t,x,0,0) \leq a$ if $t \in [0,T]$, $x \leq -M \Rightarrow f(t,x,0,0) \geq b$ if $[0,T]$, and $a \leq y_T \leq b$, $y_T = \int_0^T y(t) \, dt$.

Then there exists $N_y \in Z^+$ such that

$$L_n(x) - N_n(x) = Q_n y \qquad (x \in X_n, Q_n y \in Y_n), \qquad (3.6)$$

has a solution $x_n \in X_n$ for each $n \geq N_y$ such that $x_{n_j} \to x$ in X for some $\{n_j\} \subset \{n\}$, $x(t)$ satisfies (3.1), and one has the estimate

$$|x(t)| \leq M + \sqrt{T} \frac{\pi T p_0(\sqrt{T}A + \|y\| + BM\sqrt{T})}{2\pi^2 p_0(p_0 - D) - T[p_0(TB + \pi C) + \pi p_1 D]}.$$

$$(3.E)$$

Furthermore, if x is the only solution for a given $y \in C$, then $x_n \to x$ in X.

Proof. Since Theorem 3.2 can be applied to the problem concerning the periodic motions of a satellite in its elliptic orbit where the estimate (3.E) plays an essential role, we outline here the proof of Theorem 3.2 by following the arguments of [235].

To establish the existence of the set $G = B(0,r) = \{x \in X : |x|_2 < r\}$ for which conditions (b), (c), and (d) of Theorem 3.1 hold, let $x \in X$ be a solution of

$$(px')' = -\lambda f(t,x,x',x'') - \lambda y \quad \text{for some} \quad \lambda \in (0,1]. \tag{3.7}$$

Integrating (3.7) from 0 to T and using the boundary conditions, we get

$$\int_0^T \{f(t,x,x',x'') - y\} \, dt = 0. \tag{3.8}$$

It follows from (3.8) and (C3) that there exists $t_0 \in [0,T]$ such that $|x(t_0)| < M$. If we write $x(t) = a_0 + u(t)$ with $a_0 = (1/T) \int_0^T x \, dt$, then $\int_0^T u \, dt = 0$, $x'(t) = u'(t)$, $x(t) = x(t_0) + \int_0^t x'(s) \, ds$, so as in [235], one shows that

$$|x(t)| \leq M + \sqrt{T}\| u' \|, \quad \| x \| \leq M\sqrt{T} + \frac{T}{\pi} \| u' \|. \tag{3.9}$$

Now, in view of (3.9), the equality

$$-(px')'x = -\lambda f(t,x,x',x'')x - \lambda yx$$

implies that

$$p_0\| x' \|^2 \leq \int_0^T px'^2 \, dt$$

$$\leq \int_0^T \{A + B| x | + C| x' | + D| x'' | + | y |\} | u | \, dt.$$

It follows from this and the Wirtinger's inequality (see [111]) that

$$p_0\| u' \|^2 \leq (\sqrt{T} A + \| y \| + B\| x \| + C\| u' \| + D\| u'' \|)\| u \|$$

$$\leq \frac{T}{2\pi} \{A\sqrt{T} + \| y \| + B\| x \| + C\| u' \| + D\| u'' \|\}\| u' \|. \tag{3.10}$$

Now it follows from (3.7) and (C1) that

$$p_0\| u'' \| \leq \| px'' \| \leq \sqrt{T} A + B\| x \|$$

$$+ (C + p_1)\| u' \| + D\| u'' \| + \| y \|$$

and

$$\| x'' \| = \| u'' \| \le \frac{1}{p_0 - D} [\sqrt{T} A + \| y \|$$

$$+ B\| x \| + (C + p_1) \| u' \|]. \quad (3.11)$$

In view of this, it follows from (3.10) and a simple calculation that

$$p_0\| u' \|^2 \le \frac{T}{2\pi} \| u' \| \left\{ \frac{p_0}{p_0 - D} (\sqrt{T} A + \| y \|) \right.$$

$$\left. + \frac{p_0 B}{p_0 - D} \| x \| + \left[C + \frac{D(C + p_1)}{p_0 - D} \right] \| u' \| \right\}.$$

The last inequality and (3.9) imply that

$$p_0\| u' \|^2 \le \frac{T}{2\pi} \| u' \| \left[\frac{p_0}{p_0 - D} (\sqrt{T} A + \| y \|) + \frac{p_0 B}{p_0 - D} \sqrt{T} M \right.$$

$$\left. + \frac{p_0 B}{p_0 - D} \frac{T}{\pi} + \frac{C p_0 + D p_1}{p_0 - D} \| u' \| \right]$$

and thus combining the terms we get the looked for estimate

$$\| u' \| \le \frac{\pi T p_0 [\sqrt{T} A + \| y \| + B M \sqrt{T}]}{2\pi^2 p_0(p_0 - D) - T[p_0(TB + \pi C) + \pi p_1 D]} \equiv A_0. \quad (3.12)$$

In view of (3.9), the inequality (3.12) implies the following estimates:

$$| x(t) | \le M + \sqrt{T} \cdot A_0 \ \forall\, t \in [0,T], \qquad \| x \| \le M\sqrt{T} + \frac{T}{\pi} A_0,$$

$$(3.13)$$

which will be used in Theorem 3.2, where A_0 is defined by the right-hand side of (3.12). Since $\| x' \| = \| u' \| \le A_0$, it follows from (3.13) and (3.11) that $\| x'' \| \le A_1$ with $A_1 > 0$, a constant independent of x and $\lambda \in (0,1]$. Since $x'(t_1) = 0$ for some $t_1 \in (0,1)$, one easily shows that

$$| x'(t) | \le \| x'' \| \le A_1 \qquad \forall\, t \in [0,T]. \quad (3.14)$$

Now, in view of (C1), we obtain from the equality

$$- p x'' = - \lambda f(t,x,x',x'') - \lambda y + p' x'$$

the inequality

$$p_0| x'' | \le A + B| x | + C| x' | + D| x'' | + | y | + p_1| x' |.$$

Since $p_0 < D$, it follows from this and (3.13) and (3.14) that

$$| x''(t) | \leq \frac{1}{p_0 - D} [A + B| x | + (C + p_0)| x' |$$

$$+ | y |] \leq A_2 \quad \forall\, t \in [0,1], \tag{3.15}$$

where the constant $A_2 > 0$, which can be explicitly given, is independent of x and $\lambda \in (0,1]$.

The discussion above implies the existence of a constant $M_2 > 0$ such that if $x \in X$ is a solution of (3.7) for some $\lambda \in (0,1]$, then $| x |_2 \leq M_2$, so (b) of Theorem A holds for any $r > M_2$.

Now let $r > \max\{M_1, M_2\}$, let $G = B(0,r) = \{x \in X : | x |_2 \leq r\}$, and let $Qu = (1/T) \int_0^T u \, dt$ for $u \in Y$. Then Q is a projection of Y onto $N(L)$. Note that (c) of Theorem A holds, for if $x \in N(L) \cap \partial G$, then $x(t)$ is a constant function, say $x(t) \equiv c$, and $| x |_2 = | c | = r \geq M_1 \geq M$. Hence (C3) implies that

$$QNx - Qy = \frac{1}{T} \int_0^T \{f(t,x,0,0) - y\}\, dt \neq 0 \,\forall\, x \in N(L) \cap \partial G.$$

$$\tag{3.16}$$

To verify Theorem 3.1(d), let the bilinear form $[\cdot,\cdot]$ on $Y \times X$ be defined by $[v,x] = \int_0^T vx \, dt$; then one can show that $[\cdot,\cdot]$ is continuous and that (3.3) holds. We now claim that (i) of (C4) implies (d1), while (ii) of (C4) implies (d2) of Theorem A. Indeed, if $x \in N(L) \cap \partial G$, then $| x |_2 = | c | = r \geq M_1$. Thus it follows from (i) of (C4) that if $c = r$, then

$$\int_0^T \{f(t,c,0,0) - y(t)\}\, dt \geq T(a - y_T) \geq 0$$

and, consequently,

$$[QNx - Qy,\, x]$$

$$= \frac{1}{T} \int_0^T \left(\int_0^T \{f(t,c,0,0) - y\}\, dt \right) c\, dt = (a - y_T)c \geq 0.$$

On the other hand, if $c = -r$, then $c < -M_1$ and $\int_0^T \{f(t,c,0,0) - y\}\, dt \leq T(b - y_T) \leq 0$, and consequently, $[QNx - Ny,\, x] \geq 0$ [i.e., (d1) holds]. Similar argument shows that (ii) implies (d) of Theorem 3.1.

Thus all the hypotheses of Theorem 3.1 have been verified, and consequently, in either case, the conclusions of Theorem 3.2 follow from Theorem 3.1.   Q.E.D.

Remark 3.1 On various occasions one may need the estimate (3.E) when $T = 2\pi$ and $p(t) \equiv 1$, which in this case reduces to the estimate

$$|x(t)| \le M + \frac{2\pi(A + BM) + \sqrt{2\pi}\|y\|}{1 - 2B - C - D} \qquad \text{for} \quad t \in [0,2\pi].$$

$$(3.E')$$

Remark 3.2 As before we note in passing that condition (C2) of Theorem 3.2 holds if, for example, f is independent of x'' or f satisfies:

(C2a) There is a constant $k \in [0,p_0)$ such that

$$|f(t,p,r,q) - f(t,p,r,\bar{q})|$$
$$\le k|q - \bar{q}| \qquad \forall \quad t \in [0,T] \quad \forall \quad p,r,q,\bar{q} \in R.$$

3.3 Hilbert Space Approach

In this section we reformulate the periodic boundary value problem as an operator equation with operators L and N acting between two Hilbert spaces since in that case, as will be shown, we can weaken conditions on $f(x,p,r,q)$. We note that going over the proof of Theorem 3.2 we see that if that theorem is formulated in the Hilbert space setting, its conclusions remain valid for $f(t,x,x',x'')$ satisfying conditions stated below, which are weaker than (C2a).

Indeed, let $Y \equiv H = L^2(0,T)$ and let $X = H_0 \equiv \{x \in W_2^2(0,T) : x^{(j)}(0) = x^{(j)}(T), j = 0,1\}$, where $W_2^2(0,T)$ is the Sobolev space, whose inner product and norm are given by

$$(u,v)_2 = \sum_{i=0}^{2} (u^{(i)},v^{(i)}), \quad \|u\|_2 = \left(\sum_{i=0}^{2} \|u^{(i)}\|^2\right)^{1/2} \quad \forall \quad u,v \in W_2^2.$$

Then by the solution of (3.1) we mean a function x in H_0 that satisfies (3.1) for t in $(0,T)$ a.e. We shall refer to such a function x as the *generalized solution* of (3.1). Now, instead of the continuity condition imposed on f: $[0,T] \times R^3 \to R$ in (C1), it suffices to assume that f is a Carathéodory function and then, in view of the growth conditions imposed on f in (C1), the theory of the Nemytskii operators implies that $N: H_0 \to H$, defined by $Nx = f(t,x,x',x'')$ for $x \in H_0$, is continuous and maps bounded sets in H_0 into bounded sets in H (see [1]).

Now it was noted before that one of the basic problems in applying Theorem 3.1 to semilinear differential equations is to be able to impose reasonable conditions on the nonlinear function f [in our case on the function $f(t,x,x',x'')$] so that the operator $L - \lambda N: X \to Y$ is A-proper.

In case $X = H_0 \subset W_2^2(0,T)$ and $Y = H = L^2(0,T]$ we may assume that f satisfies either (C2b) or (C2c) if $p(t) \equiv 1$, where:

(C2b) There is $k \in [0,1)$ such that for $t \in [0,T]$ a.e. we have

$$[f(t,p,r,q) - f(t,p,r,\bar{q})][q - \bar{q}] \geq -k|\, q - \bar{q}\,|^2 \qquad \forall \quad p,r,q,\bar{q} \in R.$$

(C2c) There is a function $\alpha: (0,\infty) \to [0,1)$ such that

$$[f(t,p,r,q) - f(t,p,r,\bar{q})][q - \bar{q}]$$
$$\geq -\alpha(s)\,|\, q - \bar{q}\,|^2 \qquad \forall \quad p,r,q,\bar{q} \in R \quad \text{with} \quad |\, q - \bar{q}\,| \geq s.$$

It will be shown below that when (C2b) or (C2c) holds, then $L - \lambda N$: $H_0 \to H$ is A-proper w.r.t. $\Gamma_L = \{X_n,Y_n,Q_n\}$ for each $\lambda \in [0,1]$, where $Q_n: H \to Y_n$ an orthogonal projection for each $n \in Z^+$.

Thus we will obtain the following Hilbert space variant of Theorem 3.2.

Theorem 3.3   Suppose that the Carathéodory function $f: [0,T] \times R^3 \to R$ in (C1) satisfies (C2b) or (C2c). If (C4) of Theorem 3.2 also holds, the periodic boundary value Problem (3.1) is feebly a-solvable w.r.t. Γ_L and, in particular, (3.1) has a generalized solution $x \in H_0$. If x is the only solution, (3.1) is strongly a-solvable.

To prove Theorem 3.3, it suffices to establish the following:

Lemma 3.1   Suppose that in addition to (C1), f satisfies

(C2b) There exists a constant $\alpha \in [0, p_0)$ such that for $t \in [0,1]$ a.e. $[f(t,p,r,q) - f(t,p,r,\bar{q})][q - \bar{q}] \geq -\alpha|\, q - \bar{q}\,|^2$ for all $p,r,q,\bar{q} \in R$.

(C2c) There exists a function $\alpha: (0,\infty)$ into $[0,p_0)$ such that $[f(t,p,r,q) - f(t,p,r,\bar{q})][q - \bar{q}] \geq -\alpha(s)\,|\, q - \bar{q}\,|^2$ for $t \in [0,1]$ a.e. and all $p,r,q,\bar{q} \in R$ with $|\, q - \bar{q}\,| \geq s$.

Then the mapping $T_\lambda \equiv L - \lambda N: H_0 \to H$ is A-proper for each $\lambda \in (0,1]$.

Proof.   When the function f satisfies condition (C2b), the A-properness of T_λ was proved.

We now claim that T_λ is also A-proper w.r.t. Γ for each $\lambda \in (0,1]$ when f satisfies condition (C2c). Our proof is based on the argument used to prove [235, Lemma 3]. First we use the following fact, which follows from (C2c):

Let $u,v \in X$, let $\alpha > 0$, and suppose that $\|\, u'' - v''\,\|^2 = \int_0^1 |\, u'' - v''\,|^2\, dt \geq a^2$. Let $E = \{t \in [0,1] \,|\, |\, u''(t) - v''(t)\,| \geq \tfrac{1}{2}a\}$ and set $F = [0,1] - E$.

Then $\| u'' - v'' \|^2 = \int_F | u'' - v'' |^2 \, dt + \int_E | u'' - v'' |^2 \, dt$, and it follows from (C2c) that

$$\int_0^1 [f(t,u,u',u'') - f(t,u,u',v'')][u'' - v''] \, dt$$

$$= \left(\int_F + \int_E \right) [f(t,u,u',u'') - f(t,u,u',v'')][u'' - v''] \, dt$$

$$\geq - p_0 \int_F | u''(t) - v''(t) |^2 \, dt - \alpha(\tfrac{1}{2}a) \int_E | u''(t) - v''(t) |^2 \, dt.$$

$$(3.17)$$

Now it follows from

$$\int_F | u'' - v'' |^2 \, dt \leq \tfrac{1}{4}a^2 \leq \tfrac{1}{4}\| u'' - v'' \|^2$$

that

$$\int_E | u'' - v'' |^2 \, dt \geq \tfrac{3}{4}\| u'' - v'' \|^2.$$

This and (3.17) imply that

$$(f(t,u,u',u'') - f(t,u,u',v''), u'' - v'') \qquad (3.18)$$
$$\geq - p_0\| u'' - v'' \|^2 + \tfrac{3}{4}(p_0 - \alpha(\tfrac{1}{2}a))\| u'' - v'' \|^2.$$

To prove the A-properness of T_λ, let $\lambda \in (0,1]$ be fixed and let $\{x_{n_j} \mid x_{n_j} \in X_{n_j}\}$ be any bounded sequences such that $g_{n_j} \equiv Q_{n_j}T_\lambda(x_{n_j}) \to g$ for some g in Y. For simplicity set $n_j \equiv j$ for each $j \in Z^+$ and note that because $\{x_j\}$ is bounded and X is reflexive, we may assume that $x_j \rightharpoonup x_0$ for some x_0 in X. Then since W_2^2 is compactly embedded into $C^1([0,1])$, it follows that $x_j(t) \to x_0(t)$ and $x_j'(t) \to x_0'(t)$ uniformly for $t \in [0,1]$. Hence it is not hard to show that $f(t,x_j,x_j',x_0'') \to f(t,x_0,x_0',x_0'')$ in $L_2 = Y$. Now to prove that $x_j \to x_0$ in X, it suffices to show that $x_j'' \to x_0''$ in Y. Since $\text{dist}(x,X_j) \to 0$ for each x in X, there exists a sequence $\{w_j \mid w_j \in X_j\}$ such that $w_j \to x_0$ in X. This and the boundedness of $\{T_\lambda(x_j)\}$ imply that $(T_\lambda(x_j), Kw_j - Kx_0) \to 0$ since $Kw_j \to Kx_0$ in Y. Moreover, $(T_\lambda(x_j), Kx_j - Kw_j) = (Q_jT_\lambda(x_j), Kx_j - Kw_j) \to 0$ since $Q_jT_\lambda(x_j) \to g$ and $Kx_j - Kw_j \rightharpoonup 0$ in Y. It follows from this that

$$(T_\lambda(x_j), Kx_j - Kw_0)$$
$$= (T_\lambda(x_j), Kx_j - Kw_j) + (T_\lambda(x_j), Kw_j - Kx_0) \to 0.$$

Consequently, we see that

$$A_j \equiv (T_\lambda(x_j) - T_\lambda(x_0), Kx_j - Kx_j) \to 0. \qquad (3.19)$$

Now since $Kx = -x'' + x$ for $x \in X$, we can write A_j in the form

$$A_j = \int_0^1 p \mid x_j'' - x_0'' \mid^2 dt + \lambda(f(t,x_j,x_j',x_j'') \tag{3.20}$$
$$- f(t,x_j,x_j',x_0''), x_j'' - x_0'') + \phi(x_j - x_0)$$

where the function ϕ is given by

$$\phi(x_j - x_0) = \lambda(f(t,x_j,x_j',x_0'') - f(t,x_0,x_0',x_0''), x_j'' - x_0'')$$
$$+ (T_\lambda x_j - T_\lambda x_0, x_j - x_0) + (p'(x_j' - x_0), (x_j'' - x_0'')).$$

It is not hard to show that $\phi(x_j - x_0) \to 0$ since $x_j \rightharpoonup x_0$ in X and $x_j \to x_0$ in $C^1([0,1])$.

We now claim that $x_j'' \to x_0''$ in L_2. If not, there exists $a > 0$ such that $\| x_j'' - x_0'' \| \ge a$ for all $j \in Z^+$. Using (3.18) with $u = x_j$ and $v = x_0$ and (3.20) we see that

$$A_j - \phi(x_j - x_0) \ge p_0\| x_j'' - x_0'' \|^2 + \lambda(f(t,x_j,x_j',x_j'')$$
$$- f(t,x_j,x_j',x_0''), x_j'' - x_0'')$$
$$\ge [p_0(1 - \lambda) + \tfrac{3}{4}\lambda(p_0 - \alpha(\tfrac{1}{2}a))] \| x_j'' - x_0'' \|^2.$$

This leads to a contradiction since $A_j - \phi(x_j - x_0) \to 0$ by (3.19) and the property of ϕ, while the right-hand side is bounded below by $[p_0(1 - \lambda) + \tfrac{3}{4}\lambda(p_0 - \alpha(\tfrac{1}{2}a)]a^2 > 0$ for a fixed $\lambda \in (0,1]$. This contradiction shows that $x_j'' \to x_0''$ in L_2. Hence $x_j \to x_0$ in X and by continuity of T_λ and the completeness of $\{Y_n,Q_n\}$, we see that $T_\lambda(x_0) = Lx_0 - \lambda Nx_0 - g$. Q.E.D.

4. VARIATIONAL BPs AND ABPs FOR PDEs AND ODEs
4.1 ABPs for Elliptic PDEs in Divergence Form

In this section we establish the existence of variational ABPs for elliptic equation of the form

$$\mathcal{A}u = \lambda\mathcal{B}u, \quad u \in V, \qquad \mathring{W}_2^m \subseteq V \subseteq W_2^m. \tag{4.1}$$

where $\mathcal{A}$ and $\mathcal{B}$ are formal differential operators on Q of the form

$$\mathcal{A}u = \sum_{|\alpha|\le m} (-1)^{|\alpha|}D^\alpha A_\alpha(x,u,Du,\ldots,D^m u) \tag{4.2}$$

$$\mathcal{B}u = \sum_{|\beta|\le m-1} (-1)^{|\beta|}D^\beta B_\beta(x,u,Du,\ldots,D^m u) \tag{4.3}$$

with $\mathcal{A}$ and $\mathcal{B}$ being asymptotically linear in the sense to be defined below. To formulate the hypotheses that we impose on $\mathcal{A}$ and $\mathcal{B}$ we introduce the vector space R^{S_m} whose elements are $\xi = \{\xi_\alpha \mid |\alpha| \leq m\}$ and divide each such ξ into two parts, $\xi = (\eta,\zeta)$, where $\eta = \{\eta_\beta \mid |\beta| \leq m - 1\} \in R^{S_{m-1}}$ is a lower order part of ξ and $\zeta = \{\zeta_\alpha \mid |\alpha| = m\}$ is the part of ξ corresponding to mth-order derivatives. We also set $\xi(u) = \{D^\alpha u \mid |\alpha| \leq m\}$. Thus $\mathcal{A}_\alpha$ and $\mathcal{B}_\beta$ are functions of $Q \times R^{S_m}$ into R^1, on which we impose the following condition:

A(1) For each $|\alpha| \leq m$, $A_\alpha(x,\xi)$ satisfies the Carathéodory condition and there exists a constant $c_0 > 0$ such that

$$|A_\alpha(x,\xi)| \leq c_0(1 + \sum_{|\alpha| \leq m} |\xi_\alpha|^{p_{\alpha\beta}})$$

with $p_{\alpha\beta} \leq 1$ for $|\alpha| = |\beta| = m$, while in the lower-order cases the exponents may have larger upper bounds (see [33]).

A(2) If $\xi = (\eta,\zeta)$ is divided into components ζ_α of order $|\alpha| = m$ and the components η_α of order $|\alpha| \leq m - 1$, then for $x \in Q$ and $\eta \in R^{S_{m-1}}$,

$$\sum_{|\alpha| = m} [A_\alpha(x,\eta,\zeta) - A_\alpha(x,\eta,\zeta')][\zeta_\alpha - \zeta'_\alpha] > 0 \quad \text{for} \quad \zeta \neq \zeta'.$$

A(3) There exist constants $c_1 > 0$ and c_2 such that

$$\sum_{|\alpha| \leq m} A_\alpha(x,\xi)\xi_\alpha \geq c_1 \left(\sum_{|\alpha| \leq m} |\xi_\alpha|^2 \right) - c_2$$

$$\text{for all} \quad x \in Q \quad \text{and} \quad \xi \in R^{S_m}.$$

Remark 4.1 Instead of A(2) and A(3) one may assume: A(2'): There exist constants $c_3 > 0$ and c_4 such that for all x in Q, each pair ξ and ξ' in R^{S_m}, and some integer $k < m$ we have

$$\sum_{|\alpha| \leq m} [A_\alpha(x,\xi) - A_\alpha(x\xi')][\xi_\alpha - \xi'_\alpha]$$

$$\geq c_3 \left(\sum_{|\alpha| \leq m} |\xi_\alpha - \xi'_\alpha|^2 \right) - c_4 \left(\sum_{|\alpha| \leq k} |\xi_\alpha - \xi'|^2 \right).$$

B(1) For $|\beta| \leq m - 1$, the functions $B_\beta(x,\xi)$ satisfy the Carathéodory conditions and the inequality

$$|B_\beta(x,\xi)| \leq c_5[h(x) + \sum_{|\gamma| \leq m} |\xi_\gamma|^{q_{\beta\gamma}}] \quad [h \in L_2(Q)],$$

where $0 \leq q_{\beta\gamma} \leq n + 2(m - |\beta|)/n - 2(m - |\gamma|)$ for $n > 2(m - |\gamma|)$ and $q_{\beta\gamma}$ are arbitrary nonnegative numbers if $n \leq 2(m - |\gamma|)$.

To define the "variational AB problem" for equation (4.1) relative to a given closed subspace V of W_2^m with $V \supseteq \overset{\circ}{W}_2^m$ we first note that in virtue of assumptions A(1) and B(1) and the results about Nemytsky operators (see [136,280]), the generalized forms

$$a(u,v) = \sum_{|\alpha| \leq m} \langle A_\alpha(x, \xi(u)), D^\alpha v \rangle \tag{4.4}$$

$$b(u,v) = \sum_{|\beta| \leq m-1} \langle B_\beta(x, \xi(u)), D^\beta v \rangle \tag{4.5}$$

are well defined on W_2^m and for a given subspace V of W_2^m, one can associate with $a(u,v)$ and $b(u,v)$ in a unique way bounded continuous mappings T and C of V into V such that

$$(Tu,v) = a(u,v), \ (Cu,v) = b(u,v) \qquad \text{for} \quad u,v \in V. \tag{4.6}$$

Definition 4.1 Let V be a closed subspace of W_2^m with $\overset{\circ}{W}_2^m \subseteq V$. A real number $\lambda \in R$ is said to be a *variational ABP relative to V* of (4.1) if λ is an ABP for the equation

$$Tu = \lambda Cu \qquad (u \in V). \tag{4.7}$$

To apply Theorem IV.6.2 or Corollary IV.6.2 to equation (4.7), we first note that since V is a separable Hilbert space there exists a projectionally complete scheme $\Gamma_0 = \{X_n, P_n\}$ for (V,V) for which the following assertion holds.

Lemma 4.1 (a) If assumptions A(1) and A(2)–A(3) or A(1) and A(2') hold, $T: V \to V$ is A-proper w.r.t. Γ_0.

(b) If assumption B(1) holds, $C: V \to V$ is compact.

Proof. (a) First, if A(1) and A(2)–A(3) hold, then by the results of Browder [33] the map T satisfies condition (S), so T is A-proper, as was shown in Proposition I.1.4. On the other hand, if A(1) and A(2') hold, then

$$(Tu - Tv, u - v) \geq c_3 \| u - v \|_{m,2}^2$$

$$- c_4 \| u - v \|_{k,2}^2 \qquad \text{for} \quad u,v \in V$$

and this, as was shown by the author in Lemma 2 [228], implies that T is A-proper w.r.t. Γ_0.

(b) The fact that $C: V \to V$ is compact was shown earlier.   Q.E.D.

For Theorem IV.6.2 to be applicable to (4.7), we also impose the following additional conditions:

A(4) There exist functions $a_{\alpha\beta}(x) \in L^{\infty}(Q)$ for $|\alpha| \le m$ and $|\beta| \le m$ and a continuous function $\phi\colon R^{+} \to R^{+}$ with $\phi(t)/t \to 0$ as $t \to \infty$ such that

$$|A_{\alpha}(x,u,\dots,D^m u)$$
$$- \sum_{|\beta| \le m} a_{\alpha\beta}(x)D^{\beta}u| \le \phi(\|u\|_{m,2}) \qquad \text{for} \quad u \in V \quad \text{and} \quad |\alpha| \le m.$$

B(2) There exist functions $b_{\beta\gamma} \in L^{\infty}(Q)$ for $|\beta| \le m - 1$ and $|\gamma| \le m$ and a continuous function $\psi\colon R^{+} \to R^{+}$ with $\psi(t)/t \to 0$ as $t \to \infty$ such that

$$|B_{\beta}(x,u,\dots,D^m u) - \sum_{|\gamma| \le m} b_{\beta\gamma}(x)D^{\gamma}u| \le \psi(\|u\|_{m,2})$$

$$\text{for} \quad u \in V \quad \text{and} \quad |\beta| \le m - 1.$$

Now assumptions A(4) and B(2) imply that the bilinear forms

$$l(u,v) = \sum_{|\alpha|,|\beta| \le m} \langle a_{\alpha\beta}(x)D^{\beta}u, D^{\alpha}v \rangle \qquad (u,v \in V) \tag{4.8}$$

$$b(u,v) = \sum_{|\beta| \le m-1, |\gamma| \le m} \langle b_{\beta\gamma}(x)D^{\gamma}u, D^{\beta}v \rangle \qquad (u,v \in V) \tag{4.9}$$

determine bounded linear mappings $L,B\colon V \to V$ such that

$$l(u,v) = (Lu,v), \quad b(u,v) = (Bu,v) \qquad \text{for} \quad u,v \in V. \tag{4.10}$$

Lemma 4.2 Condition A(4) implies that L is the asymptotic derivative of T, while B(2) implies that B is the asymptotic derivative of C (i.e., $L = T_{\infty}$ and $B = C_{\infty}$).

Proof. We first show that

$$\|Tu - Lu\|_{m,2}/\|u\|_{m,2} \to 0 \qquad \text{as} \quad \|u\|_{m,2} \to \infty \quad (u \in V). \tag{4.11}$$

Indeed, since for all u and v in V we have the equality

$$(Tu - Lu, v) = a(u,v) - l(u,v)$$
$$= \sum_{|\alpha| \le m} \langle A_{\alpha}(x,n,\dots,D^m u) - \sum_{|\beta| \le m} a_{\alpha\beta}(x)D^{\beta}u, D^{\alpha}v \rangle,$$

it follows from the latter and condition A(4) that

$$|(Tu - Lu, v)| \le \sum_{|\alpha| \le m} \int |A_{\alpha}(x,u,\dots,D^m u)$$
$$- \sum_{|\beta| \le m-1} a_{\alpha\beta}D^{\beta}u\| D^{\alpha}v\| \, dx$$
$$\le K\phi(\|u\|_{m,2})\| v\|_{m,2}$$

for some constant $K > 0$. Since $\phi(t)/t \to 0$ as $t \to \infty$ and

$$\| Tu - Lu \| = \sup_{0 \neq v \in V} | (Tu - Lu, v) | / \| v \|_{m,2} \leq K\phi(\| u \|_{m,2}) \qquad (4.12)$$

for each fixed u in V, we obtain (4.11) from (4.12).

In a similar way, using condition B(2), one shows that $\| Cu - Bu \|_{m,2} / \| u \|_{m,2} \to 0$ as $\| u \|_{m,2} \to \infty$ (i.e., $C = C_\infty + P$ with $C_\infty = B$), and $\| Pu \|_{m,2} / \| u \|_{m,2} \to 0$ as $\| u \|_{m,2} \to \infty$.   Q.E.D.

Now, if we set $T = L + N: V \to V$, then $\| Nu \|_{m,2} / \| u \|_{m,2} \to 0$ as $\| u \|_{m,2} \to \infty$ by Lemma 4.2 and, in virtue of condition A(3), $(Tu,u) \geq c_1 \| u \|_{m,2}^2 - K_0$, where K_0 is some positive constant. Hence by Lemma IV.6.1 with $K = I$, the map $L (\equiv T_\infty)$ is injective and A-proper w.r.t. Γ_0.

In view of the discussion above, the operators $T,C: V \to V$ satisfy all conditions of Theorem IV.6.2 or Corollary IV.6.2, and therefore the following new result concerning the ABPs for (4.1) follows from Theorem 6.2.

Theorem 4.1   Let $\mathcal{A}$ and $\mathcal{B}$ be the formal differential operators satisfying conditions A(1)–A(4) and B(1)–B(2), respectively. Then each eigenvalue of

$$Lu = \lambda Bu \qquad (u \in V) \qquad (4.13)$$

of odd multiplicity is a variational ABP for (4.1).

4.2   BPs and ABPs for Semilinear Elliptic Equations

Let L be a differential operator of order $k \in Z^+$ for the form

$$Lu = \sum_{|\alpha| \leq k} a_\alpha(x)D^\alpha u \qquad (x \in Q, a_\alpha \in L^\infty(Q)) \qquad (4.14)$$

with L ordinary when $n = 1$ and partial when $n > 1$. We make the following basic hypothesis on L:

(F) For a given closed subspace V of $W_2^k(Q)$ the linear operator L given by (4.14) is a homeomorphism of V onto L_2.

In this section we apply Theorems IV.6.1 and IV.6.2 to obtain the existence of BPs and for ABPs for the problem

$$Lu + f(x,D^\gamma u) = \lambda\{Mu + g(x,D^\eta u)\} \qquad (4.15)$$

$$(u \in V, | \gamma | \leq k, | \eta | < k - 1)$$

where M is a linear operator of order $l \leq k - 1$ of the form

$$Mu = \sum_{|\alpha| \leq l} b_\alpha(x)D^\alpha u \qquad [b_\alpha \in L^\infty(Q)] \qquad (4.16)$$

and f and g are functions of $Q \times R^{S_{k-1}}$ into R^1 which satisfy the following conditions:

(C') For each $u \in W_2^k(Q)$, the map $u \mapsto F(u) \equiv f(x, D^\gamma u)$ for $|\gamma| \le k$ yields a continuous bounded operator from W_2^k to L_2 such that $T \equiv L + F: W_2^k \to L_2$ is A-proper w.r.t. the projectional scheme Γ_1 constructed below.

(C") For each $u \in W_2^{k-1}$ the map $u \mapsto G(u) = (x, D^\eta u)$ for $|\eta| \le k - 1$ is continuous and bounded.

Theorem 4.2   Suppose that L satisfies condition (F) and suppose also that in addition to conditions (C') and (C"), the functions f and g satisfy either the hypothesis

(D') $F(0) = G(0) = 0$ and $\| Fu \|_2 / \| u \|_{k,2} \to 0$ as $u\|_{k,2} \to 0$ and $\| Gu \|_2 / \| u \|_{k-1,2} \to 0$ as $\| u \|_{k-1,2} \to 0$, or the hypothesis

(E') $\| Fu \|_2 / \| u \|_{k,2} \to 0$, $\| Gu \|_2 / \| u \|_{k,2} \to 0$ as $\| u \|_{k,2} \to \infty$.

Then each eigenvalue of the linearized problem

$$Lu = \lambda Mu \qquad (u \in V) \tag{4.17}$$

of odd multiplicity is a BP for equation (4.15) if (D') holds and an ABP if (E') holds.

Proof.   Since V is a separable Hilbert space, we can choose a sequence of finite-dimensional subspaces $\{X_n\} \subset V$ such that $\mathrm{dist}(u, X_n) = \inf_{w \in X_n} \| u - w \|_{k,2} \to 0$ for each u in V. If we set $Y_n = L(X_n) \subset L_2$ for each n and let $P_n: V \to X_n$ and $Q_n: L_2 \to Y_n$ be the orthogonal projections, then the scheme $\Gamma_1 = \{X_n, P_n; Y_n, Q_n\}$ is projectionally complete for (V, L_2) since L is a homeomorphism of V onto L_2. Moreover, it is easy to see that L, considered as a map from V to L_2, is A-proper w.r.t. Γ_1, while $L + F$ is A-proper by hypothesis. Now it follows from condition (C") and the Sobolev embedding theorem that the map G, considered as a map from V to L_2, is compact and so is the linear map $M: V \to L_2$ defined by (4.16). Consequently, $T \equiv L + F: V \to L_2$ is A-proper w.r.t. Γ_1 and $C \equiv M + G: V \to L_2$ is compact.

Furthermore, if condition (D') holds, then $L, M \in L(V, L_2)$ are Fréchet derivatives at 0 of T and C, respectively. On the other hand, if condition (E') holds, L and M are the asymptotic derivatives of T and C, respectively. Consequently, the first assertion of Theorem 4.2 follows from Theorem IV.6.1, while the second from Theorem IV.6.2.   Q.E.D.

We illustrate the generality of Theorem 8 by the following examples, although more general examples can be given.

Let Q be a bounded domain in R^n. For simplicity we assume that ∂Q is of class C^∞, although this condition can be relaxed somewhat (see Browder [20]). We consider the problem

$$\Delta^2 u + f(x,D^\gamma u)$$
$$= \lambda(-\Delta u + g(x,D^\gamma u)) \qquad (u \in W_2^4 \cap \mathring{W}_2^2, |\gamma| \le 3), \quad (4.18)$$

where Δ denotes the n-dimensional Laplacian, f and g are continuous functions of $Q \times R^{S_3}$ into R^1 which satisfy condition (C″) for $k = 4$. If we set $V = W_2^4 \cap \mathring{W}_2^2$ and define the linear operator $L: V \to L_2$ by $Lu = \Delta^2 u$, then V is a closed subspace of W_2^4 and L is a homeomorphism of V onto L_2 (see [20]). It is easy to see that the linear map $M: V \to L_2$, defined by $Mu = -\Delta u$ for $u \in V$, is compact. Now the compactness of $F,G: V \to L_2$, given by $F(u) \equiv f(x,D^\gamma u)$ and $(Gu) \equiv g(x,D^\gamma u)$, follows from condition (C″) for $k = 4$. Thus, by Theorem 4.2, we have

Corollary 4.1   If in addition to the condition above we assume that F and G satisfy either (D′) or (E′) with $k = 4$, then each eigenvalue of the equation

$$\Delta^2 u + \lambda \Delta u = 0 \qquad (u \in W_2^4 \cap \mathring{W}_2^2) \tag{4.19}$$

of odd multiplicity is a BP for (4.18) if (D′) holds and an ABP for (4.18) if (E′) holds.

4.3   Bifurcation Problems for Ordinary Differential Operators

As our next example, we consider a bifurcation problem for the quasilinear ordinary differential equation of the form

$$u^{(m)}(x) + f(x,u^{(1)}, \ldots ,u^{(m-1)})$$
$$= \lambda g(x,u,u^{(1)}, \ldots ,u^{(m-1)}) \qquad (u \in \bar{B}(0,r) \subset W_2^m[a,b]) \tag{4.20}$$

$$W_i(u) \equiv \sum_{i=0}^{m-1} [\alpha_{ij}u^{(j)}(a) + \beta_{ij}u^{(j)}(b)] = 0 \qquad (1 \le i \le m), \tag{4.21}$$

where α_{ij}, β_{ij} are constants and the functions $f(x,z_0,z_1, \ldots ,z_{m-1})$, $g(x,z_0,z_1, \ldots ,z_{m-1})$, $\partial f(x,z_0, \ldots ,z_{m-1})/\partial z_j$ and $\partial g(x,z_0, \ldots ,z_{m-1})/\partial z_j$ are defined and continuous for $a \le x \le b$ and $|z_j| \le \delta$ for some $\delta > 0$ and $0 \le j \le m - 1$ with $f(x,0, \ldots ,0) = g(x,0, \ldots ,0) = 0$. First note that since the embedding of $W_2^m[a,b]$ into $C^{m-1}[a,b]$ is compact, there exist a constant $\eta > 0$ (depending on m and a,b) such that $\| u \|_{m,2} \ge \eta |u|_{m-1}$ for all u in W_2^m with $|u|_{m-1}$ denoting the norm of u in $C^{m-1}[a,b]$. This implies that if we set $V = \{u \in W_2^m \mid W_i(u) = 0$ for i

$= 1, \ldots, m\}$ and let $B(0,r) = \{u \in V \mid \parallel u \parallel_{m,2} \le r, r = \delta/\eta\}$, the mappings $F: B(0,r) \subset V \to L_2[a,b]$ and $G: B(0,r) \subset V \to L_2[a,b]$, given by $F(u) \equiv f(x,u,u^{(1)}, \ldots, u^{(m-1)})$ and $G(u) \equiv g(x,u,u^{(1)}, \ldots, u^{(m-1)})$ for $u \in B(0,r)$, are well defined and compact. Moreover, F and G have compact Fréchet derivatives $F_0, G_0 \in L(V, L_2)$ at $0 \in V$ given by

$$F_0 v = \sum_{j=0}^{m-1} \frac{\partial f(x,0, \ldots, 0)}{\partial z_j} v^{(j)}, \tag{4.22}$$

$$G_0 v = \sum_{j=1}^{m-1} \frac{\partial g(x,0, \ldots, 0)}{\partial z_j} v^{(j)} \quad (v \in V).$$

Suppose that the homogeneous equation $u^{(m)} = 0$ has only the trivial solution $u(x) \equiv 0$ satisfying the boundary condition (4.21). Then, as is well known, the linear mapping L defined on V by $Lu = u^{(m)}$ is a homeomorphism of V onto L_2 and thus A-proper w.r.t. to a suitably chosen projectionally complete scheme $\Gamma_2 = \{X_n, P_n; LX_n, Q_n\}$ for (V, L_2).

In view of the discussion above, an immediate consequence of Theorem 4.2 (see Remark IV.6.2) is the following:

Corollary 4.2 If in addition to the conditions on $u^{(m)}$, f, and g above, we assume that G_0 is not the null operator and that the problem

$$u^{(m)} + \sum_{j=0}^{m-1} \frac{\partial f(x,0, \ldots, 0)}{\partial z_j} u^{(j)} = 0, \qquad W_j(u) = 0 \tag{4.23}$$

has only the trivial solution $u(x) = 0$, then each eigenvalue of

$$u^{(m)} + \sum_{j=0}^{m-1} \frac{\partial f(x,0, \ldots, 0)}{\partial z_j} u^{(j)}$$

$$= \lambda \sum_{j=0}^{m-1} \frac{\partial g(x,0, \ldots, 0)}{\partial z_j} u^{(j)} \quad (u \in V) \tag{4.24}$$

of odd multiplicity is a BP for (4.20)–(4.21).

We complete this section by applying Corollary 4.2 to a more concrete example of (4.20)–(4.21) which is furnished by the one-dimensional boundary value problem arising in connection with the longitudinal bending of a compressed rod, say, of length 1. In this situation it is known (see [136]) that the bending $u(x)$ is determined by the boundary value problem

$$-u''(x) = \lambda \rho(x) u(x) \sqrt{1 - (u')^2} \tag{4.25}$$

$$u(0) = u(1) = 0.$$

Passing to an equivalent problem involving nonlinear compact integral operators, it was shown by Krasnoselskii [136] that each eigenvalue of

$$-u'' = \lambda\rho(x)u, \qquad u(0) = u(1) = 0 \tag{4.26}$$

is a bifurcation point for (4.25).

The purpose of this discussion is to show how this result of Krasnoselskii for equation (4.25) follows from Corollary 4.2 without passing to an equivalent integral eigenvalue problem as was done in [136].

Let $W_2^2 \equiv W_2^2[0,1]$ and let $\mathring{W}_2^1 = \{u \in W_2^1 \mid u(0) = u(1) = 0\}$. Since W_2^2 is compactly embedded into $C^1 \equiv C^1[0,1]$, there exists a constant $\eta_1 > 0$ such that $|u|_1 \leq \eta_1 \|u\|_{2,2}$ for all u in W_2^2 and therefore $|u'| \leq |u|_1 \eta_1 \|u\|_{2,2}$. Let $V = W_2^2 \cap \mathring{W}_2^1$, $B(0,r) = \{u \in V \mid \|u\|_{2,2} \leq r, r = 1/\eta_1\}$ and define $T: V \to L_2[0,1]$ by $Tu = -u''$ for $u \in V$ and $C: B(0,r) \subset V \to L_2$ by $Cu = \rho(x)u\sqrt{1 - (u')^2}$. Then V is a closed subspace of W_2^2, T is a linear homeomorphism of V onto L_2 and $C: B(0,r) \subset V \to L_2$ is well defined and compact with $C(0) = 0$. Thus the bifurcation problem for (4.25) reduces to the bifurcation problem for the equation

$$Tu = \lambda Cu \qquad (u \in B(0,r) \subset V). \tag{4.25'}$$

Since $T \in L(V,L_2)$, its Fréchet derivative is $T_0 = T$, while as is not hard to see, C has a Fréchet derivative $C_0 \in L(V,L_2)$ given by $C_0 u = \rho u$, which is compact as a map of V to L_2. Thus, by Corollary 4.2, every eigenvalue of (4.26) of odd multiplicity is BP for equation (4.25). Since each eigenvalue of (4.26) is simple and thus odd, Krasnoselskii's result follows from Corollary 4.2.

Finally, let us remark that as the last three examples show, even those cases where the classical bifurcation results are applicable (via integral equations), the bifurcation and asymptotic bifurcation results based on A-proper mapping theory derived in Section IV.6 offer an alternative and a more direct approach to the bifurcation problems involving certain differential operators.

Appendix

The purpose of this appendix is to collect a number of known analytical tools that were used in this monograph and which are scattered in various books. This is done for the convenience of the reader. We consider the following topics:

1. Upper and lower limits of functions
2. Finite element methods for ODEs and elliptic PDEs
3. Lebesgue measure and integral and some important convergence theorems
4. Lebesgue spaces
5. Weak and weak* convergence
6. Sobolev spaces and Sobolev embedding theorem

Now we describe briefy the material contained in each topic and indicate the books and their authors where a detailed discussion and proofs can be found.

1. UPPER AND LOWER BOUNDS AND LIMITS OF FUNCTIONS [108,304]

Let S be a given set and let f be a function whose domain $D(f) = S$. Then f is a *real-valued* function on S if its range $R(f) \subset \mathbb{R}$. The *least upper bound* of $f(x)$ on S is defined to be the least upper bound of the set of numbers $R(f) = f(S)$. We denote it by "l.u.b. of $f(x)$ on S". The *greatest lower bound* of $f(x)$ on S [i.e., g.l.b. of $f(x)$ on S] is defined similarly.

If S is a one-dimensional space, a monotonic single-valued function f on S has a right-hand limit at each right-hand accumulation point of S, and a left-hand limit at each left-hand accumulation point of S.

In case c is in the closure $\overline{S}$ of S where f is defined, $N(c,\delta)$ an open neighborhood of c and $SN(c;\delta) = S \cap N(c;\delta)$, then the l.u.b. $f(x)$ on $SN(c;\delta)$ is a function $g(\delta)$ which is single-valued and nondecreasing for $\delta > 0$, and hence has a limit at $\delta = 0$. This limit is called the *upper limit* of $f(x)$ *at* c and is denoted by $\lim_{x=c} \sup f(x)$ or $\overline{\lim}_{x=c} f(x)$, that is, $\lim_{x=c} \sup f(x) = $ g.l.b. [l.u.b. $f(x)$ for $x \in SN(c;\delta)] = \lim_{\delta \to 0}$[l.u.b. $f(x)$ for $x \in SN(c;\delta)$]. A similar definition holds for the *lower limit*, denoted by $\lim_{x=c} \inf f(x)$ or $\underline{\lim}_{x=c} f(x)$.

Note that the upper and lower limits always exist, finite or infinite, at every point of the closure of the domain of the function. We certainly have

$$\lim_{x=c} \inf f(x) \leq \lim_{x=c} \sup f(x).$$

It can be shown that $\lim_{x=c} \sup f(x)$ is a finite number a iff:

(i) $\epsilon > 0$: $\exists N(c)$ such that $x \in N(c)$ implies that $f(x) < a + \epsilon$.
(ii) $\epsilon > 0$: $\exists N(c;\epsilon)$ such that $f(x) \not< a - \epsilon$.

On the other hand, $\lim_{x=c} \sup f(x) = -\infty$ iff to each $\epsilon > 0 \; \exists \; N(c)$ such that x in $N(c) \Rightarrow f(x) < -1/\epsilon$. Also, $\lim_{x=c} \sup f(x) = \lim_{x=c} \inf f(x)$ iff $\lim_{x=c} \inf(x) = \lim_{x=c} \sup f(x)$.

The following inequalities involving upper and lower bounds and upper and lower limits of sums and differences of functions are sometimes useful.

(1a) $\underline{\lim} f + \underline{\lim} g \leq \underline{\lim}(f + g) \leq \left\{ \dfrac{\underline{\lim} f + \overline{\lim} g}{\overline{\lim} f + \underline{\lim} g} \right\} \leq \overline{\lim}(f + g) \leq$
$\overline{\lim} f + \overline{\lim} g$.

(1b) $\underline{\lim} f - \overline{\lim} \leq \underline{\lim}(f - g) \leq \left\{ \dfrac{\underline{\lim} f + \overline{\lim} g}{\overline{\lim} f + \underline{\lim} g} \right\} \leq \overline{\lim}(f - g) \leq$
$\overline{\lim} f - \underline{\lim} g$,

where all upper and lower limits are taken at the same point a in $\overline{S}$.

2. FINITE ELEMENT METHOD FOR ODEs AND ELLIPTIC PDEs [313,314,316]

It is interesting to note that the projection method (i.e., the abstract operator version of the Galerkin method) for the approximation solvability of linear and semilinear ODEs and PDEs in suitable function spaces can be realized by the finite element method.

Basic ideas   Suppose that we want to solve an elliptic boundary value problem via the Galerkin or projection method. If we use a polynomial Galerkin scheme, then as a rule (see [313,314,316] and others) the corresponding matrices are *not* sparse matrices (i.e., most of the entries are different from zero). This is a typical shortcoming of polynomial bases. In contrast to this unfavorable situation, both the difference method and the method of finite elements produce *sparse matrices*. Compared with the difference method, the method of finite elements is much more flexible. For example, one can vary the mesh size of the triangulation in different subregions depending on the subtle behavior of the solution. Today, the method of finite elements is widely used in engineering and in the natural sciences (see [316]). This method represents one of the important *achievements* of the modern numerical mathematics.

2.1   Prototype for Finite Elements in $\mathbb{R}^1$

Let $\alpha < \beta \leq \gamma$. We start with

$$w(x) = \begin{cases} 0 & \text{if } x = \alpha, \gamma \\ 1 & \text{if } x = \beta \end{cases}$$

and extend w to a function $\mathbb{R} \to \mathbb{R}$ via linear interpolation. Thus way we obtain

$$w(x) = \begin{cases} 0 & \text{if } x \notin [\alpha, \beta] \\ (x - \alpha)/(\beta - \alpha) & \text{if } x \in [\alpha, \beta] \\ (\gamma - x)/(\gamma - \beta) & \text{if } x \in [\beta, \gamma] \end{cases}$$

(Figure 1). We now consider a triangulation J of the compact interval $[a,b]$; that is, if we choose nodes a_i such that $a = a_0 < a_1 < a_2 < \cdots < a_k = b$, the mesh of T is defined through $h(J) = \max_i(a_{i+1} - a_i)$. We set

$$w_j(a_i) = \begin{cases} 1 & \text{if } i = j \\ 0 & \text{if } i \neq j \end{cases}$$

and extend w_j to a function $w_j: [a,b] \to \mathbb{R}$ via linear interpolation. The functions $w_0, w_1, \ldots, w_k$ are called *finite elements*. Furthermore, we set

$$X(J) = \text{span}\{w_0, w_1, \ldots, w_k\} \quad \text{and} \quad \dot{X}(J) = \text{span}\{w_1, \ldots, w_{k-1}\}.$$

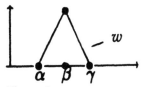

α β γ
w

Figure 1.

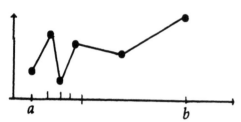

Figure 2.

Then $X(J)$ consists of all piecewise linear, continuous functions w: $[a,b]$ $\to \mathbb{R}$ with arbitrary prescribed values $w(a_0), \ldots, w(a_k)$ at nodes (Figure 2). Moreover, we get

$$\mathring{X}(J) = \{w \in X(J)\colon w(a) = w(b) = 0\}.$$

Obviously, we obtain the *generalized orthogonality* relations

$$\int_a^b w_i(x)w_j(x) \, dx = 0 \quad \text{if } |i - j| \geq 2.$$

These relations are responsible for the appearance of *sparse matrices* in connection with the Galerkin–finite element method.

Now let (J_n) be a sequence of triangulations of the interval $[a,b]$, where the mesh size goes to zero as $n \to \infty$ [i.e., $h(J_n) \to 0$ as $n \to \infty$]. Let $1 \leq p < \infty$ and set $h_n \equiv h(J_n)$ as well as $X_n = X(J_n)$ and $\mathring{X}_n = \mathring{X}(J_n)$. It is not hard to show that

(a) $\{X_n\}$ forms a Galerkin–finite element scheme in $X = W_p^1(a,b)$.
(b) For all $u \in W_p^2(a,b)$, $\text{dist}_x(u,X_n) \leq \text{const } h_n \| u \|_{2,p}$

(i.e., the accuracy of approximation is proportional to the mesh size h_n).

(c) $\{\mathring{X}_n\}$ forms a Galerkin–finite element scheme in $\mathring{X} \equiv \mathring{W}_p^1(a,b)$.
(d) For all $u \in W_p^2(a,b) \cap \mathring{X}$, $\text{dist}_x(u,\mathring{X}) \leq \text{const } h_n \| u \|_{2,p}$.

The constants in (b) and (d) depend on $(b - a)$ and p.

2.2 Piecewise Linear Finite Elements in $\mathbb{R}^2$

We want to generalize the previous results from $\mathbb{R}^1$ to $\mathbb{R}^2$. Let G be a bounded polygonal region in $\mathfrak{R}^2$. We choose a triangulation J of G with nodes $N_1, \ldots, N_k$ (Figure 3). Moreover, we set

$$h(J) = \begin{cases} \text{maximal diameter of the} \\ \text{triangles of } J \end{cases}$$

$\delta(J) \equiv$ minimal angle of the triangles of J.

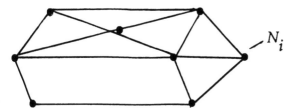

Figure 3.

The following elementary result is decisive for the construction of finite elements (see [313,314]).

(L) Let T be a triangle with vertices P_1,P_2,P_3 (Figure 4). Then there exists a unique linear function $w(\zeta,\eta) = a + b\zeta + c\eta$ with prescribed values $w(P_1),w(P_2),w(P_3)$.

Construction (C) We now prescribe the values $w(N_1), \ldots ,w(N_k)$ at the nodes and extend w uniquely to a function $w: \overline{G} \to \mathbb{R}$ via linear interpolation according to (L). Then w has the following properties:

(i) w is continuous on $\overline{G}$.
(ii) w is piecewise differentiable, and $w \in W_p^1(G), 1 \leq p < \infty$.

The set of all functions w constructed by (C) above is denoted by $X(J)$. In particular, if we set $w_j(N_i) = 1$ if $i = j$ and $w_j(N_i) = 0$ if $i \neq j$, and if we construct $w_j: \overline{G} \to \mathbb{R}$ by (C), the finite elements w_j form a basis of the space $X(J)$; that is,

$$X(J) = \text{span}\{w_i, \ldots ,w_k\}.$$

Furthermore, we set

$$\mathring{X}(J) = \{w \in X(J) : w = 0 \text{ on } \partial G\}.$$

Obviously, $\mathring{X}(J)$ consists of all $w \in X(J)$ with $w(N_i) = 0$ for all nodes $N_i \in \partial G$. If the nodes N_i and N_j do not belong to the neighboring triangles,

$$(w_i,w_j)_{1,2} \equiv \int_G \left(w_i w_j + \sum_{i=1}^{2} D_r w_i D_r w_i \right) dx = 0.$$

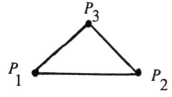

Figure 4.

This generalized orthogonality relation produces *sparse matrices* in connection with the Galerkin–finite element method.

Finally, we consider a sequence $\{J_n\}$ of triangulations of G, where the mesh size goes to zero as $n \to \infty$; that is,

$$\lim_{n \to \infty} h(J_n) = 0.$$

Moreover, we assume that $\{J_n\}$ is *nondegenerate*; that is, the minimal angles of the triangles are bounded below:

$$\inf_n \delta_n > 0.$$

We set $h_n = h(J_n)$ as well as $X_n = X(J_n)$ and $\mathring{X}_n = \mathring{X}_n(J_n)$. Let $1 \le p < \infty$. Then the following hold:

(a) $\{X_n\}$ is a Galerkin–finite element scheme in $= W_p^1(Q)$.
(b) For all $u \in W_p^2(Q)$, $\text{dist}_x(u, X_n) \le \text{const } h_n \| u \|_{2,p}$.
(c) $\{\mathring{X}_n\}$ is the Galerkin–finite element scheme in $\mathring{X} = \mathring{W}_p^1(G)$.
(d) For all $u \in W_p^2 \cap \mathring{X}$, $\text{dist}_x(u, \mathring{X}_n) \le \text{const } h_n \| u \|_{2,b}$.

The constants in (b) and (d) depend only on G, b, and $\inf_n \delta_n$.

(I) Let T be a triangle of fixed triangulation J_n of G. For $m = 0,1$, we construct the operator $\Pi : W_p^2(G) \to W_p^m(G)$ in the following way. Let $u \in W_p^2(Q)$ be given. This implies that $u \in C(\overline{G})$ by the Sobolev embedding theorem. Thus it is meaningful to set $(\Pi u)(N_i) = u(N_i)$ and to extend this to a function $\Pi u \colon \overline{G} \to \mathbb{R}$ via linear interpolation according to (L) above. Note the following.

(α) The construction of Πu is unique by (L), and we have $\Pi u \in W_p^m(Q) \cap X_n$.
(β) $\Pi u = u$ on a triangle T if u is a linear function on T.
(γ) Π is affinely invariant on T.

Consequently, for $m = 0,1$, one has

$$\| u - \Pi u \|_{W_p^1(T)} \le C(h_n^2 + h_n^2/\rho_n)^p \| u \|_{W_p^2(T)},$$

where C is a constant. Summing over all triangles $T \in J_n$, we find that

$$\| u - \Pi u \|_{W_p^1(G)}^p \le C(h_n^2 + h_n^2/\rho_n)^p \| u \|_{W_p^2}^p (G).$$

(II) We now consider the sequence $\{J_n\}$ of triangulation. We set $\Pi_n = \Pi$. From $\inf_n \delta_n > 0$ we obtain the key relation

$$\sup h_n/\rho_n < \infty.$$

Thus we get

$$\| u - \Pi_n u \|_{W_p^1(G)} \le \text{const } h_n \| u \|_{W_p^2(G)}.$$

Since $\Pi_n u \in X_n$, we obtain (b), that is,

$$\text{dist}_x(u, X_n) \leq \text{const } h_n \| u \|_{W_2^2(G)}.$$

(III) Let $u \in W_p^1(G)$ be given. The set $C^\infty(\overline{G})$ is dense in $W_2^1(G)$. Thus for each $\epsilon > 0$, there exists a $v \in C^\infty(G)$ with $\| u - v \|_{1,p} < \epsilon$. This implies that

$$\| u - \Pi_n v \|_{1,p} \leq \| u - v \|_{1,p} + \| v - \Pi_n v \|_{1,p}$$
$$\leq \epsilon + \text{const. } h_n \| v \|_{2,p} \leq 2\epsilon$$

for all $n \geq n_0(\epsilon)$. Since $\Pi_n v \in X_n$,

$$\lim_{n \to \infty} \text{dist}_x(u, X_n) = 0.$$

Therefore, $\{X_n\}$ is a Galerkin–finite element scheme in $W_p^1(G)$.

2.3 Generalization to Piecewise Quadratic Finite Elements

The previous results can easily be extended to piecewise polynomial functions of degree $k \geq 2$. To explain this we consider the special case $k = 2$. Consider Fig. 5 and let T be a triangle with vertices P_1, P_2, P_3. Denote by P_4, P_5, P_6 the midpoints of the sides of T (Figure 5). The following elementary result will be crucial.

(Q) There exists a unique quadratic polynomial

$$w(\xi, \eta) = a + b\xi + c\eta + d\xi^2 + e\eta^2 + f\xi\eta$$

with prescribed values $w(P_1), \ldots, w(P_6)$.

Let J be a triangulation of G (Figure 3). For each triangle $T \in J$, we prescribe the values $w(P_1), \ldots, w(P_6)$. Using (Q), we extend these values to a function $w: \overline{G} \to \mathbb{R}$. Then w has the following properties:

(i) w is continuous on $\overline{G}$.
(ii) w is piecewise differentiable, and $w \in W_p^1(G)$, $1 \leq p < \iota$.

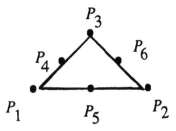

Figure 5.

Let $X^{(2)}(J)$ denote the set of all these piecewise quadratic polynomials. Moreover, we set

$$\mathring{X}^{(2)}(J) = \{w \in X^{(2)}(T) : w = 0 \text{ on } \partial G\}.$$

We now consider a sequence $\{J_n\}$ of triangulations with $h_n \to 0$ as $n \to \infty$ and $\inf_n \delta_n > 0$. Let $1 \le p < \infty$. Set $Y_n = X^{(2)}(J_n)$ and $\mathring{Y}_n = \mathring{X}^{(2)}(J_n)$. Then the following hold:

(a) $\{Y_n\}$ is a Galerkin–finite element scheme in $X = W_p^1(G)$.
(b) For all $w \in W_p^3(T)$, $\text{dist}_x(u, Y_n) \le \text{const } h_n^2 \| u \|_{3,p}$.
(c) $\mathring{Y}_n$ is a Galerkin–finite element scheme in $\mathring{X} = \mathring{W}_p^1(G)$.
(d) For all $u \in W_p^3(G) \cap \mathring{X}$, $\text{dist}_x(u, \mathring{Y}_n) \le \text{const } h_n^2 \| u \|_{3,p}$.

The constants in (b) and (d) depend only on G, p, and $\inf_n \delta_n$.

Note that in contrast to the piecewise linear finite elements above, we find here a higher-order accuracy of the approximation according to the appearance of h_n^2. Roughly speaking, it follows that we have the following situation in the case where $0 \le m \le k + 1$:

2.4. In the Sobolev space $W_p^m(G)$, the accuracy of approximation of $W_p^{k+1}(G)$-function by piecewise polynomial functions of degree k is proportional to

$$h^{k+1-m},$$

where h denotes the mesh size of the triangulation.

This result remains true for all bounded polyhedral regions in $\mathbb{R}^N$, $N \ge 1$, provided that $(k + 1)p > N$ and $1 \le p < \infty$.

3. LEBESGUE MEASURE, INTEGRAL, AND SOME IMPORTANT CONVERGENCE THEOREMS [108,304]

We recall that in the generalization of elementary geometrical notions (e.g., volumes of cuboids in $\mathbb{R}^N$) one associates in analogy with each set $M \subset \mathbb{R}^N$ a number $\text{meas}(M)$, called the *Lebesgue measure* of M and denoted by $\mu(M)$. These sets are called *measurable*. For example, open and closed sets are always measurable. Note that $\mu(M) = 0$ iff to each $\epsilon > 0$ there exists a countable number of N-dimensional intervals M_i and that $\cup_{i=1}^\infty M_i$ cover M and $\mu(\cup_{i=1}^\infty M_i) < \epsilon$.

A property P holds *almost everywhere* (a.e.) on M if P holds on all of M except for a set of measure zero.

3.1 Measurable Functions

$f: M \subseteq \mathbb{R}^N \to X$ is called a step function iff f is piecewise constant; that is, M is measurable and has finitely many pairwise disjoint subsets M_1,

... ,$M_n \subseteq M$ such that $f(x) = a_i$ for all $x \in M_i$, $M(A_i) < \infty$, $i = 1$, ... ,n; and $f(x) = 0$ otherwise.

The integral of the step function is defined by

$$\int_M f(x) \, dx \equiv \sum_{i=1}^{n} a_i \mu(M_i).$$

$f: N \subseteq \mathbb{R}^N \to X$, with values in the Banach space X, is called *measurable* iff the following hold:

(i) The domain of definition M is measurable.
(ii) There exists a sequence $\{f_n\}$ of step functions $f_n: M \to X$ such that

$$\lim_n f_n(x) = f(x) \qquad \text{for almost all} \quad x \in M.$$

Example $f: M \subset \mathbb{R}^N \to X$ is almost everywhere continuous $\Rightarrow f$ is measurable.

If $f, g, f_k: M \subset \mathbb{R}^N \to X$ are measurable, $k = 1, 2, 3, \ldots$; $\varphi, \psi: M \to \mathbb{R}$ are measurable, and $\lim_k f_k(x) \equiv h(x)$ exists for all $x \in M$, then $\varphi f + \psi g$, $x \to \| f(x) \|$, and h are measurable on M.

$f: M \subset \mathbb{R}^N \to X$ is measurable iff $x \to \langle g, f(x) \rangle$ is measurable on M for all $g \in X^*$.

If f is changed on a set of measure zero, it does not change the measurability of f.

Let $f: M \times U \to Y$, $(x, u) \to f(x, u)$, $M \subset \mathbb{R}^N$, be measurable with U and Y Banach spaces. If f satisfies the Carathéodory condition [i.e., $x \to f(x, u)$ is measurable on M for all $u \in U$, and $u \to f(x, u)$ is continuous on U for $x \in M$ a.e.], then $x \mapsto u(x)$ is measurable on $M \Rightarrow x \mapsto f(x, u)$ is measurable on M (cf. [136,280,304]).

We now extend the classical Riemann integral. $M \subset \mathbb{R}^N \to X$ is called *integrable* over M if M is measurable, there exists a sequence of step functions $f_n: M \to X$ so that $\lim_m f_n(x) = f(x)$ for $x \in M$ a.e. (i.e., f is measurable) and to each $\epsilon > 0$ there is $n_0(\epsilon)$ with $\int_M \| f_n(x) - f(x) \| \, dx < \epsilon \ \forall \ n, m \geq n_0(\epsilon)$.

(3.1a) The *Lebesgue integral* of an integrable function on M is defined by

$$\int_M f(x) \, dx \equiv \lim_n \int_M f_n(x) \, dx.$$

This integral is meaningful, the limit always exists, and the limit is independent of the choice of step functions in (3.ii).

Example $f: M \subset \mathbb{R}^N \to X$ is continuous a.e., $\mu(M) < \infty$, $\sup_M \| f(x) \| < \infty \Rightarrow \int_N f(x) \, dx$ exists.

(3.1b) *Majorant criterion.* All the following integrals exist and we have the estimates

$$\left\| \int_M f \, dx \right\| \leq \int_M \| f(x) \| \, dx \leq \int_M g \, dx$$

provided that the following two conditions hold:

(i) $\| f(x) \| \leq g(x)$ for $x \in M$ a.e., and $\int_M g \, dx$ exists.
(ii) $f : M \subset \mathbb{R}^N \to X$ is measurable.

(3.1c) *Norm criterion* (absolute integrability). Let $f: M \subseteq \mathbb{R}^N \to X$ be measurable. Then

$$\int_M f(x) \, dx \quad \text{exists iff} \quad \int_M \| f(x) \| \, dx \quad \text{exists.}$$

(3.1d) *Majorized convergence.* We have

$$\lim_n \int_M f_n \, dx = \int_M \lim_n f_n(x) \, dx,$$

where all integrals and limits exist, provided that the following conditions hold:

(i) $\| f_n(x) \| \leq g(x)$ for a.e. $x \in M$ for all n and $\int_M g \, dx$ exists.
(ii) $\lim_n f_n(x)$ exists for a.e. $x \in M$, where $f_n : M \subset \mathbb{R}^N \to X$ is measurable for all n.

This theorem of Lebesgue describes one of the most important properties of the integral and will be used frequently. The key condition is the majorant condition (i).

(3.1e) *Generalized majorized convergence.* Theorem (3.1d) remains true if we replace (i) with the following more general assumption:

$$\| f_n(x) \| \leq g_n(x) \qquad \text{for a.e.} \quad x \in M \quad \text{and all} \quad n \in Z^+.$$

All functions $g_n, g: M \to R^1$ are integrable and we have convergence $g_n \to g$ a.e. on M as $n \to \infty$ along with

$$\int_M g_n \, dx \to \int_M g \, dx \qquad \text{as} \quad n \to \infty.$$

(3.1f) *Monotone convergence.* We have

$$\lim_n \int_M f_n \, dx = \int_M \lim_n f_n(x) \, dx,$$

where all the integrals and limits exist, provided that the following two conditions are satisfied:

(i) $f_n: M \subseteq \mathbb{R}^N \to \mathbb{R}$ is integrable for all $n \in Z^+$.

(ii)   The sequence $\{f_n\}$ is *monotone* increasing (or monotone decreasing) and $\sup_n | \int_M f_n(x) \, dx | < \infty$.

(3.1g) *Lemma of Fatou.* We have

$$\int_M \varliminf_n f_n(x) \, dx \le \varliminf_n \int_M f_n(x) \, dx,$$

where all integrals exist, provided that the following two conditions hold:

(i) $f_n: M \subseteq \mathbb{R}$ is nonnegative and integrable for all $n \in Z^+$.

(ii) $\varliminf_n \int_M f_n(x) \, dx < \infty$.

This lemma generalizes the well-known relation

$$\varliminf_n a_n + \varliminf_n b_n \le \varliminf_n (a_n + b_n)$$

for nonnegative real numbers a_n and b_n.

(3.1h) *Absolute continuity of the integral.* Let $f: M \subset \mathbb{R}^N \to X$ be integrable. Then, for each $\epsilon > 0$, there exists a $\delta(\epsilon) > 0$ such that

$$\left\| \int_H f \, dx \right\| \le \int_H \| f(x) \| \, dx < \epsilon$$

holds for all subsets H of M with $\mu(H) < \delta(\epsilon)$.

(3.1i) *Absolute continuity and the convergence theorem of Vitali.* We have

$$\lim_n \int_M f_n(x) \, dx = \int_M \lim_n f_n(x) \, dx,$$

where all integrals and the left-hand limit exist, provided that four conditions are satisfied:

(i) $f_n: M \subseteq \mathbb{R}^N \to \mathbb{R}$ is integrable for all $n \in Z^+$.

(ii) $\lim_n f_n(x)$ exists for a.e. $x \in M$ and is finite.

(iii)   Equicontinuity: For each $\epsilon > 0$, there exists $\delta(\epsilon) > 0$ such that

$$\sup_n \int_H | f_n(x) | \, dx < \epsilon$$

holds for all subsets H of M with $\mu(H) < \delta(\epsilon)$.

(iv)   If $\mu(M) = \infty$, then for each $\epsilon > 0$, there exists a subset S of M such that $\mu(S) < \infty$ and $\sup_n \int_{M+S} | f_n(x) | < \epsilon$.

(3.1j) *Convergence theorem of Vitaly–Hahn–Saks.* Theorem (3.1i) remains true if we replace (iii) with the following weaker condition.

(iii*) $\lim_n \int_H f_n \, dx$ exists and is finite for al measurable subsets of M.

In addition, (iii*) implies (iii).

(3.1k) *Bounded convergence (BC) theorem.* Let f be a Lebesgue measurable function defined on a closed and bounded interval $[a,b]$ with norm $\| f \|_p$ and $\| f \|_\infty = $ ess sup$\{|f(x)| \mid a \le x < b\}$.

The bounded convergence theorem is a useful result from Lebesgue integration theory, which using the L_p norm notation, can be stated as follows:

BC theorem   If $f, f_1, f_2, \ldots$, is a sequence of measurable functions and K is a positive constant such that $\| f_n \|_\infty \le K$, $n = 1, 2, 3, \ldots$, and $\lim_n f_n(x) = f(x)$ a.e. $x \in [a,b]$, then $\lim_n \| f_n - f \|_p = 0$ for any $p \in (0,\infty)$.

Stated in this form the BC theorem can be extended to apply to sequences of functions that are uniformly bounded in the L_r-norm, $0 < r \le \infty$. Indeed, if $r \in (0,\infty]$ and f, $\{f_n\}$ are measurable functions such that $\| f_n \|_r \le K$ for $n = 1, 2, \ldots$, and $\lim_n f_n(x) = f(x)$ a.e. $x \in [a,b]$, then $\lim_n \| f_n - f \|_p = 0$ for any $p \in (0,r)$.

The last BC theorem is also valid if we replace $([a,b],m)$ by a complete finite measure space (X,μ) if we define $\| f \|_p = (\int_X | f |^p \, d\mu)^{1/p}$ and similarly extend the symbol $\| f \|_\infty$. The proof goes through without any change (see [304]). For further discussion of the BC theorem when $([s,b],m)$ is replaced by a nonfinite measure space (X,μ) (see [304]). This result is very powerful, yet its proof depends on three well-known results of real variables: Fatou's lemma, Egoroff's theorem, and Hölder's inequality.

(3.1l) *Convergence theorem of Egorov.* Each convergent sequence of measurable functions on a set of finite measure is *uniformly convergent* up to a small set; that is, let $\{f_n\}$ be a sequence of measurable functions $f_n: M \subset \mathbb{R}^N \to X$ with $\mu(M) < \infty$ such that $f_n(x) \to f(x)$ as $n \to \infty$ for $x \in M$ a.e.. Then the function $f: M \to X$ is measurable, and for each $\delta > 0$, there exists a subset M_δ of M such that $f_n(x) \to f(x)$ uniformly as $n \to \infty$ on $M - M_\delta$ and $\mu(M_\delta) < \delta$.

4. LEBESGUE SPACES $L_p(Q)$ ([304])

Let $Q \subset \mathbb{R}^N$ be a bounded domain with smooth boundary ∂Q and $1 \le p < \infty$. The Lebesgue space $L_p(Q)$ consists of measurable functions $u: Q \to \mathbb{R}$ such that $\| u \|_p \equiv (\int_Q | u(x) |^p \, dx)^{1/p} < \infty$.

$L_p(Q)$ is a separable Banach space with $C^\infty(Q)$ dense in $L_p(Q)$; $L_p(Q)$ is uniformly convex and thus reflexive for $p \in (1,\infty)$. The space $L_2(Q)$ is called a *Hilbert space* with the scalar product $(u,v) = \int_Q uv\, dx$.

4.1 Hölder Inequality and Other Inequalities

Let $1 < p_1 < p_2 < \cdots < p_N < \infty$ be given with $\sum_{i=1}^{N} p_i^{-1} = 1$, $u_i \in L_{p_i}(Q)$ for all i. Then

$$\left| \int_Q \prod_{i=1}^{N} u_i(x)\, dx \right| \leq \prod_{i=1}^{N} \left[\int_Q |u_i(x)|^p\, dx \right]^{1/p},$$

where all integrals exist.

In working with integral equations and elsewhere, one often has the following inequalities for nonnegative real numbers $\zeta_i, \ldots, \zeta_n$ (a, b, $c > 0$ and depend only on N, s, and r):

$$a\left(\sum_{i=1}^{N} \zeta_i^s \right)^{1/s} \leq \sum_{i=1}^{N} \zeta_i \leq b\left(\sum_{i=1}^{N} \zeta_i^s \right)^{1/s} \quad (1 \leq s < \infty) \tag{4.1a}$$

$$\left(\sum_{i=1}^{N} \zeta_i \right)^r \leq c \sum_{i=1}^{N} \zeta_i^r, \quad 0 < r < \infty, \tag{4.1b}$$

$$\zeta\eta \leq \frac{\zeta^p}{p} + \frac{\eta^q}{q}; \quad \prod_{i=1}^{N} \zeta_i \leq \sum_{i=1}^{N} \zeta^{p_i}/p_i \quad \text{(Young's inequality)} \tag{4.1c}$$

$$\sum_{i=1}^{N} \zeta_i\eta_i \leq \left(\sum_{i=1}^{N} \zeta_i^p \right)^{1/p} \left(\sum_{i=1}^{N} \eta_i^q \right)^{1/q} \quad \text{(Hölder's inequality)} \tag{4.1d}$$

Example From $u \in L_q(Q)$, $v_i \in L_{p_i}(Q)$, $1 \leq q$, $p_i < \infty$, and $|w(x)| \leq |u(x)| + \sum_{i=1}^{N} |v_i(x)|^{p_i/q}$ for $x \in Q$, it follows that $w \in L_q(Q)$ for measurable $w: Q \to \mathbb{R}$ and

$$\| w \|_q \leq \text{const} \left[\left(\| u \|_q + \sum_{i=1}^{N} \| v_i \|_{p_i}^{(p_i/q)} \right) \right]$$

(4.1e) $L_p(Q) \subset L_q(Q)$ for $1 \leq q < p$; $\| u \|_q \leq c_{p,q} \| u \|_p \ \forall\, u \in L_p(Q)$; in particular: $L_2(Q) > L_3(Q) \geq L_4(Q) \geq$.

4.2 Dual Space

Let $p^{-1} + q^{-1} = 1$, $1 < p, q < \infty$. For a fixed $u \in L_q(Q)$, the element $\bar{u}$ defined by $\bar{u}(v) = \int_Q uv\, dx$ for all $v \in L_p(Q)$ is a linear continuous functional on $L_p(Q)$, $\bar{u} \in (L_p(Q))^*$ and $\| \bar{u} \| = \| u \|_q$. All elements $\bar{u} \in$

$(L_p(Q))^*$ can be obtained in this way, where u is uniquely determined by $\tilde{u}$. If we identify $\tilde{u}$ with u, we can write

$$\langle u,v \rangle_X \equiv \langle \tilde{u},v \rangle_X \equiv \int_Q uv\, dx, \qquad \text{where} \quad X \equiv L_p(Q).$$

The equality $(L_p(Q))^* = L_q(Q)$ is to be understood in this sense.

(4.2a) *Convergent subsequences.* (i) $u_n \to u$ in $L_p(G)$ as $n \to \infty$, $1 \le p < \infty \Rightarrow$ there exists a subsequence $\{u_{n'}\}$ with $u_{n'}(x) \to u(x)$ for a.e. $x \in Q$.

(ii) $u_n \to u$ in $L_p(G)$ as $n \to \infty$, $1 \le p < \infty$ and $u_n(x) \to v(x)$ for a.e. $x \in Q \Rightarrow u(x) = v(x)$ for a.e. $x \in Q$.

5. WEAK AND WEAK* CONVERGENCE [313,314]

Let X be a Banach space with X^* and X^{**} denoting its first and second dual spaces, respectively.

A sequence $\{x_n\} \subset X$ is said to converge *strongly* to x_0 in X iff $\| x_n - x_0 \| \to 0$ as $n \to \infty$; in this case we write $x_n \to x_0$ in X.

A sequence $\{x_n\} \subset X$ is said to converge *weakly* to x_0 in X iff $f(x_n) \to f(x_0)$ in $\mathbb{R}$ for each $f \in X^*$; in this case we write $x_n \rightharpoonup x_0$ in X.

A sequence $\{f_n\} \subset X^*$ is said to converge weakly to f_0 in X^* iff $F(f_n) \to F(x_0)$ in $\mathbb{R}$ for each $F \in X^{**}$; in this case we write $f_n \rightharpoonup f_0$ in X^*.

A sequence $\{f_n\} \subset X^*$ is said to *converge weakly** to f in X^* iff $f_n(x) \to f(x)$ in $\mathbb{R}$ for all $x \in X$; and we write $f_n \rightharpoonup f_0$ weakly*.

Clearly, if $\{f_n\} \subset X^*$ is such that f_n converges weakly f_0 in X^*, then f_n converges weakly* to f_0. The converse is true if X is reflexive (i.e., $X = X^{**}$).

A Banach space S is said to be sequentially weakly complete if every weakly convergent sequence of X converges weakly to an element in X.

It is clear that when X is reflexive, X is weakly complete and, moreover, every bounded set in X is sequentially weakly compact. But the space $C(\overline{G})$ is not weakly complete and a bounded set in $C(\overline{G})$ is not sequentially weakly compact.

6. SOBOLEV SPACES $W_p^m(Q)$, $\mathring{W}_p^m(Q)$ [1,87,270,314]

Let $1 \le p < \infty$, $m = 0, 1, 2, \ldots$; and let $Q \subset \mathbb{R}^N$ ($N \ge 1$) be a bounded domain with smooth boundary. By a multi-index $\alpha = (\alpha_1, \ldots, \alpha_N)$, we understand an N-tuple of nonnegative integers $\alpha_1, \ldots, \alpha_N$. We set $|\alpha| = \alpha_1 + \cdots + \alpha_N$ and $D_u^\alpha = D_1^{\alpha_1} \cdots D_N^{\alpha_N} u$, that is, $D^{\alpha_u} = \partial^{|\alpha|_n}/\partial \zeta_1^{\alpha_1} \cdots \partial \zeta_N^{\alpha_N}$, where $x = (\zeta_1, \ldots, \zeta_N)$, and for $\alpha = 0$ we set $D^0 u = u$.

(6.1) Let u, v be two locally integrable functions defined on Q. We say that $D^\alpha u = v$ in the *weak sense* (and call v the αth *weak* or *generalized derivative* of u) if, for every $\varphi \in C_0^\infty(Q)$ (set of infinitely differentiable functions with compact support in Q),

$$\int_G u D^\alpha \varphi = (-1)^{|\alpha|} \int_G v\varphi \, dx.$$

It is easy to show that if u has a weak derivative $D^\alpha u$, and if $D^\alpha u$ has a weak derivative $D^\beta(D^\alpha u)$, then u has a weak derivative $D^{\beta+\alpha}u$ and $D^{\beta+\alpha}u = D^\beta(D^\alpha u)$. One can, also, derive other standard properties:

(6.2) The Sobolev space $W_p^m(G)$ for $p \in [1,\infty)$ is the set of all functions $u \in L_p(Q)$, which have weak (or generalized) derivatives up to order m such that $S^\alpha u \in L_p(G)$ for $|\alpha| \le m$. We set $W_p^0(G) = L_p(G)$. The space $W_p^m(G)$ together with the norm $\| \cdot \|_{m,p} \equiv (\sum_{|\alpha| \le m} \int_G |D^\alpha u|^p)^{1/p}$ is a real Banach space provided that we identify any two functions which differ on a set of N-dimensional Lebesgue measure zero. The space $W_p^m(G)$ is separable for $p \in [1,\infty)$, for $p \in (1,\infty)$ it is uniformly convex and thus reflexive, and $C^\infty(G)$ is dense in $W_p^m(G)$ for ∂G sufficiently smooth.

The space $\mathring{W}_p^m(G)$ denotes the closure of $C_0^\infty(G)$ in $W_p^m(G)$. Explicitly, this means the following. Let $u \in W_p^m(G)$. Then u belongs to $W_p^m(G)$ iff there exists a sequence $\{u_n\} \subset C_0^\infty(G)$ with $\| u_n - u \|_{m,p} \to 0$ as $n \to \infty$. $\mathring{W}_p^m(G)$ is a Banach space with the norm $\| \cdot \|_{m,p}$ and $\mathring{W}_2^m(G)$ is a Hilbert space with the scalar product $(\cdot,\cdot)_{m,2}$.

Suppose that $X \subset Y$. The embedding operator $E: X \to Y$ assigns to each $x \in X$ the "same" element $x \in Y$. The embedding is continuous (compact) when E is continuous (compact).

(6.3) *Sobolev embedding theorem* (see [87,313,314]. Suppose $G \subset \mathbb{R}^N$ is a bounded domain with the boundary ∂G sufficiently smooth (e.g., $\partial G \in C^{0,1}$ means that ∂G is piecewise smooth).

Then: (1) The embeddings $W_p^k(G) \subset W_q^j(G)$, $\mathring{W}_p^k \subset \mathring{W}_p^k$ are *continuous* if $d \equiv p^{-1} - N^{-1}(k - j) \le q^{-1}$ and completely continuous (i.e., compact) if $d < q^{-1}$.

(2) The embeddings $W_p^k(G) \subset C^{j+\alpha}(\overline{G}) \subset C^j(\overline{G})$ are compact for $p^{-1} - N^{-1}(k - j - \alpha) < 0$, where $C^{j+\alpha}(\overline{G})$ is the Hölder space defined by $\{f \in C^j(G): \| f \|_{j+\alpha} < \infty\}$, where for $\alpha \in (0,1]$, we define the Hölder constant $H_\alpha(f)$ by

$$H_\alpha(f) = \sup\{(f(p) - f(q))/\| p - q \| : p,q \in G, p \ne q\}$$

and set

$$\| f \|_{j+\alpha} = \sum_{|i|=j} \sup |D^i f(p)| + \sum_{|i|=j} H_\alpha(D^i f).$$

We note in passing that if β is such that $\theta < \beta < \alpha \le 1$, then bounded sets in $C^{j+\alpha}(G)$ are relatively compact in $C^{j+\beta}(G)$.

In (1) there is no need to assume any condition on ∂G when $\mathring{W}_p^k(G) \subset \mathring{W}_p^j(G)$. It should be noted that the Sobolev embedding theorem is essential for the study of ODEs and PDEs.

References

1. Adams, R. A., *Sobdev Spaces,* Academics Press, New York (1975).

2. Agmon, S., *Lectures on Elliptic Boundary Value Problems,* Mathematical Studies, Van Nostrand Reinhold, Princeton, N.J. (1967).

3. Aldarwish, A., and Langenbach, A., Lösung der asymptotischen Bifur-kationsprobleme für parametrische Gleichungen, Math. Nachr. 65 (1975), 47–58.

4. Alexander, J. C., and Fitzpatrick, P. M., Galerkin approximations in sev-eral parameter bifurcation problems, Math. Proc. Cambridge Philos. Soc. 87 (1980), 489–900.

5. Asplund, A., Averaged norms, Israel J. Math. 5 (1967), 227–233.

6. Altman, M., A fixed point theorem in Banach spaces, Bull. Acad. Polon. Sci. Cl. III 5 (1957), 19–22.

7. Amann, H., Ein Existenz und Eindeutigkeit für die Hammerstensche Glei-chung in Banach–Räuanen, Math. Z. 111 (1969), 175–190.

8. Amann, N., Fixed points of asymptotically linear maps in ordered Banach spaces, J. Funct. Anal. 14 (1973), 162–171.

9. Amann, H., Ambrosetti, A., and Mancini, G., Elliptic equations with non-invertable Fredholm linear part and bounded nonlinearities, Math. Z. 158 (1979), 179–194.

10. Banaš, J., and Goebel, K., *Measure of Noncompactness in Banach Spaces*, vol. 60, Marcel Dekker, New York (1980).

11. Belluce, L. P., and Kirk, W. A., Fixed point theorems for certain classes of nonexpansive mappings, Proc. AMS 20 (1969), 141–146.

12. Benavides, T. D., Some properties of the set and ball-measure of noncompactness and applications, J. London Math. Soc. 34 (1986), 120–128.

13. Benavides, T. D., Set-contractions and ball-contractions in L^p-spaces, J. Math. Anal. Appl. 159 (1991), 505–506.

14. Beurling, A., and Livingston, A. E., A theorem on duality mappings in Banach spaces, Ark. Mat. 4 (1962), 405–411.

15. Borysovich, Yu. G., Zviagin, V. G., and Sapranov, Yu. I., Nonlinear Fredholm transformations and the Leray–Schauder theory, Uspehi Mat. Nauk 12 (1977), 3–54.

16. Brezis, H. R., Équations et inéquations non-linéaires dans les espaces vectoriels et dualité, Ann. Inst. Fourier (Grenoble) 18(1) (1968), 115–175.

17. Brezis, H. R., Quelgues properties des opérateurs monotones et des semi-groups non linéaires, Lect. Notes Math. 543 (1976), 58–82.

18. Brezis, H., and Nirenberg, L., Characterization of the ranges of some nonlinear operators and applications to BV problems, Ann. Scuola Norm. Sup. Pisa 5 (1978), 225–326.

19. Brouwer, L. E. J., Über Abbildung von Mannigfaltigkeiten, Math. Ann. 71 (1912), 97–115.

20. Browder, F. E., On the spectral theory of elliptic differential operators, Math. Ann. 142 (1961), 22–130.

21. Browder, F. E., Topological methods for nonlinear elliptic equations of arbitrary order, Pacific J. Math. 17 (1966), 17–31.

22. Browder, F. E., Functional analysis and partial differential equations II, Math. Ann. 145 (1962), 81–226.

23. Browder, F. E., Semicontractive and semiaccretive mappings in Banach spaces, Bull. AMS 74 (1968), 660–665.

24. Browder, F. E., Nonlinear equations of evolution and nonlinear accretive operators in Banach spaces, Bull. AMS 73 (1967), 867–874.

25. Browder, F. E., Existence and uniqueness theorems for solutions of nonlinear BV problems, Proc. Symp. Appl. Math. AMS 17 (1965), 24–49.

26. Browder, F. E., Mapping theorems for noncompact nonlinear operators in Banach spaces, Proc. Natl. Acad. Sci. USA 54 (1965), 337–342.

27. Browder, F. E., The solvability of nonlinear functional equations, Duke Math. J. 30 (1963), 557–566.

28. Browder, F. E., Nonexpansive nonlinear operators in a Banach space, Proc. Natl. Acad. Sci. USA 54 (1965/66), 1041–1044.

29. Browder, F. E., Nonlinear accretive mappings in Banach spaces, Bull. AMS 73 (1967), 470–476.

30. Browder, F. E., Nonlinear elliptic BV problems, Bull. AMS 69 (1963), 862–874.

31. Browder, F. E., Fixed point theorems for nonlinear semicontractive mappings in Banach spaces, Arch. Rational Mech. Anal. 21 (1966), 259–269.

32. Browder, F. E., Nonlinear elliptic BV problems and the generalized topological degree, Bull. AMS 78 (1970), 999–1005.

33. Browder, F. E., Existence theorems for nonlinear partial differential equations, Proc. Symp. Pure Math. AMS 16 (1970), 1–60.

34. Browder, F. E., On a theorem of Beurling and Livingston, Canad. J. Math 17 (1965), 367–372.

35. Browder, F. E., Nonlinear eigenvalue problems and Galerkin approximations, Bull. AMS 74 (1968), 651–656.

36. Browder, F. E., Nonlinear operators and nonlinear equations of evolution in Banach spaces, Proc. Symp. Pure Math. 18, part II, AMS, Providence, R.I., (1976); in mimeographed form, the book appeared in 1970.

37. Browder, F. E., The degree mapping and its generalizations, Contemp. Math. AMS 21 (1983), 15–40.

38. Browder, F. E., and De Figueiredo, D., J-Monotone nonlinear operators in Banach spaces, Proc. K. Ned. Akad. Amsterdam 28 (1966), 412–420.

39. Browder, F. E., and Hess, P., Nonlinear mappings of monotone type in Banach spaces, J. Funct. Anal. 11 (1972), 251–274.

40. Browder, F. E., and Nussbaum, R. D., The topological degree for noncompact compact mappings in Banach spaces, Bull. AMS 74 (1968), 671–676.

41. Browder, F. E., and Petryshyn, W. V., The topological degree and Galerkin approximations for noncompact operations in Banach spaces, Bull. AMS 74 (1968), 641–648.

42. Browder, F. E., and Petryshyn, W. V., Approximation methods and the generalized topological degree for nonlinear mappings in Banach spaces, J. Funct. Anal. 3 (1969), 217–245.

43. Calderon, A. P., Intermediate spaces and interpolation, Conference at Warsaw (1960).

44. Calvert, B., and Webb, J. R. L., An existence theorem for quasimonotone operators, Accad. Nat. Lincei Ser. 8 50 (1971), 362–368.

45. Calvert, B., and Gupta, C. P., Nonlinear elliptic BV problems in L^P-spaces and the sum of ranges of accretive operators, Nonlinear Anal. 2 (1978), 1–26.

46. Canfora, A., La teoría del grado topological per una classe di operatóri non compatti in spàzi di Hilbert, Ricerche Mat. 28 (1979), 109.

47. Cesari, L., Alternative methods in nonlinear analysis, in *"International Conference on Differential Equations"* (H. A. Antosiewicz, ed.), Academic Press, New York (1975), 95–148.

48. Cesari, L., in *Functional Analysis and Differential Equations* (L. Cesari, R. Kannan, and J. D. Schieur, eds.), vol. 19, Marcel Dekker, New York (1976).

49. Chandler, C., and Gibson, A. G., A two-Hilbert space formulation of multiscattering theory, Proc. Conf. on Math Methods and Applications of Scattering Theory, Washington, D.C. (1979), 134–148.

50. Chang, S. H., Periodic solutions of $(a(t)x')' + f(t,x) = p(t)$, Q. J. Math. 27 (1976), 497–508.

51. Conjura, E., and Petryshyn, W. V., Extension of nonlinear densely defined operators, rates of convergence for the error and the residual, and application to differential equations, J. Math. Anal. Appl. 64 (1978), 651–694.

52. Conti, G., and De Pacscali, E., Remark on surjectivity of quasibounded P-compact maps, Rend. Ist. Mat. Univ. Trieste 8 (1976), 167–171.

53. Crandall, M. G., and Pazy, A., Semi-groups of nonlinear contractions and dissipative maps, J. Funct. Anal. 3 (1969), 376–418.

54. Crandal, M. G., and Rabinowitz, P. H., Bifurcation from simple eigenvalues, J. Funct. Anal. 8 (1971), 321–340.

55. Cronin, J., *Fixed Point and Topological Degree in Nonliner Analysis,* Mathematical Surveys, AMS, Providence, R.I. (1964).

56. Dancer, E. N., Some remarks on a theorem of Kachurowskii, J. Math. Anal. Appl. 62 (1978), 525–529.

57. Dancer, E. N., Bifurcation theory in real Banach spaces, Proc. London Math. Soc. 23 (1971), 699–734.

58. Danes, J., On densifying and relating mappings and their application to nonlinear functional analysis, Proc. Summer School, Berlin (1974), 15–56.

59. Darbo, G., Punti uniti in transformazióni a condomìnio non compatto, Rend. Sem. Mat. Univ. Padova, 24 (1955), 84–92.

60. Deimling, K., Remarks on P_γ-compact operators, Math. Ann. 189 (1970), 185–190.

61. Deimling, K., Fixed points of generalized P-compact operators, Math. Z. 115 (1970), 188–196.

62. Deimling, K., *Nonlinear Functional Analysis*, Springer-Verlag, New York, 1985.

63. Deimling, K., Zeros of accretive operators, Manuscripta Math. 13 (1974), 365–374.

64. Dolph, C. L., and Minty, G. J., *On Nonlinear Integral Equations of the Hammerstein Type*, Univ. of Wisconsin Press, Madison, Wis. (1964), 99–154.

65. Dugundji, J., An extension of Tietze's theorem, Pacific J. Math. 1 (1951).

66. Dunford, N., and Schwartz, J. T., *Linear Operators*, Interscience, New York, part I (1958), part II (1963).

67. Dubinskii, Yu. A., Nonlinear elliptic and parabolic equations of arbitrary order, Uspehi Mat. Nauk 139 (1968), 45–90.

68. Džiškariani, A. V., The least square and Boknow–Galerkin methods, Ž. Vyčisl. Mat. i Mat. Fiz. 8 (1968), 1110–1116.

69. Edelstein, M., The construction of asymptotic center with a fixed point property, Bull. AMS 78 (1972), 206–208.

70. Edmunds, D. E., and Webb, J. R. L., A Leray–Schauder theorem for a class of nonlinear operators, Math. Ann. 182 (1969), 207–212.

71. De Figueiredo, D. G., On the range of nonlinear operators with linear asymptotes which are not invertible, Comment. Math. Univ. Carolin. 15 (1974), 415–428.

72. De Figueiredo, D. G., Fixed point theorems for nonlinear operators and Galerkin approximations, J. Differential Equations 3 (1967), 271–281.

73. Filippov, V. M., *Variational Principles for Nonpotential Operators*, Transl. Math. Monographs, vol. 77, AMS, Providence, R.I. (1989).

74. Fitzpatrick, P. M., A generalized degree for uniform limits of A-proper mappings, J. Math. Anal. Appl. 35 (1971), 536–552.

75. Fitzpatrick, P. M., Surjectivity results for nonlinear mappings from a Banach space to its dual, Math. Z. 204 (1973), 177–188.

76. Fitzpatrick, P. M., On the structure of the set of solutions of equations involving A-proper mappings, Trans. AMS 189 (1974), 107–131.

77. Fitzpatrick, P. M., Existence results for equations involving noncompact perturbations of Fredholm mappings with applications to differential equations, J. Math. Anal. Appl. 61 (1978), 157–177.

78. Fitzpatrick, P. M., The stability of parity and global bifurcation via Galerkin approximations, J. London Math. Soc. (2) 38 (1988), 153–165.

79. Fitzpatrick, P. M., On nonlinear perturbations of linear second order elliptic boundary value problems, Math. Proc. Cambridge Philos. Soc. (1978), 143–157.

80. Fitzpatrick, P. M., Homotopy, linearization, and bifurcation, Nonlinear Anal. 12 (1988), 171–184.

81. Fitzpatrick, P. M., A-proper mappings and their uniform limits, Ph.D. thesis, Rutgers University, New Brunswick, N.J. (1971).

82. Fitzpatrick, P. M., and Petryshyn, W. V., Galerkin methods in the constructive solvability of nonlinear Hammerstein equations with applications to differential equations, Trans. AMS, 38 (1978), 477–500.

83. Fitzpatrick, P. M., and Petryshyn, W. V., Some applications of A-proper mappings, Nonlinear Anal. 3 (1979), 525–537.

84. Fitzpatrick, P. M., and Pejsachowicz, J., An extension of the Leray–Schauder degree for fully nonlinear elliptic problems, Proc. Symp. Pure Math. 45 (I), AMS, Providence, R.I. (1986), 425–439.

85. Fitzpatrick, P. M., Massabo, I., and Pejsachovicz, J., On the covering dimension of the set of solutions of some nonlinear equations, Trans. AMS 296 (1986), 777–798.

86. Förste, J., Zur Diffusion in einen fallenden Flüssigkeitefilm, Monatsber. Dtsch. Akad. Wiss. Berlin 8(H) 12 (1968).

87. Friedman, A., *Partial Differential Equations*, Holt, Rinehart and Winston, New York (1969).

88. Friedrichs, K. O., Spectral Theorie halbbeschränkter Operatoren, I–III, Math. Ann. 109 (1934) 465–487, 685–713; 110 (1935) 777–779.

89. Frum-Ketkov, R. L., On mappings of the sphere of a Banach space, Soviet Math. Dokl. 8 (1967), 1004–1006.

90. Fučik, S., Fredholm alternative for nonlinear operators in Banach spaces and its applications to differential and integral equations, Časopis Pěst. Mat. 96 (1971), 371–390.

91. Fučik, S., and Kufner, A., *Nonlinear Differential Equations*, Elsevier, New York (1980).

92. Fučik, S., Kučera, M., and Nečaš, J., Ranges of nonlinear asymptotically linear operators, J. Differential Equations 17 (1975), 375–394.

93. Funk, P., *Variationsrechnung und ihre Anwendung in Physik and Technik,* Springer-Verlag, Berlin (1962).

94. Furi, M., and Vignoli, A., On α-nonexpansive mapping and fixed points, Rend. Accad. Naz. Lincei, 482 (1970), 135–198.

95. Furi, M., Martelli, M., and Vignoli, A., On the solvability of nonlinear operator equations in normal spaces, Ann. Mat. Pura Appl. (4) 124 (1980), 321–343.

96. Furi, M., Martelli, M., and Vignoli, A., Stably-solvable operators in Banach spaces, Atti Accad. Naz. Rend. Cl. Sci. Fis. Mat. Natur. 60 (1976), 21–26.

97. Goebel, K., An elementary proof of the fixed point theorem of Browder and Kirk, Michigan Math. J. 16 (1969), 381–383.

98. Gaines, R. E., and Mawhin, J. L., *Coincidence Degree and Nonlinear Differential Equations,* Lecture Notes 568, Springer-Verlag, Berlin (1977).

99. Gajewski, H., Über eine Klasse nichtlinearen Gleichungen mit monotonen Operatoren, Math. Nachr. 40 (1969), 357–366.

100. Gajewski, H., Gröger, K., and Zacharias, K., *Nichtlineare Operator-Gleichungen und Operatordifferential-Gleichunger,* Akademic-Verlag, Berlin (1974).

101. Gohberg, I. C., Goldstien, L. S., and Markus, A. S., Investigation of some properties of bounded linear operators in connection with their q-norms, Uch. Zap. Kishinev. Univ. Tr. 29 (1957), 29–36.

102. Göhde, D., Zum Prinzip der kontraktiven Albildung, Math. Nachr. 30 (1965), 251–258.

103. Goldenstein, L. S., and Markus, A. S., On a measure of noncompactness of bounded sets and linear operators. *Studies in Algebra and Mathemtical Analysis,* Izdat. Karta Moldav., Kishinev, Moldavia (1965).

104. Gärding, L. Dirichlet's problem for linear elliptic partial differential equations, Math. Scand. 1 (1953), 55–72.

105. Granas, S., On a class of nonlinear mappings in Banach spaces, Bull. Acad. Polon. Sci. Cl. III 59 (1957), 867–870.

106. Granas, S., On a class of nonlinear mappings in Banach spaces, Bull. Acad. Polon. Sci. Cl. III 9 (1975), 867–870.

107. Granas, A., Guenther, R. B., and Lee, J. W., On the theorem of S. Berstein, Pacific J. Math. 75 (1978), 67–82.

108. Graves, L. M., *The Theory of Functions of Real Variables,* McGraw-Hill, New York (1956).

109. Hagen-Torn and Michlin, S. G., On the solvability of the Ritz nonlinear system, Dokl. Akad. Nauk SSSR 73 (1950), 1121–1124.

110. Hale, J. K., Ordinary differential equations, Wiley-Interscience, New York (1969).

111. Hardy, G. H., Littlewood, J. E., and Polya, G., *Inequalities*, Cambridge Univ. Press, London (1964).

112. Heinz, E., An elementary analytic theory of the degree of a mapping in N-dimensional space, J. Math. Mech. 8 (1959), 231–247.

113. Hess, P., On nonlinear mappings of monotone type homotopic to odd mappings, J. Funct. Anal. 11 (1972), 138–167.

114. Hess, P., On a theorem of Landesman and Lazer, Indiana Univ. Math. J. 2 (1974), 822–830.

115. Hetzer, G., Some remarks of Φ_+-operators and on the coincidence degree for Fredholm equations with noncompact linear perturbation, Ann. Soc. Sci. Bruxelles Ser. 189 (1975), 497–508.

116. Hetzer, G., A continuation theorem and the variational solvability of quasilinear elliptic BV problems, Nonlinear Anal. 4 (1980), 773–780.

117. Jascova, G. N., and Jakovlev, M. N., Some condition for the stability of the Petrov–Galerkin method, Trudy Mat. Inst. Steklov. 66 (1962), 182–189.

118. Kachurovskii, R. I., On the Fredholm theory for nonlinear operator equations, Dok. Akad. Nauk SSR 192 (1970), 969–972.

119. Kachurovskii, R. I., Nonlinear monotone operators in Banach spaces, Uspehi Mat. Nauk 23 (1968), 121–168.

120. Kačanow, L. M., *Foundation of the Theory of Plasticity* (Engl. Transl.), American Elsevier, New York (1971).

121. Kaniel, S., Quasi-compact nonlinear operators in Banach spaces and applications, Arch. Rational Mech. Anal 20 (1965), 259–278.

122. Kannan, R., and Schur, J., Boundary value problems for even order nonlinear differential equations, Bull. AMS, 82 (1) (1976).

123. Kantorovich, L. V., Functional analysis and applied mathematics, Uspehi Mat. Nauk 28 (1948), 89–185.

124. Kadec, M. I., Spaces isomorphic to a locally uniformly convex space, Izv. Vysš. Učebn. Zaved. Mat. 13 (1959), 51–57.

125. Kartsatos, A. G., and Parrot, M. E., Global solutions of functional evolution equations involving locally defined Lipschitzian perturbations, J. London Math. Soc. 27 (1983), 306–316.

126. Kato, T., Accretive oeprators and nonlinear evolution equations in Banach spaces, Proc. Symp. Pure Math., vol. 28(I) AMS, Providence, R.I. (1968), 138–161.

127. Kato, T., Demicontinuity, hemicontinuity, and monotonicity, Bull, AMS, 73 (1967), 886–889.

128. Kato, T., Nonlinear semigroups and evolutions equations, J. Math. Soc. Japan, 19 (1967), 508–519.

129. Keller, J. B., and Antman, S. (eds.), *Bifurcation Theory and Nonlinear Eigenvalue Problems*, W. A. Benjamin, New York (1969).

130. Kellogg, R. B., Li, T-Y., and Yorke, J., A constructive proof of the Brouwer fixed point theorem and computation results, SIAM, Num. Anal. 13 (1976), 473–483.

131. Kirk, W. A., A fixed point theorem for mappings which do not increase distances, Amer. Math. Monthly 72 (1965), 1004–1006.

132. Kirk, W. A., On nonlinear mappings of strongly semicontractive type, J. Math. Anal. Appl. 27 (1969), 409–412.

133. Kirk, W. A., Mappings of generalized contractive type, J. Math. Anal. Appl. 32 (1970), 567–572.

134. Kirk, W. A., and Morales, C., Condensing maps and the Leray–Schauder boundary conditions, Nonlinear Anal. 3 (1979), 533–538.

135. Kirk, W. A., and Morales, C., Fixed point theorems for local strong pseudo-contractions, Nonlinear Anal. 4 (1980), 363–368.

136. Krasnoselskii, M. A., *Topological Methods in the Theory of Nonlinear Integral Equations*, Pergamon Press, Oxford (1964).

137. Krasnoselskii, M. A., Two remarks on the method of successive approximations, Uspehi Mat. Nauk 10 (1955), 123–127.

138. Krasnoselskii, M. A., and Sobolevskii, P. E., Structure of the set of solutions of an equation of parabolic type, Ukrainian Math. J. 16 (1964), 319–333.

139. Krasnoselskii, M. A., Vainikko, G. M., Zabreiko, P. P., Ruticky, Ja. B., and Stecenko, V. J., *Approximate Solution of Operator Equations*, Nauka, Moscow (1969).

140. Krauss, E., A degree for operations of monotone type, Math. Nachr. 114 (1983), 53–62.

141. Kravchuk, M. P., *Application of the Method of Moments for the Solvability of Linear Differential and Integral Equations*, vols. 1 and 2, Ukrainian Akademic Press, Kiev (1932/36).

142. Kröger, H., and Perne, R., A theorem on A-proper mappings and its application to scattering theory, J. Math. Phys. 23 (1982), 715–719.

143. Krylov, M. M., *Fundamental Problems in Mathematical Theory and Technology: Scientific Study in Applied Mathematics,* Ukrainian Akademic Press, Kiev (1932).

144. Kuratowski, K., Sur les espaces complet, Fund. Math. 15 (1930), 301–309.

145. Ladyženskaya, O. A., On integral estimates, convergence, approximation methods, and solution of in functionals for elliptic equations, Vestnik Leningrad Univ. 13 (7) (1958), 60–69.

146. Laško, A. D., On the convergence of the methods of Galerkin type, Dokl. Akad. Nauk SSSR 120 (1958), 242–244.

147. Lami Dozo, E., Opérateurs non-expansifs, P-compacts et propriétés géométriques de la norme, Doctoral thesis, Université Libre de Bruzelles (1969/70).

148. Landesman, E. M., and Lazer, A. C., Nonlinear perturbations of linear elliptic BV problems at resonance, J. Math. Mech. 19 (1970), 609–623.

149. Langebach, A., Über nichtlineare Gleichungen mit differenzierbarten Regularizatoren und Verzweingungsprobleme, Math. Nachr. 34 (1967).

150. Lax, P. D., and Milgram, A. N., Parabolic equations, in *Contributions to the Theory of PDE's,* Ann. Math. Studies 33, Princeton Univ. Press, Princeton, N.J. (1954), 167–190.

151. Leray, J., Topologie der espaces abstraits de M. Banach, C. R. Acad. Sci. Paris 200 (1935), 1082–1084.

152. Leray, J., and Lions, J. L., Quelques résultats de Višik sur les probléms elliptiques non linéaires par les méthodes de Minty–Browder, Bull. Soc. Math. Fr. 93 (1965), 97–107.

153. Leray, J., and Schauder, J., Topologie et équations functionnelles, Ann. Sci., Ecol. Norm. Sup. Sér. 15 (1934), 45–78.

154. Lions, J. L., *Quelgues méthodes des résolution des probléms aux limites non linéaires,* Dunod Gauthier-Villons, Paris (1969).

155. Lloyd, N. G., *Degree Theory,* Cambridge Univ. Press, Cambridge (1978).

156. Luchka, A. Yu., *Projection-Iterative Methods for the Solvability of Differential and Integral Equations,* Naukova Dumka, Kiev (1980).

157. Luchka, A. Yu., and Luchka, T. F., *On the Origin and the Development of Direct Methods of Mathematical Physics,* Naukova Dumka, Kiev (1985).

158. Massabo, I., and Nistri, P., A topological degree for multivalued A-proper maps in Banach spaces, Boll. Un. Mat. Ital. 13-B (1976), 672–685.

159. Martelli, M., and Vignoli, A., Eigenvectors and surjectivity for α-Lipschitz mappings in Banach spaces, Ann. Mat. Pura Appl. (4) 94 (1972), 1–9.

160. Marti, J. T., *Introduction to the Theory of Bases*, No. 18, Springer-Verlag, Berlin (1970).

161. Martin, R. H., Jr., *Nonlinear and Differential Equations of Evolution in Banach Spaces*, Wiley Interscience, New York, (1976).

162. Martyniuk, A. E., Some new application of the Galerkin type methods, Mat. Sb. 49 (1959), 85–108.

163. Mawhin, J., Equivalence theorems for nonlinear operator equations and coincidence degree for some mappings in locally topological vector spaces, J. Differential Equations 12 (1972), 610–636.

164. Medvedev, V. A., On the convergence of the Bubnuv–Galerkin method, Prikl. Mat. Meh. 27 (1963), 1148–1151.

165. Mikhlin, S. G., *Variationsmethoden der mathematischen Physik*, Akademie-Verlag, Berlin (1962).

166. Mikhlin, S. G., *The Numerical Performance of Variational Methods*, Walters-Noordhoff, Groningen, The Netherlands (1971).

167. Mikhlin, S. G., On the method of Ritz for nonlinear problems, Dokl. Akad. Nauk SSSR 142 (1962), 792–793.

168. Mikhlin, S. G., On the stability of certain computational methods, Dokl. Akad. Nauk SSSR 157 (1964), 271–273.

169. Milojević, P. S., The solvability of operator equations with asymptotic quasibounded nonlinearity, Proc. AMS 76 (1979), 293–298.

170. Milojević, P. S., Solvability of oeprator equations involving nonlinear perturbations of Fredholm maps with nonnegative index and applications, vol. 957, Springer, (1982), 212–228.

171. Milojević, P. S., A generalization of the Leray–Schauder theorem and surjectivity results for multivalued A-proper and pseudo-A-proper mappings, Nonlinear Anal. 1 (1977), 263–276.

172. Milojević, P. S., On the solvability and continuation type results for nonlinear equations with applications, I, Proc. 3rd International Symposium on Topology and Its Application, Belgrade (1977), 468–485.

173. Milojević, P. S., On the solvability and continuation type results for nonlinear operator equations with applications, II, Canad. Math. Bull. 25 (1982), 98–109.

174. Milojević, P. S., On the index and the covering dimension of the solution set of semilinear equations, Proc. Symp. Pure Math. AMS, 45(2) (1986), 183–205.

175. Milojevič, P. S., and Petryshyn, W. V., Continuation and surjectivity theorems for uniform limits of A-proper mappings with applications, J. Math. Anal. Appl. 62 (1978), 368–400.

176. Milojevič, P. S., and Petryshyn, W. V., Continuation theorems and the approximation-solvability of equations involving multivalued A-proper mappings, J. Math. Anal. Appl. 6 (1977), 658–692.

177. Minty, G., On the "monotonicity" method for the solution of nonlinear equations in Banach spaces, Proc. Natl. Acad. Sci. USA 50 (1963), 1038–1041.

178. Minty, G., Monotone (nonlinear) operators in Hilbert space, Duke Math. J. 29 (1962), 241–346.

179. Morales, C., Pseudo-contractive mappings and the Leray–Schauder boundary conditions, Comment. Math Univ. Carolin. 24 (1979), 745–756.

180. Morales, C., Nonlinear equations involving m-accretive operators, J. Math. Anal. Appl. 97 (1983), 329–336.

181. Morawetz, C. (ed.), *Frontiers of Mathematical Sciences*, Comm. Pure Appl. Math. 39, Suppl. 5 (1986).

182. Nashed, M. Z., Generalized inverse mapping theorems and related applications of generalized inverse in nonlinear analysis, in *Nonlinear Equations in Abstract Spaces* (V. Lakshmikantham, ed.), Academic Press, New York (1978), 217–252.

183. Nashed, M. Z., Differentiability and related properties of operators, in *Nonlinear Functional Analysis and Applications* (L. Rall, ed.), Academic Press, New York (1971), 103–309.

184. Nashed, M. Z., A retrospective and prospective survey of generalized inverses of operators, Math. Res. Center Rep. 1125, Univ. of Wisconsin, Madison, Wis. (1971).

185. Nashed, M. Z., and Wong, J. S., Some variants of a fixed point theorem of Krasnoselsky and applications to nonlinear integral equations, J. Math. Mech. 18 (1969), 767–777.

186. Nečaš, J., On the range of nonlinear operators with linear asymptotes which are not invertible, Comment. Math. Univ. Carolin. 14 (1973), 63–72.

187. Nečaš, J., Fredholm alternative for nonlienar operators and applications to PDE's and integral equations, Časopis Pěst. Mat. 9 (1972), 65–71.

188. Nirenberg, L., An application of generalized degree to a class of nonlinear problems, Trois, Coll. Anal. Functionalle, Liege (1970) (Math. Vander, 57–74).

189. Nirenberg, L., *Topics in Nonlinear Functional Analysis*, Lecture Series, New York University, New York (1974).

190. Nirenberg, L., Variational and topological methods in nonlinear problems, Bull. AMS 4(3) (1981), 267–302.

191. Nussbaum, R. D., The fixed point index and fixed point theorems for k-set-contractions, Ph.D. thesis, University of Chicago (1969).

192. Nussbaum, R. D., The fixed point index for local condensing maps, Ann. Mat. Pura Appl. 89 (1971), 217–258.

193. Nussbaum, R. D., Degree theory for local condensing maps, J. Math. Anal. Appl. 37 (1972), 741–766.

194. Nussbaum, R. D., The radius of the essential spectrum, Duke Math. J. 38 (1970), 473–478.

195. Nussbaum, R. D., Estimates for the number of solutions of operator equations, Applicable Anal. 1 (1971), 183–200.

196. Nussbaum, R. D., Global bifurcation theorem with application to functional differential equations, J. Funct. Anal. 19 (1975).

197. O'Neil, T., and Thomas, J. W., On the equivalence of multiplicity and the generalized topological degree, Trans. AMS 167 (1972), 333–344.

198. Pascali, D., A bifurcation theory involving A-proper mappings, Libertas Math. 7 (1987), 47–58.

199. Petrov, G. I., Application of the Galerkin method to the stability problems of viscous fluid, Appl. Math. Mech. 3 (1940).

200. Petry, W., Existence theorems for a class of nonlinear operator equations. J. Math. Anal. Appl. 43 (1973), 250–260.

201. Petry, W., Ein Leray–Schauder Satz mit Anwendunger auf verallgemeinern Hammersteinsche Gleichungen, Math. Nachr. 48 (1970), 49–68.

202. Petryshyn, W. V., On a fixed point theorem for nonlinear P-compact operators in Banach spaces, Bull. AMS 72 (1966), 329–334.

203. Petryshyn, W. V., Fixed point theorems involving P-compact, semicontractive, and accretive operators not defined on all of Banach space, J. Math. Anal. Appl. 23 (1968), 336–354; see also 15 (1966), 228–242.

204. Petryshyn, W. V., Projection methods in nonlinear numerical functional analysis, J. Math. Mech. 17 (1967), 352–372.

205. Petryshyn, W. V., Remarks on the approximation-solvability of nonlinear functional equations, Arch. Rational Mech. Anal. 26 (1967), 43–49.

206. Petryshyn, W. V., Iterative construction of fixed points of contractive type

mappings in Banach spaces, in *Numerical Analysis of POE's* (C.I.M.E. 2o, Ciclo, Ispra, 1967), Edizione Cremonese, Rome (1968), 307–339.

207. Petryshyn, W. V., On the approximation-solvability of nonlinear equations (C.I.M.E. 2o, Ciclo, Ispra, 1967), Edizione Cremonese, Rome (1968), Math. Ann. 177 (1968), 156–165.

208. Petryshyn, W. V., On the projective solvability and the Fredholm alternative for equations involving linear A-proper operators, Arch. Rational Mech. Anal. 30 (1968), 270–284.

209. Petryshyn, W. V., Further remarks on nonlinear P-compact operators in Banach space, Proc. Natl. Acad. Sci. USA 55 (1966), 684–687.

210. Petryshyn, W. V., Nonlinear equations involving noncompact operators, Proc. Symp. on Nonlinear Functional Analysis, AMS, 18, part I (1970), 206–233.

211. Petryshyn, W. V., Solvability of linear and quasilinear elliptic boundary value problems via the A-proper mapping theory, Numer. Funct. Anal. Optim. 2 (1980), 543–635.

212. Petryshyn, W. V., Existence theorems for semilinear abstract and differential equations noninvertible linear parts and noncompact pertubations, in *Nonlinear Equations in Abstract Spaces* (V. Lakshmikantham, ed.), Academic Press, New York (1978), 217–252.

213. Petryshyn, W. V., On existence theorems for nonlinear equations involving noncompact mappings, Proc. Natl. Acad. Sci. USA 67 (1970), 326–330.

214. Petryshyn, W. V., Some further results on periodic solutions of certain higher nonlinear differential equations, Nonlinear Anal. 8 (1984), 1055–1069.

215. Petryshyn, W. V., Direct and iterative methods for the solution of linear operator equations in Hilbert space, Trans. AMS 105 (1962), 136–175.

216. Petryshyn, W. V., On a class of K-p.d. and non-K-p.d. operators and operator equations, J. Math. Anal. Appl. 10 (1965), 1–24.

217. Petryshyn, W. V., On the extension and the solution of nonlinear operator equations, Illinois J. Math. 10 (1966), 255–274.

218. Petryshyn, W. V., Nonlinear Friedrichs' extension of K-strongly stable operators and the error estimate for the Galerkin method, Nonlinear Anal. 17 (1991).

219. Petryshyn, W. V., Remarks on condensing and k-set-contractive mappings, J. Math. Anal. Appl. 39 (1972), 717–747.

220. Petryshyn, W. V., The solvability of nonlinear equations involving abstract and differential equations, Proc. Functional Analysis Methods in Numerical

Analysis, Special session, AMS, St. Louis, Mo., Jan. 1977 (M. Z. Nashed, ed.), No. 701, Springer-Verlag, New York (1979), 209–247.

221. Petryshyn, W. V., Fixed point and surjectivity theorems via the A-proper mappings theory with application to differential equations, in Proc. Conference, (E. Fadell and G. Fournier, eds.), Sherbrook, Quebec, 1980, No. 886, Springer-Verlag, New York, 1981.

222. Petryshyn, W. V., Invariance of domain theorem for locally A-proper maps and its applications, J. Funct. Anal. 5 (1970), 137–159.

223. Petryshyn, W. V., Stability theory for linear A-proper mappings, Proc. Centennial Congress of Shevdrenko Scientific Society, Math.-Phys. Section, New York, 16 (1973).

224. Petryshyn, W. V., Approximation-solvability of periodic boundary value problem via the A-proper mapping theory, Proc. Symp. Pure Math. AMS 45/2 (1986), 261–282.

225. Petryshyn, W. V., On the approximation-solvability of equations involving A-proper and pseudo-A-proper mappings, Bull. AMS. 81 (1975), 223–312.

226. Petryshyn, W. V., Solvability of semilinear elliptic BV problems at resonance via the A-proper type operator theory, Annalele stiutifice ale Universitatu, Al. I. Cuza Iasi, Suppl. XXV, s.I a (1979), 133–152.

227. Petryshyn, W. V., Using degree theory for densely defined A-proper maps in the solvability of semilinear equations with unbounded and noninvertible linear part, Nonlinear Anal. 4 (1980), 251–281.

228. Petryshyn, W. V., Bifurcation and asymptotic bifurcation for equations involving A-proper maps with application to differential equations, J. Differential Equations 28 (1978), 124–154.

229. Petryshyn, W. V., Antipodes theorem for A-proper mappings and its applications to mappings of the modified type (S) and (S$_+$) and to mappings with the pm-property, J. Funct. Anal. 7 (1971), 165–211.

230. Petryshyn, W. V., Fixed point theorems for various classes of 1-set and 1-ball constructive mappings in Banach spaces, Trans. AMS 182 (1973), 323–352.

231. Petryshyn, W. V., On the relationship of A-properness to mappings of monotone type with application to elliptic equations, in *Fixed Point Theory and Its Applications* (S. Swahninatham, ed.), Academic Press, New York (1976).

232. Petryshyn, W. V., Fredholm theory for abstract and differential equations with noncompact nonlinear perturbations of Fredholm maps, J. Math. Anal. Appl. 72 (1979), 472–499.

233. Petryshyn, W. V., On nonlinear equations involving pseudo-proper mappings and their uniform limits with applications, J. Math. Anal. Appl. 38 (1972), 672–720.

234. Petryshyn, W. V., *Topological Degree for Densely Defined A-Proper Operators and Semilinear Equations*, Monograph, 305 pp. Cambridge University Press (to appear).

235. Petryshyn, W. V., and Yu, Z. S., Periodic solutions of nonlinear second-order ODE which are not solvable for the highest order derivative, J. Math. Anal. Appl. 89 (1982), 462–488.

236. Petryshyn, W. W., and Tucker, T. S., On the functional equations involving nonlinear P-compact operators, Trans. AMS 35 (1969), 343–373.

237. Petryshyn, W. V., and Fitzpatrick, P. M., On 1-set and 1-ball contractions with application to perturbation problems for nonlinear bijective maps and Fredholm maps, Boll. Un. Mat. Ital. 7 (1973), 102–124.

238. Petryshyn, W. V., and Fitzpatrick, P. M., A degree theory, fixed point theorems, and mapping theorems for multivalued noncompact mappings, Trans. AMS, 194 (1974), 756–767.

239. Pohožaev, S. I., The solvability of nonlinear equations with odd operators, Functional Anal. Priložen. 1 (1967), 66–67.

240. Pohožaev, S. I., On an approach to nonlinear equations, Dokl. Akad. Nauk SSSR 247 (1979), 912–916.

241. Polskii, N. I., On the convergence of certain approximate methods of analysis, Ukrainian Math. J. 7 (1955), 56–70.

242. Polskii, N. I., Projective methods in applied mathematics, Dokl. Akad. Nauk SSSR 143 (1962), 787–790.

243. Potter, A. J. B., Nonlinear A-proper mappings of the analytic type, Canad. J. Math. 25 (1973), 468–474.

244. De Prima, C. R., and Petryshyn, W. V., Remarks on the strict monotonicity and surjectivity properties of duality maps defined on real normed space, math. Z. 123 (1971), 49–55.

245. Rabinowitz, P. H., Some aspects of nonlinear eigeuvalue problems, Rocky Mountain J. Math. 3 (1973), 161–202.

246. Rabinowitz, P. H., On bifurcation theory from infinity, J. Differential Equations 14 (1973), 462–475.

247. Reich, S., Product formulas, nonlinear semigroups, and accretive operators, J. Funct. Anal. 36 (1980), 147–168.

248. Reich, S., Remarks on fixed points, Atti Accad. Lincei 52 (1972), 689–697.

249. Reinermann, J., Fixtpunkzätze von Krasnoselski-Typ, Math. Z. 119 (1971), 339–344.

250. Rothe, E., Zur Theorie der topologischem Ordnung und der Vektorfelder in Banachschen Räumen, Compositio Math. 5 (1937).

251. Rothe, E., On the Cesari index and the Browder–Petryshyn degree, in *Intern. Symp. Dynam. Systems, Gainsville, 1970*, Academic Press, New York (1972), 295–312.

252. Rockafellar, R. T., Local boundedness of nonlinear maximal monotone operators, Michigan Math. J. (1963), 397–407.

253. Sadovskii, B. N., Limit-compact and condensing operators, Mat. Sb. 27 (1972), 85–155.

254. Sanches, L. A., A note on a differential equations of the dynamics of wires, Boll. Un. Mat. Ital. 16-A (5) (1978), 391–397.

255. Schaefer, H. H., Über the Methode der a-priori Schranken, Math. Ann. 129 (1955), 415–416.

256. Schauder, J., Der Fitzpunksatz in Funktionalräumen, Studia Math. 20 (1930), 171–180.

257. Schauder, J., Invariannz des Gebietes in Funktionalräumen, Studia Math. 1 (1929), 123–139.

258. Schumann, R., and Zeidler, E., The finite difference method for quasilinear elliptic equations of order $2m$, Numer. Funct. Anal. Optim. 1(2) (1979), 161–194.

259. Shinbrot, M., A fixed point theorem and some applications, Arch. Rational Mech. Anal. 19 (1964), 255–271.

260. Schöneberg, R., A degree theory for semicondensing vector fields in infinite dimensional Banach spaces and applications, Nonlinear Anal. 4 (1980), 393–405.

261. Schwartz, J. T., *Nonlinear Functional Analysis,* Gordon and Breach, New York (1969).

262. Skrypnik, I. V., *Nonlinear Elliptic Equations of Higher Order,* Naukova Dumka, Kiev (1973).

263. Skrypnik, I. V., *Solvability and Properties of Solutions of Nonlinear Elliptic Equations,* vol. 9, Itogi Nauki i Tekniki, Moscow (1976).

264. Skrypnik, I. V., On quasilinear elliptic equations of higher order with continuous generalized solutions, Differencial'nye Uravnenija 6 (1978), 1104–1119.

265. Skrypnik, I. V., Topological characteristics of general nonlinear elliptic operators, Dokl. Akad. Nauk SSSR 239 (1978), 538–541.

266. Skrypnik, I. V., Coercivity inequality for a pair of linear elliptic operators, Dokl. Akad. Nauk SSSR 239 (1978), 275–278.

267. Skrypnik, I. V., Quasilinear Dirichlet problem on domains with porous boundary, Dokl. Akad. Nauk Ukrainian RSR Ser. A 2 (1982), 21–26.

268. Skrypnik, I. V., *Nonlinear Elliptic Boundary Value Problems*, Texte zur Mathematik, vol. 91, Teubner, Leipzig (1986).

269. Skrypnik, I. V., *Methods for Investigation of Nonlinear Elliptic Boundary Value Problems*, Nauka, Moscow (1990).

270. Sobolev, S., *Application of Fundamental Analysis to Mathematical Physics*, Univ. Press, Leningrad (1950).

271. Sobolevskii, P. E., On the equations with operators forming an acute angle, Dokl. Akad. Nauk SSSR 116 (1957), 754–757.

272. Stuart, C. A., Some bifurcation theory for k-set-contractions, Proc. London Math. Soc. 27 (1973), 531–550.

273. Stuart, C. A., Self-adjoint square roots of positive self-adjoint bounded linear operators, Proc. Edinburgh Math. Soc. 18 Ser II (1972/73), 77–79.

274. Stuart, C. A., and Toland, J. F., The fixed point index of a linear k-set-contraction, J. London Math. Soc. (2) 6 (1973), 317–320.

275. Toland, J. F., Global bifurcation theory via Galerkin method, Nonlinear Anal. 1 (1977), 305–317.

276. Toland, J. F., Bifurcation and asymptotic bifurcation for noncompact non-symmetric gradients operators, Proc. Royal Soc. Edinburgh 73A(8) (1974/75), 137–147.

277. Treves, F., *Basic Linear Partial Differential Equations*, Academic Press, New York (1975).

278. Troesch, B. A., Integral inequalities for two functions, Arch. Rational Mech. Anal. 24 (1967), 129–140.

279. Tucker, T. S., Stability of nonlinear computing schemes, SIAM J. Numer. Anal. 6 (1966), 72–81.

280. Vainberg, M. M., *Variational methods for the Study of Nonlinear Operators*, GITTL, Moscow (1956); Engl. transl., Holden-Day, San Francisco (1964).

281. Vainberg, M. M., On some new principle in the theory of nonlinear equations, Uspehi Mat. Nauk 15 (1960), 243–244.

282. Vainikko, G., Necessary and sufficient condition for the stability of the Galerkin–Petrov method, Tartu Riikl. Ül. Toimetised 117 (1965), 141–147.

283. Vainikko, G., A perturbed Galerkin method and the general theory of approximate methods for nonlinear equations, Ž. Vyčisl. Mat. i Mat. Fiz. 7 (1967), 723–751.

284. Varga, R. S., Functional analysis and approximation theory in numerical analysis, SIAM Reg. Conf. Ser. Appl. Math. 3 (1971), 1–76.

285. Vidossich, G., On Peano phenomenon, Boll. Un. Mat. Ital. (4) 3 (1970), 32–42.

286. Vignoli, A., On quasibounded mappings and nonlinear functional equation, Atti Accad. Naz. Rend. Cl. Sci. Fis. Mat. Natur. 50 (1971), 114–117.

287. Vignoli, A., On α-contractions and surjectivity, Boll. Un. Mat. Ital. 4 (1971), 446–455.

288. Višik, M. I., Quasilinear strongly elliptic systems of differential equations in divergence form, Trudy Moskov. Mat. Obšč. 12 (1962), 125–184.

289. Webb, J. R. L., On a characterization of k-set-contractions, Acad. Naz. Lincei Ser. 8 50 (1971), 358–361.

290. Webb, J. R. L., Fixed point theorems for nonlinear semicontractive operators in Banach spaces, J. London Math. Soc. 9 (1969), 683–688.

291. Webb, J. R. L., Remarks on k-set-contractions, Boll. Un. Mat. Ital. 4 (1971), 614–629.

292. Webb, J. R. L., Mapping and fixed point theorems for nonlinear operators in Banach spaces, Proc. London Math. Soc. 20 (1970), 451–468.

293. Webb, J. R. L., A fixed point theorem and applications to functional equations in Banach spaces, Boll. Un. Mat. Ital. (4) 4 (1971), 775–788.

294. Webb, J. R. L., Existence theorems for sums of k-ball-contractions and accretive operators via A-proper mappings, Nonlinear Anal. 5 (1981), 891–896.

295. Webb, J. R. L., On the property of duality mappings and the A-properness of accretive operators, Bull. London Math. Soc. 13 (1981), 235–238.

296. Webb, J. R. L., Mappings of accretive and pseudo-A-proper type, J. Math. Anal. Appl. 85 (1982), 146–152.

297. Webb, J. R. L., A-properness and fixed points of weakly inward mappings, J. London Math. Soc. 27 (1983), 141–149.

298. Webb, J. R. L., Approximation-solvability of nonlinear analysis, Spring School on Nonlinear Analysis, Prague (1982), 234–257.

299. Webb, J. R. L., Topological degree and A-proper operators, Linear Algebra and Its Appl. (1986), 1–16.

300. Webb, J. R. L., and Welsh, S. C., Topological degree and global bifurcation, Proc. Symposium on Pure Mathematics, 45 (II), AMS Providence, R.I. (1986), 527–531.

301. Webb, J. R. L., and Welsh, S. C., Existence and uniqueness of IV problem for a class of second-order differential equations, J. Differential Equations 82 (1989).

302. Weinberg, J. A., Topological degree of locally A-proper maps, J. Math. Anal. Appl. 65 (1978), 66–79.

303. Welsch, S. C., Global results concerning bifurcation for Fredholm maps of index zero with a transversality condition, Nonlinear Anal. 12 (1988), 1137–1148.

304. Wheeden, R. L., and Zygmund, A., *Measure and Integral: Introudction to Real Analysis,* Marcel Dekker, New York (1977).

305. Willem, M., On a results of Rothe about the Cesari index and A-proper mappings, Boll. Un. Mat. Ital. 17-A (1980), 178–182.

306. Wong, S. F., The topological degree of A-proper maps, Canad. J. Math. 23 (1971), 403–412.

307. Wong, S. E., Le degré topologique de certain applications non-compacts nonlinéaires, Ph.D. dissertation, University of Montreal (1969).

308. Yichum, Z., Surjectivity of perturbed maximal monotone mappings, Kexue Tongbao 29 (1984), 857–860.

309. Yichum, Z., and Guanghong, Y., The generalized topological degree for nonlinear mappings of monotone type, Chinese Ann. Math. 10A (1989), 67–71.

310. Yood, B., Properties of linear transformations preserved under addition of a completely continuous transformation, Duke Math. J. 18 (1951), 599–612.

311. Zarantonello, E., Solving functional equations by contractive averaging, Tech. Rep. 160, U.S. Army Research Center, Madison, Wis. (1960).

312. Zarantonello, E., The closure of the numerical range contains the spectrum, Bull. AMS 70 (1964), 781–787.

313. Zeidler, E., *Nonlinear Functional Analysis and Its Applications,* vol. I, *Fixed Point Theorems,* Springer-Verlag, Berlin (1986).

314. Zeidler, E., *Nonlinear Functional Analysis and Its Applications,* vol. II(B), *Nonlinear Monotone Operators,* Springer-Verlag, Berlin (1990).

315. Zdanov, H. A., On the convergence of some variants of the method of Galerkin, Dokl. Akad. Nauk SSSR 115 (1957), 223–225.

316. Zienkiewicz, O. C., *The Finite Element Method in Engineering Science*, 2nd ed., McGraw-Hill, New York (1971); German transl., Carl Hanser, Munich (1975).

Supplementary Literature Relevant to the Material Contained in This Book

Altman, M., *Contractors and Contractor Directions: Theory and Applications*, Marcel Dekker, New York (1965).

Berežanskii, Yu. M. Selfadjoint operators in spaces of functions of infinitely many variables, Transl. Math. Monographs 63, AMS, Providence, R.I. (1986).

Berger, M. S., *Nonlinearity and Functional Analysis*, Academic Press, New York (1977).

Bogolubov, N. N., *On Some Statistical Methods in Mathematical Physics*, Ukrainian Akademic Press, Kiev (1945).

Cioranescu, I., *Geometry of Banach Spaces, Duality Mappings, and Nonlinear Problems*, Kluwer Academic, Dordrecht, The Netherlands (1990).

Cronin, J., *Differential Equations: Introduction and Qualitative Theory*, Marcel Dekker, New York (1980).

De Lafferiere, B., and Petryshyn, W. V., New positive fixed point and eigenvalue results for P_γ-compact maps and some and some applications, Nonlinear Anal. 13 (1989), 1427–1440.

Eisenack, G., and Fenske, C., *Fixpunktheorie*, Biograph. Inst. Mannhein/Wien/ Zürich, B. I. Wissenschafts-verlag, Zürich (1978).

Fitzpatrick, P. M., and Pejsachowicz, J., Orientation and Leray–Schauder theory for fully nonlinear BV problems, Memoirs AMS (to appear).

Fitzpatrick, P. M., and Petryshyn, W. V., Connected components of solutions of nonlinear eigenvalue problems, Houston J. Math. 8 (1982), 477–500.

Gossez, J. P., Nonlinear elliptic BV problems for equations with rapidly or slowly increasing coefficients, Trans. AMS 190 (1970), 163–205.

Guo, D., *Nonlinear Functional Analysis* (Chinese with English Introduction), Shandong Science and Technology Press, Shandong (1985).

Istratescu, V. I., *Fixed Point Theory; An Introduction*, D. Reidel;, Dordrecht, The Netherlands (1981).

Ize, J., Massabo, I., and Vignoli, A., Degree theory for equivariant maps, the general S^1-action, Memoirs AMS (to appear).

Jeggle, H., *Nichtlinear Funktionalanlysis*, Teubner, Stuttgart (1979).

Joshi, M. C., and Bose, R. K., *Some Topics in Nonlinear Functional Analysis*, Wiley, New York (1985).

Kannan, R., Periodic solution of pendulum-type equations, J. Differential Equations 59 (1985), 123–144.

Kazdan, U. L., and Warner, F. W., Remarks on some quasilinear elliptic equations, Comm. Pure Appl. Math. 38 (1975), 144–165.

Krylov, M. M., and Bogolubuv, M. M., On the Rayleigh principle with theory of differential equations of mathematical physics and about one of the Euler's methods in variable calculus (Ukrainian), Trudy Fiz.-Mat. Section Ukrainian Acad. Sci. 3 (1926), 38–57.

Kryszewski, W., Przeraolski, B., and Werenski, S., Remarks on approximation methods in degree theory, Trans. AMS. 316 (1989), 97–114.

Landes, R., and Mustonen, V., Boundary value problems for strongly nonlinear second order elliptic equations, Boll. Un. Mat. Ital. 4-B (1985), 15–32.

McKenna, P. J., Nonselfadjoint semilinear problems at resonance in the alternative method, J. Differential Equations 33 (1979), 275–293.

Milojević, P. S. (ed.), *Nonlinear Functional Analysis*, Marcel Dekker, New York (1989).

Mitropolsky, Yu. A., and Khoma, H. P., *Mathematical Foundation of Asymptotic Methods of Nonlienar Mechanics*, Naukova Dumka, Kiev (1983).

Pascali, D., and Sburlan, S., *Nonlinear Mappings of Monotone Type*, Editura Acad. Bucaresti, Romania, Sijthoff en Norrdhoff, Alphen aan den Rijn, The Netherlands.

Perestiuk, N. A., and Samoilenko, A. M., *Differential Equations with Impulse Effect*, Visčă Skola, Kiev (1987).

Petryshyn, W. V., Fredholm alternative for nonlinear A-proper maps with applications to nonlinear elliptic BV problems, J. Funct. 18 (1975), 288–317.

Petryshyn, W. V., Approximation-solvability of periodic boundary value problems via the A-proper mapping theory, Proc. Symp. Pure Math. 45(II) (1986), 261–282.

Petryshyn, W. V., and Yu, Z. S., Existence theorems for higher order nonlinear periodic BV problems, Nonlinear Anal. 6 (1982), 943–969.

Petryshyn, W. V., and Yu, Z. S., Nonresonance and existence for nonlienar BV problems, Houston J. Math. 9 (1983), 511–536.

Pohežayev, S. I., On a constructive emthod in variational calculus, Dokl. Akad. Nauk USSR 198 (1988), 1330–1334.

Rankin, S. M., III, Boundary value problems for partial functional differential equations, Pacific J. Math. 9 (1980), 459–468.

Vainberg, M. M., and Trenogin, V. A., The method of Lyapunov and Schmidt in the theory of nonlinear equations and their further development, Russian Math. Surveys 17 (1962), 1–60.

Webb, J. R. L., Boundary value problems for strongly nonlinear elliptic equations, J. London Math. Soc. 21 (1980), 123–132.

Notation

$\Gamma = \{X_n, V_n; E_n, W_n\}$	approximation scheme
$D \subset X$	subset
X^*	dual space to X
sup	supremum, least upper bound
inf	infimum, greatest lower bound
lim sup ($\overline{\lim}$)	limit superior
lim inf ($\underline{\lim}$)	limit inferior
R^1	real numbers
R^n	n-dimensional euclidean space
Z_+	positive integers
$\| \cdot \|$	norm
$\left. \begin{array}{l} E_1 \cup E_2 \\ \cup_{i=1}^n E_i \end{array} \right\}$	union
$E_1 \cap E_2$	intersection
$E_1 \setminus E_2$	difference
$\overline{E}$	closure
∂E	boundary
$[E]$	linear span
$T^{-1}(E)$	preimage of the set E
$B(x,r)$	open ball with center x and radius r
$x_n \rightarrow x$	strong convergence
$x_n \rightharpoonup x$	weak convergence

$T: D(T) \subset X \to Y$	map from the set D into the space Y
$D(T)$	domain of definition of T
$L(X,Y)$	space of continuous linear operators
$N(L)$	kernel or null space of L
$R(L)$	range of L
L^*	adjoint of L
L^{**}	adjoint of L^* (second adjoint of L)
L^+	generalized inverse of L
L^{-1}	inverse of L
$\oplus$	direct topological sum
$[a,b]$	close interval
(a,b)	open interval
$[a,b)$	semiopen interval
$\varnothing$	empty set
L^p	
C^m	
L_∞	classes of functions
C^∞	
C_0^∞	
ℓ^p	classes of sequences
$\|\cdot\|_p$	L^p-norm
$\neq$	not equal
$\geq$	greater than or equal to
$\leq$	less than or equal to
$\mathscr{A} \Rightarrow \mathscr{B}$	$\mathscr{A}$ imples $\mathscr{B}$
$\beta(Q)$	ball-measure of noncompactness
$\gamma(Q)$	set-measure of noncompactness
$\int_Q f(x)\,dx$	Lebesgue integral in $\mathbb{R}^n$
$W_p^m(Q)$	Sobolev space
a.e.	almost everywhere
$\Phi(Q)$	denotes either $\beta(Q)$ or $\gamma(Q)$
$\|\cdot\|_{m,p}$	norm in W_p^m
$\overset{\circ}{W}{}_p^m(Q)$	closure of C_0^∞ in $W_p^m(Q)$
$H^m(Q)$	Hilbert space $W_2^m(Q)$
$(\cdot,\cdot)_{m,2}$	inner product in H^m
$H^0(\equiv W_2^0(Q))$	identical with $L^2(Q)$
F_x'	Fréchet derivative of F at x
T_∞'	asymptotic derivative (or derivative at ∞) of T
T surjective	mapping onto, i.e., $T(D) = Y$
T injective	one-to-one mapping

T bijective	surjective and injective
$X \times Y$	cross product
$\text{dist}(x,E)$	distance from the point x to the set E
$\text{co}(E)$	convex hull of E
$\sigma(T)$	spectrum of T
$\sigma_e(T)$	essential specrum of T
$\text{ind}(T)\ (=\ i(E))$	Fredholm index
$\alpha = (\alpha_1, \ldots, \alpha_n)$	multiindex of nonnegative integers
$\lvert \alpha \rvert$	identical to $\lvert \alpha \rvert = \alpha_1 + \alpha_2 + \cdots + \alpha_n$
$D^{\alpha}u = D_1^{\alpha_1} \cdots D_n^{\alpha_n}u$	partial derivative of u of order $\lvert \alpha \rvert = \alpha_1 + \cdots + \alpha_n$
$\text{deg}_{\text{B}}(T,D,y)$	Brouwer degree
$\text{deg}_{\text{LS}}(T,D \cdot y)$	Leray–Schauder degree
$\text{deg}_{\beta}(T,D,y)$	degree of k-ball-contraction vector field
$\text{Deg}(T,D,y)$	generalized degree for A-proper map
(f,x)	value of $f \in X^*$ at $x \in X$
Π_i	product
$\sum_i$	sum
$V \subsetneqq W$	imbedding of space V into W
R^{sm}	space of all derivatives of order less than or equal to m
$R^{s'm}$	space of all derivatives of order m
$\text{Re } a$	real part of the complex number a
$J_{\psi}(x)$	Jacobian determinant of ψ at x
$I_D(x)$	inward set on D at x in D
$x \in D$	x belongs to D
$x \notin D$	x does not belong to D
ODEs	ordinary differential equations
PDEs	partial differential equations
BP	bifurcation point
ABP	asymptotic bifurcation point
Δ	Laplace operator
$\Phi_0(X,Y)$	space of Fredholm operators of index 0
$\sqrt{}$	square root sign

Index